Developments in Clay Science – Volume 6

Natural and Engineered Clay Barriers

Developments in Clay Science – Volume 6

Natural and Engineered Clay Barriers

Edited by

Christophe Tournassat
Water, Environment and Ecotechnology Division, French Geological Survey (BRGM), Orléans, France
Earth Sciences Division, Lawrence Berkeley National Laboratory, Berkeley, CA, USA

Carl I. Steefel
Earth Sciences Division, Lawrence Berkeley National Laboratory, Berkeley, CA, USA

Ian C. Bourg
Department of Civil and Environmental Engineering, Princeton University, Princeton, NJ, USA
Earth Sciences Division, Lawrence Berkeley National Laboratory, Berkeley, CA, USA

Faïza Bergaya
Centre de Recherche sur la Matière Divisée, Centre National de la Recherche Scientifique (CNRS), Orléans, France

ELSEVIER

AMSTERDAM • BOSTON • HEIDELBERG • LONDON • NEW YORK • OXFORD
PARIS • SAN DIEGO • SAN FRANCISCO • SINGAPORE • SYDNEY • TOKYO

Elsevier
Radarweg 29, PO Box 211, 1000 AE Amsterdam, Netherlands
The Boulevard, Langford Lane, Kidlington, Oxford OX5 1GB, UK
225 Wyman Street, Waltham, MA 02451, USA

British Library Cataloguing in Publication Data
A catalogue record for this book is available from the British Library

Library of Congress Cataloging-in-Publication Data
A catalog record for this book is available from the Library of Congress

ISBN: 978-0-08-100027-4
ISSN: 1572-4352

For information on all Elsevier publications
visit our website at http://store.elsevier.com

Contents

List of Contributors

Pierre M. Adler, Sorbonne Universités, UPMC Univ. Paris 06, UMR-7619 METIS, Paris cedex, France. E-mail: pierre.adler@upmc.fr

Scott Altmann, Andra, Châtenay-Malabry, France.

A. Amann-Hildenbrand, Energy and Mineral Resources Group, Institute of Geology and Geochemistry of Petroleum and Coal, Aachen, Germany. E-mail: alexandra.amann@emr.rwth-aachen.de

Faïza Bergaya, Centre de Recherche sur la Matière Divisée, Centre National de la Recherche Scientifique (CNRS), Orléans, France. E-mail: f.bergaya@cnrs-orleans.fr

Olivier Bildstein, Atomic Energy and Alternative Energies Commission, Nuclear Energy Division, Cadarache, Saint Paul-lez-Durance, France. E-mail: olivier.bildstein@cea.fr

Mikhail Borisover, Agricultural Research Organization, Institute of Soil, Water and Environmental Sciences, The Volcani Center, Bet Dagan, Israel. E-mail: vwmichel@volcani.agri.gov.il

Ian C. Bourg, Department of Civil and Environmental Engineering, Princeton University, Princeton, NJ, USA; Earth Sciences Division, Lawrence Berkeley National Laboratory, Berkeley, CA, USA. E-mail: bourg@princeton.edu

Jordi Cama, Institute of Environmental Assessment and Water Research, IDAEA, CSIC, Barcelona, Spain. E-mail: jordi.cama@idaea.csic.es

Francis Claret, Water, Environment and Ecotechnology Division, French Geological Survey (BRGM), Orléans, France. E-mail: f.claret@brgm.fr

Philippe Cosenza, University of Poitiers, CNRS, UMR 7285 IC2MP-HydrASA, ENSIP, Poitiers, France. E-mail: philippe.cosenza@univ-poitiers.fr

R. Cuss, British Geological Survey, Nottingham, UK. E-mail: rjcu@bgs.ac.uk

James A. Davis, Earth Sciences Division, Lawrence Berkeley National Laboratory, Berkeley, CA, USA. E-mail: jadavis@lbl.gov

C. Davy, Ecole Centrale de Lille/LML UMR CNRS 8107, Cité Scientifique, Villeneuve d'Ascq Cedex, France. E-mail: catherine.davy@ec-lille.fr

Ghislain de Marsily, Sorbonne Universités, UPMC Univ. Paris 06, UMR-7619 METIS, Paris cedex, France. E-mail: gdemarsily@aol.com

Jiwchar Ganor, Department of Geological and Environmental Sciences, Ben-Gurion University of the Negev, Beer Sheva, Israel. E-mail: ganor@bgu.ac.il

Eric C. Gaucher, TOTAL, E&P, Pau, France. E-mail: eric.gaucher@total.com

Julio Gonçalvès, University Aix-Marseille, UMR-7330 CEREGE, Aix en Provence, France. E-mail: goncalves@cerege.fr

J. Harrington, British Geological Survey, Nottingham, UK. E-mail: jfha@bgs.ac.uk

E. Jacops, Energy and Mineral Resources Group, Institute of Geology and Geochemistry of Petroleum and Coal, Aachen, Germany; SCK•CEN, Belgian Nuclear Research Centre, Expert Group, Waste & Disposal, Mol, Belgium; KU Leuven, Department of Earth & Environmental Sciences, Heverlee, Belgium. E-mail: ejacops@sckcen.be

B.M. Krooss, Energy and Mineral Resources Group, Institute of Geology and Geochemistry of Petroleum and Coal, Aachen, Germany. E-mail: bernhard.krooss@emr.rwth-aachen.de

N. Maes, SCK•CEN, Belgian Nuclear Research Centre, Expert Group, Waste & Disposal, Mol, Belgium. E-mail: nmaes@SCKCEN.be

Virginie Marry, Sorbonne Universités, UPMC Univ. Paris 06, UMR 8234 PHENIX, Paris, France; CNRS, UMR 8234 PHENIX, Paris, France. E-mail: virginie.marry@upmc.fr

Aliaksei Pazdniakou, Sorbonne Universités, UPMC Univ. Paris 06, UMR-7619 METIS, Paris cedex, France. E-mail: aliaksei.pazdniakou@upmc.fr

A. Revil, Department of Geophysics, Colorado School of Mines, Green Center, Golden, CO, USA; ISTerre, CNRS, Université de Savoie, Le Bourget du Lac, France. E-mail: arevil@mines.edu

Benjamin Rotenberg, Sorbonne Universités, UPMC Univ. Paris 06, UMR 8234 PHENIX, Paris, France; CNRS, UMR 8234 PHENIX, Paris, France. E-mail: benjamin.rotenberg@upmc.fr

Jonny Rutqvist, Earth Sciences Department, Lawrence Berkeley National Laboratory, Berkeley, CA, USA. E-mail: Jrutqvist@lbl.gov

F. Skoczylas, Ecole Centrale de Lille/LML UMR CNRS 8107, Cité Scientifique, Villeneuve d'Ascq Cedex, France. E-mail: Frederic.Skoczylas@ec-lille.fr

Carl I. Steefel, Earth Sciences Division, Lawrence Berkeley National Laboratory, Berkeley, CA, USA. E-mail: cisteefel@lbl.gov

Christophe Tournassat, Water, Environment and Ecotechnology Division, French Geological Survey (BRGM), Orléans, France; Earth Sciences Division, Lawrence Berkeley National Laboratory, Berkeley, CA, USA. E-mail: c.tournassat@brgm.fr

Agnès Vinsot, Andra, LSMHM, Bure, France. E-mail: agnes.vinsot@andra.fr

Acknowledgments

The editors, Christophe Tournassat, Carl I. Steefel, Ian C. Bourg, and Faïza Bergaya would like to acknowledge all of the authors of this volume for their nice contributions. They also thank all of the reviewers for their important help and contributing insights to improve the chapters.

Christophe Tournassat is especially grateful to Faiza Bergaya, the instigator of this volume, for her endless support and motivation. He is also particularly grateful to Carl Steefel, for his invitation and warm welcome in the Earth Science Division of the Berkeley National Laboratory, where this volume became a living project. He would also like to express his gratitude to Ian Bourg for his welcome in Berkeley and for the discussions about clay mineral properties (and other themes), and to all colleagues and friends from BRGM, LBNL, and other places, who accepted to contribute to this volume. This work would have not been feasible without the full support from BRGM (C. Truffert, C. King, and F. Claret). L'Institut Carnot funded the visit of C. Tournassat at the Lawrence Berkeley National Laboratory. C. Tournassat would also like to thank warmly S. Gaboreau and N. Marty for providing the images of the cover.

Faiza Bergaya is grateful to CNRS for giving her the opportunity, as Director Emeritus, to ensure continuity of her work as Series Editor of the *Developments in Clay Science*. She also thanks S. Bonnamy, the Director of CRMD laboratory, for providing all facilities for her research activities.

The contribution of C. Steefel was supported by the Director, Office of Science, Office of Basic Energy Sciences, Chemical Sciences, Geosciences, and Biosciences Division, US Department of Energy under Contract No. DE-AC02-05CH11231.

The authors of Chapter 5 would like to acknowledge the thoughtful review by N. Michau and J.-E. Lartigue, and fruitful discussions with other colleagues including X. Bourbon, B. Cochepin, Y. Linard, I. Munier (ANDRA), Ph. Blanc, S. Gaboreau, S. Grangeon, C. Lerouge, N. Marty (BRGM), C. Bataillon, D. Féron, P. Frugier, M. Libert, and M. Schlegel (CEA).

Christophe Tournassat
Carl I. Steefel
Ian C. Bourg
Faïza Bergaya
January 2015

Introduction

Christophe Tournassat[a,b], Carl I. Steefel[b], Ian C. Bourg[b,c] and Faïza Bergaya[d]

[a] *Water, Environment and Ecotechnology Division, French Geological Survey (BRGM), Orléans, France;* [b] *Earth Sciences Division, Lawrence Berkeley National Laboratory, Berkeley, CA, USA;* [c] *Department of Civil and Environmental Engineering, Princeton University, Princeton, NJ, USA;* [d] *Centre de Recherche sur la Matière Divisée, Centre National de la Recherche Scientifique (CNRS), Orléans, France*

In recent years, the scientific community has seen a remarkable surge of interest in the properties of clays as they apply in a variety of natural and engineered settings. In part, this renewed interest is traceable to the very property that, in the past, had relegated clay-rocks to a minor status in hydrology, namely their low hydraulic conductivity. While clay-rocks might be largely bypassed by contaminant plumes in groundwater aquifers and by saline fluids in sedimentary basins, their low permeability allows them to play key roles in several important subsurface energy-related applications, including the long-term storage of nuclear wastes in geologic repositories and CO_2 sequestration in subsurface geologic formations. In these applications, the low transmissivity of clay-rich geologic formations or engineered clay barriers provides at least part of the basis for isolation of radionuclide contaminants and CO_2 from the biosphere. Clay materials are an important part of the multibarrier systems for nuclear waste storage under consideration worldwide, but their performance must be demonstrated on the timescale of hundreds to thousands of years (Altmann, 2008; Busch et al., 2008; Chapman and Hooper, 2012; Armitage et al., 2013; Neuzil, 2013). The low permeability of clay-rich shales also explains why hydrocarbon resources are not easily exploited from these formations, thus requiring in many cases special procedures such as hydraulic fracturing in order to extract them.

In addition to their low permeability, clay minerals have other properties of interest in these applications, including their very high adsorption capacity (*Chapter 2, in this volume*). The strong adsorption and resulting retardation of many contaminants by clay minerals make them ideal for use in natural or engineered barrier systems, particularly where there is a desire to improve confidence in the safety of waste isolation beyond reliance on slower transport rates alone. In addition, the high pH/redox buffering capacity (*Chapter 3, in this volume*) and slow dissolution kinetics of clay minerals (*Chapter 4, in this volume*), along with the slow diffusive mass transport in clay-rich media (*Chapter 6, in this volume*), make clay-rocks and engineered clay barriers remarkably stable under the chemical perturbations generated by high partial pressure of CO_2 or by the presence of concrete, steel, and other exogenous materials (*Chapter 5, in this volume*).

While clay materials offer some striking benefits in these and other applications, their properties and behavior under relevant conditions remain only partly understood. With the exception of the work by Bredehoft and Papadopolous (1980), Bredehoft et al. (1983), and Neuzil (1982, 1986, 1993, 1994), the hydrodynamics of clay-rocks had, until these last two decades, attracted only limited attention from hydrogeologists. As discussed by Neuzil (2013), flow through clay-rich formations may not be adequately described by Darcy's Law. In fact, engineered clay barriers and clay-rocks show a remarkable array of macroscale properties such as high swelling pressure, very low permeability, semipermeable membrane properties, and a strong coupling between geochemical, mechanical, and osmotic properties (Malusis et al., 2003; Malusis and Shackelford, 2004). These properties are thought to arise from the distinct geochemical, transport, and mechanical properties of the interlayer (nano)pores of swelling clay minerals such as Na^+-montmorillonite and other smectites (*Chapters 8–10, in this volume*). Clay-rocks typically show a nonlinear dependence of the flow field on the pore pressure, particularly at low pressure gradients and flow rates where threshold behavior prevails. Much of this anomalous behavior is traceable to chemical, electrical potential, and thermal gradients that result in nonconjugate driving forces for hydrodynamic flow and molecular diffusion. The prediction of gas migration through clay barriers (e.g., CO_2 from carbon sequestration storage, or H_2 generated by radiolysis or corrosion of steel containers) is a difficult challenge as well because of the complex interplay of the gas transport processes with the mechanical properties and the pore structure of clay-rocks (*Chapter 7, in this volume*). Even where hydrodynamic flow through clay-rocks is limited or suppressed altogether, diffusion offers another possible means for transport that must be evaluated. This task is rendered difficult by the incomplete understanding of the microstructure and surface electrostatics of clay-rich materials, such that multiple models exist with very different underlying concepts/hypotheses on the diffusion and semipermeable properties of the clay nanopores (*Chapter 6, in this volume*).

The development of predictive mesoscale models of water, gas, and solute mass fluxes in nanoporous media is in fact a long-standing challenge in the geosciences. The behavior of nanoporous clay environments is complicated by the fact that the pore structure of clay materials is heterogeneous, such that water and ions can be present in bulk-liquid-like water, on external surfaces of clay particles, and in nano-scale confinement in clay interlayers (*Chapter 1, in this volume*). To understand and predict the coupling phenomena, it is often necessary to examine the physical processes at the pore scale, upscale the physical laws to the continuum scale, and compare continuum scale model predictions to geophysical or other macroscopic observables. A range of upscaling strategies has been developed to predict the various properties of interest for clay materials (*Chapters 8–11, in this volume*).

This volume opens on the surface and chemical properties of clay minerals and clay barriers (*Chapters 1–4*). Then, it focuses on mass fluxes through clay barriers (*Chapters 5–7*) and on coupled thermo–hydro–mechanical processes (*Chapters 8 and 9*). The end of the volume is focused on upscaling modeling strategies and their applications (*Chapters 10 and 11*).

A large part of the current understanding of clay barrier properties has been gained through studies conducted on radioactive waste storage systems, a fact that is reflected in most of the chapters. However, the recent breakthroughs in the field and the challenges that remain are not limited to this application. For instance, the development of recovery techniques for gas and light liquid hydrocarbons from shale has created a new series of challenges for the clay scientist community. Hopefully, this volume can provide a solid basis to the clay and nonclay scientist communities for the identification of current understanding, recent breakthroughs, and the challenges that remain in the field of clay barriers.

Note on Terminology and Abbreviations

For the purpose of consistency of clay terminology, the abbreviations used in all chapters of this volume follow the terminology of the *Handbook of Clay Science* (Bergaya and Lagaly, 2013). The most used abbreviations are Bent for bentonite, Sm for smectite, Mt for montmorillonite, Kaol for kaolinite, and I-Sm for illite-smectite, the clays and clay minerals most frequently encountered in clay barriers.

REFERENCES

Altmann, S., 2008. Geo'chemical research: a key building block for nuclear waste disposal safety cases. J. Contam. Hydrol. 102, 174–179.

Armitage, P., Faulkner, D., Worden, R., 2013. Caprock corrosion. Nat. Geosci. 6, 79–80.

Bergaya, F., Lagaly, G., 2013. Handbook of Clay Science. Developments in Clay Science, second ed. Elsevier.

Bredehoeft, J.D., Papadopulos, S.S., 1980. A method for determining the hydraulic properties of tight formations. Water Resour. Res. 16, 233–238.

Bredehoeft, J., Neuzil, C., Milly, P., 1983. Regional Flow in the Dakota Aquifer: A Study of the Role of Confining Layers. US Geological Survey Water Supply Papers 2237, p. 45.

Busch, A., Alles, S., Gensterblum, Y., Prinz, D., Dewhurst, D.N., Raven, M.D., Stanjek, H., Krooss, B.M., 2008. Carbon dioxide storage potential of shales. Int. J. Greenhouse Gas Control 2, 297–308.

Chapman, N., Hooper, A., 2012. The disposal of radioactive wastes underground. Proc. Geol. Assoc. 123, 46–63.

Malusis, M.A., Shackelford, C.D., Olsen, H.W., 2003. Flow and transport through clay membrane barriers. Eng. Geol. 70, 235–248.

Malusis, M.A., Shackelford, C.D., 2004. Predicting solute flux through a clay membrane barrier. J. Geotech. Geoenviron. Eng. 130, 477–487.

Neuzil, C., 1982. On conducting the modified "slug" test in tight formations. Water Resour. Res. 18, 439–441.

Neuzil, C., 1986. Groundwater flow in low-permeability environments. Water Resour. Res. 22, 1163–1195.

Neuzil, C., 1993. Low fluid pressure within the Pierre Shale: a transient response to erosion. Water Resour. Res. 29, 2007–2020.

Neuzil, C., 1994. How permeable are clays and shales? Water Resour. Res. 30, 145–150.

Neuzil, C., 2013. Can shale safely host US nuclear waste? Eos, Trans. Am. Geophys. Union 94, 261–262.

Chapter 1

Surface Properties of Clay Minerals

Christophe Tournassat,[a,c] Ian C. Bourg,[b,c] Carl I. Steefel[c] and Faïza Bergaya[d]

[a] *Water, Environment and Ecotechnology Division, French Geological Survey (BRGM), Orléans, France;* [b] *Department of Civil and Environmental Engineering, Princeton University, Princeton, NJ, USA;* [c] *Earth Sciences Division, Lawrence Berkeley National Laboratory, Berkeley, CA, USA;* [d] *Centre de Recherche sur la Matière Divisée, Centre National de la Recherche Scientifique (CNRS), Orléans, France*

Chapter Outline

Developments in Clay Science, Vol. 6C. http://dx.doi.org/10.1016/B978-0-08-100027-4.00001-2

Clay minerals are defined as a group of phyllosilicates of small size, typically less than 2 μm in their larger dimension (Bergaya and Lagaly, 2013a). They have a high specific surface area (SSA), the highest known among major natural minerals. Unlike other high-SSA natural minerals such as zeolites and manganese and iron (hydr)oxides, clay minerals are frequently the main components of extended sedimentary stratigraphic layers. Illite and smectite (Sm) alone may constitute ~30% of all sedimentary rocks (Garrels and Mackenzie, 1971). Most macroscopic properties of clay media are related to physical and chemical processes that take place at clay mineral surfaces. In turn, the surface properties of clay minerals are intimately linked to their crystallographic properties.

The present chapter introduces key terms and concepts in clay science that are used in this volume. For the sake of simplicity, this introduction focuses on the clay minerals that are most prevalent in engineered and natural clay barriers such as smectite (Sm) (particularly montmorillonite (Mt)), illite, mixed layers illite−smectite (I-Sm), and kaolinite (Kaol). Broader reviews of clay mineral crystallography and surface properties can be found in several books (Grim, 1968; Güven, 1992; Bergaya and Lagaly, 2013b).

1.1 FROM SHEETS TO CLAY MINERAL LAYERS

1.1.1 Structure of Clay Mineral Layers

The fundamental structure of clay minerals consists of a sheet of edge-sharing $M_{Oct}O_6$ octahedra (M_{Oct} = Al, Mg, or Fe), the *octahedral sheet*, fused to one or two sheets of corner-sharing $M_{Tet}O_4$ tetrahedra (M_{Tet} = Si or Al), the *tetrahedral sheet(s)*. The association of one octahedral sheet and one or two tetrahedral sheets forms a *clay mineral layer*. The first criterion in classifying clay minerals is their layer type: layers with one tetrahedral sheet form the 1:1 (or tetrahedral-octahedral, TO) layer type; layers with two tetrahedral sheets (on either side of the octahedral sheet) form the 2:1 (or TOT) layer type. Smectite and illite have 2:1 layer structures, whereas Kaol has a 1:1 layer structure (Figure 1.1).

The M_{Oct} metals in the octahedral sheet of clay minerals consist predominantly either of divalent metals (Mg, Fe^{II}), in which case all octahedral sites are occupied by a metal and the clay mineral is known as *trioctahedral*, or of trivalent metals (Al, Fe^{III}), in which case only two-thirds of the octahedral sites are occupied in a honeycomb pattern and the clay mineral is known as *dioctahedral*. Dioctahedral clay minerals with Al as the main octahedral cation (including Sm, illite, and Kaol) are the predominant type of clay minerals in most sedimentary environments. Their ideal structural formulas are $Si_2Al_2O_5(OH)_4$ and $Si_4Al_2O_{10}(OH)_2$ for the TO and TOT layers, respectively (Figure 1.1).

Clay mineral structures contain three types of oxygen atoms: basal O atoms (O_b) that bridge neighboring $M_{Tet}O_4$ tetrahedra and form a plane of O atoms

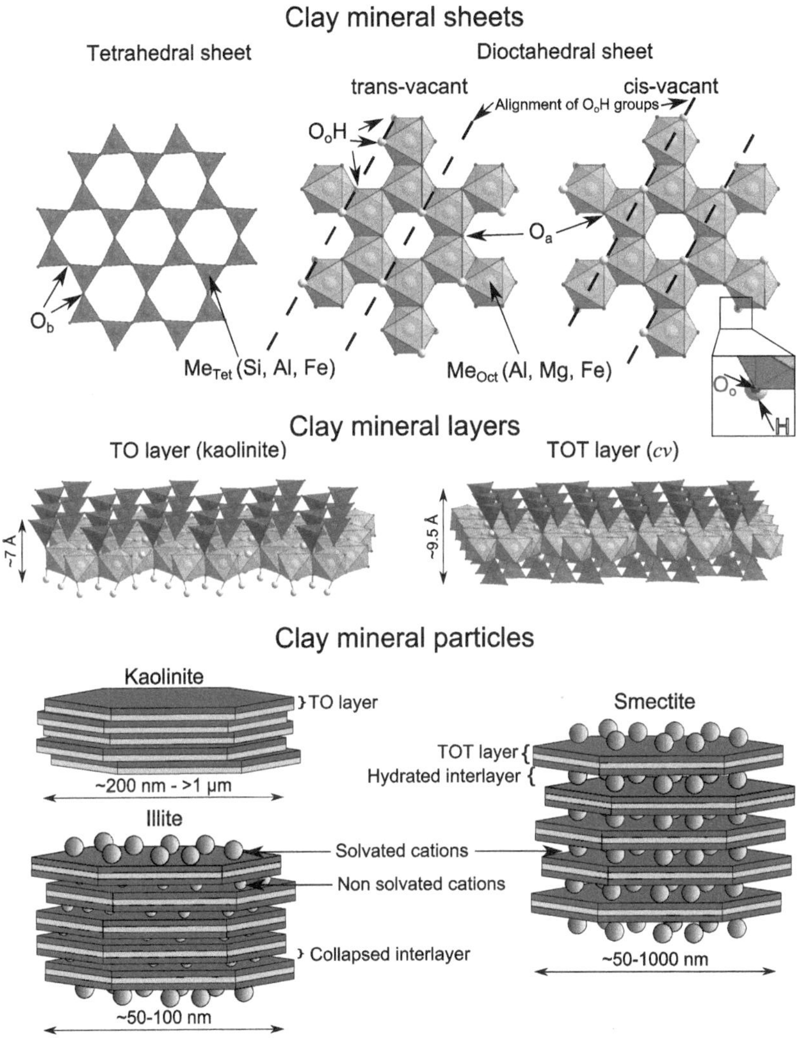

FIGURE 1.1 From top to bottom: tetrahedral and octahedral sheets, TO (Kaol) and TOT layers (*cv*-Mt), and clay mineral particles. The Kaol layer structure was taken from the COD database (Gražulis et al., 2012). The *cv*-Mt structure was taken from Tsipursky and Drits (1984).

constituting the *siloxane surface*; apical O atoms (O_a) that link $M_{Tet}O_4$ tetrahedra to $M_{Oct}O_6$ octahedra; and octahedral O atoms (O_o) that connect $M_{Oct}O_6$ tetrahedra and almost always carry a proton (OH). Each clay mineral layer has two basal surfaces. In TOT layer type clay minerals, these surfaces are both siloxane surfaces (i.e., planes of O_b atoms). In TO layer type clay minerals, one basal surface is a siloxane surface, while the other is a metal-oxide-like plane of protonated O_o atoms (Figure 1.1).

In TOT layers, each octahedral site is surrounded by two O_o atoms and four O_a atoms. However, not all octahedral sites have the same geometry with regard to the positions of their O_o anions. Specifically, one-third of the octahedral sites are known as *trans-octahedra*, because their O_o atoms are located on opposite corners of the octahedron; the remaining octahedral sites are *cis-octahedra*, because their O_o atoms are located on the same edge of the octahedron. Consequently, dioctahedral TOT layer may be either *cis-* or *trans-vacant*, depending on whether their octahedral vacancies are located on *cis-* or *trans-octahedral* sites (Figure 1.1). Montmorillonite, a common type of dioctahedral smectite, usually has a *cis*-vacant structure, whereas illite exhibits either *cis-* or *trans-vacant* structures (Drits and Zviagina, 2009; Brigatti et al., 2013).

1.1.2 Layer Charge and Charge Compensation Mechanisms

A particular feature of many TOT clay minerals is their significant negative layer charge density x (in moles of charge per mole of clay mineral, defined on the basis of a $Si_4Al_2O_{10}(OH)_2$ ideal layer formula unit). This layer charge arises from *isomorphic substitutions* of tetrahedral or octahedral metals. In dioctahedral clay minerals, the most common substitutions are of Si by Al in the tetrahedral sheet and of Al by Mg, Fe^{II}, or Fe^{III} in the dioctahedral sheet. Additional phenomena that can influence the layer charge include the presence of vacant octahedral sites in trioctahedral clay minerals (denoted by an empty square in the unit cell formula, □), the presence of trioctahedral domains in dioctahedral clay minerals, and the partial dehydroxylation of the octahedral sheet arising from oxidation/reduction reaction of octahedral iron (Manceau et al., 2000). The layer charge is not necessarily spatially uniform in the layer: the location of isomorphic substitutions can be ordered, clustered, or randomly distributed (Vantelon et al., 2001, 2003; Gates, 2005; Ngouana Wakou and Kalinichev, 2014). The resulting negative layer charge is balanced primarily by the presence of alkali and alkaline earth metals (Na^+, K^+, Ca^{2+}, Mg^{2+}) on the clay mineral basal surfaces. Among the TOT clay minerals, smectite can have a range of negative layer charges x between 0.2 and 0.6 $mol_c\ mol^{-1}$, while illite has x values between 0.6 and 0.9 $mol_c\ mol^{-1}$ (Sposito et al., 1999; Brigatti et al., 2013). The TO clay minerals, including kaolinite, have x values close to zero.

1.1.3 Aspect Ratio and Morphology of Clay Mineral Particles

A clay mineral particle is formed by the stacking of up to dozens of clay mineral layers (see Section 1.2.1). From crystallographic data, it can be easily estimated that the distance between the planes of oxygen atoms on opposite layer surfaces is 6.54 Å for Mt and 4.5 Å for Kaol. A rough estimation of the layer thickness can be obtained by adding to these values twice the ionic radius of oxygen (~1.5 Å). As a result, the layer thickness is about 7 Å for a TO layer and 9.5 Å for a TOT layer. The thickness of each layer is much smaller than its basal plane dimensions, which range from 50 to 100 nm for illite (Poinssot et al., 1999; Sayed Hassan et al., 2006), from 50 to 1000 nm for Mt (Zachara et al., 1993; Tournassat et al., 2003; Yokoyama et al., 2005; Le Forestier et al., 2010; Marty et al., 2011), and from less than 200 nm to more than 1 μm for Kaol (Djéran-Maigre et al., 1998). Consequently, clay minerals generally present a high aspect ratio with different morphologies: Kaol and well-crystallized illite have a tendency toward hexagonal and elongated hexagonal morphologies respectively, whereas Mt and less well-crystallized illite have mostly irregular platy or lath-shaped morphologies.

1.2 FROM LAYERS TO PARTICLES AND AGGREGATES

1.2.1 Layer Stacking and Hydration

Layers stack to form *clay mineral particles* as shown in Figure 1.1. The number of layers stacked in a single particle depends on the nature of the clay mineral. Illite particles typically consist of 5 to 20 stacked TOT layers. For Kaol, the number of stacked layers can range from 10 to more than 200 in a single sample (Sayed Hassan et al., 2006). For Sm (a swelling clay mineral), the layers can become completely delaminated; the number of layers per Sm particle increases with decreasing water chemical potential and also tends to increase with the valence of the charge-compensating cation (Banin and Lahav, 1968; Shainberg and Otoh, 1968; Schramm and Kwak, 1982; Sposito, 1992; Saiyouri et al., 2000). Due to the irregular morphology of clay mineral layers, the edge surfaces of different layers in a single particle may be misaligned. Moreover, translational and rotational disorder between adjacent layers makes the structure turbostratic at the scale of individual clay mineral particles.

The space between clay mineral layers (the *interlayer space*) is either empty (if $x = 0$) or occupied by cations that compensate the layer charge (if $x > 0$). In nonswelling clay minerals with $x > 0$ (such as illite), the interlayer cations are nonsolvated and consist predominantly of K^+ or NH_4^+. In swelling clay minerals such as smectite, the interlayer space contains water in variable quantity as a function of temperature, applied stress, the amount and origin of layer charge (from tetrahedral or octahedral substitutions), the water chemical potential, and the identity of the interlayer cation(s) (Cases et al., 1992; Bérend et al., 1995; Cases et al., 1997; Saiyouri et al., 2004; Holmboe et al., 2012; Ngouana Wakou and Kalinichev, 2014). Interlayer water molecules are strongly influenced by the interlayer cations and by the siloxane surface (Sposito and Prost, 1982). Interlayer cations in swelling clay minerals tend to be fully solvated except in the case of cations with low hydration energies (such as K^+, Rb^+, Cs^+, or NH_4^+) that adsorb as inner-sphere surface complexes.

The variable amount of interlayer water in Sm leads to significant swelling associated with variations in interlayer distance. At the scale of an individual Sm particle, this swelling can be characterized by X-ray diffraction measurements of the *basal reflection* d_{001} (the sum of the layer thickness and interlayer distance). The interlayer distance is sensitive to the same conditions that influence interlayer water (Bradley et al., 1937; Méring and Glaeser, 1954; Norrish, 1954; Slade et al., 1991; Sato et al., 1992; Kozaki et al., 1998; Ferrage et al., 2005, 2007a,b, 2010; Holmboe et al., 2012). At low water contents (below ~0.5 $g_{water}/g_{clay\ mineral}$) clay mineral swelling occurs in a stepwise manner with discrete stable basal spacings at $d_{001} = 11.8-12.7$ Å (one-layer hydrate), 14.5−15.7 Å (two-layer hydrate), 18.4−19 Å (three-layer hydrate), and up to 19−22 Å corresponding to the four-layer hydrate (Holmboe et al., 2012; Lagaly and Dékány, 2013). In the case of Na^+- and Li^+-Sm, swelling can proceed to much larger d_{001} values with increasing water chemical potential in a continuous rather than stepwise manner (Norrish, 1954). These two types of clay mineral swelling (stepwise and continuous) are termed *crystalline* and *osmotic swelling*. The swelling−shrinkage phenomenon has important implications for the mechanical and hydrologic properties of Sm.

1.2.2 Mixed-Layer Clay Minerals

Mixed-layer clay minerals exhibit alternating layers with contrasting structures, compositions, and basal distances, or with different layer displacement or rotation between consecutive layers (Sakharov and Lanson, 2013). Illite-smectite mixed layer minerals are very common in clay-rock stratigraphic layers. As a general rule, the surface properties of mixed layer mineral particles (hydration, adsorption) are not strictly equivalent to the simple combination of the properties of their constitutive layers.

1.2.3 Particle SSA

From the crystallographic information it is possible to calculate SSA values for individual layers. For a triclinic unit cell with dimensions $a \times b$ in the layer plane and angle γ between the $\boldsymbol{a}$ and $\boldsymbol{b}$ vectors, the *basal* SSA (Figure 1.2) of an individual layer is given by:

$$S_{layer}^{basal} = \frac{2ab \sin\gamma \times N_A}{M_{clay}} \tag{1.1}$$

where M_{clay} is the molar mass of the clay mineral unit cell (in g mol^{-1}, based on a $O_{20}(OH)_4$ structural formula for TOT clay minerals or a $O_{10}(OH)_8$ formula for

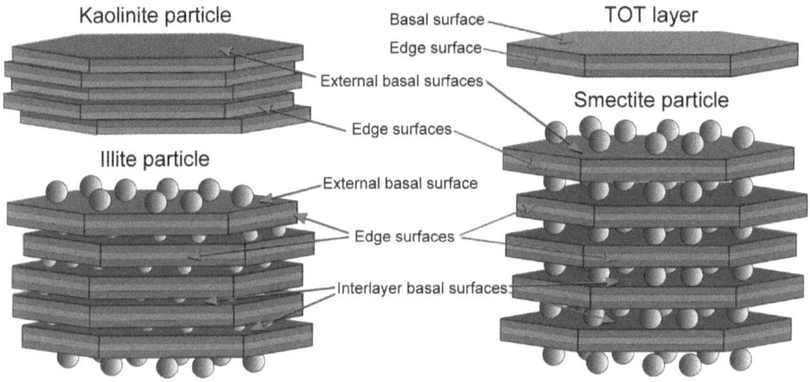

FIGURE 1.2 Positions of the edge, external basal, and internal basal surfaces on a TOT layer and in kaolinite, illite, and smectite particles.

TO clay minerals) and N_A is Avogadro's constant (6.022×10^{23} mol^{-1}). For Kaol ($a \times b = 5.15 \times 8.94$ Å^2, $\gamma = 89.8°$, Brigatti et al., 2013), illite ($a \times b = 5.20 \times 9.0$ Å^2, $\gamma = 90°$, Drits et al., 1993), and Mt ($a \times b = 5.18 \times 8.98$ Å^2, $\gamma \approx 90°$, Tsipursky and Drits, 1984), Eqn 1.1 yields $S_{layer}^{basal} = 1070, 720$, and 720–790 m^2 g^{-1}, respectively. For Mt, the range of layer chemical compositions and the nature of interlayer cation result in a range of S_{layer}^{basal} values.

Calculation of the *edge* SSA of an individual layer (Figure 1.2) requires knowledge of the perimeter, the thickness and the mass of one layer (P_{layer}, h_{layer} and m_{layer} in m, m^3 and g respectively):

$$S_{layer}^{edge} = \frac{h_{layer} P_{layer}}{m_{layer}} \tag{1.2}$$

For a layer having a regular hexagonal shape, side length of l, and thickness approximated by the layer-to-layer thickness, c^* (i.e., summing the layer thickness and the interlayer distance), the edge SSA is given by:

$$S_{hex\ layer}^{edge} = \frac{2c^*}{\sqrt{3}l} \times S_{layer}^{basal} \tag{1.3}$$

Accordingly, for $c^* = 7$ Å (for TO clay minerals) or 9.5–10 Å (for TOT clay minerals) and a particle diameter of 500 Å ($l \approx 250$ Å), the edge SSA of individual layers is roughly 20 to 30 times lower than their basal SSA.

With layer stacking, the basal SSA is split into contributions from the external and internal basal surfaces (Figure 1.2). If an average of n_c layers are stacked in one particle and if simplifying assumptions are used that all the layers have the same shape and size and that the layer edges are perfectly aligned, the estimation of the relative contribution of external and internal basal surfaces to the overall basal SSA is straightforward:

$$S_{layer}^{external\ \ basal} = \frac{S_{layer}^{basal}}{n_c}$$
$$S_{layer}^{interlayer\ \ basal} = S_{layer}^{basal}\left(1 - \frac{1}{n_c}\right) \tag{1.4}$$

In reality, Eqn (1.4) provides only an upper bound on the internal SSA and a lower bound on the external basal SSA, because the simplifying assumptions listed above are only approximately valid.

The relationship between edge and external basal SSA is obtained by combining Eqns (1.4) and (1.3):

$$S_{hex\ layer}^{edge} = \frac{2c^{*}n_c}{\sqrt{3}l} \times S_{layer}^{external\ \ basal} \tag{1.5}$$

As shown in Eqn (1.5), the edge and external basal SSA are similar if the average layer stacking is $n_c \sim l/10$ (with l expressed in angstroms). This condition is never met even for illite and Sm particles with the smallest observed layer dimensions. In short, the edge SSA of illite, Mt, and Kaol particles is always smaller (and sometimes much smaller) than their external basal SSA.

Experimental characterizations of clay mineral surfaces often include the SSA measured by N_2 gas adsorption with the Brunauer–Emmett–Teller technique (N_2-BET). The interpretation of N_2-BET surface areas of clay minerals should be carried out with caution for several reasons (Bergaya, 1995): firstly, N_2 probes only the external surfaces of clay mineral particles, i.e., it does not access the interlayer space, even in swelling clay minerals. Secondly, in addition to forming a monolayer on the external clay mineral surfaces, N_2 condenses in pores formed by the aggregation of clay mineral particles (Chiou et al., 1993; Michot and Villiéras, 2013). Thirdly, the N_2-BET surface area quantifies the sum of two SSA (external basal surface + edge surface) for surfaces that have very different properties. Finally, N_2-BET surface area is measured on dry samples, whereas the microstructure of swelling clay minerals is sensitive to water content. Despite these caveats, the N_2-BET surface area can provide useful information, for example, on the external SSA of nonswelling clay minerals such as illite and Kaol.

Measurements of the relative contributions of edge and external basal surface area to the total external surface area can be achieved by statistical analysis of particle morphology using atomic force microscopy and transmission electron microscopy (TEM) techniques (Nadeau, 1985; Bickmore et al., 2001; Cadene et al., 2005). Alternatively, the derivative isotherms summation (DIS) method can distinguish between different clay mineral surfaces (edge vs. external basal surfaces) in a single gas adsorption measurement based on differences in adsorption energy (Michot et al., 1990; Villiéras et al., 1992, 1997; Michot and Villiéras, 2013). The few studies that compared microscopic imaging and DIS methods in the case of illite and Mt particles yielded satisfactory agreement (Table 1.1) with a slight overestimation of edge surface area obtained by the DIS method (Reinholdt et al., 2013). In the case of swelling clay minerals, the total SSA (internal + external) of the clay mineral particles can be measured using adsorbents that induce clay mineral swelling, such as ethylene glycol monoethyl ether (EGME). For nonswelling clay minerals, the EGME-accessible surface area is close to the N_2-BET surface area (i.e., the external surface area); for swelling clay minerals, the EGME-accessible surface area is often commensurate with the total SSA calculated from crystallographic considerations (Srodon and McCarty, 2008) (Eqn (1.1), Table 1.1), but sometimes shows notable differences that depend on experimental conditions (Chiou and Rutherford, 1997; Michot and Villiéras, 2013). Equation (1.1) can be used to estimate the internal surface area, providing that the external surface area is known. However, this equation cannot be applied directly in the case of mixed layer clay minerals if the relative proportion of swelling and nonswelling interlayer spaces is not precisely known.

As shown by Eqn (1.4), an alternative route toward quantifying the proportion of internal and external basal surfaces in swelling clay minerals consists in determining the number of layers per stack. In certain cases, n_c can be determined in hydrated conditions. For example, light scattering and anion exclusion measurements have shown that n_c in Sm dispersions depends on the salinity, the nature of exchangeable cation, and the history of the clay mineral (for example, n_c shows significant hysteresis during cation exchange experiments) (Sposito, 1992; Verburg et al., 1995; Bourg and Sposito, 2011), while X-ray diffraction measurements and TEM characterization have shown that n_c in clay mineral pastes also depends on the solid–water ratio (Saiyouri et al., 2000; Melkior et al., 2009; Muurinen, 2009).

1.2.4 Nature of the External Basal Surfaces of Clay Mineral Particles

In nonswelling clay minerals, the nature of the clay mineral layers that form the external basal surfaces of each particle (the outer surface layer, OSL) can be different from the nature of the layers in the core of the particle. For example, three different types of Kaol OSL have been described: a 7 Å Kaol TO layer as described in Section 1.1.1; an uncharged (pyrophyllite-like) TOT layer on one side of the Kaol particle (such that the stacking sequence in the particle goes TOTO...TOT), and a charged TOT layer (Sm) on one or both sides of the Kaol particle (Ma and Eggleton, 1999a). Similarly, illite particles may terminate with a Kaol layer (Tsipursky et al., 1992). This heterogeneity in particle composition has little or no influence on SSA but may profoundly influence the surface charge and surface chemistry of

TABLE 1.1 Representative Specific Surface Area (SSA) Values and Cation Exchange Capacity (CEC) for Illite and Mt Minerals as a Function of Selected Measurement Conditions

	Main Exchangeable Cation	CEC (mol kg^{-1})	SSA (m^2 g^{-1})						References
						DIS		AFM	
			Crystallographic	EGME	N_2-BET	Basal	Edge	Edge	
Mt Swy-1, <0.2 μm fraction	Ca	0.86 (0.90[a])	755	781					Srodon and McCarty (2008)
Mt Swy-1, <0.2 μm fraction	Ca	0.78	"		24				Chiou and Rutherford (1997)
Mt Swy-1	Na	0.66	"		19				Chiou and Rutherford (1997)
Mt Swy-1, <2 μm fraction	Na	0.84–0.9	"		35				Baeyens and Bradbury (1997)
Mt Swy-2, <2 μm fraction	Na		750–780	737	45.4	21.7	25.4		Le Forestier et al. (2010)
Mt MX80, <2 μm fraction	Na		780			26.6	8.5	7–9	Tournassat et al. (2003)
Illite du Puy	Na	0.22	–	95	97				Missana et al. (2009)
Illite du Puy	Na		–			133	38	21–30	Sayed Hassan et al. (2006)

[a]*Value corrected from the water content of the clay mineral dried at 110°C.*

the particles. For mixed layer minerals the question of the nature of the OSL is even more acute (Sakharov and Lanson, 2013) and the charge density of the external basal surfaces is not usually known.

1.2.5 Charge Balance at the Scale of a Clay Mineral Particle

An important feature of clay mineral particles is their intrinsic surface charge density σ_{in} ($\text{mol}_c\,\text{kg}^{-1}$). This surface charge density is the sum of two contributions: the net structural surface charge density σ_0 ($\text{mol}_c\,\text{kg}^{-1}$) and the net proton surface charge density σ_{H} ($\text{mol}_c\,\text{kg}^{-1}$) (Sposito, 1998):

$$\sigma_{\text{in}} = \sigma_0 + \sigma_{\text{H}} \tag{1.6}$$

Charge balance on a clay mineral particle imposes that Δq, the sum of the adsorbed ion charge densities q_i of all species except surface-complexed H^+ and OH^- ions, equals the opposite of the intrinsic surface charge density:

$$\sigma_0 + \sigma_{\text{H}} + \Delta q = 0 \tag{1.7}$$

The three types of clay mineral surfaces in Figure 1.2 contribute differently to the charge balance relation (1.7). If $S_{layer}^{edge} \ll S_{layer}^{basal}$ and all layers in a particle are similar, the net structural surface charge density depends only on the layer charge x, the molar mass of a clay mineral unit cell (M_{clay}), and the average number of layers per particle (in the case of nonswelling clay minerals):

$$\sigma_0(\text{swelling clay minerals}) = \frac{x}{M_{clay}} \tag{1.8}$$

$$\sigma_0(\text{non-swelling clay minerals}) = \frac{1}{n_c}\frac{x}{M_{clay}} \tag{1.9}$$

Net proton surface charge is much more localized, because it arises primarily from proton and hydroxyl complexation by oxide-type surface functional groups that exist only on edge surfaces and, in the case of TO clay minerals, on the octahedral basal surface (Tombácz and Szekeres, 2006). Proton adsorption by cation exchange on basal surfaces also contributes to σ_{H} at low pH values (Bourg et al., 2007).

1.2.6 From Particles to Aggregates and Porous Media

When packed together, clay mineral particles form *aggregates* and their external surfaces delineate *interparticle spaces*. Assemblies of aggregates delineate *interaggregate spaces* (Bergaya and Lagaly, 2013a). The sum of the volumes occupied by interlayer, interparticle, and interaggregate spaces, normalized to the total volume of the porous medium, is the *porosity* of the porous medium (θ). If the dry bulk density (ρ_{dry}) of the porous medium and the mean layer density

(ρ_{layer}) of the clay minerals are known, the calculation of the porosity is straightforward:

$$\theta = 1 - \frac{\rho_{dry}}{\rho_{layer}} \tag{1.10}$$

In compacted clay, pores are heterogeneous in size and their aspect ratio (the ratio of the longest to the shortest pore dimension) tends to be very large due to the high aspect ratio of the clay mineral layers and particles (for example, the aspect ratio is 200 for a Sm with an interlayer distance of ~1 nm (three-layer hydrate) and a layer characteristic length of 200 nm). A key measure of pore size is that of the smallest pore dimension, usually referred to as the pore width. If the clay mineral is homogeneous and the basal surfaces are perfectly parallel, the mean width of the pores ($\overline{d_{pore}}$) can be calculated according to:

$$\overline{d_{pore}} = \frac{2\theta}{\rho_{dry} \times SSA_{particle}} \tag{1.11}$$

where $SSA_{particle}$ is the sum of the SSAs of edge surfaces, external basal surfaces, and water-accessible (non-collapsed) interlayers. In the case of Mt, the mean width of the pore becomes similar to the interlayer distance of the three-layer hydrate at a clay mineral dry density of ~1.5 kg dm^{-3}, because of the large total SSA of the particles (Figure 1.3). In the case of illite, the mean pore width remains larger than that of the three-layer hydrate up to ~2.5 kg dm^{-3}, because the interlayer spaces are collapsed and the overall SSA is approximately ten times lower than that of Mt particles.

In clay-rocks, the presence of nonclay minerals (e.g., quartz, carbonates, pyrite) has a large influence on the pore size distribution and the structure of the pore network (Keller et al., 2013). The pore size distribution of clay-rocks is generally bimodal or more complex: interaggregate spaces can be as wide as few micrometers whereas interlayer spaces typically have a thickness on the order of 1 nm (Keller et al., 2011). Following the IUPAC nomenclature in use for porous materials, pores can be classified in three size categories (Rouquerol

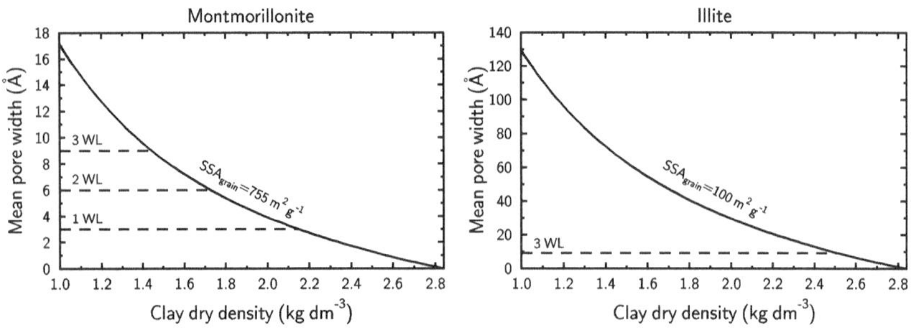

FIGURE 1.3 Mean pore width calculated according to Eqn (1.11) for two different clay minerals: a pure Mt (left) and a pure illite (right). Horizontal dashed lines marked 1, 2, and 3WL indicate the interlayer distances of the one-, two-, and three-layer hydrates, respectively.

et al., 1994): micropores have widths smaller than 2 nm, mesopores have widths between 2 and 50 nm, and macropores have widths larger than 50 nm.

The structure of the pore network is a key controlling factor for the fluid properties in clay media. For example, the nanometer-scale pore systems of gas shales are an important control on hydrocarbon storage capacity and fluid transmissivity to fracture networks (Chalmers et al., 2012). In water-saturated conditions, the pores in clay material and clay-rocks are usually fully saturated as evidenced by the agreement between the porosity values derived from water loss measurements, density measurement (wet, dry, and grain densities), and water diffusion-accessible porosity measurements (Fernández et al., 2014). These results imply that the pore network is fully connected, though a full characterization of the pores responsible for the connectivity between the macropores, i.e., the characterization of the pore throats, is still pending. Porosity imaging at the nanometer scale and segmentation of the total porosity pose a real challenge to unraveling the structure of the pore networks (Hemes et al., 2013).

1.3 SURFACE PROPERTIES OF BASAL SURFACES

1.3.1 Cation Exchange and Cation Exchange Capacity

As noted above, negative layer charge arising from isomorphic substitutions is balanced primarily by cations located in the interlayer space or on external basal surfaces. Unless they are located in collapsed interlayer spaces of non-swelling clay minerals, these cations are *readily exchangeable* by other cations present in solution. The exchange of charge-compensating cations between the clay mineral surface and bulk liquid water, a *cation exchange* reaction, can be described with the following expression:

$$z_j X_{z_i} C_i + z_i C_j^{z_j+} \rightleftharpoons z_i X_{z_j} C_j + z_j C_i^{z_i+} \tag{1.12}$$

where X^- represents a generic negatively charged exchange site. For practical purposes, the selectivity of the reaction described by Eqn (1.12) can be quantified using a variety of empirical *ion exchange selectivity coefficients*, K_{ex}, such as those of Vanselow (1932), Gapon (1933), and Gaines and Thomas (1953):

$$\begin{aligned}
{}^{V}K_{\text{ex}}^{\text{i}\to\text{j}} &= \frac{(x_{C_j})^{z_i}}{(x_{C_i})^{z_j}} \frac{\left\{C_i^{z_i+}\right\}^{z_j}}{\left\{C_j^{z_j+}\right\}^{z_i}} \\
{}^{G}K_{\text{ex}}^{\text{i}\to\text{j}} &= \frac{E_{C_j}}{E_{C_i}} \frac{\left\{C_i^{z_i+}\right\}^{1/z_i}}{\left\{C_j^{z_j+}\right\}^{1/z_j}} \\
{}^{GT}K_{\text{ex}}^{\text{i}\to\text{j}} &= \frac{(E_{C_j})^{z_i}}{(E_{C_i})^{z_j}} \frac{\left\{C_i^{z_i+}\right\}^{z_j}}{\left\{C_j^{z_j+}\right\}^{z_i}}
\end{aligned} \tag{1.13}$$

where x_{C_i} denotes the fractional contribution of cation C_i to the total number of moles of exchangeable cations; E_{C_i} denotes the fractional contribution of C_i to the total number of moles of charge of exchangeable cations; brackets denote the activities of C_i and C_j in aqueous solution; and ${}^{V}K_{\text{ex}}^{\text{i}\rightarrow\text{j}}$, ${}^{G}K_{\text{ex}}^{\text{i}\rightarrow\text{j}}$, and ${}^{GT}K_{\text{ex}}^{\text{i}\rightarrow\text{j}}$ are the Vanselow, Gapon, and Gaines–Thomas selectivity coefficients, respectively. The selectivity coefficients defined by Eqns (1.13) are not thermodynamic equilibrium constants and, therefore, they may vary with the aqueous geochemistry, with the occupancy of the exchanger (clay mineral) phase, and with other conditions such as the width of the interlayer space (Bourg and Sposito, 2011). For example, in Na^+–K^+ exchange reactions on illite or Mt, $K_{\text{ex}}^{\text{Na}\rightarrow\text{K}}$ decreases as x_K (or, equivalently, E_K) increases (Jensen, 1973).

The value of the selectivity coefficient for a given exchange reaction quantifies the affinity of a surface for a given cation over another cation. The value of the selectivity coefficient depends on the choice of model, except for reactions between monovalent cations (where ${}^{V}K_{\text{ex}}^{\text{i}\rightarrow\text{j}} = {}^{G}K_{\text{ex}}^{\text{i}\rightarrow\text{j}} = {}^{GT}K_{\text{ex}}^{\text{i}\rightarrow\text{j}}$) or for homovalent exchange reactions between multivalent cations (where ${}^{V}K_{\text{ex}}^{\text{i}\rightarrow\text{j}} = {}^{GT}K_{\text{ex}}^{\text{i}\rightarrow\text{j}}$). Experimental data on binary exchange reactions on Mt are broadly consistent with ${}^{V}K_{\text{ex}}^{\text{i}\rightarrow\text{j}} \sim 1$ for strongly-solvated cations that adsorb as fully solvated outer-sphere surface complexes (Li^+, Na^+, divalent cations), whereas larger ${}^{V}K_{\text{ex}}^{\text{i}\rightarrow\text{j}}$ values are obtained for weakly-solvated cations that adsorb as partly-solvated inner-sphere surface complexes (K^+, Rb^+, Cs^+) (Sposito et al., 1983; McBride, 1994; Steefel et al., 2003; Bourg and Sposito, 2011; Bergaya et al., 2013).

In the case of swelling clay minerals, ion exchange occurs simultaneously on internal and external basal surfaces. The size of the clay particle and of the interlayer space may influence ion exchange selectivity (Laird and Shang, 1997). For example, the affinity for Ca^{2+} relative to Na^+ may be greater in Sm interlayer space than on external surfaces. In the case of nonswelling clay minerals such as illite, most K^+ cations in the collapsed interlayer space are not readily exchangeable. In this case, the exchangeable cations include only those located on the external surfaces. A small fraction of the interlayer K^+ cations, located near the edges of the particle (at the so-called *frayed edge sites*), may be exchangeable for other weakly-hydrated cations (Rb^+, Cs^+, NH_4^+) (Comans et al., 1991; Poinssot et al., 1999; Bradbury and Baeyens, 2000).

As noted above, the total specific amount of charge that is balanced by exchangeable cations is referred to as the CEC of the clay mineral. The part of the CEC value that is due to the compensation of the layer structural charge is termed permanent CEC because it does not depend on external conditions. For Mt, the value of the CEC is usually commensurate with the layer charge due to isomorphic substitutions. The permanent CEC of illite is far lower than the cumulated charge of its constitutive layers but is roughly equivalent to the charge of its OSL. The permanent charge of Kaol is close to zero except for Kaol samples exhibiting a Sm OSL (Ma and Eggleton, 1999b).

1.3.2 Protonation and Deprotonation of Oxygen Atoms on Basal Surfaces

It is generally considered that the oxygen atoms on the siloxane surface (O_b) cannot be protonated in the water acidity range, although recent surface force measurements on Kaol particles challenge this hypothesis (Gupta and Miller, 2010). Conversely, the $>Al_2{-}OH$ functional groups on the octahedral basal surface of Kaol have a well-established ability to gain or loose protons when exposed to liquid water that contributes to the observed pH-dependent surface charge of Kaol (Brady et al., 1996; Tombácz and Szekeres, 2006).

1.3.3 Electrostatic Potential, Cation Condensation, and Anion Exclusion

The negative charge of clay mineral basal surfaces is screened by cation adsorption and anion exclusion near the clay mineral surface in a region known as the *electrical double layer* (EDL). Measurements of anion exclusion and electrophoretic mobility in aqueous dispersions of clay mineral particles indicate that the EDL has a thickness on the order of several nanometers with a strong dependence on ionic strength (Sposito, 1992). The EDL can be conceptually subdivided into a *Stern layer* containing inner- and outer-sphere surface complexes and a *diffuse layer* containing ions that interact with the surface through long-range electrostatics (Henderson and Boda, 2009; Lee et al., 2010).

The diffuse layer composition and structure is the matter of intensive research because it governs many macroscopically observed phenomena including swelling, osmosis, and particle aggregation (Leroy et al., 2006). A simplified yet powerful description of the diffuse layer can be achieved in the framework of the *modified Gouy–Chapman model*, which uses the Poisson equation and the Boltzmann distribution and makes the following three simplifying hypotheses: that the solution phase is a uniform continuum characterized solely by its dielectric permittivity ε ($78.3 \times 8.85419 \cdot 10^{-12}$ F m^{-1} for water at 298 K); that the surface charge σ (in m^{-2}) is uniformly distributed in the interfacial plane; and that the potential of mean force $W_i(x)$ applying to an ion i at a distance x from the surface is determined only by the mean electrostatic potential at the same position, $\psi(x)$. Additionally, the modified Gouy–Chapman model postulates that the ions cannot approach the surface closer than a distance a. If clay mineral basal surfaces are modeled as infinite planes, the Poisson equation can be written as follows:

$$\frac{d^2\psi(x)}{dx^2} = -\frac{1}{\varepsilon}\sum_i z_i F c_i(x) \qquad (1.14)$$

and the Boltzmann equation is given by:

$$c_i(x) = c_{i0}\exp\left(\frac{-z_i F \psi(x)}{RT}\right) \qquad (1.15)$$

where $c_i(x)$ and c_{i0} are the volumetric concentrations of ion i at a distance x and at infinite distance from the surface (mol m^{-3}), F is Faraday's constant (96 485 C mol^{-1}), R is the ideal gas constant (8.314 J K^{-1} mol^{-1}), T is absolute temperature (K), and the electrostatic potential is conventionally defined as zero at an infinite distance from the surface. In the case of an isolated plane, Eqn (1.14) can be solved analytically for simple electrolyte compositions. In the case of a 1:1 electrolyte such as NaCl, the electrostatic potential is related to the ionic strength ($c_{Na0} = c_{Cl0} = 1000I$)[1] through the following equation:

$$\frac{F\psi(x)}{RT} = 4\text{arctanh}\left[\tanh\left(\frac{F\psi(a)}{4RT}\right) \times \exp(-\kappa(x-a))\right] \tag{1.16}$$

where κ is the inverse of the Debye length (κ^{-1}):

$$\kappa = \sqrt{\frac{2F^2 1000I}{\varepsilon RT}} \tag{1.17}$$

and the electrostatic potential at the distance of closest approach is given by:

$$\frac{F\psi(a)}{RT} = -2\text{arcsinh}\left(\frac{F^2|\sigma|}{2\kappa\varepsilon RTN_A}\right) \tag{1.18}$$

Equations (1.16) and (1.18) are strictly valid only for 1:1 electrolytes. Other analytical solutions to Eqn (1.14) exist for n:n symmetric electrolytes with $n > 1$ (e.g., $CaSO_4$), for 2:1 electrolytes (e.g., $CaCl_2$) and for 2:1:1 mixed electrolytes (e.g., $CaCl_2$/NaCl) (Chen and Singh, 2002). In practice, however, Eqns (1.16) and (1.18) are often used regardless of the type of electrolyte.

The electrostatic potential and Na/Cl concentration profiles obtained from Eqns (1.16) and (1.18) are shown in Figure 1.4 for $T = 298$ K, $\sigma = 6.2 \cdot 10^{17}\ m^{-2} = 0.62\ nm^{-2} = -0.1\ C\ m^{-2}$ (a value characteristic of Mt basal surfaces; see Table 1.1) and $I = 0.1$ or $I = 0.01$. The characteristic thickness of the diffuse layer (i.e., the length scale associated with the decay of $\psi(x)$) depends on ionic strength and is related to the Debye length through the exponentially decaying term of Eqn (1.16). The characteristic distance associated with cation adsorption is significantly shorter than the characteristic thickness of the diffuse layer, a phenomenon known as counterion condensation (Sposito, 2004).

In the modified Gouy–Chapman model, cation adsorption and anion exclusion at the clay mineral–water interface are fully determined by long-range electrostatic interactions between ions and the charged surface. This purely long-range electrostatic approach, with a single value of a for all ions, cannot explain observed differences in adsorption selectivity between cations of same valence. These differences can be ascribed to the presence of a Stern layer of specifically adsorbed cations (inner- or outer-sphere surface complexes)

1. The adimensional form of ionic strength is used here.

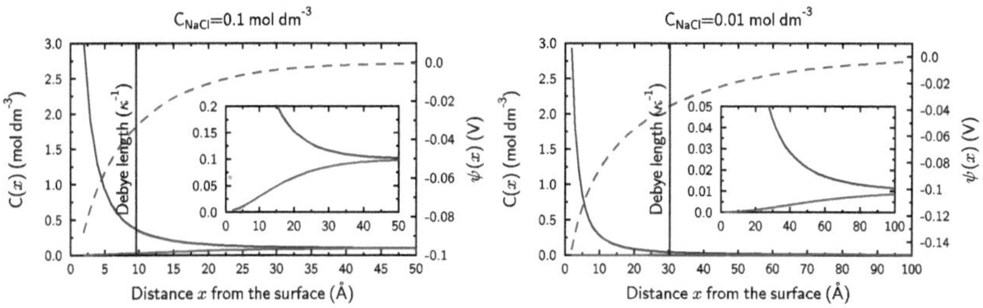

FIGURE 1.4 Modified Gouy–Chapman model prediction of the electrostatic potential, ψ (red dashed line), and Na^+ and Cl^- concentrations (blue (dark gray in print version) and green (light gray in print version) lines, respectively) as a function of distance from an external basal surface in contact with an infinite volume of NaCl solution at a temperature of 298 K. The uniform surface charge is $-0.1\ C\ m^{-2}$ and the bulk NaCl concentration is $0.1\ mol\ dm^{-3}$ (left figure) or $0.01\ mol\ dm^{-3}$ (right figure). The distance of closest approach a is set to 1.9 Å.

on the clay mineral surface (Dzombak and Hudson, 1995; Leroy et al., 2007; Tournassat et al., 2009; Appelo et al., 2010; Tournassat et al., 2011).

Equation (1.16) is highly nonlinear and difficult to combine with other equations to study coupled processes. Consequently, a simpler linearized version of Eqn (1.16) is sometimes used, known as the Debye–Hückel approximation:

$$\psi^{DH}(x) = \psi(a)\exp(-\kappa(x-a)) \tag{1.19}$$

The Debye–Hückel approximation is accurate only if the electrostatic potential at the surface is small, typically $\psi(a) \ll 2RT/F$. At 25°C, the condition is fulfilled only for $\psi(a) \ll 50$ mV.

Equations (1.16) and (1.18) are no longer valid if the distance between two clay mineral basal surfaces is less than approximately ten Debye lengths, because the diffuse layers on opposing clay mineral surfaces influence each other (Figure 1.5, top). This is typically the case in conditions relevant to Sm-rich clay-rocks or engineered clay barriers: for example, the mean pore width in Sm compacted to a dry bulk density of $1.5\ kg\ dm^{-3}$ is ~9 Å (Figure 1.3) while the Debye length of a $0.1\ mol\ L^{-1}$ NaCl solution is ~10 Å (Figure 1.4). Equations (1.16) and (1.18) also are not strictly applicable in the case of multi-ionic or multivalent solution compositions. In these cases, the Poisson–Boltzmann equation must be solved numerically for the geometry and electrolyte of interest. Alternatively, a much simpler *mean potential model* (often improperly referred to as a Donnan model) can be used. On the mean potential model, the diffuse layer is defined as a discrete region of volume V_{DL} with a uniform electrostatic potential ψ_M imposed by the following charge-balance relation:

$$\frac{\sigma A}{N_A} - \sum_i z_i V_{DL} c_{i0} \exp\left(\frac{-z_i F \psi_M}{RT}\right) = 0 \tag{1.20}$$

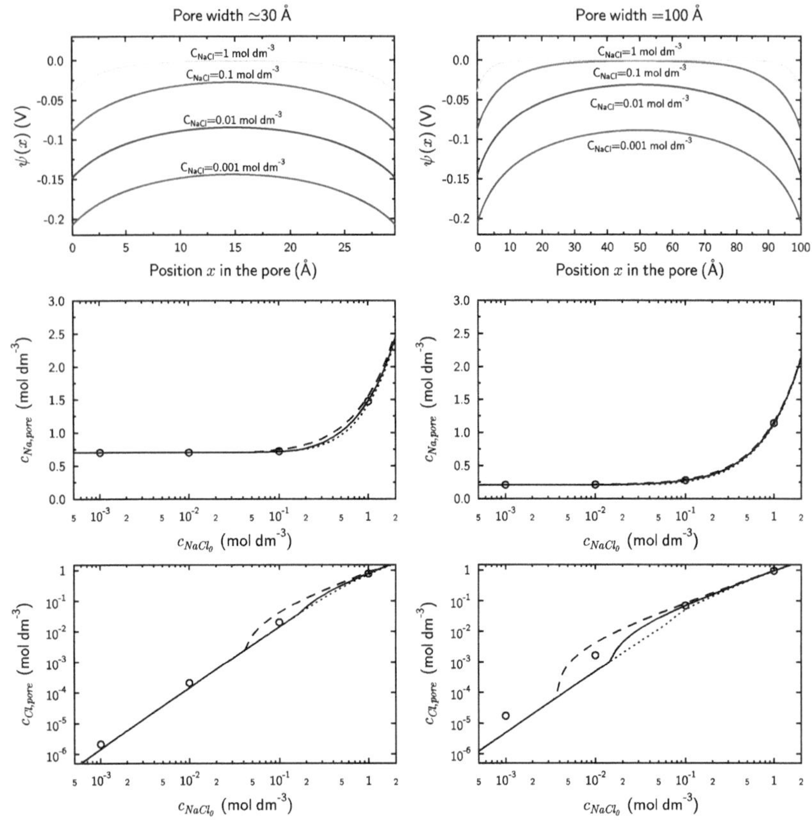

FIGURE 1.5 Top figure: electrostatic potential between two parallel clay mineral surfaces ($\sigma = 0.1$ C m^{-2}) as a function of external NaCl concentration and pore width (left: 30 Å; right: 100 Å) according to the numerical resolution of the Poisson–Boltzmann equation. Middle figure: comparison of the mean Na concentration calculated from (1) the integration of the Poisson–Boltzmann potential profile (circles) and (2) the mean potential model using a diffuse layer thickness of 1, 2, or 4 times the Debye length (dash, plain, and dotted lines respectively). Bottom figure: same as middle figure but for the mean Cl concentration.

where A is the value of the total surface area (m^2). The volume V_{DL} can, equivalently, be defined as a diffuse layer thickness $d_{DL} = V_{DL}/A$. With a pore width of l (m) between two clay mineral surfaces of surface charge σ, and considering a NaCl solution at equilibrium with a concentration c_{NaCl0} (in mol m^{-3}), Na^+ and Cl^- total concentrations in the pore ($c_{Na,\ pore}$, $c_{Cl,\ pore}$) are given by:

$$c_{Na,\ pore} = \frac{c_{NaCl0}}{l} \times \left((l - 2d_{DL}) + 2d_{DL} \exp\left(\frac{-F\Psi_M}{RT} \right) \right)$$

$$c_{Cl,\ pore} = \frac{c_{NaCl0}}{l} \times \left((l - 2d_{DL}) + 2d_{DL} \exp\left(\frac{F\Psi_M}{RT} \right) \right) \quad (1.21)$$

$$\frac{F\Psi_M}{RT} = \operatorname{asinh}\left(\frac{\sigma}{N_A \times 2d_{DL} \times c_{NaCl0}} \right)$$

if $l > 2d_{DL}$ and

$$c_{Na,\ pore} = c_{NaCl0} \exp\left(\frac{-F\psi_M}{RT}\right)$$
$$c_{Cl,\ pore} = c_{NaCl0} \exp\left(\frac{F\psi_M}{RT}\right) \tag{1.22}$$

otherwise. Figure 1.5 shows that a d_{DL} value corresponding to two Debye lengths (i.e., a mean potential volume extending two Debye lengths from each surface) leads to a strict equality of the mean potential model and the Poisson–Boltzmann model in the case of monovalent ions, as long as the diffuse layers from each surface do not interact. If the diffuse layers overlap, a mean potential model applied to the entire pore volume has the tendency to underestimate the mean anion concentration.

1.4 SURFACE PROPERTIES OF EDGES

1.4.1 Structure of Edges

At the edges of the layers, the clay mineral structure is cleaved in a manner that is not precisely established. It is usually assumed that the edges have a TO or TOT structure similar to the structure of the inner part of the layer with surface metals (Si, Al, and their substituting atoms) having the same coordination as in the inner structure and with most external oxygen atoms being under-coordinated (White and Zelazny, 1988). However, experimental results and atomistic simulations suggest that octahedral and tetrahedral metals on clay edges can adopt a different stoichiometry and coordination than in the bulk mineral structure, at least in some conditions (Carroll and Walther, 1990).

1.4.2 Protonation/Deprotonation, Edge Surface Charge, and Electrostatic Potential

Clay mineral edges carry a pH-dependent net proton surface charge (σ_H) that arises from the acid–base reactivity of edge surface functional groups. This acid base reactivity has been studied extensively using acid–base titrations of clay mineral dispersions (Charlet et al., 1993; Wanner et al., 1994; Zysset and Schindler, 1996; Baeyens and Bradbury, 1997; Avena, 2002; Tombacz and Szekeres, 2004, 2006; Duc et al., 2005a,b, 2006, 2008; Tertre et al., 2006). In these experiments, the consumption or release of a proton by/from the surface is measured as a function of acid/base additions at various salt background concentrations. If the initial surface charge of the titrated mineral is known, these measurements can be used to calculate the point of zero net proton charge (p.z.n.p.c., where σ_H is the net proton charge). In the case of solids with no structural charge ($\sigma_0 = 0$), the p.z.n.p.c. is equal to the point of zero net charge (p.z.n.c.) and the titration curve is directly related to the intrinsic charge of the surface (Sposito, 1998). Using these constraints, it is in principle possible to construct a surface charge and potential model by defining a set of surface reactions and proton affinity constants in a way similar to what can be done for oxides (Stumm et al., 1970; Schindler and Gamsjager, 1972; Sposito, 1984; Schindler and Stumm, 1987; Hiemstra et al., 1989; Davis et al., 1990; Dzombak and Morel, 1990; Sverjensky, 1993; Hiemstra et al., 1996), e.g.,

$$> SOH^{x+} \rightleftharpoons SO^{(x-1)+} + H^{+} \quad \log K_H \tag{1.23}$$

with:

$$K_H = \frac{(>SO^{(x-1)+})(H^+)}{(>SOH^{x+})} \tag{1.24}$$

where values in parentheses are activities.

In surface complexation models, the activity of a surface species is usually assumed to be equal to the ratio of the surface species concentration over the total concentration of site [$>S_{tot}$], times a Boltzmann factor that provides an activity coefficient correction:

$$(>SOH^{x+}) = \frac{[>SOH^{x+}]}{[>S_{tot}]}\exp\left(\frac{-xF\psi_s}{RT}\right) \tag{1.25}$$

$$\left(>SO^{(x-1)+}\right) = \frac{[>SO^{(x-1)+}]}{[>S_{tot}]}\exp\left(\frac{-(x-1)F\psi_s}{RT}\right) \tag{1.26}$$

where ψ_s is the surface potential at the protonation/deprotonation site, a quantity that is not measurable and must be predicted by a model. Oxide surface complexation models make the hypothesis that the charge is homogeneously distributed on a flat and infinite surface. For clay mineral edge surfaces, this hypothesis is not valid because the edge surface is very different from a flat infinite surface. Furthermore, for Sm or illite, permanent structural charge is distributed on the basal surfaces while variable charge is localized on the edges and, for Kaol, basal and edge surfaces carry different types of functional groups (Brady et al., 1996). This specificity of clay mineral surfaces makes the modeling of potentiometric titration data challenging, because the mutual influence of basal and edge electrostatic potentials must be taken into account to correctly predict σ_H as a function of pH and ionic strength (Secor and Radke, 1985; Chang and Sposito, 1996; Bourg et al., 2007; Delhorme et al., 2010) (Figure 1.6).

1.4.3 Surface Complexation, Cation and Anion Exchange

Similarly to oxide surfaces, clay mineral edge surfaces can bind inorganic or organic cations, anions, and molecules through short-range interactions with specific surface sites (Goldberg and Criscenti, 2008; Lagaly and Dékány, 2013).

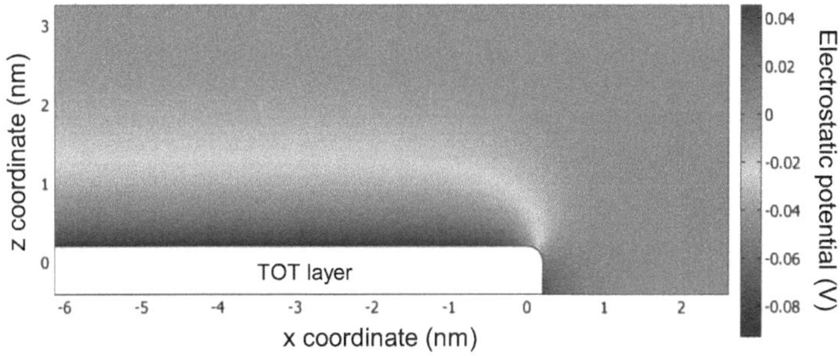

FIGURE 1.6 Mutual influences of the edge and basal surface charge on the electrostatic potential developing in the vicinity of the edge surface of a Mt layer. *Figure adapted from Tournassat et al. (2013).* The color scale corresponds to the potential value (in V) calculated at an ionic strength of 0.1 for a uniform basal charge density of -0.109 C m^{-2} and an edge charge density of 0.1 C m^{-2}.

Non-specific interactions also take place that compensate the charge of the edge surface and that explain the pH-dependent CEC of clay minerals (Sposito, 2004). Those processes are detailed in the following Chapter 2.

1.5 SUMMARY

The microstructural and surface properties of clay minerals introduced in the present chapter are generally well characterized, and the underlying mechanisms well understood, thanks to extensive research efforts in clay science since the 1950s. Still, important fundamental issues remain to be elucidated that impact the interpretation of a number of macroscopic observations. In particular, significant uncertainties remain with regards to the nature of the OSL of clay mineral particles in natural samples, the detailed microstructure of clay media at in-situ conditions (pressure, saturation), and the atomic-level structure of particle edges. Future breakthroughs in these areas will significantly advance current understanding of clay mineral surface properties.

REFERENCES

Appelo, C.A.J., Van Loon, L.R., Wersin, P., 2010. Multicomponent diffusion of a suite of tracers (HTO, Cl, Br, I, Na, Sr, Cs) in a single sample of Opalinus Clay. Geochim. Cosmochim. Acta 74, 1201–1219.

Avena, M.J., 2002. Acid-base behavior of clay surfaces in aqueous media. Encycl. Surf. Colloid Sci. 37–63.

Baeyens, B., Bradbury, M.H., 1997. A mechanistic description of Ni and Zn sorption on Na-montmorillonite. Part I: titration and sorption measurements. J. Contam. Hydrol. 27, 199–222.

Banin, A., Lahav, N., 1968. Particle size and optical properties of montmorillonite in suspension. Isr. J. Chem. 6, 235–250.

Bérend, I., Cases, J.-M., François, M., Uriot, J.-P., Michot, L., Masion, A., Thomas, F., 1995. Mechanism of adsorption and desorption of water vapour by homoionic montmorillonites: 2. the Li^+, Na^+, K^+, Rb^+ and Cs^+ exchanged forms. Clays Clay Miner. 43, 324–336.

Bergaya, F., 1995. The meaning of surface area and porosity measurements of clays and pillared clays. J. Porous Mater. 2, 91–96.

Bergaya, F., Lagaly, G., 2013a. Chapter 1-General introduction: clays, clay minerals, and clay science. In: Bergaya, F., Lagaly, G. (Eds.), Handbook of Clay Science. Part A. Fundamentals, Developments in Clay Science. Elsevier, pp. 1–19.

Bergaya, F., Lagaly, G., 2013b. Handbook of Clay Science, second ed. Developments in Clay Science. Elsevier.

Bergaya, F., Lagaly, G., Vayer, M., 2013. Chapter 2.11-Cation and anion exchange. In: Bergaya, F., Lagaly, G. (Eds.), Handbook of Clay Science. Part B. Techniques and Applications, Developments in Clay Science. Elsevier, pp. 333–359.

Bickmore, B.R., Bosbach, D., Hochella, Jr., M.F., Charlet, L., Rufe, E., 2001. In situ atomic force microscopy study of hectorite and nontronite dissolution: implications for phyllosilicate edge surface structures and dissolution mechanisms. Am. Mineral. 86, 411–423.

Bourg, I.C., Sposito, G., Bourg, A.C.M., 2007. Modeling the acid-base surface chemistry of montmorillonite. J. Colloid Interface Sci. 312, 297–310.

Bourg, I.C., Sposito, G., 2011. Ion exchange phenomena. In: Handbook of Soil Science, second ed. CRC Press, Boca Raton (Chapter 16).

Bradbury, M.H., Baeyens, B., 2000. A generalised sorption model for the concentration dependent uptake of caesium by argillaceous rocks. J. Contam. Hydrol. 42, 141–163.

Bradley, W.F., Grim, R.E., Clark, G.F., 1937. A study of the behavior of montmorillonite on wetting. Z. Kristallogr. 97, 260–270.

Brady, P.V., Cygan, R.T., Nagy, K.L., 1996. Molecular controls on kaolinite surface charge. J. Colloid Interface Sci. 183, 356–364.

Brigatti, M.F., Galán, E., Theng, B.K.G., 2013. Chapter 2-structure and mineralogy of clay minerals. In: Bergaya, F., Lagaly, G. (Eds.), Handbook of Clay Science, Developments in Clay Science. Elsevier, pp. 21–81.

Cadene, A., Durand-Vidal, S., Turq, P., Brendle, J., 2005. Study of individual Na-montmorillonite particles size, morphology, and apparent charge. J. Colloid Interface Sci. 285, 719–730.

Carroll, S.A., Walther, J.V., 1990. Kaolinite dissolution at 25°, 60°, and 80°C. Am. J. Sci. 290, 797–810.

Cases, J.-M., Bérend, I., Besson, G., François, M., Uriot, J.P., Thomas, F., Poirier, J.E., 1992. Mechanism of adsorption and desorption of water vapor by homoionic montmorillonite. 1. The sodium-exchanged form. Langmuir 8, 2730–2739.

Cases, J.-M., Bérend, I., François, S.M., Uriot, J.P., Michot, L.J., Thomas, F., 1997. Mechanism of adsorption and desorption of water vapour by homoionic montmorillonite: 3. the Mg^{2+}, Ca^{2+}, Sr^{2+} and Ba^{2+} exchanged forms. Clays Clay Miner. 45, 8–22.

Chalmers, G.R., Bustin, R.M., Power, I.M., 2012. Characterization of gas shale pore systems by porosimetry, pycnometry, surface area, and field emission scanning electron microscopy/transmission electron microscopy image analyses: examples from the Barnett, Woodford, Haynesville, Marcellus, and Doig units. AAPG Bull. 96, 1099–1119.

Chang, F.-R.C., Sposito, G., 1996. The electrical double layer of a disk-shaped clay mineral particle: effect of electrolyte properties and surface charge density. J. Colloid Interface Sci. 178, 555–564.

Charlet, L., Schindler, P.W., Spadini, L., Furrer, G., Zysset, M., 1993. Cation adsorption on oxides and clays: the aluminum case. Aquat. Sci. 55, 1015–1621.

Chen, Z., Singh, R.K., 2002. General solution for Poisson–Boltzmann equation in semiinfinite planar symmetry. J. Colloid Interface Sci. 245, 301–306.

Chiou, C.T., Rutherford, D.W., Manes, M., 1993. Sorption of nitrogen and ethylene glycol monoethyl ether (EGME) vapors on some soils, clays, and mineral oxides and determination of sample surface areas by use of sorption data. Environ. Sci. Technol. 27, 1587–1594.

Chiou, C.T., Rutherford, D.W., 1997. Effects of exchanged cation and layer charge on the sorption of water and EGME vapors on montmorillonite clays. Clays Clay Miner. 45, 867–880.

Comans, R.N.J., Haller, M., De Preter, P., 1991. Sorption of cesium on illite: non-equilibrium behaviour and reversibility. Geochim. Cosmochim. Acta 55, 433–440.

Davis, J.A., Kent, D.B., Hochella, M.F., White, A.F., 1990. Mineral-water interface geochemistry. Rev. Mineral. 23, 177–260.

Delhorme, M., Labbez, C., Caillet, C., Thomas, F., 2010. Acid-base properties of 2:1 clays. I. Modeling the role of electrostatics. Langmuir 26, 9240–9249.

Djéran-Maigre, I., Tessier, D., Grunberger, D., Velde, B., Vasseur, G., 1998. Evolution of microstructures and of macroscopic properties of some clays during experimental compaction. Mar. Pet. Geol. 15, 109–128.

Drits, V.A., Weber, F., Salyn, A.L., Tsipursky, S.I., 1993. X-ray identification of one-layer illite varieties: application to the study of illites around uranium deposits of Canada. Clays Clay Miner. 41, 389.

Drits, V.A., Zviagina, B.B., 2009. Trans-vacant and cis-vacant 2: 1 layer silicates: structural features, identification, and occurrence. Clays Clay Miner. 57, 405–415.

Duc, M., Gaboriaud, F., Thomas, F., 2005a. Sensitivity of the acid-base properties of clays to the methods of preparation and measurement: 1. Literature review. J. Colloid Interface Sci. 289, 139–147.

Duc, M., Gaboriaud, F., Thomas, F., 2005b. Sensitivity of the acid-base properties of clays to the methods of preparation and measurement: 2. Evidence from continuous potentiometric titrations. J. Colloid Interface Sci. 289, 148–156.

Duc, M., Thomas, F., Gaboriaud, F., 2006. Coupled chemical processes at clay/electrolyte interface: a batch titration study of Na-montmorillonites. J. Colloid Interface Sci. 300, 616–625.

Duc, M., Carteret, C., Thomas, F., Gaboriaud, F., 2008. Temperature effect on the acid-base behaviour of Na-montmorillonite. J. Colloid Interface Sci. 327, 472–476.

Dzombak, D.A., Morel, F.M.M., 1990. Surface Complexation Modeling-Hydrous Ferric Oxide. New York.

Dzombak, D.A., Hudson, R.J.M., 1995. Ion-exchange—the contributions of diffuse layer sorption and surface complexation. In: Huang, C.P., O'Melia, C., Morgan, J.J. (Eds.), Aquatic Chemistry. Interfacial and Interspecies Processes, pp. 59–94.

Fernández, A.M., Sánchez-Ledesma, D.M., Tournassat, C., Melón, A., Gaucher, E.C., Astudillo, J., Vinsot, A., 2014. Applying the squeezing technique to highly consolidated clayrocks for pore water characterisation: lessons learned from experiments at the Mont Terri Rock Laboratory. Appl. Geochem. 49, 2–21.

Ferrage, E., Lanson, B., Sakharov, B.A., Drits, V.A., 2005. Investigation of smectite hydration properties by modeling experimental X-ray diffraction patterns: Part I. Montmorillonite hydration properties. Am. Mineral. 90, 1358–1374.

Ferrage, E., Kirk, C.A., Cressey, G., Cuadros, J., 2007a. Dehydration of Ca-montmorillonite at the crystal scale. Part 2. Mechanisms and kinetics. Am. Mineral. 92, 1007–1017.

Ferrage, E., Lanson, B., Sakharov, B.A., Geoffroy, N., Jacquot, E., Drits, V.A., 2007b. Investigation of dioctahedral smectite hydration properties by modeling of X-ray diffraction profiles: influence of layer charge and charge location. Am. Mineral. 92, 1731–1743.

Ferrage, E., Lanson, B., Michot, L.J., Robert, J.-L., 2010. Hydration properties and interlayer organization of water and ions in synthetic Na-smectite with tetrahedral layer charge. Part 1. Results from X-ray diffraction profile modeling. J. Phys. Chem. C 114, 4515–4526.

Gaines, G.L. Jr., Thomas, H.C., 1953. Adsorption studies on clay minerals. II. A formulation of the thermodynamics of exchange adsorption. J. Chem. Phys. 21, 714–718.

Gapon, E.N., 1933. On the theory of exchange adsorption in soils. J. Gen. Chem. USSR (Engl. Transl.) 3, 144–160.

Garrels, R.M., Mackenzie, F.T., 1971. Evolution of Sedimentary Rocks. WW Norton & Co, New York.

Gates, W.P., 2005. Infrared spectroscopy and the chemistry of dioctahedral smectites. In: Kloprogge, T. (Ed.), Vibrational Spectroscopy of Layer Silicates and Hydroxides, CMS Workshop Lectures, vol. 13. The Clay Minerals Society, Aurora, CO, pp. 125–168.

Goldberg, S., Criscenti, L.J., 2008. Modeling Adsorption of Metals and Metalloids by Soil Components. Biophysico-chemical processes of heavy metals and metalloids in soil environments. John Wiley & Sons, Hoboken, NJ. Modeling adsorption of metals and metalloids by soil components 215–264.

Gražulis, S., Daškevic, A., Merkys, A., Chateigner, D., Lutterotti, L., Quirós, M., Serebryanaya, N.R., Moeck, P., Downs, R.T., Le Bail, A., 2012. Crystallography open database (COD): an open-access collection of crystal structures and platform for world-wide collaboration. Nucleic Acids Res. 40, D420–D427.

Grim, R.E., 1968. Clay Mineralogy, second ed.

Gupta, V., Miller, J.D., 2010. Surface force measurements at the basal planes of ordered kaolinite particles. J. Colloid Interface Sci. 344, 362–371.

Güven, N., 1992. Clay-Water Interface and Its Rheological Implications. Clay Minerals Society.

Hemes, S., Desbois, G., Urai, J.L., De Craen, M., Honty, M., 2013. Variations in the morphology of porosity in the Boom Clay Formation: insights from 2D high resolution BIB-SEM imaging and mercury injection porosimetry. Neth. J. Geosci. 92, 275–300.

Henderson, D., Boda, D., 2009. Insights from theory and simulation on the electrical double layer. Phys. Chem. Chem. Phys. 11, 3822–3830.

Hiemstra, T., De Wit, J.C.M., Van Riemsdijk, W.H., 1989. Multisite proton adsorption modelling at the solid/solution interface of (hydr)oxides: a new approach. II. Application to various important (hydr)oxides. J. Colloid Interface Sci. 133, 105–117.

Hiemstra, T., Venema, P., Van Riemsdijk, W.H., 1996. Intrinsic proton affinity of reactive surface groups of metal (hydr)oxides: the bond valence principle. J. Colloid Interface Sci. 184, 680–692.

Holmboe, M., Wold, S., Jonsson, M., 2012. Porosity investigation of compacted bentonite using XRD profile modeling. J. Contam. Hydrol. 128, 19–32.

Jensen, H.E., 1973. Potassium–calcium exchange equilibria on a montmorillonite and a kaolinite clay. Agrochimica 17, 181–190.

Keller, L.M., Holzer, L., Wepf, R., Gasser, P., 2011. 3D geometry and topology of pore pathways in Opalinus clay: Implications for mass transport. Appl. Clay Sci. 52, 85–95.

Keller, L.M., Holzer, L., Schuetz, P., Gasser, P., 2013. Pore space relevant for gas permeability in Opalinus clay: statistical analysis of homogeneity, percolation, and representative volume element. J. Geophys. Res. Solid Earth 118, 2799–2812.

Kozaki, T., Fujishima, A., Sato, S., Ohashi, H., 1998. Self-diffusion of sodium ions in compacted montmorillonite. Nucl. Technol. 121, 63–69.

Lagaly, G., Dékány, I., 2013. Chapter 8-Colloid Clay science. In: Bergaya, F., Lagaly, G. (Eds.), Handbook of Clay Science, Developments in Clay Science. Elsevier, pp. 243–345.

Laird, D.A., Shang, C., 1997. Relationship between cation exchange selectivity and crystalline swelling in expanding 2:1 phyllosilicates. Clays Clay Miner. 45, 681–689.

Lee, S.S., Fenter, P., Park, C., Sturchio, N.C., Nagy, K.L., 2010. Hydrated cation speciation at the muscovite (001)-water interface. Langmuir 26, 16647–16651.

Le Forestier, L., Muller, F., Villieras, F., Pelletier, M., 2010. Textural and hydration properties of a synthetic montmorillonite compared with a natural Na-exchanged clay analogue. Appl. Clay Sci. 48, 18–25.

Leroy, P., Revil, A., Coelho, D., 2006. Diffusion of ionic species in bentonite. J. Colloid Interface Sci. 296, 248–255.

Leroy, P., Revil, A., Altmann, S., Tournassat, C., 2007. Modeling the composition of the pore water in a clay-rock geological formation (Callovo-Oxfordian, France). Geochim. Cosmochim. Acta 71, 1087–1097.

Ma, C., Eggleton, R.A., 1999a. Surface layer types of kaolinite: a high-resolution transmission electron microscope study. Clays Clay Miner. 47, 181–191.

Ma, C., Eggleton, R.A., 1999b. Cation exchange capacity of kaolinite. Clays Clay Miner. 47, 174–180.

Manceau, A., Drits, V.A., Lanson, B., Chateigner, D.L., Wu, J., Huo, D., Gates, W.P., Stucki, J.W., 2000. Oxidation-reduction mechanism of iron in dioctahedral smectites: 2. Crystal chemistry of reduced garfield nontronite. Am. Mineral. 85, 153–172.

Marty, N.C.M., Cama, J., Sato, T., Chino, D., Villiéras, F., Razafitianamaharavo, A., Brendlé, J., Giffaut, E., Soler, J.M., Gaucher, E.C., Tournassat, C., 2011. Dissolution kinetics of synthetic Na-smectite. An integrated experimental approach. Geochim. Cosmochim. Acta 75, 5849–5864.

McBride, M.B., 1994. Environmental Chemistry of Soils. Oxford University Press, New York Oxford.

Melkior, T., Gaucher, E.C., Brouard, C., Yahiaoui, S., Thoby, D., Clinard, C., Ferrage, E., Guyonnet, D., Tournassat, C., Coelho, D., 2009. Na^+ and HTO diffusion in compacted bentonite: effect of surface chemistry and related texture. J. Hydrol. 370, 9–20.

Méring, J., Glaeser, R., 1954. Sur le rôle de la valence des cations échangeables dans la montmorillonite. Bull. Soc. Fr. Mineral. Cristallogr. 77, 519–530.

Michot, L., François, M., Cases, J.-M., 1990. Surface heterogeneity studied by a quasi-equilibrium adsorption procedure. Langmuir 6, 637–643.

Michot, L.J., Villiéras, F., 2013. Chapter 2.10-Surface area and porosity. In: Bergaya, F., Lagaly, G. (Eds.), Handbook of Clay Science, Developments in Clay Science. Elsevier, pp. 319–332.

Missana, T., Alonso, U., García-Gutiérrez, M., 2009. Experimental study and modelling of selenite sorption onto illite and smectite clays. J. Colloid Interface Sci. 334, 132–138.

Muurinen, A., 2009. Studies on the Chemical Conditions and Microstructure in the Reference Bentonites of Alternative Buffer Materials Project (ABM) in Äspö. Posiva Oy, Helsinki (Finland).

Nadeau, P.H., 1985. The physical dimensions of fundamental clay particles. Clay Miner. 20, 499.

Ngouana Wakou, B.F., Kalinichev, A.G., 2014. Structural arrangements of isomorphic substitutions in Smectites: molecular simulation of the swelling properties, interlayer structure, and dynamics of hydrated Cs-Montmorillonite revisited with new clay models. J. Phys. Chem. C.

Norrish, K., 1954. The swelling of montmorillonite. Faraday Discuss. Soc. 18, 120–134.

Poinssot, C., Baeyens, B., Bradbury, M.H., 1999. Experimental and modelling studies of caesium sorption on illite. Geochim. Cosmochim. Acta 63, 3217–3227.

Reinholdt, M.X., Hubert, F., Faurel, M., Tertre, E., Razafitianamaharavo, A., Francius, G., Prêt, D., Petit, S., Béré, E., Pelletier, M., Ferrage, E., 2013. Morphological properties of vermiculite particles in size-selected fractions obtained by sonication. Appl. Clay Sci. 77, 18–32.

Rouquerol, J., Avnir, D., Fairbridge, C.W., Everett, D.H., Haynes, J.H., Pernicone, N., Ramsay, J.D.F., Sing, K.S.W., Unger, K.K., 1994. Recommendations for the characterization of porous solids (Technical Report). Pure Appl. Chem. 66, 1739–1758.

Saiyouri, N., Hicher, P.Y., Tessier, D., 2000. Microstructural approach and transfer water modelling in highly compacted unsaturated swelling clays. Mech. Cohesive-Frict. Mater. 5, 41–60.

Saiyouri, N., Tessier, D., Hicher, P.Y., 2004. Experimental study of swelling in unsaturated compacted clays. Clay Miner. 39, 469–479.

Sakharov, B.A., Lanson, B., 2013. Chapter 2.3-X-ray identification of mixed-layer structures: modelling of diffraction effects. In: Bergaya, F., Lagaly, G. (Eds.), Handbook of Clay Science, Developments in Clay Science. Elsevier, pp. 51–135.

Sato, T., Watanabe, T., Otsuka, R., 1992. Effects of layer charge, charge location, and energy change on expansion properties of dioctahedral smectites. Clays Clay Miner. 40, 103–113.

Sayed Hassan, M., Villieras, F., Gaboriaud, F., Razafitianamaharavo, A., 2006. AFM and low-pressure argon adsorption analysis of geometrical properties of phyllosilicates. J. Colloid Interface Sci. 296, 614–623.

Schindler, P., Gamsjager, H., 1972. Acid-base reactions of the TiO_2 (anatase)-water interface and the point of charge of TiO_2 suspensions. Kolloid Z. Pollymere 250, 759–763.

Schindler, P.W., Stumm, W., 1987. The surface chemistry of oxides, hydroxides and oxide minerals. In: Stumm, W. (Ed.), Aquatic Surface Chemistry. Wiley Interscience, New York, pp. 311–338.

Schramm, L.L., Kwak, J.C.T., 1982. Influence of exchangeable cation composition on the size and shape of montmorillonite particles in dilute suspension. Clays Clay Miner. 30, 40–48.

Secor, R.B., Radke, C.J., 1985. Spillover of the diffuse double layer on montmorillonite particles. J. Colloid Interface Sci. 103, 237–244.

Shainberg, I., Otoh, H., 1968. Size and shape of montmorillonite particles saturated with Na/Ca ions (inferred from viscosity and optical measurements). Isr. J. Chem. 6, 251–259.

Slade, P.G., Quirk, J.P., Norrish, K., 1991. Crystalline swelling of smectite samples in concentrated NaCl solutions in relation to layer charge. Clays Clay Miner. 39, 234–238.

Sposito, G., Prost, R., 1982. Structure of water adsorbed on smectites. Chem. Rev. 82, 553–573.

Sposito, G., Holtzclaw, K.M., Jouany, C., Charlet, L., 1983. Cation selectivity in sodium-calcium, sodium-magnesium, and calcium-magnesium exchange on Wyoming bentonite at 298 K. Soil Sci. Soc. Am. J. 47, 917–921.

Sposito, G., 1984. The Surface Chemistry of Soils. Oxford University Press, New York.

Sposito, G., 1992. The diffuse-ion swarm near smectite particles suspended in 1:1 electrolyte solutions: modified Gouy-Chapman theory and quasicrystal formation. In: Güven, N., Pollastro, R.M. (Eds.), Clay Water Interface and Its Rheological Implications. Clay minerals society, pp. 127–156.

Sposito, G., 1998. On points of zero charge. Environ. Sci. Technol. 32, 2815–2819.

Sposito, G., Skipper, N.T., Sutton, R., Park, S.-H., Soper, A.K., 1999. Surface geochemistry of the clay minerals. Proc. Natl. Acad. Sci. U.S.A. 96, 3358–3364.

Sposito, G., 2004. The Surface Chemistry of Natural Particles. Oxford University Press, New York.

Srodon, J., McCarty, D.K., 2008. Surface area and layer charge of smectite from CEC and EGME/ H_2O-retention measurements. Clays Clay Miner. 56, 155–174.

Steefel, C.I., Carroll, S., Zhao, P., Roberts, S., 2003. Cesium migration in Hanford sediment: a multisite cation exchange model based on laboratory transport experiments. J. Contam. Hydrol. 67, 219–246.

Stumm, W., Huang, C.P., Jenkins, S.R., 1970. Specific chemical interactions affecting the stability of dispersed systems. Croat. Chem. Acta 42, 223–244.

Sverjensky, D.A., 1993. Physical surface complexation models for sorption at the mineral-water interface. Nature 364, 776–780.

Tertre, E., Castet, S., Berger, G., Loubet, M., Giffaut, E., 2006. Surface chemistry of kaolinite and Na-montmorillonite in aqueous electrolyte solutions at 25 and 60°C: experimental and modeling study. Geochim. Cosmochim. Acta 70, 4579–4599.

Tombácz, E., Szekeres, M., 2004. Colloidal behavior of aqueous montmorillonite suspensions: the specific role of pH in the presence of indifferent electrolytes. Appl. Clay Sci. 27, 75–94.

Tombácz, E., Szekeres, M., 2006. Surface charge heterogeneity of kaolinite in aqueous suspension in comparison with montmorillonite. Appl. Clay Sci. 34, 105–124.

Tournassat, C., Neaman, A., Villiéras, F., Bosbach, D., Charlet, L., 2003. Nanomorphology of montmorillonite particles: Estimation of the clay edge sorption site density by low-pressure gas adsorption and AFM observations. Am. Mineral. 88, 1989–1995.

Tournassat, C., Chapron, Y., Leroy, P., Boulahya, F., 2009. Comparison of molecular dynamics simulations with triple layer and modified Gouy-Chapman models in a 0.1 M NaCl–montmorillonite system. J. Colloid Interface Sci. 339, 533–541.

Tournassat, C., Bizi, M., Braibant, G., Crouzet, C., 2011. Influence of montmorillonite tactoid size on Na-Ca cation exchange reactions. J. Colloid Interface Sci. 364, 443–454.

Tournassat, C., Grangeon, S., Leroy, P., Giffaut, E., 2013. Modeling specific pH dependent sorption of divalent metals on montmorillonite surfaces. A review of pitfalls, recent achievements and current challenges. Am. J. Sci. 313, 395–451.

Tsipursky, S.I., Drits, V.A., 1984. The distribution of octahedral cations in the 2:1 layers of dioctahedral smectites studied by oblique-texture electron diffraction. Clay Miner. 19, 177–193.

Tsipursky, S.I., Eberl, D.D., Buseck, P.R., 1992. Unusual tops (bottoms?) of particles of 1M illite from the Silverton caldera (CO). In: Proceedings of the American Society of Agronomy Annual Meeting, Minneapolis, pp. 381–382.

Vanselow, A.P., 1932. The utilization of the base-exchange reaction for the determination of activity coefficients in mixed electrolytes. J. Am. Chem. Soc. 54, 1307–1311.

Vantelon, D., Pelletier, M., Michot, L.J., Barrès, O., Thomas, F., 2001. Fe, Mg and Al distribution in the octahedral sheet of montmorillonites. An infrared study in the OH-bending region. Clay Miner. 36, 369–379.

Vantelon, D., Montarges-Pelletier, E., Michot, L.J., Briois, V., Pelletier, M., Thomas, F., 2003. Iron distribution in the octahedral sheet of dioctahedral smectites. An FeK-edge X-ray absorption spectroscopy study. Phys. Chem. Miner. 30, 44–53.

Verburg, K., Baveye, P., McBride, M.B., 1995. Cation-exchange hysteresis and dynamics of formation and breakdown of Montmorillonite quasi-crystals. Soil Sci. Soc. Am. J. 59, 1268–1273.

Villiéras, F., Cases, J.-M., François, M., Michot, L.J., Thomas, F., 1992. Texture and surface energetic heterogeneity of solids from modeling of low pressure gas adsorption isotherms. Langmuir 8, 1789–1795.

Villiéras, F., Michot, L.J., Bardot, F., Cases, J.-M., François, M., Rudzinski, W., 1997. An improved derivative isotherm summation method to study surface heterogeneity of clay minerals. Langmuir 13, 1104–1117.

Wanner, H., Albinson, Y., Karnland, O., Wieland, E., Wersin, P., Charlet, L., 1994. The acid/base chemistry of montmorillonite. Radiochim. Acta 66/67, 157–162.

White, G.N., Zelazny, L.W., 1988. Analysis and implications of the edge structure of dioctahedral phyllosilicates. Clays Clay Miner. 36, 141–146.

Yokoyama, S., Kuroda, M., Sato, T., 2005. Atomic force microscopy study of montmorillonite dissolution under highly alkaline conditions. Clays Clay Miner. 53, 147–154.

Zachara, J.M., Smith, S.C., McKinley, J.P., Resch, C.T., 1993. Cadmium sorption on specimen and soil smectites in sodium and calcium electrolytes. Soil Sci. Soc. Am. J. 57, 1491–1501.

Zysset, M., Schindler, P.W., 1996. The proton promoted dissolution kinetics of K-montmorillonite. Geochim. Cosmochim. Acta 60, 921–931.

Chapter 2

Adsorption of Inorganic and Organic Solutes by Clay Minerals

Mikhail Borisover[a] and James A. Davis[b]

[a] *Agricultural Research Organization, Institute of Soil, Water and Environmental Sciences, The Volcani Center, Bet Dagan, Israel;* [b] *Earth Sciences Division, Lawrence Berkeley National Laboratory, Berkeley, CA, USA*

Chapter Outline

Developments in Clay Science, Vol. 6C. http://dx.doi.org/10.1016/B978-0-08-100027-4.00002-4

2.1 INTRODUCTION

Adsorption of inorganic and organic solutes by clay minerals, including competitive or synergetic effects in multisolute systems, affects straightforwardly the rates and mechanisms of the chemical transfer through the clay barriers. Some effects of adsorption interactions on diffusion through the clay barriers are discussed in Chapters 6, 8, and 10 in this volume. Clay mineral adsorption of ions, molecules, and low- and high-molecular mass compounds may have also other impacts by changing clay mineral hydration and swelling, by affecting clay mineral porosity and coating surfaces, by contributing to a development of clay mineral-associated bioorganic phases, and by modifying surface tension at clay mineral–water interface.

General properties of basal and edge surfaces of clay minerals and their chemical structure responsible for cation exchange, anion exclusion, protonation, deprotonation, and clay swelling are described in Chapter 1 in this volume. Therefore, the goal of this chapter is to summarize the major mechanisms controlling clay mineral interactions with inorganic and organic species, cations, anions, and molecules and to discuss their relationship with the surfaces and interfaces of clay minerals. This chapter will emphasize the role of the interlayer space and edges, the contributions of electrostatic and covalent interactions into inorganic solute adsorption by clay minerals, the effects of surface and bulk water in adsorption of organic solutes. To make the content of this chapter relevant to the engineered and environmental applications of clay minerals, the interactions of clay mineral surfaces with natural organic matter (NOM) and natural nanoparticles of other minerals will be evaluated as well as their influence on adsorptive properties of clay minerals and the role of dissolved organic matter (DOM) and wetting–drying processes on ability of clay minerals to interact with solutes.

2.2 CLAY MINERALS AND SURFACE FUNCTIONAL GROUPS

Negative charge arising from isomorphic substitutions in clay mineral structures is balanced mostly by cations located in the interlayer space or on external basal surfaces (Chapter 1 in this volume). These cations occupy what can be referred to as cation exchange sites, although the cations remain at least partly solvated and form loose, outer-sphere (OS) complexes with the surfaces in the interlayer space and external basal planes. This ion exchange behavior is mostly independent of pH.

At the edges of clay mineral layers the clay mineral structure is broken, resulting in the formation of surface hydroxyl groups analogous to those that form on the surfaces of simple Al and Si oxides and (hydr)oxides. It is usually assumed that clay mineral edges have a tetrahedral–octahedral (TO) or tetrahedral–octahedral–tetrahedral (TOT) structure and chemical composition similar to the structure of the inner part of the layer, with surface metals (Si, Al, and their substituting atoms) having the same coordination, with external oxygen atoms being undercoordinated. These sites can then undergo surface protonation/deprotonation and surface complexation reactions with cations and anions, following the typical pH-dependent adsorption behavior observed on oxide and (hydr)oxide minerals. In addition, the undercoordinated nature of the external oxygen atoms results in a greater lability for the external metal atoms, allowing metal substitution reactions to occur at the edges of clay mineral layers for metal ions of similar ionic radius and coordination chemistry.

2.3 INORGANIC SOLUTE ADSORPTION–DESORPTION MECHANISMS

Cation adsorption onto clay minerals can be classified into several categories depending on the underlying mechanisms by which the cation bonds with the clay mineral. For transition metal cations, lanthanides, and actinides, it is known that at low pH adsorption on most clay minerals is ionic strength dependent, and becomes less ionic strength pH *dependent* at pH values greater than 6. These trends can be explained in terms of the two types of surface charge on clay minerals. Fixed charge sites (sometimes called constant potential sites) dominate adsorption at low pH values. Cation adsorption at such fixed charge sites is caused by OS surface complex formation, which is driven by electrostatic and hydrogen-bonding (H-bond) forces and energies. Thus, for this type of adsorption, the strength of the bonding is influenced by ionic strength, because the ionic strength affects the degree to which the constant surface charge is distributed into solution. At greater pH values, edge sites on the clay minerals become more negatively charged and are the predominant contributors to cation adsorption because of the stronger, covalent chemical interactions (i.e., inner-sphere (IS) surface complexation).

These general adsorption trends have been observed for transition metal cations, lanthanides, and actinides for a range of clay minerals and cationic adsorbates (Turner et al., 1996; Schlegel et al., 2001b; Vico, 2003; Rabung et al., 2005; Bradbury and Baeyens, 2005, 2009a; Gu et al., 2010; Dähn et al., 2011).

In contrast, adsorption of alkaline earth and alkali cations is always dominated by electrostatic bonding, and is thus only weakly dependent on pH and highly dependent on ionic strength.

Anion adsorption to clay minerals is generally weak relative to cations, due to charge repulsion from the fixed negative charge (Vinsova et al., 2004; Jan et al., 2007). However, some anions are capable of forming strong covalent bonds with the edge sites of clay minerals (Goldberg, 2002). As in the case of anion surface complexation on the surfaces of metal oxides, anion adsorption decreases with increasing pH for clay minerals.

2.3.1 Interlayer Adsorption

As described in more detail in Chapter 1 (this volume), montmorillonite (Mt) carries a permanent negative surface charge resulting from the substitution of lattice cations with cations of lower charge. The structural negative surface charge, known as the cation exchange capacity (CEC), is neutralized in the interlayer space by electrostatically bound cations held close to the surface that can exchange with other cations in solution. Cation exchange is characterized by (mostly) pH-independent adsorption that tends to become more important at lower pH values and at lower background electrolyte concentrations (Dähn et al., 2011). For example, Sr adsorption to Mt is ionic strength dependent, but independent of pH; similar behavior is observed for Sr^{2+} adsorption by illite except at pH >9 (Missana et al., 2008).

Xu and Harsh (1990a,b, 1992) were among the first to develop a quantitative model that described the relative binding of monovalent cations to clay minerals with structural charge deficiencies. Prior to their work, quantitative models of cation exchange had considered only "purely electrostatic" interactions between cations and the clay mineral surface and were only able to predict selectivity sequences for hard (acid)—hard (base) interactions. Xu and Harsh (1990a,b) developed a model based on the hard and soft acid and base principle, taking into account both electrostatic and covalent interactions. According to the model, the thermodynamic exchange constant could be obtained from the differences in absolute electronegativity and absolute softness of the exchanging cations in binary exchange. The authors tested the applicability of the model to Li^+, K^+, Rb^+, and Cs^+ exchange with Na^+ on clay minerals and ion exchange resins and got an excellent fit to the experimental data. McBride (1980) later related the changes in selectivity coefficient values to surface structural disorder. Auboiroux et al. (1998) later extended the model to describe divalent cation selectivity during cation exchange.

Teppen and Miller (2006) reexamined the cation exchange conceptual framework, and reasoned that computational molecular mechanics leads to a conclusion that, for two cations of equal valence, the more weakly hydrated cation tends to partition into the "subaqueous" interlayer space. Thus, Mt "selects" Cs^+ over K^+ because of the selectivity of the solution phase for the more strongly hydrated cation, K^+. According the authors this concept is a better explanation of the energetics of cation exchange selectivity than hypotheses that suggest more favorable interactions between Mt surfaces and selected cations (see discussion in Section 1.3.1). The concept was confirmed in subsequent work by Rotenberg et al. (2009). This partitioning concept rationally explains the alkali and alkaline earth selectivity sequences as well as the selectivities of Mt for larger organic cations over smaller cations. However, Fukushi et al. (2014) found that the rate of Cs^+ desorption from ion exchange sites of Mt was also strongly controlled by particle size.

Micas are a special case for adsorption of Cs^+, an important component of low-level nuclear wastes (McKinley et al., 2004). Layers are bound together by charge-compensating interlayer K^+, which is weakly hydrated and fits into the ditrigonal cavities on the surfaces of adjacent layers. The monovalent cation, Cs^+, weakly hydrated and similar in size to K^+, is strongly retained when substituted for K^+. However, McKinley et al. (2004) observed that Cs^+ was preferentially adsorbed to mica edges, steps on mica surfaces, or fractured regions in sediments from the Hanford (USA) waste site; edge site adsorption is covered in the following section. Microscopic studies showed that Cs^+ was strongly bound to frayed edge sites (FESs) in preference to sites on basal planes. Cs^+ was found to penetrate the interior of mica grains, forming discrete zones of concentration, particularly in muscovite grains. Localization of adsorbed Cs^+ in areas where K^+ was depleted suggested that weathering reactions had caused the formation of FESs within the micas.

2.3.2 Edge Site Adsorption

In addition to the electrostatic cation exchange sites described above, additional important adsorption sites are the "edge" sites, located at the edges of clay layers. They are generally present as surface hydroxyl groups coordinated to the metal atoms of the octahedral and tetrahedral sheets, and like oxide surface hydroxyl groups, these groups can be protonated, deprotonated, and form surface complexes.

2.3.2.1 Cation Surface Complexation

X-ray absorption spectroscopy has been extremely useful in describing the molecular scale details of surface complexes formed by cations on various clay minerals (Schlegel and Manceau, 2006; Schlegel and Descostes, 2009; Dähn et al., 2011). Schlegel et al. (2001b) were among the first to effectively use extended X-ray absorption fine structure (EXAFS) spectroscopy to differentiate the various bonding mechanisms on hectorite (Hect) and relate changes in bonding to kinetic measurements. In 0.01 M $NaNO_3$ solution, significant amounts of Zn^{2+} were adsorbed within the first 5 min at pH 4 and 6.5. But then, Zn^{2+} adsorption slowly increased at pH 6.5 but decreased at pH 4. In 0.3 M $NaNO_3$ solution, Zn^{2+} adsorption was slower in the first 5 min, and in contrast to the lower ionic strength results, Zn^{2+} adsorption subsequently increased at both pH values. EXAFS spectra were consistent with edge-shared Zn and Mg octahedra and corner-shared Zn octahedra and Si tetrahedral. The results indicate that IS Zn^{2+} surface complexes formed at edge sites of Hect layers. In 0.01 M $NaNO_3$ solution, a continuous evolution with increasing reaction time from predominantly OS to predominantly IS complexes was observed, suggesting that Zn^{2+} initially adsorbed as exchangeable OS complexes on interlayer sites, and then migrated to layer edges to form IS surface complexes. At pH 4, the decrease in Zn^{2+} adsorption with time was due to an initial cation exchange with Na^+, followed by exchange off of the solid with Mg^{2+} that was slowly released by dissolution of the Hect.

Dähn et al. (2003) and Schlegel and Manceau (2006) showed that Zn^{2+} and Ni^{2+} adsorbed by Mt samples resulted in the formation of similar IS complexes at edge sites. Schlegel et al. (1999) observed similar IS metal complexes for Co^{2+} adsorbed on Hect. Dähn et al. (2011) gave spectroscopic evidence for the existence of two distinct groups of edge sites of Mt that are predominant at different adsorption densities of metal ions (Figure 2.1). The EXAFS spectra for Zn exhibited a pronounced angular dependency at low adsorbed Zn^{2+} and suggested that the surface complexes were predominantly coordinated to the edges of Al-octahedral sheets. In contrast, a second group of edge sites could be identified at high adsorbed Zn^{2+} whose spectra only showed a weak angular dependency. In addition, these authors illustrated where Zn could be substituted into the Mt structure (see Section 2.3.3).

More recently, Churakov and Dähn (2012) used ab initio molecular dynamics (MD) calculations to simulate the uptake of Zn^{2+} on edge surfaces of Mt and to examine consistency of the results with EXAFS spectra for adsorbed Zn^{2+} (Figure 2.2). The experiments were carried out in a high ionic strength (NaCl) background solution that prevented Zn^{2+} from attaching to ion exchange sites in the interlayer space, thus restricting Zn^{2+} adsorption to edge sites. The MD simulation results suggest that at low loading (~ 0.4 sites/nm^2 of edge surface area) Zn^{2+} becomes structurally incorporated into the outermost *trans*-octahedra on (010) and (110) edges of the Mt, preferentially substituting for Al at *trans*-symmetric sites. At medium loading (up to 2–3 sites/nm^2 of edge surface area), Zn^{2+} forms mono- and bidentate IS surface complexes attached to the octahedral layer of (010) and (110) edge sites. The results provide good justification for the strong and weak site modeling concept for surface complexation on 2:1 clay minerals, e.g., Mt and illite (Dähn et al., 2011).

Numerous authors have studied transition metal adsorption on Mt over the last several decades. Gu et al. (2010) provide a good example of such a study, comparing adsorption of five different transition metal ions, Cd^{2+}, Cu^{2+}, Ni^{2+}, Pb^{2+}, and Zn^{2+}, onto Mt as a function of pH and ionic strength. A two-site surface complexation model was used to fit the adsorption data (Figure 2.3). The results showed that in an acidic pH range, 3–6 for Cd^{2+}, Cu^{2+}, Ni^{2+}, and

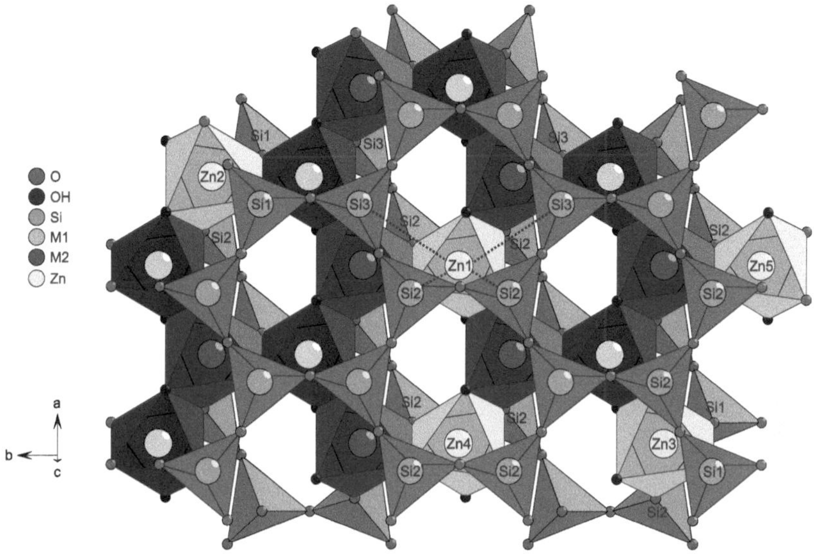

FIGURE 2.1 Illustration of Zn^{2+} incorporated in Mt structure (Zn1), and Zn^{2+} located at "strong" edge sites (Zn2–Zn5). M1 is the trans octahedron, with OH groups located at opposing corners, whereas M2 is the cis octahedron, with the OH groups located on the same edge. Zn2 and Zn3 show possible surface complexes formed at *cis*-vacant-like edge sites. Zn4 and Zn5 show two possible surface complexes formed at *trans*-vacant-like edge sites. *Reprinted with permission from Dähn et al. (2011).*

Zn^{2+}, and 3–4.5 for Pb^{2+}, adsorption was greatly affected by ionic strength, while in a higher pH range adsorption was not affected. In the lower pH range, the cations were mainly bound by formation of OS surface complexes on the permanently charged basal surface sites, while in the higher pH range adsorption occurred mainly on variably charged edge sites by formation of IS surface complexes. Adsorption affinity to the permanently charged sites was $Pb^{2+} > Cu^{2+} > Ni^{2+} \sim Zn^{2+} \sim Cd^{2+}$, while variable charged site affinity was $Pb^{2+} \sim Cu^{2+} > Zn^{2+} > Cd^{2+} > Ni^{2+}$. Gaskova and Bukaty (2008) examined experimental data for adsorption of U(VI) and Eu^{3+} on Mt, illite, and kaolinite (Kaol) over a range of chemical conditions. Because of the range of conditions, these authors were able to demonstrate that both ion exchange and surface complexation processes were responsible for the adsorption processes and calibrated model parameters to describe these data.

Bradbury and Baeyens (2005) were among the first to conduct good studies of metal adsorption competition on Mt. In particular, they focused on the influence of adsorption competition between metals of different charge under conditions dominated by edge site surface complexation, using varying combinations and concentrations of Co^{2+}, Ni^{2+}, Zn^{2+}, Eu^{3+}, Nd^{3+}, Am(III), Th(IV), and U(VI). For metals adsorbed at trace concentrations in a

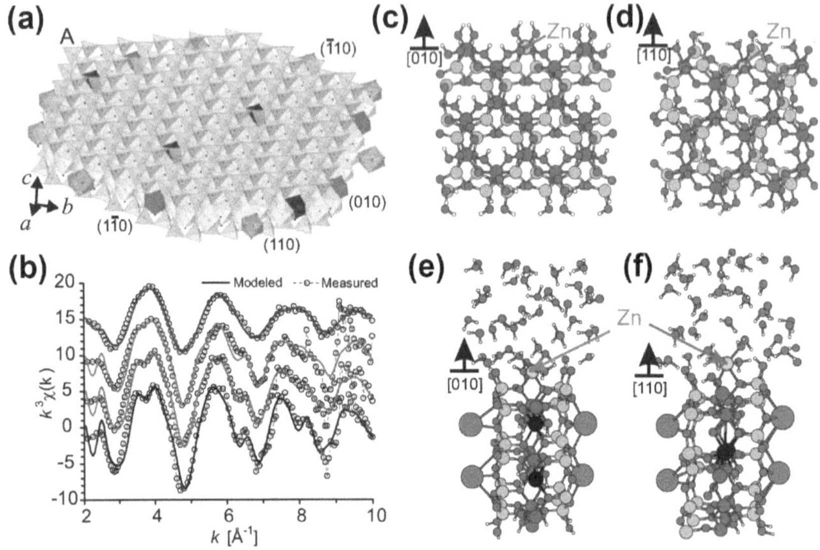

FIGURE 2.2 (a) Schematic view of adsorption sites on (010) and (110) edge planes of Mt at low and medium loading (red (very dark gray in print version) and green (dark gray in print version) octahedra, respectively). Structural incorporation of Zn^{2+} is illustrated as black polyhedral; *trans-* and *cis-*Al octahedra are shown in blue (very light gray in print version) and yellow (light gray in print version) colors, respectively. Silica tetrahedra are orange. (b) Best fit of measured EXAFS spectra. (c–f) Snapshots from atomistic simulations of Zn^{2+} adsorption on (110) and (010) edge surfaces. Oxygen atoms are red (very dark gray in print version); Al green (light gray in print version); H white; Si yellow (pale gray in print version); K pink (gray in print version); and Zn blue (light gray in print version). (c) complexation, using varying combinations and concentrations of Co(II), Ni(II), Zn(II), Eu(III), and (d) are the dominant sorption complexes at low loading. (e) and (f) are the dominant adsorption complexes at medium loading. *Reprinted with permission from Churakov and Dähn (2012).*

background electrolyte containing a competing metal and at pH >6, metals with similar chemistry (valence state, hydrolysis behavior) competed with one another, but metals with different chemistry did not compete. For example, Eu^{3+}, Nd^{3+}, and Am(III) exhibited adsorption competition effects, as did Ni^{2+}, Co^{2+}, and Zn^{2+}. To explain the competitive effects, the authors proposed that multiple sets of strong sites exist on the edge sites. When considering a single adsorbent such as Mt, and a range of adsorbing cations, linear free energy relationships (LFER) exist between the aqueous hydrolysis constant of the cations and surface complexation constants for strong edge sites (Bradbury and Baeyens, 2005, 2009b). Although the values of such surface complexation constants are dependent on the modeling approach used (in particular, the electrical double-layer model used), the trend in free energies suggests some level of thermodynamic significance.

As reported in other papers (Charlet et al., 2007; Gehin et al., 2007) a recent paper by Soltermann et al. (2014) illustrates the potential complexity

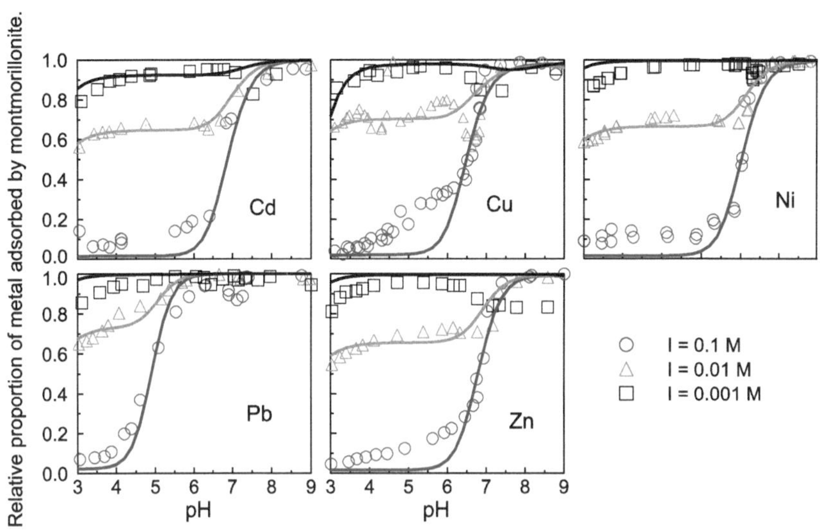

FIGURE 2.3 Relative proportions of adsorbed metal cations onto Na^+-Mt at different ionic strengths (indicated as I). Symbols are experimental data and lines are fitted using parameters in the surface complexation plus ion exchange model of Gu et al. (2010). *Reprinted with permission from Gu et al. (2010).*

of cation adsorption competition on Mt. The authors studied competitive adsorption of Fe(II) and Zn^{2+} in anoxic systems with an ionic strength of 0.1 at pH 6.4, conditions where edge site complexation should predominate. These authors found that when Zn^{2+} was at trace concentrations, Fe(II) was an effective competing cation at micromolar concentrations. In reduced systems, Fe(II) can often be present at concentrations as high as 1 mM, and hence, may be an important cation occupying edge sites. However, when Fe(II) was at trace concentrations, Zn^{2+} was not an effective competing cation for adsorption. Spectroscopic results suggested that the reason for the behavior of Fe(II) at trace concentrations was due to at least partial oxidation of the adsorbed Fe(II) to Fe(III), suggesting that certain strong-binding edge sites can catalyze surface oxidation to Fe(III).

Coppin et al. (2002) investigated the adsorption of the complete lanthanide series on Kaol and Na^+-Mt over the pH range 3–9 at two different ionic strengths, 0.025 and 0.5 M. At 0.5 M, both clay minerals exhibited the same pH dependence for the lanthanide adsorption edges, with a large increase in the distribution coefficient, K_D, above pH 5.5. At 0.025 M, the measured K_D was influenced by the CEC of the minerals. A fractionation was observed due to the selective adsorption between the rare earth elements (REE) at high ionic strength, with the heavy REE being more adsorbed than the light REE. The results can be interpreted in terms of the surface chemistry of the clay minerals, where two types of surface sites coexist: the permanent structural charge sites and variable charge, pH-dependent sites.

An important application in the consideration of adsorption on clay minerals is that clay-rocks are of strong interest to many nuclear waste management programs. Clay minerals such as illite, Mt, illite–smectite (I-Sm) mixed layers, and Kaol are important components in such rock types and often make up ≥50% of the total rock mass. Bradbury and Baeyens (2009a) studied the adsorption characteristics of Ni^{2+} and Co^{2+}, Eu^{3+}, and Sn(IV) on Na^{+}-illite, with $NaClO_4$ as the background electrolyte under anoxic conditions. Like Mt, the structure of illite is constituted from the repetition of tetrahedron–octahedron–tetrahedron layers, but illite is a nonexpanding mineral because the interlayer space is mainly occupied by poorly hydrated K^{+} cations. Adsorption was studied at trace concentrations such that reactions with strongly binding surface sites predominated. Self-consistent model parameters were derived for a surface complexation (SC) model for the adsorbing cations and a cation exchange (IX) model for major cations. Furthermore, the combined SC + IX model was consistent with previous models derived for Mt by the same authors.

The adsorption properties of minerals may be considerably altered by the presence of inorganic and organic coatings that are typically found in soils. Bellenger and Staunton (2008) investigated adsorption of trace amounts of Cs^{+} on Kaol, Mt, and illite, with varying additions of humic or fulvic acids. The presence of organic coatings decreased Cs^{+} adsorption on illite, with smaller effects observed for Mt and Kaol. Fan et al. (2014) examined the differences between Cs^{+} adsorption on illite (nonexpanding) and vermiculite (intermediate expansion) and the effect of humic acid (HA) addition on Cs^{+} adsorption. The results showed that Cs^{+} adsorption was inhibited to some extent in the ternary clay + HA + Cs^{+} system due to (1) blocked access of Cs^{+} to FES by HA, and (2) reduced availability of the interlayer site in vermiculite. In some cases, OS complexes of Cs^{+} were formed because of the HA site blocking. Interactions of clay mineral surfaces in soils and sediments with NOM and nanoparticles of other minerals and their effect on adsorption of organic solutes are presented further in Sections 2.5 and 2.6.2.1, respectively.

Rabung et al. (2005) studied adsorption of Cm^{3+} and Eu^{3+} at trace concentrations onto Ca^{2+}-Mt and Na^{+}-illite, using both batch adsorption experiments and time-resolved laser fluorescence spectroscopy (TRLFS). The results indicated the existence of OS complexes at pH <4 with Na^{+}-illite. However, for Ca^{2+}-Mt, OS complexes did not form at significant levels, due to Ca^{2+} competition for cation exchange sites. TRLFS spectra indicated the formation of IS surface complexes at pH >5 for both clay minerals. Adsorption of Nd^{3+} and Eu^{3+} onto K^{+}-illite was studied by Sinitsyn et al. (2000) as a function of ionic strength and the concentrations of the lanthanides. The shape of the adsorption isotherms indicated the existence of a multiplicity of energetically distinct surface types, reflecting the multiphase nature of natural illitic materials. Surface site types included amphoteric silanol and aluminol sites, basal planar surfaces, and FES. The FES edges were only active at low ionic strengths.

Sepiolite (Sep) is a natural, magnesium hydrosilicate clay mineral with the formula, $Mg_4Si_6O_{15}(OH)_2 \cdot 6H_2O$. Its unique fibrous structure allows absorption of organic and inorganic ions, and because of its low cost, Sep offers great potential as an ion exchanger for removing heavy metals ions from wastewaters. Bektas et al. (2004) studied adsorption of Pb^{2+} onto Sep under variable pH and temperature conditions. Equilibrium was well described by the Langmuir equation and thermodynamic parameters were derived from the temperature dependence. Vico (2003) examined Cu^{2+} and Zn^{2+} adsorption onto Sep as a function of pH and ionic strength, and with competitive adsorption of Ca^{2+}. Surface complexation with edge sites was proposed to explain the strong pH dependence of Cu^{2+} and Zn^{2+} adsorption, but some ion exchange with variable Na^+ was also noted. In addition, substitution for Mg^{2+} at the edges of the octahedral sheets may occur.

2.3.2.2 Anion Surface Complexation

On oxide surfaces, strongly adsorbing anions typically form IS surface complexes by displacing hydroxyl or water groups from the surface, forming bonds with the metal atom of the oxide, e.g., with Fe in Fe(III) oxides and hydroxides. There have been few spectroscopic studies of the structures of anionic surface complexes formed at clay mineral surfaces. Recently, Charlet et al. (2012) analyzed the structure of arsenate surface complexes formed at the edge sites of Ca^{2+}-Mt particles. In this structure, AsO_4^{3-} anions share two oxygen atoms with the apical oxygen atoms of Al (and Mg) edge site octahedra in a binuclear, bidentate fashion (Figure 2.4). This is similar to the structures formed when arsenate forms surface complexes at oxide and oxyhydroxide surfaces (Sherman and Randall, 2003).

In a series of papers, Manning and Goldberg (1996, 1997) and Goldberg (2002) studied competitive adsorption and surface complexation of a number of anions, including arsenate, arsenite, phosphate, and molybdate, and the clay minerals, Kaol, Mt, and illite. Each of these anions is believed to form surface complexes on the edge sites of Mt analogous to the structure described above. As(V) and As(III) adsorption have very different pH dependencies in these systems. For a given set of electrolyte conditions, As(V) adsorption density is greatest near pH 5, and As(III) adsorption is maximum near pH 9 (Figure 2.5). At the arsenic concentration of 20 μM and high clay mineral contents used in those experiments, there was very little adsorptive competition between As(V) and As(III). Much greater adsorptive competition between As(V) and phosphate was observed in the experiments of Manning and Goldberg (1996), suggesting that the presence of phosphate may enhance As(V) transport through clay barriers. Molybdate had a much smaller impact on As(V) adsorption. This could be explained by the fact that phosphate and As(V) have

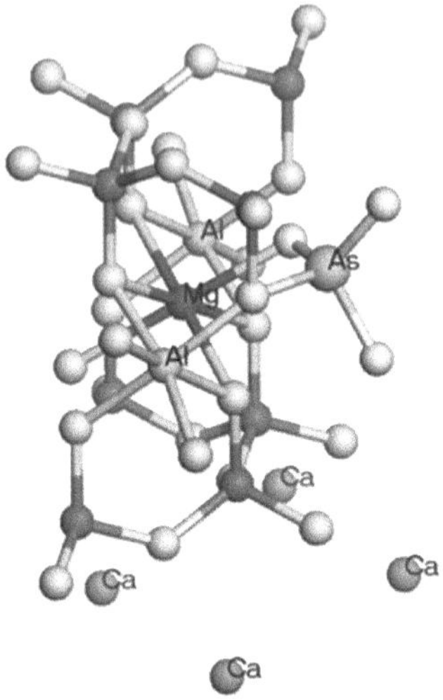

FIGURE 2.4 Structure of the surface complex formed by As(V) at the edge of a Ca^{2+}-Mt layer. *Reprinted with permission from Charlet et al. (2012).*

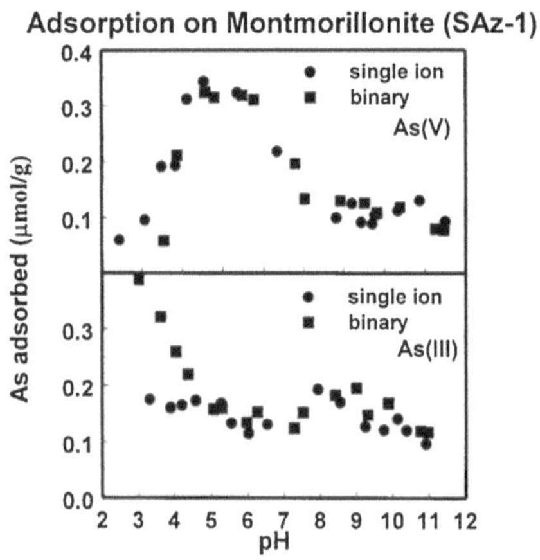

FIGURE 2.5 As(V) and As(III) adsorption on Mt as a function of pH in 0.01 M NaCl solution (initial composition) containing 20 μM arsenic. *Reprinted with permission from Goldberg (2002).*

similar adsorption pH dependencies, whereas molybdate was only strongly adsorbed at pH 4 (Manning and Goldberg, 1996).

Manning and Goldberg (1997) observed oxidation of adsorbed As(III) on Kaol and illite mineral surfaces at pH values $\geq$7. It was not clear in these experiments whether the oxidant was dissolved oxygen or trace amounts of manganese oxides present in the mineral samples. Lin and Puls (2000) also observed oxidation of adsorbed As(III) in a study of As adsorption by six different clay mineral samples, including halloysite (Hal), Kaol, illite,

illite/Mt, and chlorite (Chl). Aging of the arsenic-contacted clay mineral dispersions for two months increased both As(V) and As(III) adsorption and As(III) oxidation. Lenoble et al. (2002) determined that Fe(III)-pillared Mt was as strong as goethite as an adsorbent for As(V) and As(III), thus making it an effective material to add to clay barriers where As attenuation is important.

2.3.3 Surface Precipitation and Metal Substitution Reactions in the Clay Mineral Layer

Using EXAFS spectroscopy, Scheidegger et al. (1997) studied the molecular structure of Ni that was retained on the surfaces of pyrophyllite, Kaol, gibbsite, and Mt. The uptake of Ni from solution was a two-step process, with fast uptake in the first hour followed by much slower retention. The spectroscopic results were consistent with the presence of polynuclear Ni surface complexes, with Ni−Ni bond distances shorter than in $Ni(OH)_2(s)$. Analysis of the data strongly suggested the formation of mixed Ni/Al hydroxide phases, with a structure similar to that of takovite ($Ni_6Al_2(OH)_{16}CO_3 \cdot H_2O$). It was hypothesized that Al^{3+} was released from the mineral surfaces by dissolution and then incorporated into the mixed Ni/Al hydroxide-like phases, and that Al^{3+} release was the rate-determining step in the formation of the mixed phases.

Thompson et al. (1999) studied Co^{2+} adsorption and surface precipitation on Kaol over a period of several months. Rapid surface adsorption and precipitation accounted for early Co^{2+} uptake, but then slower precipitation caused continued Co^{2+} uptake. Co^{2+} adsorbed on Kaol as hydroxy-bridged polymers or multimers (at the Co^{2+} concentration studied = 100 μM), and subsequently some of the adsorbed Co^{2+} was transformed to a hydrotalcite-like precipitate that contained aluminum ions derived from Kaol dissolution.

Schlegel et al. (2001a) were among the first to demonstrate nucleation and epitaxial growth of Zn-phyllosilicate at the edges of Sm minerals under controlled laboratory conditions and at an environmentally relevant temperature. Using polarized extended X-ray absorption fine structure (P-EXAFS) spectroscopy, the authors showed that Zn^{2+} retained by Hect was structurally attached to the edges of hectorite layers and that the size and structure of the Zn^{2+} surface complexes varied with dissolved Si concentration and reaction time. At lower dissolved Si, small polymers were formed containing an average of two to three zinc cations, located in structural continuity with the Hect octahedral sheet. At high dissolved Si, the structure resembled a Zn-phyllosilicate, with very few magnesium cations included in the structure.

In an EXAFS study of Zn^{2+} adsorption on Mt at pH 7, Lee et al. (2004) also found that different types of surface complexes or surface precipitates were formed depending on the reaction time at a total Zn^{2+} concentration of 1 mM. For samples aged up to 11 days, the Zn−Zn contribution in the second shell EXAFS suggested formation of multinuclear surface complexes or surface precipitates. For samples aged 20 days and more, Zn−Zn and Zn−Si contributions in the second shell suggested formation of mixed metal coprecipitates such as a Zn phyllosilicate-like phase. Formation of these mixed metal solids probably accounts for the slow continuous adsorption reaction at longer aging times. Sequestration of zinc in this manner likely reduces the soil solution concentration, mobility, and toxicity of Zn^{2+} in soils and sediments.

Following on the earlier Schlegel et al. (2001a) study, Schlegel and Manceau (2006) investigated Zn^{2+} adsorption by Mt for a different set of chemical conditions, reaction times, and used P-EXAFS to identify Zn-bonding. At low initial Si concentration (70 μM), Zn^{2+} adsorption occurred rapidly during the first hour of reaction, and then leveled off at 168 h. The adsorption rate was consistent with Zn^{2+} adsorption on pH-dependent edge sites. At high initial Si concentration (500 μM), the initial adsorption rate was similar, but Zn^{2+} adsorption continued to increase substantially during the first 168 h of reaction, and was paralleled by Si uptake with a Si/Zn uptake ratio of 1.5, suggesting formation of a hydrous Zn-silicate. The spectroscopic results indicated a mixture of tetrahedrally and octahedrally coordinated zinc complexes. The proportion of tetrahedral zinc was lower in the high Si concentration samples and decreased with reaction time. At low Si concentration and 216 h of reaction, the spectra indicated the formation of zinc mononuclear IS complexes at the edges of Mt layers with both types of zinc coordination. At high Si concentration, Zn-silicate nuclei at the layer edges were detected after only 4 h of reaction. At 313 h of reactions, Zn^{2+} mononuclear complexes were no longer detected, supporting precipitation of a Zn-silicate. Thus, clay minerals may act as nucleating surface for the heterogeneous precipitation of Zn-silicate at dissolved Si concentrations relevant to natural systems.

2.4 ORGANIC SOLUTE ADSORPTION MECHANISMS

2.4.1 Adsorption of Organic Molecules by Clay Mineral Surfaces

2.4.1.1 Interactions Occurring within a Clay Mineral Phase and Interfaces

Interactions of organic compounds with clay minerals are variously affected by organic adsorbate structure, by types of clay minerals and nature of exchangeable cations, by the clay mineral charge density and by the solution properties (such as pH, ionic strength, the presence of organic cosolvents etc.). Extensive review of earlier publications presented by Lagaly et al. (2013) covered mechanisms, kinetics, and structure of the clay mineral intercalates formed by neutral organic molecules, cations, polymers, and proteins. Schoonheydt and Johnston (2013) examined clay interactions with water, organic cations and cationic proteins. Yuan et al. (2013) discussed the literature on interactions of organic cations, proteins, and nonionic organic compounds with clay minerals, with the focus on pollution control. A detailed summary of the factors involved and the mechanisms proposed for the uptake of nonionic compounds by complexes of clay minerals with quaternary ammonium cations (QAC) was also presented by Yuan et al. (2013). In addition, the review published by Cornejo et al. (2008) discussed the mechanisms of pesticide–clay mineral interactions. Due to the wide chemistry of organic pesticides involving both neutral compounds of variable chemical structure, organic cations and anions, the major mechanisms responsible for pesticide adsorption by clay minerals are meaningful also for adsorption of organic compounds from other classes not necessarily assigned to the pesticide group. Therefore, only a brief summary of various types of interactions occurring with participation of organic molecules within clay mineral phases and interfaces is provided in Table 2.1, with indicating specific surfaces of clay minerals and groups of atoms involved. The variety of interaction mechanisms of organic molecules with clay minerals provides the options for tuning adsorbate-clay mineral interactions to a certain molecular adsorbate structure. Further, this section is focused on bulk water (as a solvent) and water associated with clay minerals (by possibly multiple mechanisms which make it thermodynamically different from water molecules in an external phase) that play profound roles in adsorption of organic compounds by clay minerals.

2.4.1.2 The Significance of Solute Interactions in a Solution Phase Equilibrated with an Adsorbed State

The liquid-phase adsorption involves a transfer of adsorbing molecules from a solution to an adsorbed state. The equilibrium adsorption of organic molecules may be described by a coefficient K_d of a compound distribution between an adsorbed state and a solution phase (i.e., an adsorbed concentration over a solution concentration). A quantification of the contributions of various interactions to the K_d values is not trivial. On an empirical level, it can be carried out by using the LFER analysis exploring some solute-related descriptors of intermolecular forces (e.g., the E, S, A, B, L^{16}, and V in Eqn. (2.1)):

$$\log K_d = c + eE + sS + aA + bB + l \log L^{16} \left[or\ vV \right] \tag{2.1}$$

where E is the excess molar refraction which describes polarizability contributions from n- and π-electrons; S is the dipolarity/polarizability; A and B are the H-bond acidity and basicity, respectively; L^{16} is the compound gas/n-hexadecane partitioning coefficient strongly related to the volume V of organic compounds and associated with dispersion interactions and the cavity formation in the bulk phases (Pool, 2003). So, the LFER examination of aqueous adsorption of a series of organic compounds on an n-hexadecyl trimethylammonium (HDTMA)-exchanged Mt suggested that, in comparison to water,

TABLE 2.1 Types of Interactions of Nonionic Organic Compounds with Clay Mineral Adsorbents

Types of Interactions	Clay Mineral Surfaces and Groups of Atoms Involved
Nonspecific van der Waals (i.e., dispersion, induction, and multipole) interactions	Any surfaces in the clay phase (i.e., siloxane, OH surfaces and edges), interlayer-, surface-, and exchangeable cation-associated water, organic moieties in organic cation-modified clays
Electrostatic interactions with charges	Negatively charged clay mineral surface, positively charged counterions, and variously (pH-dependent) charged edge surfaces
H-bond donating and accepting	H-bond donation from OH surfaces, edge-associated hydroxyls, interlayer water, cation- and edge-associated water molecules; H-bond accepting by neutral and negatively charged siloxanes surfaces, edges and clay mineral-associated water. In organic cation-exchanged clays, both H-donating and accepting are possible depending on the functional groups present
Proton transfer reactions in the clay phase	Surface acidic OH groups, cation-associated water or acidic functional groups associated with organoclay fragments are capable of donating protons
Electron donor–electron acceptor interactions	Exchanged cations of transition metals and structural Lewis acid sites as electron acceptors; siloxanes as electron donors; donating/accepting moieties in organic cation-modified clays

the organoclay (OC) adsorbent is more polarizable, less polar, and cohesive and has stronger H-bond acceptor and weaker H-bond donor capabilities (Zhu et al., 2011a).

Therefore, multiple solute interactions occurring in a solvent phase may compensate, to a certain extent, adsorbate interactions within clay minerals and interfaces. A possibly little contribution of a certain type of interactions, e.g., H-bond formation, to a compound adsorption from a liquid phase does not necessarily indicate an insignificance of such interactions in any of coequilibrating phases.

Many nonpolar (low-polar) organic compounds, e.g., hydrocarbons and their halogen-substituted derivatives, demonstrate a tendency to escape from aqueous phases to a hydrophobic phase. This tendency often associated with "hydrophobic adsorption" or "hydrophobic binding" does not involve "hydrophobic bonds" between an adsorbate and an adsorbent and per se is not linked to a certain type of adsorption site. Thermodynamically, the hydrophobic effect responsible for reducing aqueous solubilities of nonpolar molecules and their expelling from aqueous solutions is associated (at room temperature) with the entropy loss upon hydration of nonpolar moieties (Chandler, 2005).

2.4.1.3 The Role of Clay Mineral-Associated Water in Organic Molecule—Clay Mineral Interactions

Adsorptive properties of clay minerals toward nonionic organic molecules are strongly related to the presence of adsorbed water in clay mineral. The literature on clay mineral−water interactions is broad (e.g., see in Yariv, 1992; more recently, Schoonheydt and Johnston, 2013; Lagaly and Dekany, 2013). Clay mineral−water interactions are influenced by the nature of exchangeable cations, their valence and size that determine the cation hydration potential (Schoonheydt and Johnston, 2013). In the presence of water,

- some clay minerals swell and open the interlayer spaces leading to more accessible surfaces potentially able to interact with organic compounds;
- adsorbed water may bridge organic molecules via H-bond formation or participate in proton transfer reactions;
- water competes for clay adsorption sites and prevents direct organic adsorbate interactions with exchangeable cations, OH and siloxane surfaces.

The actual availability of siloxane surfaces in clay minerals for aqueous adsorption of organic molecules is increased when the surface charge density is reduced (Jaynes and Boyd, 1991). Hydration of clay surfaces can be suppressed by replacing typically present Na^+ and Ca^{2+} cations with other less-hydrated inorganic cations such as K^+ or Cs^+ or organic cations. As a result, the aqueous adsorption of pesticides, nitro-substituted aromatic compounds, and other nonionized organic contaminants by clay minerals is enhanced (Aggarwal et al., 2006; Liu et al., 2009). Thus, it was concluded that maintaining the expanded clay mineral structure with a reduced hydration contributes to enhanced adsorption of organic molecules from water. Moderate heating or air-drying of the clay adsorbents (such as Na^+-, K^+-, and QAC-exchanged Mt) was suggested to hinder the following clay mineral rehydration thus suppressing water-adsorbate competition and enhancing aqueous adsorption of certain organic molecules (Chappell et al., 2005; Borisover et al., 2012).

The competitive role of water in interactions of organic compounds with clay minerals becomes obvious from a general decrease of organic vapor adsorption (described by the adsorbed state/gas phase K_d values) with increasing relative air humidity ($RH = p_W/p_W^{sat} \times 100$, where p_W and p_W^{sat} are the partial water vapor pressure and saturated vapor pressure of water, respectively)

(Goss and Eisenreich, 1996; Goss et al., 2003). A "differential stoichiometry" in mutual relations between adsorbed organic (A) and water (W) molecules may be quantified directly from the equilibrium vapor K_d versus RH data (Borisover, 2013):

$$(\partial n_W/\partial n_A)_{T,P,RH} = (\partial \ln K_d/\partial \ln RH)_{T,P,RH,n_A} \tag{2.2}$$

where n_A and n_W correspond to adsorbed amounts of organic and water molecules, respectively, at given temperature T and pressure P. Table 2.2 provides some examples calculated by means of Eqn (2.2) which show how different organic molecules, *n*-hexane and diethyl ether, affect the hydration status of various clay minerals at various RH (at "infinitely low" adsorbed concentrations of organic molecules A). So, an *n*-hexane molecule expels between two and four molecules of water from Ca^{2+}-bentonite (Bent) as compared to 0.7–1.4 water molecules expelled from both Kaol within the same 40–80% RH range thus suggesting a greater hydration of a microenvironment accommodating *n*-hexane molecules in Ca^{2+}-Bent as compared with Kaol.

The derivatives $(\partial n_W/\partial n_A)_{T,P,RH}$ can be also related to the molecular descriptors of organic adsorbates within the LFER approach. The dipolarity/polarizability, H-bond basicity, and the molar volume of organic adsorbates contribute to expelling clay mineral-bound water molecules (i.e., to the negative values of the derivative in Eqn. (2.2)) whereas, in contrast, polarizability contributions from *n*- and π-electrons increase the clay mineral hydration upon introducing the molecules of organic compounds to the adsorbed state (Borisover, 2013).

The effect of the full clay mineral hydration on organic compound–clay mineral interactions is obtained by comparing, on the activity-based scale, the organic vapor adsorption isotherms on dehydrated clay mineral adsorbents and the aqueous adsorption isotherms. Such a comparison demonstrated the strong water–organic adsorbate competition for multiple nonpolar organic molecules on various organic cation-exchanged Sm (e.g., Lee et al., 1990; Kukkadapu and Boyd, 1995). The vapor adsorption study may be difficult (or even impossible) for multiple organic compounds of interest, due to their low volatility. In this case the aqueous adsorption compared with the adsorption on a dehydrated clay mineral adsorbent from a relatively inert medium, e.g., saturated hydrocarbon, may reveal adsorbate-water interplays on the background of clay mineral swelling. Such a comparison is shown in Figure 2.6 for adsorption of nitrobenzene and *m*-nitrophenol on Na^{+}-Bent and its HDTMA-form exchanged by 90%. Water-adsorbate competition is adsorbate-dependent, becomes clearly suppressed on the HDTMA-exchanged Bent as compared with its Na^{+}-form and practically disappears for *m*-nitrophenol adsorption on HDTMA-clay mineral thus suggesting a link between an adsorbate-water competition and an adsorbate capability of specific interactions (Borisover et al., 2008).

TABLE 2.2 Derivatives $(\partial n_W/\partial n_A)_{T,P,RH}$ for *n*-Hexane and Diethyl Ether Vapors Adsorbed by Variously Hydrated Clay Minerals

Clay Mineral Adsorbent	$(\partial n_W/\partial n_A)_{T,P,RH}$[a] at Various RH (Indicated in Percent)			
	40	80	40	80
	n-Hexane		Diethyl Ether	
Ca^{2+}-Kaol	−0.7	−1.4	−1.5	−2.9
Na^{+}-Kaol	−0.7	−1.4	−1.3	−2.6
Ca^{2+}-Bent	−1.9	−3.7	na[b]	na

[a]*Calculated by using Eqn (2.2) from the experimental data obtained at 70°C by Goss (1993).*
[b]*na means not available.*

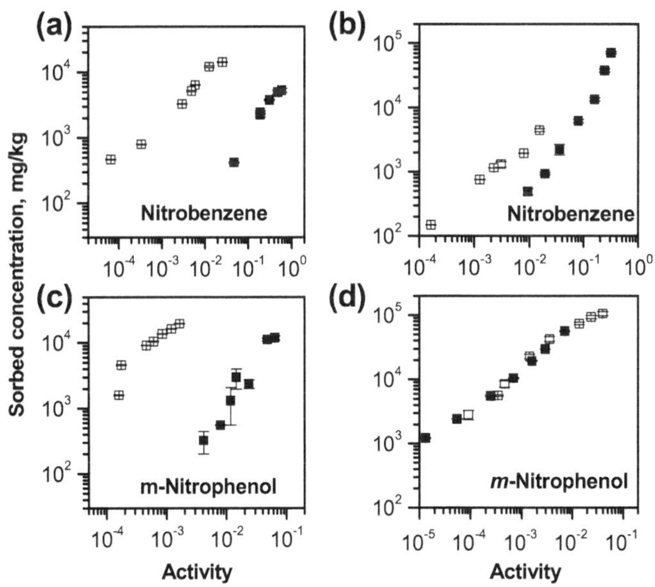

FIGURE 2.6 Adsorption isotherms determined for nitrobenzene and *m*-nitrophenol on Na^{+}-bentonite (a) and (c) and its *n*-HDTMA-exchanged form (90% of cation exchange capacity), (b) and (d): adsorbed concentration are plotted against a compound activity (referred to a pure (subcooled) liquid state). The open squares represent the data obtained on a dehydrated adsorbent in *n*-hexadecane; the filled squares represent the data obtained from water. *This figure is adapted with permission from Borisover et al. (2008). Copyright (2008) American Chemical Society.*

2.4.2 Adsorption of Charged Organic Species by Clay Mineral Surfaces

2.4.2.1 Organic Cations

Organic cations are effectively adsorbed on clay minerals with isomorphous substitution via cation exchange. This is the long-studied clay mineral adsorption mechanism established for multiple organic cations including dyes, surfactants, proteins (Cornejo et al., 2008; Lagaly et al., 2013; Schoonheydt and Johnston, 2013; Yuan et al., 2013). A brief summary of other processes contributing to adsorption of organic cations by clay minerals is provided in Table 2.3 (e.g., see also Yuan et al., 2013).

However, despite a longtime research, there is a lack of sufficient consistent data on adsorption of organic cations by clay minerals at low adsorbent loadings where adsorbate–adsorbate interactions may be considered unimportant (Droge and Goss, 2013a). Therefore, by examining adsorption of a large series of organic cations on Bent, Kaol, and illite, Droge and Goss (2013a) demonstrated that the K_d values describing the distribution of organic cations between a clay mineral adsorbent (at low loading) and an aqueous

TABLE 2.3 Mechanisms and Processes Controlling Interactions of Organic Cations and Anions with Clay Minerals

Charged Entity	Important Interaction Mechanisms
Organic cations	The cation exchange due to the negatively charged siloxane surfaces and pH-dependent edge sites
	Lateral (van der Waals) interactions between organic cations (adsorbed both via ion exchange or by any other mechanisms)
	Electron donor (negatively charged siloxane surface)–electron acceptor (organic cation) interactions
	H-bond with interlayer water and oxygen on the clay mineral surface
Organic anions	The anion exchange at positively charged edge sites (at low pH), on the clay minerals enriched with organic cations above the CEC, the clay-cationic polymer complexes, the clays pillared with poly(hydroxo metal) cations
	Ligand exchange (with the structural OH groups)
	The H-bond with the surface- and edge-associated OH groups, the coordinated or surface-adsorbed water
	Bridging to clay mineral surfaces via multivalent cations

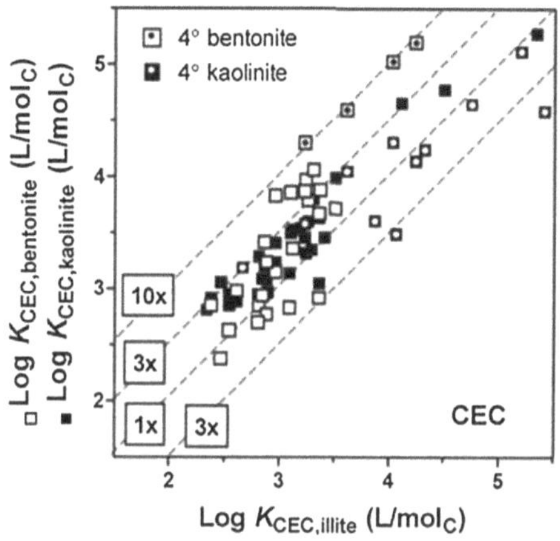

FIGURE 2.7 The cation exchange capacity (CEC)-normalized distribution coefficients K_{CEC} of a series of organic cations on bentonite, kaolinite, and illite (at the presence of 15 mmol of NaCl and pH 6). Quaternary ammonium cations (4°) are marked by a contrasting dot. *This figure is reprinted with permission from Droge and Goss (2013a). Copyright (2013) American Chemical Society.*

phase (of a given ionic composition) and normalized by CEC of clay minerals varied generally only within a factor of ±3 for (Figure 2.7). One conclusion from this work was that despite organic cations may undergo multiple possible interactions with clay minerals (Table 2.3), the accounting for the CEC becomes sufficient to describe a variability between adsorption coefficients (at low adsorbate loadings) on the three different clay minerals characterized by different microenvironments of negative charges.

Van der Waals lateral interactions of QAC exchanged on swelling and nonswelling clay minerals are important and result in a unique S-like shape of the QAC adsorption isotherms (Xu and Boyd, 1995). Clay adsorption of QAC with long aliphatic chains, at sufficiently high uptakes (but below a CEC) may occur with a partial cation exchange (Xu and Boyd, 1995; Sheng et al., 1998). Organic cations adsorbed (together with their anion) by "a hydrophobic bonding" mechanism may relatively easily be desorbed as distinct to the organic cations exchanged (Sheng et al., 1998). A hydrophobic effect expelling nonpolar moieties from water contributes to any mechanism of organic cation adsorption by clay minerals from aqueous solutions. The significant amount of organic cations in the clay mineral exceeding the CEC may be obtained also by adsorbing the micelles formed by the long-chain QAC (Nir et al., 2013).

2.4.2.2 Organic Anions

Adsorption of organic anions by negatively charged clay minerals is generally limited and relatively weak but there are several mechanisms that may yet contribute to interactions of organic anions with clay minerals (Table 2.3; Cornejo et al., 2008; Yuan et al., 2013). Adsorption of organic anions may be significantly enhanced on organic cation-modified clay minerals when the extent of the loading of organic cations exceeded the clay CEC. The positive charge developed by the clay mineral-loaded organic cations provides an option for the anion exchange. Organic anions may be accumulated on clay minerals via the positively charged QAC-based micelles used to accommodate anions (Nir et al., 2013). Polymeric organic polycations adsorbed on clay mineral may also provide the exchange sites for organic anions both when the polycation adsorbed amount is below and above the clay mineral CEC (Yuan et al., 2013). Clays pillared with poly(hydroxo metal) cations can promote the adsorption of anions.

Due to the electrostatic repulsion between anions and clay mineral surfaces, the negative adsorption increasing the concentration of an organic anion in a solution was also reported (Cornejo et al., 2008). Electrostatic repulsion may occur also between adsorbed organic anions. So, in examining adsorption of perfluoroalkyl acids by Kaol, an electrostatic repulsion between adsorbed acids was suggested as a primary inhibitory factor, and the enhanced adsorption of the longer-chained acids as compared with the shorter-chained was related to the hydrophobic effect expelling CF_2 groups of the organic acids from water (Xiao et al., 2011). Ligand exchange or the replacement of the structural OH groups at the edges and also on basal surfaces is known for some inorganic anions (Bergaya et al., 2013; Lagaly and Dekany, 2013). H-bond could be expected between organic anions and H-bond donors in the clay mineral.

2.4.2.3 Organic Compounds Undergoing a Partial Ionization and Containing Both Basic and Acidic Functional Groups

When organic compounds undergo a partial ionization, the K_d describing adsorption of a compound from an aqueous medium may be represented by Eqn (2.3) (Schwarzenbach et al., 2003):

$$K_d = \theta_i \times K_{d,i} + (1 - \theta_i) \times K_{d,n} \tag{2.3}$$

where θ_i is the fraction of the ionized form in a solution, and $K_{d,i}$ and $K_{d,n}$ are the distribution coefficients for ionized and neutral forms of an adsorbing compound. If the adsorption isotherms of ionized and nonionized species may be considered linear upon their solution concentrations, then, $K_{d,i}$ and $K_{d,n}$ can be obtained from the K_d versus θ_i plot. However, the inherent nonlinearity of adsorption isotherms may strongly affect the shape of a K_d versus θ_i profile and should be taken into account in adsorption modeling (Xiao and Pignatello, 2014).

When organic molecules contain both basic and acidic functional groups, depending on pH, they behave either as cations, anions, or zwitterions thus demonstrating even more complexed behavior in their interactions with clay minerals. For example, multiple mechanisms are suggested for adsorption of enrofloxacin, a veterinary chemotherapeutic agent, by Sm and Kaol (Yan et al., 2012) thus involving the cation exchange of positively charged species, the formation of the OS complexes and the IS complex formation with the participation of a carboxylic group. Adsorption of the zwitterionic amino acids is controlled by the cation exchange and the complexation of the interlayer cations by the carboxylate groups of the zwitterion and the anionic form (Lagaly et al., 2013).

2.5 INTERACTIONS OF CLAY MINERAL SURFACES IN SOILS AND SEDIMENTS WITH NOM AND NATURAL NANOPARTICLES OF OTHER MINERALS

Due to high activities of clay mineral surfaces, in natural environments clays undergo multiple interactions with NOM including humic substances (HS), polysaccharides, proteins (and multiple other low- and high-molecular organic substances), and inorganic components such as metal oxides often present as nanoparticles. Chemistry of NOM in soils and possible relations with the presence of clay minerals are discussed by Baldock and Broos (2012). The strong interactions between clay mineral and NOM make themselves evident in close associations between clay and NOM contents in soils. Clay mineral–NOM interactions depend on the clay mineral type (TO or TOT), the functional group content and the molecular size of NOM components, the pH, the presence of inorganic monovalent or divalent cations (Baldock and Broos, 2012). The structure of organic moieties of HS is important for clay–HS interactions. For example, Balcke et al. (2002) demonstrated that the HS adsorption by Kaol correlated directly with the aromaticity of HS and inversely with their polarity expressed as the oxygen to carbon ratio. The specific surface area of clay minerals is more important than the total clay content. Yuan and Theng (2012) provided a detailed summary of mechanisms of clay–HS interactions. Although ionizable humic and fulvic acids should be repelled by negatively charged clay mineral surfaces, there are multiple attractive interactions between HS and clay mineral, e.g., cation- and water-bridging, anion and ligand exchange and other binding modes listed in Tables 2.1 and 2.3 that operate simultaneously, with importance of each mechanism changing from soil to soil (Yuan and Theng, 2012).

Possible scenarios of protein–clay mineral interactions are reviewed by Schoonheydt and Johnston (2013): positively charged proteins undergo cation exchange such that the adsorption of cationic proteins in the interlayer space of Sm is dependent both on the molar mass of the proteins and the swelling behavior of the Sm. The adsorption of negatively charged proteins is contributed by dispersive interactions with siloxane surfaces needed to overcome the electrostatic clay mineral–protein and protein–protein repulsions. Protein adsorption is pH-dependent, increases with decreasing negative charge of the protein molecules and may be associated with protein conformation changes (Lagaly et al., 2013).

Inorganic (hydr)oxide are ubiquitous in the environment and may have multiple impacts on activities of clay mineral surfaces. Polymeric (hydr) oxides of aluminum, iron, chromium, zinc, and titanium can intercalate into Sm by cation exchange (Brigatti et al., 2013). Heterocoagulation reactions between oppositely charged metal oxide nanoparticles and larger clay minerals colloids may affect reactive surface area and adsorption reactions (Qafoku, 2012). So, positively charged iron (hydr)oxide can coat the negatively charged clay mineral surfaces thus blocking adsorption sites at clay mineral surfaces but instead providing the iron oxide surfaces. Lagaly and Dekany (2013) discussed the colloidal behavior of clay mineral–oxide dispersions and concluded that it depends on pH of a solution, the point of zero charge (PZC) of oxides, the relative size of the particles, the charge density and the way how the particle were brought into contact. Heterocoagulation between the positively charged oxide particles and clay mineral particles is expected to occur at $pH < 6$, if PZC of the oxide is about 6. Interactions of smaller particles of gibbsite and iron (hydr)oxide with larger clay mineral particles may lead to recharging surfaces and changing the clay colloidal behavior (Lagaly and Dekany, 2013). Another important effect is that the cementation of clay minerals by coating with amorphous (hydr)oxides and the interlayer penetration of poly(hydroxo aluminum/iron) ions may reduce the mechanical swelling of the clay minerals (Lagaly and Dekany, 2013). Kampf et al. (2012) indicated that precipitated or coagulated metal oxides may affect adsorption of cations and anions (as in the case of iron oxides and aluminum oxides), the oxidation–reduction reactions (as for iron and manganese oxides) whereas ubiquitous essentially noncharged and poorly hydrated silicon oxide has relatively small effect on the physicochemical processes and acts as a diluent to the much more reactive clay minerals and oxides of Fe, Al, Mn. Iron oxide may form deposits on Kaol particles, and the attraction between positively charged iron oxide particles and negatively charged clay minerals contributes to the soil aggregate stability (Kampf et al., 2012). Cai et al. (2014) concluded on the attachment of titanium dioxide (TiO_2) nanoparticles to dispersed Bent and Kaol particles and attributed the increased transport of nano-TiO_2 to the clay mineral–TiO_2 interactions. The precipitation and/or adsorption of iron and aluminum oxides on clay mineral surfaces may also contribute to the development of complexed organomineral assemblages. A soft X-ray spectromicroscopic study of mineral-organic matter assemblages in soil clay fractions demonstrated a similar association of iron oxides and clay minerals with organic matter in organomineral complexes (Chen et al., 2014a).

2.6 ADSORPTION PROCESSES ON CLAYS IN NATURAL AND ENGINEERED ENVIRONMENTS

Adsorption interactions of metals, metalloids, and organic compounds with clays in natural and engineered environments may differ significantly from these observed on "pure" clay minerals. These differences are due for many possible reasons but rarely have the effects been studied from a mechanistic point of view. Some of the more important phenomenological observations made in studies of clay-rocks, clayey sediments, and engineered barriers will be reported below.

2.6.1 Interactions of Metal and Metalloid Ions with Clays in Natural and Engineered Environments

Manceau et al. (2005) studied the natural speciation of Ni(II) and Zn(II) in a clay horizon of a paddy soil (Taiwan) using EXAFS spectroscopy. Ni^{2+} was found to substitute for manganese in the manganese layer of a natural lithiophorite in the soil. In contrast, Zn^{2+} was found to fill vacant sites in the gibbsitic layer of the lithiophorite. The partitioning of Ni^{2+} and Zn^{2+} between the two layers was a result of the general preference of Ni^{2+} for slightly smaller sites. Additional characterization revealed that Zn^{2+} was predominantly bound to hydroxy-Al interlayers sandwiched between vermiculite layers in the fine soil matrix. The incorporation of Zn^{2+} in the gibbsitic layer of both lithiophorite and vermiculite helps to increase the stability of these minerals by providing positive charge to balance the negative charge from the TOT clay mineral layer and the manganese layer of lithiophorite. The results emphasize the importance of metal substitution reactions in soils and sediments, which provide a binding environment for sequestration of some trace metals and control metal mobility within soil pore-water solutions. Following on from this work, Vespa et al. (2010) investigated the speciation of Zn(II) in smelter-affected agricultural soils. While the previous work with uncontaminated soils identified clay mineral as the main sequestration phase, the polluted soil sample was more ferruginous with an average Fe/Al atomic ratio of 1.1. The Zn(II) and Fe(III) in the neoformed Sm phase in the polluted soil were derived from the weathering of ZnS, ZnO, FeS_2, and $ZnFe_2O_4$ particles from the smelter. These cations slowly dissolve and then coprecipitate with Al(III) and Si(IV) in the soil clay matrix. Although likely forming surface complexes initially, the Zn^{2+} eventually is sequestered by the metal substitution reactions. Zinc sequestration in the octahedral sheet of dioctahedral Sm is potentially irreversible, because this type of clay mineral is stable over a large pH range, and the neoformed species is analogous to the uncontaminated species which formed over time during pedogenesis.

While metal substitution reactions can be important for the overall adsorption of some metals in soils and sediments, this occurs specifically in cases where the adsorbing metal has the ionic radius and coordination chemistry appropriate to fit in the structure of the "adsorbing" clay mineral. There are also cases where adsorbed metals and actinides are presumed to remain as surface complexes over longer periods of time rather than proceeding to a metal substitution reaction, for example, in the cases of U(VI), Rh(IV), and Tc(IV) (Turner et al., 1996; Vinsova et al., 2004).

Other important issues regarding adsorption include the application of laboratory results from dispersed systems (and associated thermodynamic adsorption data or model parameters) to rocks and engineered barriers. For example, Maes et al. (2008) measured Cs^+ adsorption onto Boom clay in batch adsorption tests on dispersed systems and then tested the measured adsorption parameters in modeling the results of long-term diffusion experiments (fourteen years) on intact clay cores. The transport retardation factors obtained from batch adsorption data tended to be higher than those obtained by migration experiments. Their analysis suggested that not all adsorption sites are available in a compacted system. In contrast, van Loon and Glaus (2008) examined a similar issue for compacted Bent and hypothesized that compaction affects the thermodynamics of the adsorption process such that Cs^+ adsorption on compacted Bent should increase. The authors argued that, due to a reduction of the interlayer space, there is a less capability of the interlayer water to hydrate cations. Cations with a low hydration tendency, such as Cs^+, therefore accumulate in the interlayer space, whereas highly hydrated cations such as Na^+ tend to accumulate in the bulk water. However, in an experimental test of the hypothesis, van Loon et al. (2009) found no significant effect in comparing the adsorption of Cs^+ on crushed and intact (noncrushed) Opalinus clay, and others (Montavon et al., 2006; Chen et al., 2014b) have found that compaction does not decrease Cs^+ adsorption in natural clay-rocks or in Bent. Appelo et al. (2010) attributed the earlier reported "discrepancies" partly to neglecting the role of dead-end pores in migration experiments, but also to use an approximate isotherm in diffusion modeling. The authors demonstrated that it is important to incorporate exact Cs^+ nonlinear adsorption behavior into diffusion models (rather than a simplifying Freundlich isotherm). Without the correct sorption isotherm, Cs^+ adsorption will be calculated differently in migration and batch experiments, because it is the slope of the isotherm that is the important term in diffusion models.

Clay liners are often used to seal landfills, and the transport of toxic elements through these barriers is important to groundwater quality. As(V), As(III), and Se(IV) adsorption from landfill leachate solutions onto Mt and Kaol was studied (Frost and Griffin, 1977) because of the presence in these minerals in clay barriers. Studies have also been made of adsorption of other oxyanion contaminants to clay minerals that may comprise clay barrier or clay-rock systems. For example, Missana et al. (2009) determined that adsorption of Se(IV) on FEBEX Bent and on an illite sample was strong in the weakly acidic pH range ($pH < 7$). Adsorption of the Se(IV) was similar for each clay-rich sample and independent of ionic strength and Se concentration. Cr(VI) is more weakly adsorbed than As(V), As(III), or phosphate, and is only effectively adsorbed by illite or Kaol under more acidic conditions than Se(IV), i.e., pH 2–5 (Zachara et al., 1988; Ajouyed et al., 2011). However, the adsorption of Cr(VI) is strong enough to allow natural clays to be used as landfill liners to attenuate Cr(VI) as a contaminant (Lu et al., 2009). Taylor et al. (2000) found that Fe(II)-bearing clay minerals, e.g., Mt, illite, and vermiculite, were effective in reducing Cr(VI) to Cr(III) at the mineral surface, greatly increasing the retention of chromium and indicating that redox potential within clay barriers is important in its mobility.

2.6.2 Interactions of Organic Compounds with Clays in Natural and Engineered Environments

This section is focused on three environmental factors expected to affect organic compound–clay interactions: (1) presence of NOM and metal oxides, (2) presence of DOM, and (3) wetting–drying process.

2.6.2.1 Presence of NOM and Metal Oxides

Due to a strong water competition for clay mineral adsorption sites (see Section 2.4.1.3), the role of clay minerals as adsorbents toward nonionized organic compounds becomes diminished in wet soils and sediments, and NOM is often considered as a major natural adsorbent component controlling aqueous adsorption of multiple organic compounds (Chiou and Soup, 1985; Schwarzenbach et al., 2003). Even a small fraction of NOM on clay mineral adsorbents may control the aqueous adsorption of hydrophobic organic compounds. For example, the 0.01% of organic carbon on aluminum or iron oxide is sufficient to fully control the aqueous adsorption of various polycyclic aromatic hydrocarbons (Mader et al., 1997). The capability of NOM to cover clay mineral surfaces and block the adsorption sites may further diminish the role of clay minerals in adsorption of organic compounds by wet soils and sediments (Charles et al., 2006). The opposite scenario seems to be also possible when the NOM adsorbed an external clay mineral surfaces activates the intercalation of fulvic acids (Dubbin et al., 2014). The coating effect reducing the availability and adsorption affinity of clay mineral surfaces is recognized for iron oxide (Laird et al., 1994). Soil cementing metal oxides may reduce the access to clay mineral interlayer by inhibiting swelling (Li et al., 2003).

The relative contribution of clay mineral fractions to the organic compound adsorption by soils depends on the type of the clay mineral, the nature of the exchanged cation, the clay fraction in the soil, and organic compound distribution coefficients associated with different soil components (Li et al., 2003). Weakly hydrated exchangeable inorganic cations enhance adsorption of organic compounds by clay minerals (see Section 2.4.1.3), and therefore increase the relative clay mineral contribution to the organic compound adsorption by soils and clay mineral–NOM complexes. The contributions of a K^+-exchanged clay mineral and a model NOM to the adsorption of diuron and atrazine by a hypothetical soil of a variable Sm/NOM composition are compared in Figure 2.8 (Sheng et al., 2001). As shown in Figure 2.8, assuming

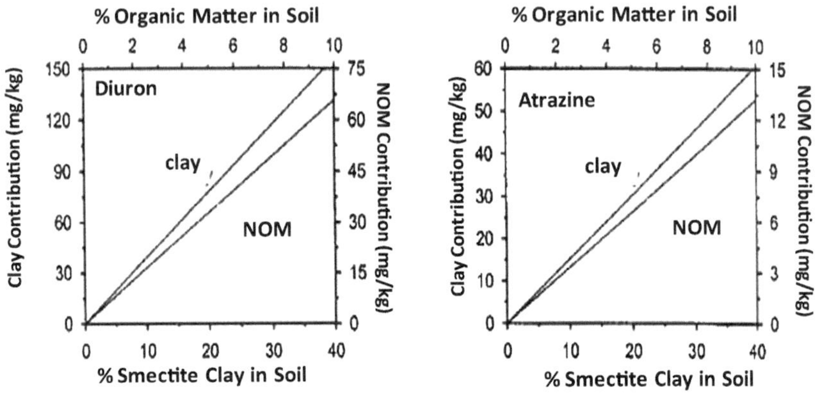

FIGURE 2.8 Adsorbed concentrations of diuron and atrazine on the clay and natural organic matter (NOM) components of the model (hypothetical) soils composed of a K^+-Sm and a muck soil organic matter (at the solution concentrations of the compounds corresponding to 10% of the aqueous solubility; Sheng et al., 2001). *This figure is adapted with permission from Sheng et al. (2001). Copyright (2001) American Chemical Society.*

the additivity of adsorption contributions from a K^+-Sm and a model NOM, the K^+-clay mineral may indeed be an adsorbent controlling the retention of the organic compounds by soils. Although typically K^+ is not the dominant exchangeable cation, the scenarios are possible when the interlayer space may be occupied with K^+ even in the presence of other cations (e.g., the cation demixing; Sheng et al., 2001). The contributions of clay mineral, NOM and metal oxide fractions to the overall soil adsorption of organic compounds may not be additive (Celis et al., 1998; Zhang et al., 2008). Specifically, clay mineral–NOM interactions can result in conformational NOM changes as compared with the bulk or isolated NOM.

Organic cation-modified clay minerals appear to be competitive with NOM toward aqueous adsorption of nonionized organic compounds. The content of organic carbon associated with clay mineral-attached cations is large, and, in addition, the organic carbon-normalized distribution coefficients for organic molecules interacting with OC may exceed significantly the values typically reported for organic molecules on soil NOM (Boyd et al., 1988; Borisover et al., 2001).

Regarding adsorption of organic cations in various natural and engineered systems containing NOM, the clay minerals may play a more distinct role than in adsorption of nonionized organic compounds. For a given ionic solution composition, the K_d values for the organic cation adsorption to the soils were predicted with the averaged deviation by -0.1 ± 0.4 log units from the measured values by relating adsorption onto soil NOM and clay mineral to the soil organic carbon content and the clay CEC, respectively (Droge and Goss, 2013b).

2.6.2.2 Dissolved Organic Matter

The presence of multiple organic and inorganic soluble components in aqueous phases of natural and engineered systems, such as in soil and aquatic environments, wastewater treatment facilities or waste disposal sites, may have various impacts on interactions of target organic compounds with clay minerals. DOM is capable of forming water-soluble complexes with organic molecules and ions or competing with organic adsorbates for adsorption sites thus reducing the organic compound affinity to adsorb (Schwarzenbach et al., 2003; Gerstl and Graber, 2011). Also, the opposite effects of DOM are possible, i.e., coadsorption of DOM–organic compound complexes onto the adsorbent matrix or DOM adsorption providing an additional domain for adsorption of a target compound (Muller et al., 2007). Diverse anthropogenic contaminants including low-molecular mass organic compounds, metals, surface-active compounds capable of forming micelles or adsorbing at natural surfaces, water-soluble polymers, and others may be involved in similar adsorption relations with organic compounds of interest thus resulting in a competition at surfaces, formation of water-soluble complexes and coadsorption.

2.6.2.3 Wetting–Drying Cycles

Wetting–drying of clay materials may have impact on ability of clay minerals to interact with organic compounds. It is not only because the removal of water decreases the water competition for adsorption sites and affects swelling/shrinking of clay minerals. Heller-Kallai (2013) summarized some studies demonstrating the changes in clay minerals due to the wetting–drying cycles, such as ordering of layer stacking accompanied by the potassium fixation in Sm, a deprotonation at the presence of soluble potassium salts and the increase in the layer charge, the variability in a specific surface area. Bergaya et al. (2013) indicated other effects of wetting–drying on clays such as the formation of the exchange sites with high selectivity for cesium ions upon the repeated wetting–drying of potassium Mt, drying-induced hysteresis in the K^+/Ca^{2+} exchange on Mt, the changes in the Bent CEC caused by extensive drying.

Dehydration of HDTMA-exchanged Mt and decrease of its basal spacing caused by drying are only partially reversible upon rehydration (Zhu et al., 2011b). Since the clay mineral properties are affected by drying/wetting, its impact on clay mineral interactions with organic molecules and ions is well expected. Indeed, drying/rewetting the potassium Sm (but not Ca^{2+}-Sm) involved a hysteretic (reduced) swelling of clay mineral particles but produced an affinity for atrazine adsorption from water by one order of magnitude greater as compared with that on the never-dried K^+-Sm (Chappell et al., 2005). In addition, at increasing hydration, a configuration of the NOM associated with clay minerals may change (e.g., by exposing the hydrophilic groups at the interfaces and hiding the hydrophobic groups in the NOM interior; Liu and Lee, 2006) thus potentially affecting the availability of clay mineral surfaces for interactions with organic compounds.

2.7 SUMMARY

Adsorption of inorganic solutes on clay minerals can occur via three main mechanisms: (1) cation exchange with fixed charge sites in interlayer spaces and along basal planes, (2) surface complexation at edge sites, and (3) metal substitution reactions with cations in the clay mineral structure. Cation adsorption at fixed charge sites is driven by electrostatic and hydrogen-bonding forces and energies and involves the formation of OS complexes. In contrast, IS surface complexation at edge sites generally involves stronger, covalent chemical interactions. Metal substitution reactions involve the exchange of adsorbing cations with Al^{3+} or Mg^{2+} cations in the clay mineral structure, generally at layer edges. These reactions can be detected with spectroscopy by determining the coordination numbers of the adsorbing metal with surrounding oxygen atoms and the number of other metal atoms that are located in close proximity. Metal substitution can occur when the adsorbing cation has a similar atomic radius to the exchanging metal cation and sufficient time allows the reaction to proceed following initial IS complex formation. Sequestration of Zn^{2+} and Ni^{2+} by this mechanism likely reduces the soil solution concentration, mobility, and toxicity of Zn^{2+} and Ni^{2+} in soils and sediments.

For transition metal cations, lanthanides, and actinides, adsorption on most clay minerals is ionic strength dependent at low pH, and becomes more ionic strength *independen*t and pH *dependent* at pH values greater than 6. This occurs via a transition from a predominant cation exchange mechanism at low pH and to predominant IS surface complexation at edge sites at higher pH. Anion adsorption is generally weak, due to charge repulsion from the fixed negative charge, but some anions (phosphate, arsenate, arsenite, molybdate) can form strong IS surface complexes with edge sites. Spectroscopic evidence exists suggesting that there are two distinct groups of edge sites on Mt that predominate adsorption at different loadings of metal ions on the surface. In addition, competitive adsorption behavior suggests that multiple sets of strong sites exist on the edge sites. When considering a single adsorbent such as Mt, and a range of adsorbing cations, LFER can be derived between the aqueous hydrolysis constant of the cations and semiempirical, IS surface complexation constants for binding at edge sites. Competitive adsorption between anions that form IS complexes at edge sites has also been demonstrated, e.g., between phosphate and arsenate, suggesting that phosphate may enhance As(V) transport through clay barriers. Fe(III)-pillared Mt can be synthesized as an effective material to add to clay barriers where strong retention of anions is important.

Several groups have proposed surface complexation combined with ion exchange models (SC + IX) that have model parameters derived from fits to experimental adsorption data. The most comprehensive of model databases has been developed by Bradbury and Baeyens (2005, 2009a,b) in terms of application to numerous metal ions and actinides, Mt, and illite. However, the database of this model has not yet been extended to adsorbing anions. Moreover, the model is clearly semiempirical in nature, and its thermodynamic significance has been questioned because the conceptual model is based on uncharged surfaces (Tournassat et al., 2013). Fundamental understanding of the electrical charge and potential of Mt surfaces has evolved, and it is now believed that the constant potential of the basal planes affects the potential experienced by ions bonded at edge sites. Thus, the state-of-the-art of SC + IX models is still immature and pseudothermodynamic. There is considerable room for improvement in this area, following on from the developments in electrical double layer theory that have been described.

Numerous data have been collected regarding interactions of organic molecules and ions with natural or modified clays. The data include the results of adsorption experiments performed at various pH values, with various exchanged cations and at differing solution ionic strength, in the presence of competitive ions and DOM in solutions, from water–organic mixtures and at variable water activity. Adsorption data analysis within phenomenological models, modern spectroscopic techniques, and molecular modeling provides insights into the major mechanisms controlling interactions of organic molecules and ions and their distribution between the clay minerals and aqueous solutions. However, despite the progress in knowledge on interactions of organic compounds with clay minerals, multiple questions need to be better understood. There is a need in predictive tools that allow estimation of the thermodynamics of adsorption interactions of organic compounds with various clay minerals. In part, an LFER-based analysis has been suggested for vapor adsorption of organic molecules on clay minerals, for aqueous adsorption of organic compounds on organic cation-modified clay minerals and for organic cation adsorption on clay mineral surfaces. However, apparently there is no yet a general approach, either empirical or theoretical, which would be effective in predicting adsorption of various organic compounds on clay mineral surfaces. Current predictive or descriptive tools refer to the adsorption/distribution coefficients obtained at a significant dilution of solutes, i.e., they do not account for the effect of saturating adsorption sites. The latter becomes important in multiple scenarios when concentrations of organic adsorbates cannot be considered low. Similarly, there are no quantitative approaches which would allow relating the molecular structure of organic adsorbates with the adsorption–desorption hysteresis. More work is needed to develop the tools for estimating organic compound–clay mineral interactions in natural and engineered environments where adsorption interactions are complicated by the presence of alternative adsorbents (i.e., NOM or metal oxide), DOM, variable solution composition and are found in a multifaceted interplay with solute transport and bio-, photo-, and chemical transformations.

REFERENCES

Aggarwal, V., Li, H., Boyd, S.A., Teppen, B.J., 2006. Enhanced sorption of trichloroethene by smectite clay exchanged with Cs^+. Environ. Sci. Technol. 40, 894–899.

Ajouyed, O., Hurel, C., Marmier, N., 2011. Evaluation of the adsorption of hexavalent chromium on kaolinite and illite. J. Environ. Prot. 2, 1347–1352.

Appelo, C.A.J., Van Loon, L.R., Wersin, P., 2010. Multicomponent diffusion of a suite of tracers (HTO, Cl, Br, I, Na, Sr, Cs) in a single sample of opalinus clay. Geochim. Cosmochim. Acta 74, 1201–1219.

Auboiroux, M., Melou, F., Bergaya, F., Touray, J.C., 1998. Hard and soft acid-base model applied to bivalent cation selectivity on a 2:1 clay mineral. Clays Clay Miner. 46, 546–555.

Balcke, G.U., Kulikova, N.A., Hesse, S., Kopinke, F.D., Perminova, I.V., Frimmel, F.H., 2002. Adsorption of humic substances onto kaolin clay related to their structural features. Soil Sci. Soc. Am. J. 66, 1805–1812.

Baldock, J.A., Broos, K., 2012. Soil organic matter. In: Huang, P.M., Lin, Y., Sumner, M.E. (Eds.), Handbook of Soil Sciences. Properties and Processes, second ed. CRC Press, Taylor & Francis Group, Boca Raton, pp. 1–11.

Bektas, N., Agim, B.A., Kara, S., 2004. Kinetic and equilibrium studies in removing lead ions from aqueous solutions by natural sepiolite. J. Hazard. Mater. 112, 115–122.

Bellenger, J.-P., Staunton, S., 2008. Adsorption and desorption of ^{85}Sr and ^{137}Cs on reference minerals, with and without inorganic and organic surface coatings. J. Environ. Radioact. 99, 831–840.

Bergaya, F., Lagaly, G., Vayer, M., 2013. Cation and anion exchange; chapter 2.11. In: Bergaya, F., Lagaly, G. (Eds.), Developments in Clay Science, Handbook of Clay Science. Techniques and Applications, second ed., Volume 5B. Elsevier, Amsterdam, pp. 333–359.

Borisover, M.D., Graber, E.R., Bercovich, F., Gerstl, Z., 2001. Suitability of dye-clay complexes for removal of non-ionic organic compounds from aqueous solutions. Chemosphere 44, 1033–1040.

Borisover, M., Gerstl, Z., Burshtein, F., Yariv, S., Mingelgrin, U., 2008. Organic sorbate-organoclay interactions in aqueous and hydrophobic environments: sorbate-water competition. Environ. Sci. Technol. 42, 7201–7206.

Borisover, M., Bukhanovsky, N., Lapides, I., Yariv, S., 2012. The potential of thermally treated organobentonites to adsorb organic compounds from water. Appl. Clay Sci. 67–68, 151–157.

Borisover, M., 2013. The effect of organic sorbates on water associated with environmentally important sorbents: estimating and the LFER analysis. Adsorption 19, 241–250.

Boyd, S.A., Mortland, M.M., Chiou, C.T., 1988. Sorption characteristics of organic compounds on hexadecyltrimethylammonium-smectite. Soil Sci. Soc. Am. J. 52, 652–657.

Bradbury, M.H., Baeyens, B., 2005. Experimental measurements and modeling of sorption competition on montmorillonite. Geochim. Cosmochim. Acta 69, 4187–4197.

Bradbury, M.H., Baeyens, B., 2009a. Sorption modelling on illite. Part I: titration measurements and the sorption of Ni, Co, Eu and Sn. Geochim. Cosmochim. Acta 73, 990–1003.

Bradbury, M.H., Baeyens, B., 2009b. Sorption modelling on illite. Part II: actinide sorption and linear free energy relationships. Geochim. Cosmochim. Acta 73, 1004–1013.

Brigatti, M.F., Galan, E., Theng, B.K.G., 2013. Structure and mineralogy of clay minerals. Chapter 2. In: Bergaya, F., Lagaly, G. (Eds.), Developments in Clay Science, Handbook of Clay Science. Fundamentals, Second ed., Volume 5A. Elsevier, Amsterdam, pp. 21–81.

Cai, L., Tong, M., Wang, X., Kim, H., 2014. Influence of clay particles on the transport and retention of titanium dioxide nanoparticles in quartz sand. Environ. Sci. Technol. 48, 7323–7332.

Celis, R., Cornejo, J., Hermosin, M.C., Koskinen, W.C., 1998. Sorption of atrazine and simazine by model associations of soil colloids. Soil Sci. Soc. Am. J. 62, 165–171.

Chandler, D., 2005. Hydrophobic effect: interfaces and the driving force of hydrophobic assembly. Nature 437, 640–647.

Chappell, M.A., Laird, D.A., Thompson, M.L., Hui, L., Teppen, B.J., Aggarwal, V., Johnston, C.T., Boyd, S.A., 2005. Influence of smectite hydration and swelling on atrazine sorption behavior. Environ. Sci. Technol. 39, 3150–3156.

Charles, S.M., Li, H., Teppen, B.J., Boyd, S.A., 2006. Quantifying the availability of clay surfaces in soils for adsorption of nitrocyanobenzene and diuron. Environ. Sci. Technol. 40, 7751–7756.

Charlet, L., Scheinost, A.C., Tournassat, C., Greneche, J.M., Gehin, A., Fernandez-Martinez, A., Coudert, S., Tisserand, D., Brendle, J., 2007. Electron transfer at the mineral/water interface: selenium reduction by ferrous iron sorbed on clay. Geochim. Cosmochim. Acta 71, 5731–5749.

Charlet, L., Bardelli, F., Parsons, C., He, J., Chakraborty, S., Gailer, J., 2012. Arsenic binding onto phyllosilicates and glutathione: soil immobilisation and human excretion mechanisms. In: Ng, Noller, Naidu, Bundschuh, Bhattacharya (Eds.), Understanding the Geological and Medical Interface of Arsenic. Taylor & Francis Group, London, pp. 59–62.

Chen, C., Dynes, J.J., Wang, J., Karunakaran, C., Sparks, D.L., 2014a. Soft X-ray spectromicroscopy study of mineral-organic matter associations in pasture soil clay fractions. Environ. Sci. Technol. 48, 6678–6686.

Chen, Z., Montavon, G., Ribet, S., Guo, Z., Robinet, J., David, K., Tournassat, C., Grambow, B., Landesman, C., 2014b. Key factors to understand in-situ behavior of Cs in callovo-oxfordian clay-rock (France). Chemical Geol. 387, 47–58.

Chiou, C.T., Shoup, T.D., 1985. Soil sorption of organic vapors and effects of humidity on sorptive mechanisms and capacity. Environ. Sci. Technol. 19, 1196–1200.

Churakov, S.V., Dähn, R., 2012. Zinc adsorption on clays inferred from atomistic simulations and EXAFS spectroscopy. Environ. Sci. Technol. 46, 5713–5719.

Coppin, F., Berger, G., Bauer, A., Castet, S., Loubet, M., 2002. Sorption of lanthanides on smectite and kaolinite. Chemical Geol. 182, 57–68.

Cornejo, J., Celis, R., Pavlovic, I., Ulibarri, M.A., 2008. Interactions of pesticides with clays and layered double hydroxides: a review. Clay Miner. 43, 155–175.

Dähn, R., Scheidegger, A.M., Manceau, A., Schlegel, M., Baeyens, B., Bradbury, M.H., Chateigner, D.L., 2003. Structural evidence for the sorption of Ni(II) atoms on the edges of montmorillonite clay minerals: a polarized X-ray absorption fine structure study. Geochim. Cosmochim. Acta 67, 1–15.

Dähn, R., Baeyens, B., Bradbury, M.H., 2011. Investigation of the different binding edge sites for Zn on montmorillonite using P-EXAFS - the strong/weak site concept in the 2SPNE SC/CE sorption model. Geochim. Cosmochim. Acta 75, 5154–5168.

Droge, S.T.J., Goss, K.U., 2013a. Sorption of organic cations to phyllosilicate clay minerals: CEC-normalization, salt dependency, and the role of electrostatic and hydrophobic effects. Environ. Sci. Technol. 47, 14224–14232.

Droge, S.T.J., Goss, K.U., 2013b. Development and evaluation of a new sorption model for organic cations in soil: contributions from organic matter and clay minerals. Environ. Sci. Technol. 47, 14233–14241.

Dubbin, W.E., Vetterlein, J.P., Jonsson, J.L., 2014. Fatty acids promote fulvic acid intercalation by montmorillonite. Appl. Clay Sci. 97–98, 53–61.

Fan, Q.H., Tanaka, M., Tanaka, K., Sakaguchi, A., Takahashi, Y., 2014. An EXAFS study on the effects of natural organic matter and the expandability of clay minerals on cesium adsorption and mobility. Geochim. Cosmochim. Acta 135, 49–65.

Frost, R.R., Griffin, R.A., 1977. Effect of pH on adsorption of arsenic and selenium from landfill leachate by clay minerals. Soil Sci. Soc. Am. J. 41, 53–57.

Fukushi, K., Sakai, H., Itono, T., Tamura, A., Arai, S., 2014. Desorption of intrinsic cesium from smectite: Inhibitive effects of clay particle organization on cesium desorption. Environ. Sci. Technol. 48, 10743–10749.

Gaskova, O.L., Bukaty, M.B., 2008. Sorption of different cations onto clay minerals: modelling approach with ion exchange and surface complexation. Phys. Chem. Earth 33, 1050–1055.

Gehin, A., Greneche, J.-M., Tournassat, C., Brendle, J., Rancourt, D., Charlet, L., 2007. Reversible surface-sorption-induced electron-transfer oxidation of Fe(II) at reactive sites on a synthetic clay mineral. Geochim. Cosmochim. Acta 71, 863–876.

Gerstl, Z., Graber, E.R., 2011. Impact of irrigation with treated wastewater on pesticides and other organic microcontaminants in soils. In: Levy, G.J., Fine, P., Bar-Tal, A. (Eds.), Treated Wastewater in Agriculture: Use and Impacts on Soil Environment and Crops. Blackwell Publishing, Oxford, U.K, pp. 382–397.

Goldberg, S., 2002. Competitive adsorption of arsenate and arsenite on oxides and clays. Soil Sci. Soc. Am. J. 66, 413–421.

Goss, K.-U., 1993. Effects of temperature and relative humidity on the sorption of organic vapors on clay minerals. Environ. Sci. Technol. 27, 2127–2132.

Goss, K.U., Eisenreich, S.J., 1996. Adsorption of VOCs from the gas phase to different minerals and a mineral mixture. Environ. Sci. Technol. 30, 2135–2142.

Goss, K.U., Buschmann, J., Schwarzenbach, R.P., 2003. Determination of the surface sorption properties of talc, different salts, and clay minerals at various relative humidities using adsorption data of a diverse set of organic vapors. Environ. Toxicol. Chem. 22, 2667–2672.

Gu, X.Y., Evans, L.J., Barabash, S.J., 2010. Modeling the adsorption of Cd (II), Cu (II), Ni (II), Pb (II) and Zn (II) onto montmorillonite. Geochim. Cosmochim. Acta 74, 5718–5728.

Heller-Kallai, L., 2013. Thermally modified Clay minerals. Chapter 10.2. In: Bergaya, F., Lagaly, G. (Eds.), Developments in Clay Science, Handbook of Clay Science. Fundamentals, second ed., Volume 5A. Elsevier, Amsterdam, pp. 411–433.

Jan, Y.-L., Tsai, S.-C., Wei, Y.-Y., Tung, N.-C., Wei, C.-C., Hsu, C.-N., 2007. Coupled mechanics, hydraulics and sorption properties of mixtures to evaluate buffer/backfill materials. Phys. Chem. Earth 32, 789–794.

Jaynes, W.F., Boyd, S.A., 1991. Hydrophobicity of siloxane surfaces in smectites as revealed by aromatic hydrocarbon adsorption from water. Clays Clay Miner. 39, 428–436.

Kampf, N., Scheinost, A.C., Schulze, D.G., 2012. Oxide minerals in soils. Chapter 22. In: Huang, P.M., Lin, Y., Sumner, M.E. (Eds.), Handbook of Soil Sciences. Properties and Processes, second ed. CRC Press Taylor & Francis Group, Boca Raton, pp. 22–31.

Kukkadapu, R.K., Boyd, S.A., 1995. Tetramethylphosphonium smectite and tetramethylammonium-smectite as adsorbents of aromatic and chlorinated hydrocarbons Effect of water on adsorption efficiency. Clays Clay Miner. 43, 318–323.

Lagaly, G., Dekany, I., 2013. Colloid clay science, chapter 8. In: Bergaya, F., Lagaly, G. (Eds.), Developments in Clay Science, Handbook of Clay Science, Fundamentals, second ed., Volume 5A. Elsevier, Amsterdam, pp. 243–345.

Lagaly, G., Ogawa, M., Dekany, I., 2013. Clay mineral–organic interactions, chapter 10-3. In: Bergaya, F., Lagaly, G. (Eds.), Developments in Clay Science, Handbook of Clay Science, Fundamentals, second ed., Volume 5A. Elsevier, Amsterdam, pp. 435–505.

Laird, D.A., Yen, P.Y., Koskinen, W.C., Steinheimer, T.R., Dowdy, R.H., 1994. Sorption of atrazine on soil clay components. Environ. Sci. Technol. 28, 1054–1061.

Lee, J.-F., Mortland, M.M., Chiou, C.T., Kile, D.E., Boyd, S.A., 1990. Adsorption of benzene, toluene, and xylene by two tetramethylammonium-smectites having different charge densities. Clays Clay Miner. 38, 113–120.

Lee, S., Anderson, P.R., Bunker, G.B., Karanfil, C., 2004. EXAFS study of Zn sorption mechanisms on montmorillonite. Environ. Sci. Technol. 38, 5426–5432.

Lenoble, V., Bouras, O., Deluchat, V., Serpaud, B., Bollinger, J.-C., 2002. Arsenic adsorption onto pillared clays and iron oxides. J. Colloid Interface Sci. 255, 52–58.

Li, H., Sheng, G., Teppen, B.J., Johnston, C.T., Boyd, S.A., 2003. Sorption and desorption of pesticides by clay minerals and humic acid-clay complexes. Soil Sci. Soc. Am. J. 67, 122–131.

Lin, Z., Puls, R.W., 2000. Adsorption, desorption and oxidation of arsenic affected by clay minerals and aging process. Environ. Geol. 39, 753–759.

Liu, Z., Lee, C., 2006. Drying effects on sorption capacity of coastal sediment: the importance of architecture and polarity of organic matter. Geochim. Cosmochim. Acta 70, 3313–3324.

Liu, C., Li, H., Teppen, B.J., Johnston, C.T., Boyd, S.A., 2009. Mechanisms associated with the high adsorption of dibenzo-p-dioxin from water by smectite clays. Environ. Sci. Technol. 43, 2777–2783.

Lu, H.-J., Luan, M., Zhang, J., 2009. A kinetic study on the adsorption of chromium (VI) onto a natural material used as landfill liner. Electron. J. Geotechnical Eng. 14, 1–10.

Mader, B.T., Goss, K.U., Eisenreich, S.J., 1997. Sorption of nonionic, hydrophobic organic chemicals to mineral surfaces. Environ. Sci. Technol. 31, 1079–1086.

Maes, N., Salah, S., Jacques, D., Aertsens, M., Van Gompel, M., De Cannière, P., Velitchkova, N., 2008. Retention of Cs in boom clay: comparison of data from batch sorption tests and diffusion experiments on intact clay cores. Phys. Chem. Earth 33, S149–S155.

Manceau, A., Tommaseo, C., Rihs, S., Geoffroy, N., Chateigner, D., Schlegel, M., Tisserand, D., Tamura, N., Chen, Z.-S., 2005. Natural speciation of Mn, Ni, and Zn at the micrometer scale in a clayey paddy soil using X-ray fluorescence, absorption, and diffraction. Geochim. Cosmochim. Acta 69, 4007–4034.

Manning, B.A., Goldberg, S., 1996. Modeling arsenate competitive adsorption on kaolinite, montmorillonite and illite. Clays Clay Miner. 44, 609–623.

Manning, B.A., Goldberg, S., 1997. Adsorption and stability of arsenic(III) at the clay mineral-water interface. Environ. Sci. Technol. 31, 2005–2011.

McBride, M., 1980. Interpretation of the variability of selectivity coefficients for exchange between ions of unequal charge on smectite. Clays Clay Miner. 28, 255–261.

McKinley, J.P., Zachara, J.M., Heald, S.M., Dohnakolva, A., Newville, M.G., Sutton, S.R., 2004. Microscale distribution of cesium sorbed to biotite and muscovite. Environ. Sci. Technol. 38, 1017–1023.

Missana, T., Garcia-Gutierrez, M., Alonso, U., 2008. Sorption of strontium onto illite/smectite mixed clays. Phys. Chem. Earth 33, S156–S162.

Missana, T., Alonso, U., Garcia-Gutierrez, M., 2009. Experimental study and modelling of selenite sorption onto illite and smectite clays. J. Colloid Interface Sci. 334, 132–138.

Montavon, G., Alhajji, E., Grambow, B., 2006. Study of the interaction of Ni^{2+} and Cs^{+} on MX-80 bentonite; effect of compaction using the "capillary method". Environ. Sci. Technol. 40, 4672–4679.

Muller, K., Magesan, G.N., Bolan, N.S., 2007. A critical review of the influence of effluent irrigation on the fate of pesticides in soil. Agric. Ecosyst. Environ. 120, 93–116.

Nir, S., El-Nahhal, Y., Undabeytia, T., Rytwo, G., Polubesova, T., Mishael, Y., Rabinovitz, O., Rubin, B., 2013. Clays, Clay Minerals, and pesticides. Chapter 5.2. In: Bergaya, F., Lagaly, G. (Eds.), Developments in Clay Science, Handbook of Clay Science. Techniques and Applications, second ed., Volume 5B. Elsevier, Amsterdam, pp. 645–662.

Pool, C.F., 2003. The Essence of Chromatography, pp. 15–18. Elsevier, Amsterdam.

Qafoku, N.P., 2012. Impacts of environmental nanoparticles on chemical, biological and hydrological processes in terrestrial ecosystems. In: Huang, P.M., Lin, Y., Sumner, M.E. (Eds.), Handbook of Soil Sciences. Resources Management and Environmental Impacts, second ed. CRC Press, Taylor & Francis Group, Boca Raton, pp. 1–4.

Rabung, T., Pierret, M.C., Bauer, A., Geckeis, H., Bradbury, M., Baeyens, B., 2005. Sorption of Eu(III) on Ca-montmorillonite and Na-illite. Part 1: batch sorption and time-resolved laser fluorescence spectroscopy experiments. Geochim. Cosmochim. Acta 69, 5393–5402.

Rotenberg, B., Morel, J.-P., Marry, V., Turq, P., Morel-Desrosiers, N., 2009. On the driving force of cation exchange in clays: Insights from combined microcalorimetry experiments and molecular simulation. Geochim. Cosmochim. Acta 73, 4034–4044.

Scheidegger, A.M., Lamble, G.M., Sparks, D.L., 1997. Spectroscopic evidence for the formation of mixed-cation hydroxide phases upon metal sorption on clays and aluminum oxides. J. Colloid and Interface Sci. 186, 118–128.

Schlegel, M.L., Manceau, A., Chateigner, D.L., Charlet, L., 1999. Sorption of metal ions on clay minerals: I. Polarized EXAFS evidence for the adsorption of Co on the edges of hectorite particles. J. Colloid and Interface Sci. 215, 140–158.

Schlegel, M., Manceau, A., Charlet, L., Chateigner, D., Hazemann, J., 2001a. Sorption of metal ions on clay minerals. III. Nucleation and epitaxial growth of Zn phyllosilicate on the edges of hectorite. Geochim. Cosmochim. Acta 65, 4155–4470.

Schlegel, M.L., Manceau, A., Charlet, L., Hazemann, J.L., 2001b. Adsorption mechanisms of Zn on hectorite as a function of time, pH, and ionic strength. Am. J. Sci. 301, 798–830.

Schlegel, M.L., Manceau, A., 2006. Evidence for the nucleation and epitaxial growth of Zn phyllosilicate on montmorillonite. Geochim. Cosmochim. Acta 70, 901–917.

Schlegel, M.L., Descostes, M., 2009. Uranium uptake by hectorite and montmorillonite: a solution chemistry and polarized EXAFS study. Environ. Sci. Technol. 43, 8593–8598.

Schoonheydt, R.A., Johnston, C.T., 2013. Surface and interface chemistry of clay minerals. Chapter 5. In: Bergaya, F., Lagaly, G. (Eds.), Developments in Clay Science, Handbook of Clay Science. Fundamentals, second ed., Volume 5A. Elsevier, Amsterdam, pp. 139–172.

Schwarzenbach, R.P., Gschwend, P.M., Imboden, D.M., 2003. Environmental Organic Chemistry. John Wiley and Sons, Inc, Hoboken.

Sheng, G., Wang, X., Wu, S., Boyd, S.A., 1998. Enhanced sorption of organic contaminants by smectitic soils modified with a cationic surfactant. J. Environ. Qual. 27, 806–814.

Sheng, G., Johnston, C.T., Teppen, B.J., Boyd, S.A., 2001. Potential contributions of smectite clays and organic matter to pesticide retention in soils. J. Agric. Food Chem. 49, 2899–2907.

Sherman, D.M., Randall, S.R., 2003. Surface complexation of arsenic(V) to iron(III) (hydr)oxides: structural mechanism from ab initio molecular geometries and EXAFS spectroscopy. Geochim. Cosmochim. Acta 67, 4223–4230.

Sinitsyn, V.A., Aja, S.U., Kulik, D.A., Wood, S.A., 2000. Acid-base surface chemistry and sorption of some lanthanides on K^+-saturated marblehead illite: I. Results of an experimental investigation. Geochim. Cosmochim. Acta 64, 185–194.

Soltermann, D., Fernandes, M.M., Baeyens, B., Miehé-Brendlé, J., Dähn, R., 2014. Competitive Fe(II)–Zn(II) uptake on a synthetic montmorillonite. Environ. Sci. Technol. 48, 190–198.

Taylor, R.W., Shen, S., Bleam, W.F., Tu, S.-I., 2000. Chromate removal by dithionite-reduced clays: evidence from direct x-ray adsorption near edge spectroscopy (XANES) of chromate reduction at clay surfaces. Clays Clay Miner. 48, 648–654.

Teppen, B.J., Miller, D.M., 2006. Hydration energy determines isovalent cation exchange selectivity by clay minerals. Soil Sci. Soc. Am. J. 70, 31–40.

Thompson, H.A., Parks, G.A., Brown, G.E., 1999. Dynamic interactions of dissolution, surface adsorption, and precipitation in an aging cobalt(II)-clay-water system. Geochim. Cosmochim. Acta 63, 1767–1779.

Tournassat, C., Grangeon, S., Leroy, P., Giffaut, E., 2013. Modeling specific pH dependent sorption of divalent metals on montmorillonite surfaces. A review of pitfalls, recent achievements and current challenges. Am. J. Sci. 313, 395–451.

Turner, G.D., Zachara, J.M., McKinley, J.P., Smith, S.C., 1996. Surface-charge properties and UO_2^{2+} adsorption of a subsurface smectite. Geochim. Cosmochim. Acta 60, 3399–3414.

van Loon, L.R., Glaus, M.A., 2008. Mechanical compaction of smectite clays increases ion exchange selectivity for cesium. Environ. Sci. Technol. 42, 1600–1604.

van Loon, L.R., Baeyens, Bradbury, M.H., 2009. The sorption behaviour of cesium on opalinus clay: a comparison between intact and crushed material. Appl. Geochem. 24, 999–1004.

Vespa, M., Lanson, M., Manceau, A., 2010. Natural attenuation of zinc pollution in smelter-affected soil. Environ. Sci. Technol. 44, 7814–7820.

Vico, L.I., 2003. Acid-base behaviour and Cu^{2+} and Zn^{2+} complexation properties of the sepiolite/water interface. Chemical Geol. 198, 213–222.

Vinsova, H., Konirova, R., Koudelkova, M., Jedinakova-Krizova, V., 2004. Sorption of technetium and rhenium on natural sorbents under aerobic conditions. J. Radioanal. Nucl. Chem. 261, 407–413.

Xiao, F., Zhang, X., Penn, L., Gulliver, J.S., Simcik, M.F., 2011. Effects of monovalent cations on the competitive adsorption of perfluoroalkyl acids by kaolinite: experimental studies and modeling. Environ. Sci. Technol. 45, 10028–10035.

Xiao, F., Pignatello, J.J., 2014. Effect of adsorption nonlinearity on the pH-adsorption profile of ionizable organic compounds. Langmuir 30, 1994–2001.

Xu, S., Harsh, J.B., 1990a. Monovalent cation selectivity quantitatively modeled according to hard/soft acid/base theory. Soil Sci. Soc. Am. J. 54, 357–363.

Xu, S., Harsh, J.B., 1990b. Hard and soft acid-base model verified for monovalent cation selectivity. Soil Sci. Soc. Am. J. 54, 1596–1601.

Xu, S., Harsh, J., 1992. Alkali cation selectivity and surface charge of 2:1 clay minerals. Clays Clay Miner. 40, 567–574.

Xu, S., Boyd, S.A., 1995. Cationic surfactant adsorption by swelling and nonswelling layer silicates. Langmuir 11, 2508–2514.

Yariv, S., 1992. Wettability of clay minerals. In: Schrader, M.E., Loeb, G. (Eds.), Modern Approaches to Wettability. Theory and Applications. Plenum Press, New York, pp. 279–326.

Yan, W., Hu, S., Jing, C., 2012. Enrofloxacin sorption on smectite clays: effects of pH, cations, and humic acid. J. Colloid Interface Sci. 372, 141–147.

Yuan, G., Theng, B.K.G., 2012. Clay-Organic interactions in soil environments. In: Huang, P.M., Lin, Y., Sumner, M.E. (Eds.), Handbook of Soil Sciences. Resources Management and Environmental Impacts, second ed. CRC Press, Taylor & Francis Group, Boca Raton, p. 2.

Yuan, G.D., Theng, B.K.G., Churchman, G.J., Gates, W.P., 2013. Clays and Clay minerals for pollution control. Chapter 5.1. In: Bergaya, F., Lagaly, G. (Eds.), Developments in Clay Science, Handbook of Clay Science. Techniques and Applications, second ed., Volume 5B. Elsevier, Amsterdam, pp. 587–644.

Zachara, J.M., Cowan, C.E., Schmidt, R.L., Ainsworth, C.C., 1988. Chromate adsorption by kaolinite. Clays Clay Miner. 36, 317–326.

Zhang, Y., Zhu, D., Yu, H., 2008. Sorption of aromatic compounds to clay mineral and model humic substance-clay complex: effects of solute structure and exchangeable cation. J. Environ. Qual. 37, 817–823.

Zhu, R., Chen, W., Shapley, T.V., Molinari, M., Ge, F., Parker, S.C., 2011a. Sorptive characteristics of organomontmorillonite toward organic compounds: a combined LFERs and molecular dynamics simulation study. Environ. Sci. Technol. 45, 6504–6510.

Zhu, J., Wang, T., Zhu, R., Ge, F., Yuan, P., He, H., 2011b. Expansion characteristics of organo montmorillonites during the intercalation, aging, drying and rehydration processes: effect of surfactant/CEC ratio. Colloids and Surf. A: Physicochemical and Eng. Aspects 384, 401–407.

Chapter 3

Chemical Conditions in Clay-Rocks

Christophe Tournassat,[a,e] Agnès Vinsot,[b] Eric C. Gaucher[c] and Scott Altmann[d]
[a] *Water, Environment and Ecotechnology Division, French Geological Survey (BRGM), Orléans, France;* [b] *Andra, LSMHM, Bure, France;* [c] *TOTAL, E&P, Pau, France;* [d] *Andra, Châtenay-Malabry, France;* [e] *Earth Sciences Division, Lawrence Berkeley National Laboratory, Berkeley, CA, USA*

Chapter Outline

Developments in Clay Science, Vol. 6C. http://dx.doi.org/10.1016/B978-0-08-100027-4.00003-6

3.1 INTRODUCTION

Until the end of the 1990s, deep geological clay formations had no direct industrial applications and little research had been conducted to characterize the chemistry of pore-water in indurated clay-rocks. The emergence of the concept of radioactive waste disposal in clay-rock formations led to a research boom in this area. The effectiveness of this concept relies on formation capacity to limit pollutant mass transport (low permeability, low diffusivity characteristics) and to reduce pollutant concentration by adsorption/precipitation phenomena. Other factors of paramount importance are the stability of clay-rock physical–chemical properties over geological time periods and favorable mechanical properties, such as fracture self-sealing in the zones damaged by excavation procedures (Landais, 2006; Delay et al., 2007; Neuzil, 2013; Bianchi et al., in press). The properties of clay-rock result from the individual properties of each of the mineral and organic constituents and the way they interact. The solution contained in the pore space, i.e. the pore water, is an important part of this system. The chemical composition of pore water (whether pristine or perturbed) affects a number of chemical, hydrological, and mechanical clay-rock properties (Rosenqvist, 1984; Moore, 1991; Altmann, 2008) as well as the stability of the engineered materials that they interact with (Savage et al., 2002; Gaucher and Blanc, 2006; Gaboreau et al., 2011; Marty et al., 2014). Determining the chemical conditions prevailing in clay-rock pore-water is thus an important step in predicting how clay-rock barriers will behave over time (Altmann, 2008). Chemical characterization operations for pore-water in clay-rocks differ significantly from those used in conventional ground water characterization. Pore waters in clay-rocks are notoriously difficult to sample directly, and only small volumes can usually be obtained that retain the principal characteristics representative of in situ conditions. However, improvements in characterization techniques have made it feasible to obtain representative pore-water samples, and development of thermodynamic databases dedicated to clay-rock environments now permit more accurate predictions of its composition by geochemical modeling. Following this first stage of development for the nuclear industry, scientific and industrial interest in clay formations increased because of the key role acknowledged for clay-rock layers in sealing deep CO_2 sequestration in saline aquifers, or in depleted oil and gas fields, where clay-rock formations play the role of cap-rock.

In CO_2 sequestration applications, the clay cap-rock, if present, must prevent fluids from escaping the sequestration reservoir. Low permeability, high acid buffering, and/or CO_2 adsorption capacities are all necessary properties for a clay cap-rock (Gaus et al., 2005; Busch et al., 2008). In addition, the integrity of the well-borehole systems in contact with the clay cap-rock, as well as the effectiveness of sealing at the wellbore/clay-rock interface, must be assessed for the sequestration time period (Gherardi et al., 2012). High-clay mineral surface areas favor trapping CO_2 in gaseous or supercritical forms (Cole et al., 2010). In addition, clay minerals (chlorite, illite, smectite) reacting with CO_2 can contribute to the mineralization of a part of the CO_2, leading to long-term trapping (Luquot et al., 2012).

In the field of oil and gas production in shale formations, the need to better understand the chemistry of natural and flow-back waters is emerging. To develop technologies to mitigate the release of radionuclides and heavy metals in the flow-back waters, it is necessary to understand how they originate in the waters produced in these contexts (Kargbo et al., 2010). Finally, the petroleum industry needs to predict overpressure during drilling of very deep boreholes. Clay mineral dehydration may play a very important role in the building of overpressure. Coupling clay-rock's chemical reactivity with mechanics might be a necessary step toward understanding and quantifying those interactions (Lahann and Swarbrick, 2011).

The first section of this chapter summarizes the current understanding of the composition of clay-rock formations that are under consideration as host-rocks for radionuclide repository facilities and/or as cap-rocks for gas sequestration. The second section gives an overview of the experimental methods that have been used so far to characterize clay-rock pore-water chemistries, highlighting the most and least commonly encountered errors and difficulties. The last section introduces aqueous equilibrium modeling techniques that investigate how mineral dissolution/precipitation and other processes control pore-water compositions. A large part of this chapter is based on the knowledge acquired by studying clay-rock formations similar to those expected to host nuclear waste repository facilities. Nonetheless, the experimental and modeling methods developed are useful for other applications as well.

3.2 CLAY-ROCK MINERALOGY, WATER CONTENT AND POROSITY

A range of mineralogical compositions and bulk physical properties are found in clay-rocks that are being investigated as possible gas sequestration cap-rocks or geological barriers for radioactive waste disposal. For instance, the Boom clay in Mol (Belgium) is a shallow marine deposit with a total porosity of about 40%, and water content of 20% (dry mass basis). It is a relatively soft clay-rock composed mainly of clay minerals (30–60%) and quartz (15–60%). Carbonates and pyrite are only minor mineral components (1–5% each). Boom clay can contain up to 5% of organic carbon (De Craen et al., 2004b; Deng et al., 2012). By contrast, clay-rock in the Boda formation (Mecsek Mountains, Hungary, Lázár and Máthé (2012)), which was deposited in shallow lacustrine environment, has very low porosity (1.5%) (Mell et al., 2006). It is a highly consolidated clay-rock whose mineral composition is mainly clay minerals (40–50%) and high albite content (20–40%). Carbonate minerals and quartz are also present (up to 10% and 5–15% respectively).

Not only do different clay formations have different properties and mineralogy, but the properties and mineralogy vary as a function of the layers in a single formation. For example, the clay-rocks in the Callovian-Oxfordian (COx) formation (Bure, France) have contrasting mineralogical compositions as a function of depth (Gaucher et al., 2004) (Figure 3.1), resulting also in changes in the bulk physical properties. For instance, the changes in the abundance ratio of carbonates and clay minerals could be correlated to

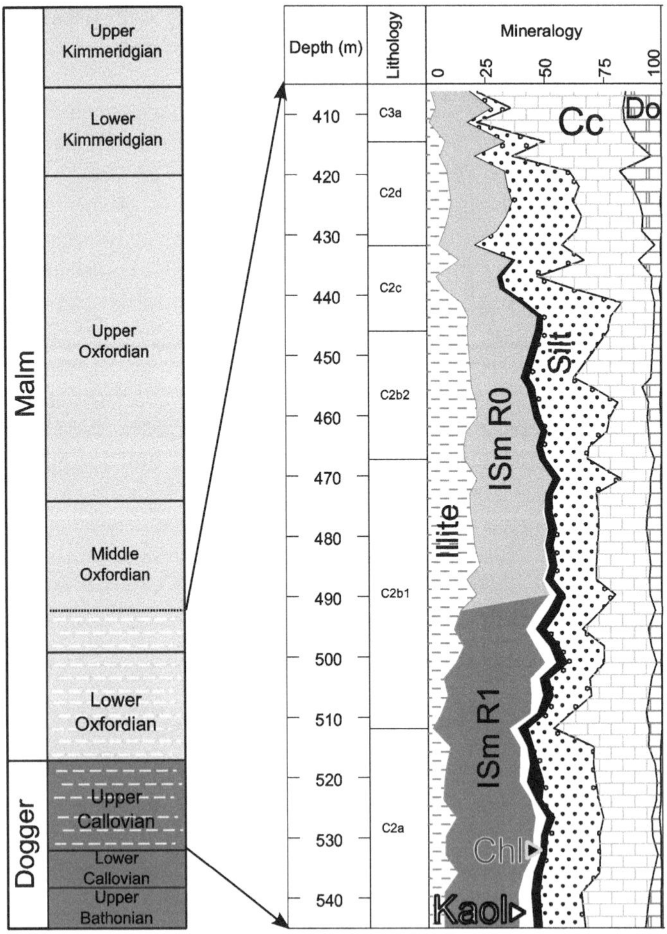

FIGURE 3.1 Lithostratigraphic column at the Bure site. Schematic distribution of the principal mineralogy: Illite, I-Sm R0 and I-Sm R1 (illite-smectite mixed layers), Chl (chlorite), Kaol (kaolinite), Silt (quartz + feldspar), Cc (calcite), Do (dolomite). *Figure adapted from Lerouge et al. (2011) reproduced with the authorization of the publisher (Elsevier).*

changes in the rock's porosity and diffusion properties (Descostes et al., 2008; Jacquier et al., 2013). The clay-rocks in the Boda formation contain hematite (7–10%), indicative of an oxidative environment. But some of the strata contain pyrite instead, evidencing alternating redox conditions within the same formation (Lázár et al., 2010).

The differences highlighted among the clay-rocks under investigation preclude the use of a generic pore-water composition applicable to all clay-rocks without consideration of site specificities. Nevertheless, a set of experimental and modeling tools have been developed that can be applied to the characterization of all sites.

3.3 INVESTIGATION METHODS FOR PORE-WATER CHEMICAL COMPOSITION CHARACTERIZATION

3.3.1 In Situ Techniques: Piezometers and Instrumented Boreholes

Collection of water seeping into deep boreholes might appear to be the most straightforward technique for obtaining representative solution samples for characterizing the chemistry of clay-rock pore-water. However, the very low-hydraulic conductivities of clay-rock formations (typically less than 10^{-12} m/s), means that years would be necessary to fill even a part of the volume of a borehole drilled from the ground level, even in the presence of hydraulic head values of several MPa. Moreover, whatever the chosen drilling technique, the drilling phase affects the composition of water accumulated in such a borehole, thus preventing collection of water having a composition representative of the in situ pristine water composition, even regarding major species.

Water sampling conditions are much more favorable in underground galleries such as those of underground research laboratories (URL) (Delay et al., 2014). Relatively short-length, small-volume boreholes may be drilled into the surrounding rock using less disturbing techniques than for deep boreholes from ground level, i.e., without mud and at low-drilling fluid pressure (several bars). These boreholes are artesian since their length is designed to reach outside the hydraulically most disturbed zone surrounding the galleries. The minimum length depends on the gallery diameter and the rock hydraulic conductivity; for indurated clay-rocks with hydraulic conductivities in the range 10^{-14}–10^{-12} m/s, it is of the order of 10 m for a 4 m-diameter gallery (Bossart et al., 2004). Under these conditions typical water production flow rates are between 3 and 50 mL day^{-1} m^{-2}, depending on the hydraulic head surrounding the borehole sampling chamber (Vinsot et al., 2008a,b). Continuous water flows within this flow rate range have been obtained in nine boreholes in the COx at Bure URL and in the Opalinus clay (OPA) at Mont Terri URL over more than five years (Vinsot et al., 2011).

Although water collection from open boreholes is often possible in URL galleries, interaction between the collected water and atmospheric air makes it impossible to determine the pH and dissolved gas concentrations of the pristine pore-water. In order to determine these parameters, two main types of URL experiments have been conducted, based on specially designed borehole equipment (Figure 3.2). The first type of experiment, conducted in a descending borehole, consists of recirculating a volume of water having a composition similar to the pore-water composition (based on geochemical modeling calculations). In this kind of experiment it is expected that recirculating solution will adjust its composition to match the equilibrium with the surrounding rock water (Vinsot et al., 2008a,b; Wersin et al., 2011a). The second type of experiment consists of recirculating a nonreactive gas, usually nitrogen (N_2) or noble gases, in an ascending borehole and measuring the composition of the collected, well-preserved seepage water (Vinsot et al., 2008a,b). In addition, it is possible to monitor changes in gas composition over time due to the degassing of volatile compounds dissolved in the pore-water (CO_2, alkanes, N_2, etc.).

In situ pore-water collection or equilibration is not completely free of experimental artifacts, which must be taken into account when attempting to estimate the composition of pristine, undisturbed pore-water. In water circulation experiments (Figure 3.2), the very low-diffusion coefficient of solute species in clay-rock porosity makes stabilization of the borehole water composition slow. Reactive transport modeling can however be used to interpret collected water composition variation over time (Appelo et al., 2008). In drainage water sampling experiments, the volumes collected are small and are representative of the solution present in a small volume of rock surrounding the borehole. Consequently, any perturbation of the chemistry of this clay-rock volume has an impact on the composition of the sampled water and can lead to measurements which are not representative of pristine water. For instance, dry drilling techniques evaporate water in the surrounding clay-rock resulting in an increased salinity of the first seepage water collected (Vinsot et al., 2013) and modified water isotope fractionation. Dry drilling with air is also a source of clay-rock and solute oxidation, causing recovery of overly high-sulfate concentrations due to the oxidation of pyrite (Pearson et al., 2003; De Craen et al., 2004b). This oxidation problem led to the development of dry drilling using N_2 or argon (Ar) gas in place of air. Experience also has shown however that, regardless of the type of experiment, the development of microorganisms in the system could not be avoided, resulting in potentially significant modifications of water composition such as dissolved organic matter (DOM) composition, sulfate and sulfide concentrations (therefore redox state), and inorganic carbon isotopic composition. Whether these microorganisms (mainly bacteria including archaea) find their origin in contamination during the drilling and equipment phases of the borehole, or also in a reviviscence of autochthonous populations from the rock,

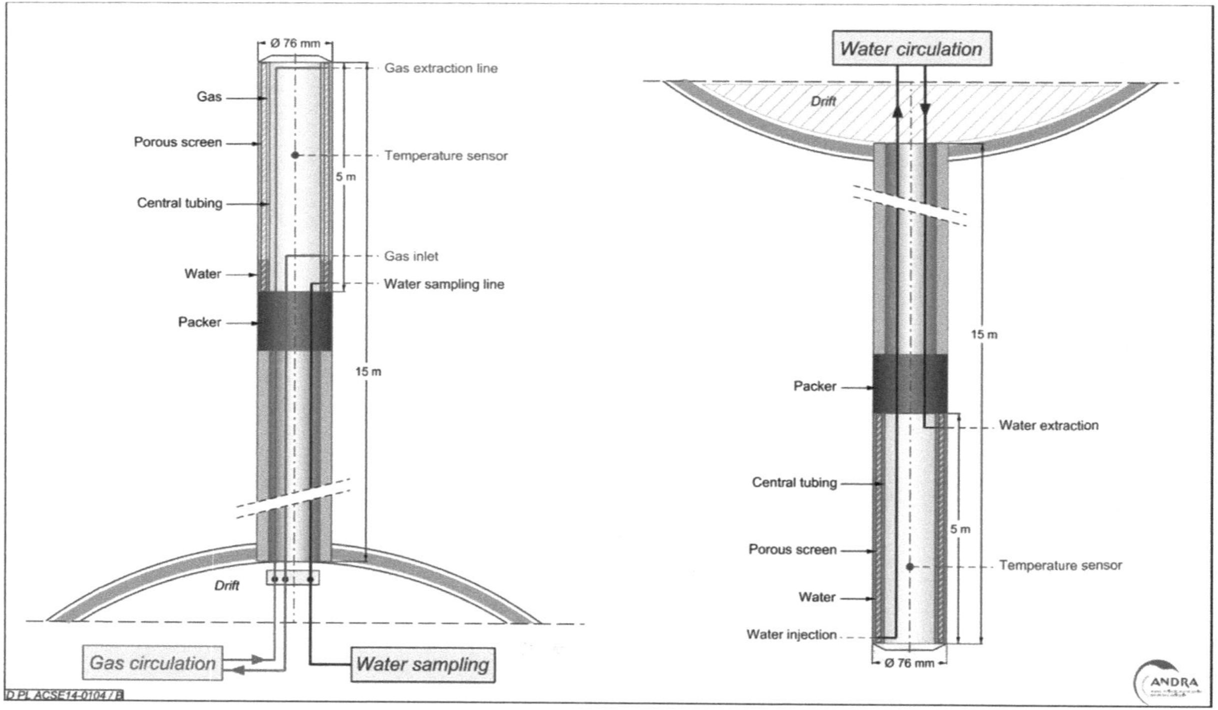

FIGURE 3.2 Schematic representation of the borehole instrumentation for a gas circulation and water sampling experiment (left) and a water circulation experiment (right).

is still under debate (Stroes-Gascoyne et al., 2007; Poulain et al., 2008; Stroes-Gascoyne et al., 2011; Urios et al., 2012).

Despite the described difficulties, measurement of pore-water composition via dedicated experiments carried out in URL boreholes provides a great deal of useful information. For example, concentration variations in a single borehole give insight into the most probable concentrations in the pore-water itself. In additionally, dedicated experiments employing carefully controlled perturbations (O_2, H_2, bacterial activity, alkaline perturbations, etc.) give very valuable information on the buffering capacity of the system as a whole (Wersin et al., 2011a,b; Vinsot et al., 2012, 2014).

3.3.2 Pore-Water Extraction from Core Samples by Squeezing

Ex situ water extraction methods (e.g., squeezing, ultracentrifugation) applied to core samples obtained from surface or URL boreholes can provide an alternative when in situ water collection is not possible for feasibility or timing reasons. Here, only squeezing will be described since it is the only method applicable to indurated clay-rocks (e.g., OPA or COx) given their low porosities (~10–20%) (Thury and Bossart, 1999; Sammartino et al., 2003; Gaucher et al., 2004; Yven et al., 2007). This technique consists of applying a very high compaction pressure to a clay-rock sample. Since the compressions of the solid phase and water phase are negligible, the change of volume is accommodated by reduced pore volume and thus expulsion of pore-water from the sample (Bufflap and Allen, 1995).

The composition of water collected by squeezing can also be affected by various artifacts (De Lange et al., 1992; Sacchi et al., 2001), and for a large part they are similar to those encountered with water samples obtained from in situ experiments. Problems due to drilling fluid interacting with the pore-water can be circumvented by drilling with air or with inert gases like N_2 or Ar. Alternatively, it is possible to remove the external part of the core that has interacted with the drilling fluid (Gaucher et al., 2004; Sanford et al., 2009). It is necessary to use chemically inert drilling fluids as much as possible, with a preference of oil-based muds over water-based muds. Note that oil-based muds often contain brines (e.g., $CaCl_2$ at 350 g L^{-1} reported in Gaucher et al., 2004) which means that the external layer of the core must be removed rapidly in order to limit diffusion of drilling mud compounds into the core. Large core diameter helps for subsampling a well-preserved central core for analysis.

One of the most problematic artifacts is oxidation of the core sample. Contrary to the in situ experiments, this problem is not limited to the drilling and the packer installation phases. It can occur during (1) sample conditioning after recovery from the core barrel and its transport to the laboratory; (2) sample conditioning in the water extraction device; (3) the water extraction procedure itself; and (4) the waiting period before of the water is analyzed. Inadequate sample handling procedures during these four phases can lead to large changes in the extracted water composition, among the most noticeable being an increased sulfate concentration due to pyrite oxidation (De Craen et al., 2004a). Samples must be stored under an inert atmosphere (N_2 or a noble gas) using different conditioning methods depending on the duration of the storage period. Common examples are inside a double envelope of heat-sealed plasticized aluminum (Gaucher et al., 2004), in sealed containers or bombs (Gaucher et al., 2006; Lassin et al., 2013), or in a glove box (De Lange et al., 1992). Inert atmosphere conditions also must be employed during the pore-water extraction procedure. Decompression and temperature changes are two commonly reported sources of artifacts that are unavoidable during core extraction, transport, and handling in the laboratory (De Lange et al., 1992; Sacchi et al., 2001). Changes in pressure and temperature potentially affect chemical equilibria and can lead to the dissolution or precipitation of mineral phases during core reequilibration under ambient conditions. This is especially true for the carbonate system. Depending on the temperature, calcite solubility increases by a factor about two to five between 1 and 500 bars but only by a few percent between 1 and 50 bars (Millero, 1982). Consequently, the pressure effect on the chemical equilibria is not so problematic for the characterization of clay-rocks from shallow formations since down to at least -500 m (corresponding to 50 bars of hydrostatic pressure) the pressure effect on solubility values is often negligible. The change of pressure due to squeezing might alter the chemical composition of the squeezed water by ultrafiltration, a phenomenon resulting from the semipermeable properties of the clay materials. Fernandez et al. (2014) have demonstrated, however, that choosing an optimal range of squeezing pressures can minimize this kind of artifact.

3.3.3 Alternative Techniques

The liquid displacement technique can be considered as an alternative technique to squeezing. It has long been used on densely packed sand and coarse-grained deposits using an immiscible fluid for flushing (Scholl, 1963). It has only recently been applied to indurated clay-rocks for at least two reasons: (1) a high fluid pressure is necessary to obtain reasonable flux in these very low permeability materials and (2) the sample preparation procedure is challenging because it must insure that there is no by-pass around the sample of the injected fluid. In the reported studies (Mäder and Gimmi, 2007; Grambow et al., 2014), the flushing fluid was not an immiscible fluid but a solution of known composition, differing from that of the pore-water. Therefore, a significant number of the water samples at the column outlet had compositions intermediate between the flushing solution and the pore-water due to diffusion processes taking place during the experiment. Coupled geochemical and transport modeling was therefore necessary to fully interpret the meaning of the sampled water compositions (Mäder and Gimmi, 2007; Grambow et al., 2014). Simple integration of the elution concentration peaks as a function of time gives information as to the total mass of a compound in a sample but this method is restricted to the analysis of solute compounds that do not interact significantly with the solid matrix (e.g., chloride and bromide anions). For instance, the concentration of various small organic molecules in the pore-water of the COx formation could be quantified using this method (Huclier-Markai et al., 2010). The small volumes of water collected (typically several microliters per day) also make it difficult to perform a comprehensive set of analyses.

New characterization techniques are always under development. For instance, Savoye et al. (2012) demonstrated that it is possible to reproduce the concept of a URL recirculation borehole experiment at the scale of a core sample in order to obtain valuable insights regarding pore-water isotopic composition and salinity. More recently Celejewski et al. (2014) proposed a method for the extraction of water from clay-rock cores using the absorption properties of cellulosic sheets. This continuous development of new approaches and techniques will certainly facilitate the investigation of the pore-water composition of clay-rocks from various origins in the near future.

3.3.4 Examples of Pore-Water Compositions

A selection of water compositions obtained from a range of different clay-rocks and clay sediments is provided in Table 3.1. While the ionic strength of clay-rock pore-water can vary widely (in the range 0.008−0.91, i.e., from fresh waters to brines), pH values most often lie in a narrow range from 6.5 to 8.5. From the pH value and the dissolved carbonate concentration (or the alkalinity), it is possible to calculate a corresponding $H_2CO_3^*$ activity (see Table 3.1). It is interesting to note that the corresponding calculated CO_2 partial pressure (p_{CO_2}) that would be measured in a gas phase at equilibrium with the pore-water in these selected examples is always higher than atmospheric p_{CO_2} ($\log p_{CO_2} \sim -3.4$). This could be due to the pH/carbonate system being controlled either within the clay-rock system or by an enriched CO_2 source outside the system. The narrow range of measured p_{CO_2} values for clay-rock or sediment layers from very different origins and locations is in line with the first of these two hypotheses.

3.3.5 Indirect Characterization of Pore-Water Chemistry

Indirect characterization of pore-water chemistry constitutes a rapid means of providing (1) prescreening as to the nature of the pore-water and (2) input parameters for geochemical modeling.

3.3.5.1 Water Content and Leaching Experiments

The water content of clay-rocks (ω in $kg_{water}\,kg^{-1}_{dry\ solid}$) is determined by gravimetric analysis. The mass of water is commonly determined as the difference between the initial mass of the sample and its mass after oven drying at

TABLE 3.1 Examples of Compositions of Water Extracted from Clay-Rocks and Clay Sediments

Location	Santa Barbara Basin, US	Chesapeake Bay, US	Mont-Terri, Switzerland	Mont-Terri, Switzerland	Bure, France	Mol, Belgium
References	Sholkovitz (1973)	Sanford et al. (2009)	Vinsot et al. (2008a)	Fernández et al. (2014)	Vinsot et al. (2008b)	Fernández (2008)
Type of rock/ sediment	Marine clay sediment (12–15 cm, core B)	Smectite layer (CBIC-28) in the ICDP-USGS deep drill hole	Indurated clay-rock—Opalinus clay (PC-C experiment)	Indurated clay-rock—Opalinus clay (HT experiment)	Indurated clay-rock—Callovian-Oxfordian clay-rock (PAC experiment, Sample EST02091F)	Boom clay (R13U m. 7.7–7.90)
Total porosity		~45%	~15%	~15%	~15%	~40%
Extraction procedure	Centrifugation (20°C)	Centrifugation (room temperature)	In situ borehole	Squeezing (75 MPa) of bore core	In situ borehole	Squeezing
Measured Values						
In situ temperature (°C)	6.4	–	13	13	25	16
pH	8.0–8.1 (at 20°C)	6.6 (at room temperature)	7.2 (at 20°C)	7.7 (at 20°C)	7.4 (at 20°C)	8.4 (at 20°C)
mmol/L						
Na	495	539	281	235	45.7	7.1
K	10.9	4.7	1.4	1.4	0.6	0.11
Ca	9.5	54.4	19.3	16.8	5.0	0.11

Continued

TABLE 3.1 Examples of Compositions of Water Extracted from Clay-Rocks and Clay Sediments—cont'd

Location	Santa Barbara Basin, US	Chesapeake Bay, US	Mont-Terri, Switzerland	Mont-Terri, Switzerland	Bure, France	Mol, Belgium
Mg	52.5	42.7	22.3	22.2	4.3	0.03
Sr	–	0.7	0.4	0.5	0.2	–
Cl	558	697	325	276	35.0	0.56
S(6)	26.3	38.6	17.0	17.4	12.5	0.44
C(4)	–	1.87	3.6	–	4.2	5.79
Alkalinity[a]	8.1	–	3.3	1.8	3.5	5.79
Calculated Values						
Calcite saturation index	–	−0.4[c]	0.2	0.4	0	0
Corrected pH[b]	–	–	7.1	7.5	7.3	8.4
Calculated p_{CO_2} (atm in log)	−2.6	−1.9	−2.1	−2.7	−2.1	−3.0
Ionic strength (–)	0.71	0.91	0.42	0.37	0.09	0.008

[a]*In mmol equivalent HCO_3^- per L.*
[b]*Assuming equilibrium with calcite at in situ temperature. Calculations were performed with PHREEQC and the Pitzer.dat database (Parkhurst and Appelo, 2013).*
[c]*Assuming a temperature of 20°C.*

110°C for 24 h. However, clay minerals can retain a nonnegligible part of their hydration water at 110°C and a higher temperature of 150°C (if possible under vacuum) or above should be preferred (Robinet et al., 2012; Bauer and Vennemann, 2014). The water content is determined on samples that have potentially been partially dehydrated and it is thus a good practice to correct this artifact by determining the saturation degree with additional measurements: namely grain density (or specific gravity), bulk dry density, and bulk wet density (Fernández et al., 2014).

Leaching experiments consist of mixing a mass *m* of dry, powdered clay-rock (g) with a volume *V* of deionized water (L) and of measuring solubilized compound concentrations after a given equilibration time. The solid–liquid separation is usually performed by centrifugation followed by filtration. Given the small size of the clay mineral particles, 0.1 μm filters (or less) should be preferred over the most commonly used 0.45 μm filters, to avoid the presence of clay mineral colloids in the filtrate. When considering a high solid to liquid ratio of 1 g/mL, the dilution ratio for the pore-water in the deionized water is 10 for a clay-rock having a water content of ~10%. This simple calculation is proof that the water composition obtained from leaching experiments cannot be representative of pore-water composition. Solutes in leached water samples originate from the dilution of chemical species present in the rock pore-water but also from the dissolution of mineral species. Mineral precipitation can also occur due, for example, to CO_2 partial pressure changes. For those reasons, it is not possible to recalculate pore-water composition by using a simple dilution correction. Finally, chemical species that are not involved in dissolution/precipitation processes can be involved in adsorption processes such as cation exchange. Following pore-water dilution and the changed ionic strength, reequilibrating the surfaces with the diluted pore-water leads to changes in the ratio of those species. For instance, as the ionic strength decreases, cation-exchange surfaces preferentially adsorb divalent species (Ca^{2+}, Mg^{2+}, Sr^{2+}) as compared to monovalent species (Na^+, K^+). Consequently, even Na^+ concentration in the pore-water cannot be obtained directly from leaching experiment results and the relative ratios of leached cation concentrations are not representative of their ratio in the pore-water either.

On the other hand, leaching experiments can provide direct quantitative information for species that do not adsorb or dissolve/precipitate i.e., anionic species such as Cl^- and Br^-. For these species, the leached concentration, $C_{leached,i}$ is, logically, linearly proportional to the ratio *m*/*V* (Bradbury and Baeyens, 1998; Pearson et al., 2003; Gaucher et al., 2006) and their total concentration in the clay-rock sample, $C_{rock,i}$, assuming that they are not present in insoluble mineral phases, can be calculated as

$$C_{rock,i} = \frac{C_{leached,i}}{m} V \tag{3.1}$$

Providing that the water content, at saturation, is known (ω in $kg_{water}\ kg_{dry\ solid}^{-1}$), it is possible to calculate the mean concentration of species i in the porosity ($\overline{C_{poro,i}}$ in mol kg_{water}^{-1}):

$$\overline{C_{poro,i}} = \frac{C_{rock,i}}{\omega} \quad (3.2)$$

This is also the case for sulfate but two processes complicate the sulfate data interpretation. The first process is the dissolution of sulfate bearing minerals. For instance, celestite, $SrSO_4$, is a mineral frequently encountered in indurated clay-rocks under investigation in France and Switzerland (Gautschi, 2001; Pearson et al., 2003; Gaucher et al., 2004). Its dissolution during leaching tests leads also to a linear increase in sulfate concentration as a function of *m*/*V*. The second process is the partial oxidation of pyrite. Sample preparation and leaching experiments are performed usually in an anoxic glove box to avoid oxidation artifacts. However, a simple mass balance calculation shows that the oxidation of only 0.5% of the pyrite inventory can increase the leached sulfate concentration by a factor of two. If oxidation occurs before the leaching test itself, the complete dissolution of pyrite oxidation by-products (e.g., gypsum) also leads to a linear increase in sulfate concentration as a function of *m*/*V*. Indeed, only samples preserved under the best conditions (e.g., conditioning and transport under liquid N_2, sample preparation under an atmosphere with less than 1 ppm O_2) led to satisfactory results for leached sulfate concentrations (Gaucher et al., 2009). At a pore scale of several nanometers, the pore-water does not have spatially uniform composition, due to cation enrichment and anion depletion in the diffuse layer. This takes place near negatively charged clay mineral surfaces (see Chapter 1 in this Volume). Only part of the porosity, that in the bigger pores, contains predominantly electroneutral pore solution, i.e., "free" or "bulk" pore-water. Consequently, $\overline{C_{poro,i}}$ is not a direct measurement of the Cl^- or Br^- anion concentration in the bulk pore-water ($C_{bulk,i}$). The ratio $\overline{C_{poro,i}}/C_{bulk,i}$ is the so-called "anion accessible porosity." Direct computation of the anion accessible porosity using $\overline{C_{poro,i}}$ values obtained from leaching experiments and $C_{bulk,i}$ values obtained from squeezing or in situ experiments repeatedly yield values close to 0.5 for clay-rock samples from the COx and OPA formations (Pearson et al., 2003; Gaucher et al., 2006; Fernández, 2008; Savoye et al., 2012; Fernández et al., 2014). Although the mechanisms responsible for the anion exclusion effect are well known (see Chapter 1, in this Volume), no one has yet been able to predict its extent based on independent measurable parameters. Comparing the properties of the two clay-rocks cited above gives a clear picture of this problem. COx and OPA clay-rocks mineral compositions are similar in nature but they differ in mineral proportions: OPA samples are richer in illite and poorer in mixed layer illite-smectite (I-Sm) phases than COx samples. Consequently, the specific surface area (SSA) of COx samples is usually higher than the SSA of OPA samples. Illite surface charge is also higher than smectite (Sm) surface charge. In addition, the pore waters of the two formations are different, the ionic strength in the OPA at Mont Terri URL being two to six times higher than that of the COx pore-water in the Bure URL. Surface charge and surface area as well as ionic strength are the most important parameters in determining the extent of the diffuse layer and its effect on pore-water composition (see Chapter 1, in this Volume). In compacted powdered montmorillonite (Mt) samples, changes in those three parameters produce large variations in $\overline{C_{poro,i}/C_{bulk,i}}$ values (Muurinen et al., 2004, 2007; Van Loon et al., 2007; Birgersson and Karnland, 2009; Tournassat and Appelo, 2011). Consequently, one should expect, a priori, different $\overline{C_{poro,i}/C_{bulk,i}}$ values for COx and OPA clay-rocks as well as different values within the same clay-rock as a function of location due to changes in mineralogy and pore-water ionic strength. This finding of an almost constant $\overline{C_{poro,i}/C_{bulk,i}}$ value for indurated clay-rock samples of different origins is still puzzling and remains a field of active research (Mazurek et al., 2011).

3.3.5.2 Cation Exchange

Cation adsorption compensates for the negative charge of the clay mineral layers. Ion exchange reactions describe the distribution of ions between adsorbed and solution phases (see Chapters 1 and 2). Since the kinetics of cation exchange reactions are usually very fast (Tang and Sparks, 1993), the cation exchanger composition of a clay-rock can be considered as an image of the cation composition of the contacting pore-water (Bradbury and Baeyens, 1998; Tournassat et al., 2007, 2008; Gaucher et al., 2009; Pearson et al., 2011). Measurement of cation exchange capacity (CEC) and cation exchanger population involves the use of an index cation, which is able to replace the cations actually adsorbed on the clay mineral. Various techniques have been developed differing in the index cation involved, on the procedure used (single step or multiple step extraction, or percolation) and on the analytical procedure. The process consists of mixing a given mass of solid sample in a solution containing the index cation so that it replaces the cations in the clay mineral, which are in turn dissolved in the contacting solution. Once the index cation has desorbed surface cations from the sample, the solution is centrifuged, filtered, and analyzed. Since the initial index cation concentration is known, the amount adsorbed can be determined and, thereby a total CEC value. By assaying the extracted cations in solution, the whole CEC can be attributed to a sum of the various exchangeable cations and the exchangeable population identified.

Various artifacts may occur during cation exchange assays. For instance, oxidation of the sample usually leads to an increase of exchangeable Ca^{2+} because the oxidation of pyrite causes a pH decrease that is buffered by carbonate mineral dissolution (mainly calcite dissolution). As for leaching experiments, this type of artifact can be avoided by handling the sample carefully. Some other artifacts are unavoidable: the presence of very soluble compounds (e.g., celestite and gypsum minerals, or simply the salt precipitated from the pore-water during sample drying) leads to overestimating their cationic constituents on the exchanger. Calcite/dolomite dissolution also leads to uncertainties in estimating Ca^{2+} and Mg^{2+} concentrations on the exchanger. Most of the time, these latter effects have limited influence compared to the variability of the measured CEC and on cation exchange populations. Otherwise, they can be quantified (1) by repeating the measurements at different solid to liquid ratio values, reaction times, and/or index cation types; (2) by analyzing the anionic composition of the final contact solution; and/or (3) by modeling the data (Dohrmann, 2006a; Gaucher et al., 2009; Koroleva et al., 2011). Finally, measurement methods have been improved and adapted to the nature of the sample in order to limit the effect of these artifacts (Dohrmann, 2006b; Dohrmann and Kaufhold, 2009; Koroleva et al., 2011).

3.3.5.3 Leaching Test and Cation Exchange Population versus Pore-Water Composition

Leaching tests and cation exchange population characterizations do not provide any direct quantitative information on pore-water composition. Nonetheless, those measurements constitute a rapid means of providing prescreening as to the nature of the pore-water: for example, pore waters with high salinity and a major Na^+ contribution to the cation balance result in cation exchange populations dominated by Na^+ and to high $\overline{C_{poro,\mathrm{Cl}}}$ values. By contrast, pore waters with low ionic strength lead to low $\overline{C_{poro,\mathrm{Cl}}}$ values and cation exchange populations dominated by Ca^{2+} and Mg^{2+} due to the exchanger preference for divalent cations over monovalent cations as the ionic strength decreases (Sposito, 1984). Leaching tests and cation exchange population information can be used in pore-water modeling approaches to obtain quantitative information on pore-water composition.

3.4 MODELING PORE-WATER COMPOSITION

Clay-rock pore-water composition models can have several distinct objectives. The first objective is to predict the composition of pore waters that cannot be extracted from the clay-rock for cost, time, or technical reasons. In that case, pore-water modeling is the only method available for estimating the pore-water composition. The second objective is to screen the solubility controls most likely to be responsible for dissolved chemical species concentrations. They are usually screened by comparing the modeled concentrations and the concentrations that are measured in seepage water or squeezed water. None of the direct characterization techniques for pore-water composition is entirely free of measurement artifacts. Modeling pore-water composition is thus a necessary step in the validation process for reference pore-water compositions. In addition, modeling approaches are necessary to validate the extrapolation of pore-water composition measured from a fully characterized site with access to water samples (URL for instance) to a less well-characterized site (industrial application site for instance). A third objective is then to predict the most probable evolution of the pore-water composition and the mineralogical assemblage as a function of perturbations (see Chapter 5, in this Volume).

3.4.1 Modeling Hypotheses, Strategies, and Ancillary Data

3.4.1.1 Applying the Gibbs's Phase Rule to Clay-Rock Systems

Diffusion processes drive exchanges between the pore-water in clay-rocks and the surrounding environment. In addition, mineral grains in clay-rocks are usually small, so they have a high SSA that enhances the rates of dissolution/precipitation kinetics. Consequently, the interaction time must be long enough to establish equilibrium conditions between the pore-water and a pool of minerals that includes also minerals with low dissolution or precipitation kinetic rates.

Once this hypothesis of local equilibrium with a pool of minerals is accepted, pore-water models rely on the Gibbs' phase rule:

$$F = C - P + 2 \tag{3.3}$$

where F is the variance or the number of degrees of freedom of the system, C is the number of independent components, and P is the number of phases in equilibrium with each other; the extra 2 is related to the pressure and temperature parameters. The objective of pore-water composition models is to reduce the number of degrees of freedom to zero, based on constraints obtained from clay-rock sample analyses. At fixed temperature and pressure, this objective is met for $P = C$, i.e., a number of phases equal to the number of components is necessary to obtain the pore-water composition. In practice, the concentrations of some constituents (the conservative tracers, e.g., Cl^- or Br^-) are not controlled by the solubility of any phases in the system, resulting rather from dilution of the sediment pore-water by meteoric water following deposition, burial, and erosion together with contributions from diffusive fluxes from or toward the surrounding stratigraphic layers (Mazurek et al., 2011). The concentrations of those components (free components) must be fixed, otherwise a degree of freedom is added to the system for each unconstrained concentration. The average concentrations of conservative tracers in the porosity can be obtained from interpreting leaching experiments and porosity/water-content measurements, however, as explained previously, the anion accessible porosity cannot yet be predicted based on clay-rock sample properties alone. Consequently, the determination of the concentrations of conservative anionic tracers is subject to the uncertainty associated with the empirical observation $\overline{C_{poro,\mathrm{Cl}}/C_{bulk,\mathrm{Cl}}} \sim 0.5$.

In Eqn (3.3), note that C represents a number of *independent* components. Elements dissolved in solution, are present in the form of distinct solute species (for instance, Ca is present in the form of Ca^{2+} and solute complexes such as

$CaCl^+$ and $CaSO_4$), but only elements can be considered as independent components since the concentrations of the different species can be obtained from (1) the reaction equilibrium constants and (2) considering a solute-species activity model. Because some elements have solute species that can be found simultaneously in various redox states (Fe, S), the redox potential of the pore-water must also be constrained, thus adding one degree of freedom to the system. As a rule, it is necessary to find n independent phases to predict the pore-water concentrations of n elements (not considering the free components) in the absence of redox-sensitive species, or $n + 1$ independent phases in the presence of one or more redox-sensitive species. The aqueous phase itself must be considered in the number of independent phases (in practice it adds a numerical constraint to the electroneutrality). Considering mineral dissolution/precipitation reactions provides the necessary additional constraints.

Cation exchange reactions are often included in pore-water modeling (Bradbury and Baeyens, 1998; Motellier et al., 2003; Pearson et al., 2003; Gaucher et al., 2006; Beaucaire et al., 2008; Gaucher et al., 2009; Pearson et al., 2011; Tremosa et al., 2012). A cation exchanger cannot really be considered as a phase in the sense of the Gibbs' phase rule. However, considering it, reduces the number of independent components if the relative occupancy of the cations on the exchanger and the exchange selectivity coefficients are known. In that case, all components that participate in cation exchange reactions can be related to the others and the number of independent components reduces to one. As an example, in a system with Na^+, Ca^{2+}, Mg^{2+}, and K^+ competing for cation exchange, Ca^{2+}, Mg^{2+}, and K^+ activities are related to the Na^+ activity according to (Bradbury and Baeyens, 1998; Tournassat et al., 2007):

$$\{Ca^{2+}\} = \frac{E_{Ca}}{E_{Na}^2} \frac{\{Na^+\}^2}{{}^{GT}K_{ex}^{Na \to Ca}} \tag{3.4}$$

$$\{Mg^{2+}\} = \frac{E_{Mg}}{E_{Na}^2} \frac{\{Na^+\}^2}{{}^{GT}K_{ex}^{Na \to Mg}} \tag{3.5}$$

$$\{K^+\} = \frac{E_{K}}{E_{Na}} \frac{\{Na^+\}}{{}^{GT}K_{ex}^{Na \to K}} \tag{3.6}$$

where ${}^{GT}K_{ex}$ is an exchange selectivity coefficient under the Gaines and Thomas convention (see Chapter 1, in this Volume), brackets denote activities, and E_i is the fractional occupancy of species i on the exchanger. In this case, Ca, Mg, and K are not qualified as independent components in the system since all of them can be expressed as a function of the component Na. Any of these elements can be considered as the unique independent component by rearranging Eqns (3.4)–(3.6) in order to redefine the other components as a function of the chosen independent one. In addition, it should be recalled that those relationships do not depend on the total amount of exchangeable cations but only on the relative occupancies of cations on the exchanger.

3.4.1.2 Selection of Minerals for Equilibrium Reactions

Once the number C of independent components has been determined, $P = C$ phases at equilibrium must be found to constrain the system entirely. The porosity of investigated clay-rock from deep geological formations is usually saturated with water (Matray et al., 2007) leaving only equilibria with mineral phases as possible constraints (equilibria with gas phases may be considered as well in unsaturated media). The selection of minerals considered in a pore-water model should obey the following guidelines (Gaucher et al., 2006): (1) the minerals must be present in the observed mineralogical assemblage, (2) the minerals must show evidence of stability/equilibrium, and (3) the minerals must have reliable thermodynamic property (enthalpies of formation, heat capacities, entropies, and Gibbs free energy) data available. Whether it is possible to consider that clay minerals are at equilibrium with a solution is still under debate (Blanc et al., 2013). This question blocked consideration of clay minerals in the earliest pore-water models (Beaucaire et al., 2000). The thermodynamic properties of clay minerals such as chlorite (Chl), illite, Sm and I-Sm mixed layers are generally not precisely known for specific clay-rocks because of their variable compositions. Thermodynamic properties of reference Chl, Sm, and illite minerals have been obtained recently from calorimetric measurements (Gailhanou et al., 2007, 2009, 2012). The solubilities deduced from those properties have been released in thermodynamic databases devoted to low temperature water–rock interactions (Blanc et al., 2012; Giffaut et al., 2014). Based on that data, Gaucher et al. (2009) and Pearson et al. (2011) conducted a sensitivity analysis on their pore model outcomes by considering various potential clay assemblages and/or various structural compositions (and solubility values) for a given clay mineral. In so doing, they demonstrated that the uncertainty on clay mineral composition and solubility has, in fact, little influence on the modeled pore-water composition (Figure 3.3). In addition, predicted pore-water compositions were very close to the composition of seepage water samples collected in the URL, further validating the use of clay mineral solubilities in predictive pore-water composition models.

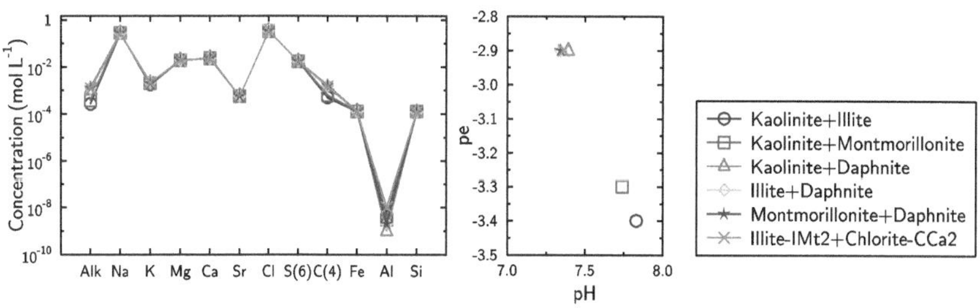

FIGURE 3.3 Comparison of Mont Terri pore-water compositions calculated with various clay mineral assemblages. *Data from Table 5 of Pearson et al. (2011).*

On the other hand, some phases may be at disequilibrium with the pore-water composition. For instance, it is usually considered that organic matter is not at equilibrium with the carbonate system. In addition, some detrital minerals, although having well-documented solubilities should not be considered in equilibrium models for low-temperature clay-rocks (< 30°C). This is especially the case for feldspars. Albite (a sodium-plagioclase feldspar) and microcline (a potassium-plagioclase feldspar) are commonly encountered as detrital minerals in clay-rock samples (Beaucaire et al., 2000; Pearson et al., 2003; Gaucher et al., 2004). Beaucaire et al. (2000, 2008) proposed those minerals as possible buffers of Na^+ and K^+ concentrations in clay-rocks. If those minerals were at equilibrium with the pore-water, it would then follow that the activity ratio $(Na^+)/(K^+)$ must be constant and be equal to the ratio of albite and microcline solubilities, $K_{albite}/K_{microcline}$. Pearson et al. (2011) and Gaucher et al. (2009) demonstrated that (1) the composition of the collected seepage or squeezing waters from three different clay-rocks (OPA, COx, and Tournemire clay-rocks) do not comply with this requirement and (2) the information on Na^+/K^+ cation exchange populations is also incompatible with the albite/microcline equilibrium. This result agrees with the results of Stefánsson and Arnórsson (2000) who concluded that feldspar equilibrium was not common at temperatures below about 50°C for a wide range of analyzed natural waters.

At last, DOM can influence the pore-water composition by complexing inorganic ions in solution, and thus, by increasing the solubility of the corresponding elements. In COx and OPA pore-water, the concentration of dissolved organic carbon (DOC) is low (in the range $1-20\ mg\ L^{-1}$) (Courdouan et al., 2007a,b) and organic compounds are mainly acidic compounds, fatty acids, aldehydes, amino acids, and lower aging humic compounds (Huclier-Markai et al., 2010), whose influence on the total concentration of major elements can be neglected. However, for trace elements, and/or for other clay-rock with higher DOC concentrations in pore-water, the effect of DOM on solubility values may be important (Liu et al., 2008).

Apart from identifying phases that are likely at or out of equilibrium with the pore-water, modelers must also verify that the chosen phases are indeed independent in the modeled system. As an example, let us consider the modeling of the following element concentrations and parameters: pH (H^+), Na, Ca, Mg, K, C(4) (for carbonate, bicarbonate, etc.), and Cl, by making the hypothesis that those are the major compounds in the pore-water, and that the cation exchange population ratios between Na, K, Ca, Mg are known. Cl concentration is assumed also to be known from an independent measurement. According to the previous section, the number of independent components is reduced to four (pH, Ca, C(4), and Cl; note that Na, Mg, or K could be considered instead of Ca). Only three independent components remain. Thus, if species are not considered as redox sensitive, three phases are needed to constrain the system. One of them is obviously water. Calcite and dolomite are frequently encountered minerals in clay-rocks and could, a priori, be used to reach the condition $P = C$. However, calcite and dolomite are related via the exchange reactions (Tournassat et al., 2008):

$$\log K_{ex}^{Ca \rightarrow Mg} = 2 \log K_{calcite} - \log K_{dolomite} \tag{3.7}$$

Consequently, calcite and dolomite are not independent phases in the selected system and the pH and the bicarbonate concentrations cannot be calculated without adding other independent phases.

3.4.2 Modeling the Pore-Water Composition in Pristine Rock

An example of a pore-water composition prediction is shown in Figure 3.4 according to a model that follows the method described in the previous sections without consideration of the DOM. The agreement between modeled and measured values is good and most important parameters can be predicted accurately, for instance pH and ionic strength. There is no reliable Al data to compare with the model prediction. The modeled Fe concentration is significantly too high. In the proposed model, the Fe concentration is controlled by siderite, whose solubility is known. However, pure siderite is seldom observed in the COx formation. The solubility of the commonly observed Fe-enriched carbonates is in turn unknown, leaving the Fe concentration prediction not as well constrained as the prediction for other elements. The consideration of

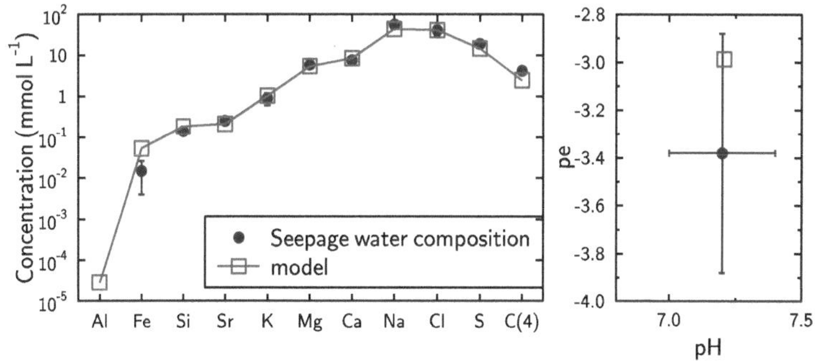

FIGURE 3.4 Top: example of a fully constrained pore-water model for the system pH, pe, Cl, S, Na, K, Ca, Mg, Sr, Si, Al, Fe in the Callovian-Oxfordian clay-rock (Bure, France). Bottom: corresponding predicted pore-water composition as compared to seepage water composition. *Adapted from Gaucher et al. (2009).*

siderite in the list of independent phases leads to an oversaturation of the modeled pore-water composition with regards to iron (hydr)oxides such as goethite or magnetite (Gaucher et al., 2009; Pearson et al., 2011). Those phases have never been observed in pristine COx samples but indirect evidence of their presence have been recently gathered using magnetic measurements (Aubourg and Pozzi, 2010; Kars et al., in press). Considering goethite or magnetite instead of siderite in the mineralogical assemblage helps to reconcile the model with the measured Fe concentrations.

Models based on the same methodology and using similar mineral phases have been applied successfully to OPA and Tournemire clay-rocks (Pearson et al., 2011; Tremosa et al., 2012), but yet, the predictions are not perfect and several shortcomings have been identified related to e.g., (1) uncertainties inherent to the use of generic exchange selectivity coefficients that are obtained experimentally on model materials; (2) uncertainty in the tabulated solubility values in thermodynamic databases; (3) imprecise knowledge of the exact chemical composition of clay mineral phases and the related uncertainty regarding their thermodynamic properties; and (4) a need to improve prediction (and in situ verification) of redox conditions. Nonetheless, available pore-water models now offer a readily available estimate of the most important pore-water parameters in clay-rock systems.

3.4.3 Additional Model Controls

The comparison between measured and predicted concentration values is not always possible and a good agreement can sometimes be fortuitous. Additional controls on the relevance of model hypotheses are thus necessary. Gaucher et al. (2010) used carbon and oxygen isotopic signatures to demonstrate that pore-water is at equilibrium with diagenetic carbonates (rather than with primary calcite) in four different clay-rock formations (Tournemire, OPA, COx, and Boom Clay) and Lerouge et al. (2010, 2011) used petrographic and strontium and oxygen isotopic data on COx clay-rocks, to show that feldspar minerals do not participate significantly in pore-water composition regulation, and that celestite does. In addition, they demonstrated that quartz, and not chalcedony, is the likely mineral responsible for controlling Si concentration.

3.4.4 Significance of Pore-Water Composition

The water mobilized by drainage or squeezing is likely to be predominantly that in larger pores and perhaps outside the first several nm surface layer, therefore the pore-water composition measured on seepage or squeezed water or that is calculated from pore-water modeling is not representative of the whole pore-water volume. Whether a pore-water volume having this electroneutral composition actually exists depends on the pore-size distribution and the ionic strength (see Chapter 1, in this Volume). Investigating pore-size distribution in OPA via imaging techniques reveals the presence of pores having equivalent radii larger than 50 nm (Keller et al., 2011, 2013). Such pore radii are much larger than the predicted size of the diffuse layer, so part of the pore volume may be filled with an electroneutral solution. Independently of whether an electroneutral solution is present in part of the porosity, the predicted pore-water composition is only representative of the solution composition at equilibrium with the mineralogical assemblage and the surfaces, far from any electrostatic influence from the surfaces. The total inventory of the solute species cannot be deduced directly from the proposed pore-water models, but additional calculation methods have been developed to predict the average composition in the diffuse layer (Leroy et al., 2007; Appelo et al., 2010).

3.5 CONCLUSION: ACHIEVEMENTS AND FUTURE CHALLENGES

This chapter focused mainly on how to characterize the major components and parameters of clay-rock pore-water. Those include pH, the major inorganic cations (mostly Na^+, Ca^{2+}, Mg^{2+}, K^+) and the major inorganic anions (Cl^-, SO_4^{2-}, bicarbonate, and carbonate anions), which are, with pe, the most important in predicting a range of properties for clay-rocks (e.g., adsorption, diffusion, and swelling). As soon as it is necessary to model the evolution of the pore-water composition and the mineralogical assemblage as a function of an external perturbation (see Chapter 5, in this Volume), it is necessary to consider the chemical reactivity of clay minerals. Consequently, clay mineral constituents (SiO_2, Al, Fe) must also be included in the characterization effort. As for Al and Fe concentrations, few measurements are available and the variability or uncertainty on those measurements is large, in part due to low Al solubility at near-neutral pH and to Fe concentration's sensitivity on redox. Indeed, while the mineral assemblage (e.g., presence of pyrite, iron hydr(oxides), or siderite) provides qualitative information on the redox environment, the quantification of the redox potential in clay-rock pore-water remains a difficult challenge (Pearson et al., 2011). Additional challenges are related to investigations (1) on natural trace element solubility controls (Grangeon et al., 2015) and (2) on natural soluble organic matter that could help us to understand and quantify the competitive or synergetic effects for adsorption on the clay-rock matrix.

REFERENCES

Altmann, S., 2008. Geo'chemical research: a key building block for nuclear waste disposal safety cases. J. Contam. Hydrol. 102, 174–179.

Appelo, C.A.J., Vinsot, A., Mettler, S., Wechner, S., 2008. Obtaining the porewater composition of a clay rock by modeling the in- and out-diffusion of anions and cations from an in-situ experiment. J. Contam. Hydrol. 101, 67–76.

Appelo, C.A.J., Van Loon, L.R., Wersin, P., 2010. Multicomponent diffusion of a suite of tracers (HTO, Cl, Br, I, Na, Sr, Cs) in a single sample of Opalinus clay. Geochim. Cosmochim. Acta 74, 1201–1219.

Aubourg, C., Pozzi, J.-P., 2010. Toward a new <250 °C pyrrhotite–magnetite geothermometer for claystones. Earth Planet. Sci. Lett. 294, 47–57.

Bauer, K.K., Vennemann, T.W., 2014. Analytical methods for the measurement of hydrogen isotope composition and water content in clay minerals by TC/EA. Chem. Geol. 363, 229–240.

Beaucaire, C., Pitsch, H., Toulhoat, P., Motellier, S., Louvat, D., 2000. Regional fluid characterisation and modelling of water-rock equilibria in the Boom clay formation and in the Rupelian aquifer at Mol, Belgium. Appl. Geochem. 15, 667–686.

Beaucaire, C., Michelot, J.L., Savoye, S., Cabrera, J., 2008. Groundwater characterisation and modelling of water-rock interaction in an argillaceous formation (Tournemire, France). Appl. Geochem. 23, 2182–2197.

Bianchi, M., Liu, H.-H., Birkholzer, J.T. Radionuclide transport behavior in a generic geological radioactive waste repository. Groundwater. http://dx.doi.org/10.1111/gwat.12171, in press.

Birgersson, M., Karnland, O., 2009. Ion equilibrium between montmorillonite interlayer space and an external solution—consequences for diffusional transport. Geochim. Cosmochim. Acta 73, 1908–1923.

Blanc, P., Lassin, A., Piantone, P., Azaroual, M., Jacquemet, N., Fabbri, A., Gaucher, E.C., 2012. Thermoddem: a geochemical database focused on low temperature water/rock interactions and waste materials. Appl. Geochem. 27, 2107–2116.

Blanc, P., Vieillard, P., Gailhanou, H., Gaboreau, S., 2013. Chapter 6—Thermodynamics of clay minerals. In: Bergaya, F., Lagaly, G. (Eds.), Handbook of Clay Science. Part A. Fundamentals, Developments in Clay Science. Elsevier, pp. 173–210.

Bossart, P., Trick, T., Meier, P.M., Mayor, J.-C., 2004. Structural and hydrogeological characterisation of the excavation-disturbed zone in the Opalinus clay (Mont Terri Project, Switzerland). Appl. Clay Sci. 26, 429–448.

Bradbury, M.H., Baeyens, B., 1998. A physicochemical characterisation and geochemical modelling approach for determining porewater chemistries in argillaceous rocks. Geochim. Cosmochim. Acta 62, 783–795.

Bufflap, S.E., Allen, H.E., 1995. Sediment pore water collection methods for trace metal analysis: a review. Water Res. 29, 165–177.

Busch, A., Alles, S., Gensterblum, Y., Prinz, D., Dewhurst, D.N., Raven, M.D., Stanjek, H., Krooss, B.M., 2008. Carbon dioxide storage potential of shales. Int. J. Greenhouse Gas Control 2, 297–308.

Celejewski, M., Scott, L., Al, T., 2014. An absorption method for extraction and characterization of porewater from low-permeability rocks using cellulosic sheets. Appl. Geochem. 49, 22–30.

Cole, D.R., Chialvo, A.A., Rother, G., Vlcek, L., Cummings, P.T., 2010. Supercritical fluid behavior at nanoscale interfaces: implications for CO_2 sequestration in geologic formations. Philos. Mag. 90, 2339–2363.

Courdouan, A., Christl, I., Meylan, S., Wersin, P., Kretzschmar, R., 2007a. Characterization of dissolved organic matter in anoxic rock extracts and in situ pore water of the Opalinus clay. Appl. Geochem. 22, 2926–2939.

Courdouan, A., Christl, I., Meylan, S., Wersin, P., Kretzschmar, R., 2007b. Isolation and characterization of dissolved organic matter from the Callovo-Oxfordian formation. Appl. Geochem. 22, 1537–1548.

De Craen, M., Van Geet, M., Wang, L., Put, M., 2004a. High sulphate concentrations in squeezed Boom Clay pore water: evidence of oxidation of clay cores. Phys. Chem. Earth Parts A/B/C 29, 91–103.

De Craen, M., Wang, L., Geet, M., van, Moors, H., 2004b. Geochemistry of Boom Clay Pore Water at the Mol Site. SCK-CEN-BLG-990, 04/MDC/P-48. SCK-CEN.

De Lange, G.J., Cranston, R.E., Hydes, D.H., Boust, D., 1992. Extraction of pore water from marine sediments: a review of possible artifacts with pertinent examples from the North Atlantic. Mar. Geol. 109, 53–76.

Delay, J., Vinsot, A., Krieguer, J.-M., Rebours, H., Armand, G., 2007. Making of the underground scientific experimental programme at the Meuse/Haute-Marne underground research laboratory, North Eastern France. Phys. Chem. Earth Parts A/B/C 32, 2–18.

Delay, J., Bossart, P., Ling, L.X., Blechschmidt, I., Ohlsson, M., Vinsot, A., Nussbaum, C., Maes, N., 2014. Three decades of underground research laboratories: what have we learned? Geol. Soc. Lond. Spec. Publ. 400, 400–401.

Deng, Y., Cui, Y.-J., Tang, A.M., Li, X.L., Sillen, X., 2012. An experimental study on the secondary deformation of Boom clay. Appl. Clay Sci. 59, 19–25.

Descostes, M., Blin, V., Bazer-Bachi, F., Meier, P., Grenut, B., Radwan, J., Schlegel, M.L., Buschaert, S., Coelho, D., Tevissen, E., 2008. Diffusion of anionic species in Callovo-Oxfordian argillites and oxfordian limestones (Meuse/Haute-Marne, France). Appl. Geochem. 23, 655–677.

Dohrmann, R., 2006a. Cation exchange capacity methodology I: an efficient model for the detection of incorrect cation exchange capacity and exchangeable cation results. Appl. Clay Sci. 34, 31–37.

Dohrmann, R., 2006b. Cation exchange capacity methodology III: correct exchangeable calcium determination of calcareous clays using a new silver-thiourea method. Appl. Clay Sci. 34, 47–57.

Dohrmann, R., Kaufhold, S., 2009. Three new, quick cec methods for determining the amounts of exchangeable calcium cations in calcareous clays. Clays Clay Miner. 57, 338–352.

Fernández, A.M., 2008. PID 3.2.1 Physical, Mineralogical and Geochemical Characterization of the Boom Clay, Callovo-Oxfordian and Opalinus Clay Rock Samples. European project FUNMIG 6th PCRD.

Fernández, A.M., Sánchez-Ledesma, D.M., Tournassat, C., Melón, A., Gaucher, E.C., Astudillo, J., Vinsot, A., 2014. Applying the squeezing technique to highly consolidated clayrocks for pore water characterisation: lessons learned from experiments at the Mont Terri rock laboratory. Appl. Geochem. 49, 2–21.

Gaboreau, S., Pret, D., Tinseau, E., Claret, F., Pellegrini, D., Stammose, D., 2011. 15 years of in situ cement-argillite interaction from Tournemire URL: characterisation of the multi-scale spatial heterogeneities of pore space evolution. Appl. Geochem. 26, 2159–2171.

Gailhanou, H., Miltenburg, J.C., van, Rogez, J., Olives, J., Amouric, M., Gaucher, E.C., Blanc, P., 2007. Thermodynamic properties of anhydrous smectite MX-80, illite IMt-2 and mixed-layer illite-smectite ISCz-1 as determined by calorimetric methods. Part I: heat capacities, heat contents and entropies. Geochim. Cosmochim. Acta 71, 5463–5473.

Gailhanou, H., van Miltenburg, J.C., van Genderen, A., Rogez, J., Greneche, J.M., Gilles, C., Gaucher, E.C., Blanc, P., 2009. Thermodynamic properties of ripidolite CCa-2. Part I. Heat capacities, heat contents and entropies. Geochim. Cosmochim. Acta 73, 4738–4749.

Gailhanou, H., Blanc, P., Rogez, J., Mikaelian, G., Kawaji, H., Olives, J., Amouric, M., Denoyel, R., Bourrelly, S., Montouillout, V., Vieillard, P., Fialips, C.I., Michau, N., Gaucher, E.C., 2012. Thermodynamic properties of illite, smectite and beidellite by calorimetric methods: enthalpies of formation, heat capacities, entropies and Gibbs free energies of formation. Geochim. Cosmochim. Acta 89, 279–301.

Gaucher, E., Robelin, C., Matray, J.M., Negrel, G., Gros, Y., Heitz, J.F., Vinsot, A., Rebours, H., Cassabagnere, A., Bouchet, A., 2004. ANDRA underground research laboratory: interpretation of the mineralogical and geochemical data acquired in the Callovian-Oxfordian formation by investigative drilling. Phys. Chem. Earth Parts A/B/C 29, 55–77.

Gaucher, E.C., Blanc, P., 2006. Cement/clay interactions—a review: experiments, natural analogues, and modeling. Waste Manage. 26, 776–788.

Gaucher, E.C., Blanc, P., Bardot, F., Braibant, G., Buschaert, S., Crouzet, C., Gautier, A., Girard, J.-P., Jacquot, E., Lassin, A., Negrel, G., Tournassat, C., Vinsot, A., Altmann, S., 2006. Modelling the porewater chemistry of the Callovian-Oxfordian formation at a regional scale. C. R. Geosci. 338, 917–930.

Gaucher, E.C., Tournassat, C., Pearson, F.J., Blanc, P., Crouzet, C., Lerouge, C., Altmann, S., 2009. A robust model for pore-water chemistry of clayrock. Geochim. Cosmochim. Acta 73, 6470–6487.

Gaucher, E.C., Lassin, A., Lerouge, C., Fléhoc, C., Marty, N.C., Henry, B., Tournassat, C., Altmann, S., Vinsot, A., Buschaert, S., others, 2010. CO_2 partial pressure in clayrocks: a general model. In: Water-Rock Interaction, vol. XIII, pp. 855–858.

Gaus, I., Azaroual, M., Czernichowski-Lauriol, I., 2005. Reactive transport modelling of the impact of CO_2 injection on the clayey cap rock at Sleipner (North Sea). Chem. Geol. 217, 319–337.

Gautschi, A., 2001. Hydrogeology of a fractured shale (Opalinus clay): implications for deep geological disposal of radioactive wastes. Hydrogeol. J. 9, 97–107.

Gherardi, F., Audigane, P., Gaucher, E.C., 2012. Predicting long-term geochemical alteration of wellbore cement in a generic geological CO_2 confinement site: tackling a difficult reactive transport modeling challenge. J. Hydrol. 420, 340–359.

Giffaut, E., Grivé, M., Blanc, P., Vieillard, P., Colàs, E., Gailhanou, H., Gaboreau, S., Marty, N., Madé, B., Duro, L., 2014. Andra thermodynamic database for performance assessment: ThermoChimie. Appl. Geochem. 49, 225–236.

Grambow, B., Landesman, C., Ribet, S., 2014. Nuclear waste disposal: I. Laboratory simulation of repository properties. Appl. Geochem. 49,, 237–246.

Grangeon, S., Vinsot, A., Tournassat, C., Lerouge, C., Giffaut, E., Heck, S., Groschopf, N., Denecke, M.A., Wechner, S., Schäfer, T., 2015. The influence of natural trace element distribution on the mobility of radionuclides. The exemple of nickel in a clay-rock. Appl. Geochem. 52, 155–173.

Huclier-Markai, S., Landesman, C., Rogniaux, H., Monteau, F., Vinsot, A., Grambow, B., 2010. Non-disturbing characterization of natural organic matter (NOM) contained in clay rock pore water by mass spectrometry using electrospray and atmospheric pressure chemical ionization modes. Rapid Commun. Mass Spectrom. 24, 191–202.

Jacquier, P., Hainos, D., Robinet, J., Herbette, M., Grenut, B., Bouchet, A., Ferry, C., 2013. The influence of mineral variability of Callovo-Oxfordian clay rocks on radionuclide transfer properties. Appl. Clay Sci. 83, 129–136.

Kargbo, D.M., Wilhelm, R.G., Campbell, D.J., 2010. Natural gas plays in the Marcellus shale: challenges and potential opportunities. Environ. Sci. Technol. 44, 5679–5684.

Kars, M., Lerouge, C., Grangeon, S., Aubourg, C., Tournassat, C., Madé, B., Claret, F. Identification of nanocrystalline goethite in reduced clay formations. Application to the Callovian-Oxfordian formation of Bure (France). American Mineralogist. http://dx.doi.org/10.2138/am-2015-5096, in press.

Keller, L.M., Holzer, L., Wepf, R., Gasser, P., 2011. 3D geometry and topology of pore pathways in Opalinus clay: implications for mass transport. Appl. Clay Sci. 52, 85–95.

Keller, L.M., Holzer, L., Schuetz, P., Gasser, P., 2013. Pore space relevant for gas permeability in Opalinus clay: statistical analysis of homogeneity, percolation, and representative volume element. J. Geophys. Res. Solid Earth 118, 2799–2812.

Koroleva, M., Lerouge, C., Mader, U., Claret, F., Gaucher, E.C., 2011. Biogeochemical processes in a clay formation in situ experiment: part B—results from overcoring and evidence of strong buffering by the rock formation. Appl. Geochem. 26, 954–966.

Lahann, R.W., Swarbrick, R.E., 2011. Overpressure generation by load transfer following shale framework weakening due to smectite diagenesis. Geofluids 11, 362–375.

Landais, P., 2006. Advances in geochemical research for the underground disposal of high-level, long-lived radioactive waste in a clay formation. J. Geochem. Explor. 88, 32–36.

Lassin, A., Marty, N., Henry, B., Trémosa, J., Gailhanou, H., Madé, B., Altmann, S., Gaucher, E.C., 2013. Equilibrium partial pressure of CO_2 in the Callovo-Oxfordian argillite as a function of relative humidity. Proc. Earth Planet. Sci. 7, 459–462.

Lázár, K., Máthé, Z., Földvári, M., 2010. Various redox conditions in Boda Claystone as reflected in the change of Fe^{2+}/Fe^{3+} ratio in clay minerals. J. Phys. Conf. Ser. 012053.

Lázár, K., Máthé, Z., 2012. Claystone as a potential host rock for nuclear waste storage. Clay minerals in nature—their characterization, modification and application. Intech 56–80.

Lerouge, C., Gaucher, E.C., Tournassat, C., Negrel, P., Crouzet, C., Guerrot, C., Gautier, A., Michel, P., Vinsot, A., Buschaert, S., 2010. Strontium distribution and origins in a natural clayey formation (Callovian-Oxfordian, Paris Basin, France): a new sequential extraction procedure. Geochim. Cosmochim. Acta 74, 2926–2942.

Lerouge, C., Grangeon, S., Gaucher, E.C., Tournassat, C., Agrinier, P., Guerrot, C., Widory, D., Flehoc, C., Wille, G., Ramboz, C., Vinsot, A., Buschaert, S., 2011. Mineralogical and isotopic record of biotic and abiotic diagenesis of the Callovian-Oxfordian clayey formation of Bure (France). Geochim. Cosmochim. Acta 75, 2633–2663.

Leroy, P., Revil, A., Altmann, S., Tournassat, C., 2007. Modeling the composition of the pore water in a clay-rock geological formation (Callovo-Oxfordian, France). Geochim. Cosmochim. Acta 71, 1087–1097.

Liu, D.J., Bruggeman, C., Maes, N., 2008. The influence of natural organic matter on the speciation and solubility of Eu in Boom Clay porewater. Radiochim. Acta 96, 711–720.

Luquot, L., Andreani, M., Gouze, P., Camps, P., 2012. CO_2 percolation experiment through chlorite/zeolite-rich sandstone (Pretty Hill Formation–Otway Basin–Australia). Chem. Geol. 294, 75–88.

Mäder, U., Gimmi, T., 2007. Species-specific transport and reactive transport modelling of a long-term core infiltration experiment on Opalinus clay. In: Clays in Natural and Engineered Barriers for Radioactive Waste Confinement, 3rd International Meeting, Lille, France, p. 20.

Marty, N.C., Munier, I., Gaucher, E.C., Tournassat, C., Gaboreau, S., Vong, C.Q., Giffaut, E., Cochepin, B., Claret, F., 2014. Simulation of cement/clay interactions: feedback on the increasing complexity of modelling strategies. Transport Porous Med. 1–21.

Matray, J.M., Savoye, S., Cabrera, J., 2007. Desaturation and structure relationships around drifts excavated in the well-compacted Tournemire's argillite (Aveyron, France). Eng. Geol. 90, 1–16.

Mazurek, M., Alt-Epping, P., Bath, A., Gimmi, T., Niklaus Waber, H., Buschaert, S., Cannière, P.D., de Craen, M., Gautschi, A., Savoye, S., others, 2011. Natural tracer profiles across argillaceous formations. Appl. Geochem. 26, 1035–1064.

Mell, P., Megyeri, J., Riess, L., Máthé, Z., Csicsák, J., Lázár, K., 2006. Sorption of Co, Cs, Sr and I onto argillaceous rock as studied by radiotracers. J. Radioanal. Nucl. Chem. 268, 405–410.

Millero, F.J., 1982. The effect of pressure on the solubility of minerals in water and seawater. Geochim. Cosmochim. Acta 46, 11–22.

Moore, R., 1991. The chemical and mineralogical controls upon the residual strength of pure and natural clays. Geotechnique 41, 35–47.

Motellier, S., Ly, J., Gorgeon, L., Charles, Y., Hainos, D., Meier, P., Page, J.-C., 2003. Modelling of the ion-exchange properties and indirect determination of the interstitial water composition of an argillaceous rock. Application to the Callovo-Oxfordian low-water-content formation. Appl. Geochem. 18, 1517–1530.

Muurinen, A., Karnland, O., Lehikoinen, J., 2004. Ion concentration caused by an external solution into the porewater of compacted bentonite. Phys. Chem. Earth Parts A/B/C 29, 119–127.

Muurinen, A., Karnland, O., Lehikoinen, J., 2007. Effect of homogenization on the microstructure and exclusion of chloride in compacted bentonite. Phys. Chem. Earth Parts A/B/C 32, 485–490.

Neuzil, C.E., 2013. Can shale safely host US nuclear waste? Eos, Trans. Am. Geophys. Union 94, 261–262.

Parkhurst, D.L., Appelo, C., 2013. Description of Input and Examples for PHREEQC Version 3—a Computer Program for Speciation, Batch-reaction, One-dimensional Transport, and Inverse Geochemical Calculations.

Pearson, F.J., Arcos, D., Boisson, J.-Y., Fernandez, A.M., Gäbler, H.-E., Gaucher, E., Gautschi, A., Griffault, L., Hernan, P., Waber, H.N., 2003. Mont Terri Project—Geochemistry of Water in the Opalinus Clay formation at the Mont Terri Rock Laboratory. Geology Series No. 5 Swiss Federal Office for Water and Geology, Bern.

Pearson, F.J., Tournassat, C., Gaucher, E.C., 2011. Biogeochemical processes in a clay formation in situ experiment: part E—equilibrium controls on chemistry of pore water from the Opalinus clay, Mont Terri underground research laboratory, Switzerland. Appl. Geochem. 26, 990–1008.

Poulain, S., Sergeant, C., Simonoff, M., Le Marrec, C., Altmann, S., 2008. Microbial investigations in Opalinus clay, an argillaceous formation under evaluation as a potential host rock for a radioactive waste repository. Geomicrobiol. J. 25, 240–249.

Robinet, J.-C., Sardini, P., Coelho, D., Parneix, J.-C., Prêt, D., Sammartino, S., Boller, E., Altmann, S., 2012. Effects of mineral distribution at mesoscopic scale on solute diffusion in a clay-rich rock: example of the Callovo-Oxfordian mudstone (Bure, France). Water Resour. Res. 48.

Rosenqvist, I., 1984. The importance of pore water chemistry on mechanical and engineering properties of clay soils. Philos. Trans. Roy. Soc. Lond. Ser. A, Math. Phys. Sci. 311, 369–373.

Sacchi, E., Michelot, J.-L., Pitsch, H., Lalieux, P., Aranyossy, J.-F., 2001. Extraction of water and solutes from argillaceous rocks for geochemical characterisation: methods, processes and current understanding. Hydrogeol. J. 9, 17–33.

Sammartino, S., Bouchet, A., Prêt, D., Parneix, J.-C., Tevissen, E., 2003. Spatial distribution of porosity and minerals in clay rocks from the Callovo-Oxfordian formation (Meuse/Haute-Marne, Eastern France)—implications on ionic species diffusion and rock sorption capability. Appl. Clay Sci. 23, 157–166.

Sanford, W.E., Voytek, M.A., Powars, D.S., Jones, B.F., Cozzarelli, I.M., Cockell, C.S., Eganhouse, R.P., 2009. Pore-water chemistry from the ICDP-USGS core hole in the Chesapeake Bay impact structure—implications for paleohydrology, microbial habitat, and water resources. Geol. Soc. Am. Spec. Pap. 458, 867–890.

Savage, D., Noy, D., Mihara, M., 2002. Modelling the interaction of bentonite with hyperalkaline fluids. Appl. Geochem. 17, 207–223.

Savoye, S., Michelot, J.-L., Matray, J.-M., Wittebroodt, C., Mifsud, A., 2012. A laboratory experiment for determining both the hydraulic and diffusive properties and the initial pore-water composition of an argillaceous rock sample: a test with the Opalinus clay (Mont Terri, Switzerland). J. Contam. Hydrol. 128, 47–57.

Scholl, D.W., 1963. Techniques for removing interstitial water from coarsegrained sediments for chemical analyses. Sedimentology 2, 156–163.

Sholkovitz, E., 1973. Interstitial water chemistry of the Santa Barbara Basin sediments. Geochim. Cosmochim. Acta 37, 2043–2073.

Sposito, G., 1984. The Surface Chemistry of Soils. Oxford University Press, New York.

Stefánsson, A., Arnórsson, S., 2000. Feldspar saturation state in natural waters. Geochim. Cosmochim. Acta 64, 2567–2584.

Stroes-Gascoyne, S., Schippers, A., Schwyn, B., Poulain, S., Sergeant, C., Simonoff, M., Le Marrec, C., Altmann, S., Nagaoka, T., Mauclaire, L., McKenzie, J., Daumas, S., Vinsot, A., Beaucaire, C., Matray, J.-M., 2007. Microbial community analysis of Opalinus clay drill core samples from the Mont Terri underground research laboratory, Switzerland. Geomicrobiol. J. 24, 1–17.

Stroes-Gascoyne, S., Sergeant, C., Schippers, A., Hamon, C., Nèble, S., Vesvres, M.-H., Barsotti, V., Poulain, S., Le Marrec, C., 2011. Biogeochemical processes in a clay formation in situ experiment: part D—microbial analyses—synthesis of results. Appl. Geochem. 26, 980–989.

Tang, L., Sparks, D.L., 1993. Cation-exchange kinetics on montmorillonite using pressure-jump relaxation. Soil Sci. Soc. Am. J. 57, 42–46.

Thury, M., Bossart, P., 1999. The Mont Terri rock laboratory, a new international research project in a Mesozoic shale formation, in Switzerland. Eng. Geol. 52, 347–359.

Tournassat, C., Gailhanou, H., Crouzet, C., Braibant, G., Gautier, A., Lassin, A., Blanc, P., Gaucher, E.C., 2007. Two cation exchange models for direct and inverse modelling of solution major cation composition in equilibrium with illite surfaces. Geochim. Cosmochim. Acta 71, 1098–1114.

Tournassat, C., Lerouge, C., Blanc, P., Brendle, J., Greneche, J.M., Touzelet, S., Gaucher, E.C., 2008. Cation exchanged Fe(II) and Sr compared to other divalent cations (Ca, Mg) in the bure Callovian-Oxfordian formation: implications for porewater composition modelling. Appl. Geochem. 23, 641–654.

Tournassat, C., Appelo, C.A.J., 2011. Modelling approaches for anion-exclusion in compacted Na-bentonite. Geochim. Cosmochim. Acta 75, 3698–3710.

Tremosa, J., Arcos, D., Matray, J.-M., Bensenouci, F., Gaucher, E.C., Tournassat, C., Hadi, J., 2012. Geochemical characterization and modelling of the Toarcian/Domerian porewater at the Tournemire underground research laboratory. Appl. Geochem. 27, 1417–1431.

Urios, L., Marsal, F., Pellegrini, D., Magot, M., 2012. Microbial diversity of the 180 million-year-old Toarcian argillite from Tournemire, France. Appl. Geochem. 27, 1442–1450.

Van Loon, L.R., Glaus, M.A., Müller, W., 2007. Anion exclusion effects in compacted bentonites: towards a better understanding of anion diffusion. Appl. Geochem. 22, 2536–2552.

Vinsot, A., Appelo, C.A.J., Cailteau, C., Wechner, S., Pironon, J., De Donato, P., De Cannière, P., Mettler, S., Wersin, P., Gäbler, H.E., 2008a. CO_2 data on gas and pore water sampled in situ in the Opalinus clay at the Mont Terri rock laboratory. Phys. Chem. Earth Parts A/B/C 33, S54–S60.

Vinsot, A., Mettler, S., Wechner, S., 2008b. In situ characterization of the Callovo-Oxfordian pore water composition. Phys. Chem. Earth Parts A/B/C 33, S75–S86.

Vinsot, A., Delay, J., La Vaissière, R., de, Cruchaudet, M., 2011. Pumping tests in a low permeability rock: results and interpretation of a four-year long monitoring of water production flow rates in the Callovo-Oxfordian argillaceous rock. Phys. Chem. Earth Parts A/B/C 36, 1679–1687.

Vinsot, F.A., Claret, Lundy, M., Wechner, S., 2012. In situ measurement of the rate of oxygen consumption by the Callovo-Oxfordian argillaceous rock. In: Montpellier 2012, Clays in Natural and Engineered Barriers for Radioactive Waste Confinement, 5th International Meeting, Montpellier France, October 22–25, pp. 302–303.

Vinsot, A., Linard, Y., Lundy, M., Necib, S., Wechner, S., 2013. Insights on desaturation processes based on the chemistry of seepage water from boreholes in the Callovo-Oxfordian argillaceous rock. Proc. Earth Planet. Sci. 7, 871–874.

Vinsot, A., Appelo, C.A.J., Lundy, M., Wechner, S., Lettry, Y., Lerouge, C., Fernandez, A.-M., Labat, M., Tournassat, C., De Canniere, P., Schwyn, B., Mckelvie, J., Dewonck, S., Bossart, P., Delay, J., 2014. In situ diffusion test of hydrogen gas in the Opalinus clay. Geol. Soc. Lond. Spec. Publ. 400, 400–412.

Wersin, P., Leupin, O.X., Mettler, S., Gaucher, E.C., Mader, U., De Canniere, P., Vinsot, A., Gabler, H.E., Kunimaro, T., Kiho, K., Eichinger, L., 2011a. Biogeochemical processes in a clay formation in situ experiment: part A—overview, experimental design and water data of an experiment in the Opalinus clay at the Mont Terri underground research laboratory, Switzerland. Appl. Geochem. 26, 931–953.

Wersin, P., Stroes-Gascoyne, S., Pearson, F.J., Tournassat, C., Leupin, O.X., Schwyn, B., 2011b. Biogeochemical processes in a clay formation in situ experiment: part G—key interpretations and conclusions. Implications for repository safety. Appl. Geochem. 26, 1023–1034.

Yven, B., Sammartino, S., Géraud, Y., Homand, F., Villiéras, F., 2007. Mineralogy, texture and porosity of Callovo-Oxfordian argillites of the Meuse/Haute-Marne region (eastern Paris Basin). Mém. Soc. Géol. Fr. 178, 73–90.

Chapter 4

Dissolution Kinetics of Clay Minerals

Jordi Cama[a] and Jiwchar Ganor[b]
[a] *Institute of Environmental Assessment and Water Research, IDAEA, CSIC, Barcelona, Spain;*
[b] *Department of Geological and Environmental Sciences, Ben-Gurion University of the Negev, Beer Sheva, Israel*

Chapter Outline

Developments in Clay Science, Vol. 6C. http://dx.doi.org/10.1016/B978-0-08-100027-4.00004-8

4.1 INTRODUCTION

This chapter reviews studies of clay mineral dissolution kinetics published during the last twenty years (1995–2014). Over this period one main goal has been the elucidation of clay mineral dissolution mechanisms. The layered structure of the clay minerals (a group of phyllosilicates) in contrast to the structure of the framework silicates has offered distinct mechanistic interpretations of their dissolution reactions. The improved microscopic techniques have rendered detailed topographic inspections at the nanoscale of the clay mineral-reacted surfaces to delve deeper into the processes controlling the dissolution reactions.

The second main goal has been the obtainment of dissolution rates under different environmental conditions with the imminent derivation of dissolution rate laws. Investigation of the effect of the primary environmental variables (pH, temperature, organic acids, solution composition, and solution saturation state), as well as the solution ionic strength, on the dissolution rates has been carried out by means of laboratory experiments, primarily using powder samples, from which the concept of reactive surface area of clay minerals has been debated.

Twenty years ago, Nagy (1995) presented an exhaustive compilation of clay mineral dissolution kinetics in a chapter entitled "Dissolution and precipitation kinetics of sheet silicates." Her review covers studies that were published from the early 1980s to 1995, along with a thorough theoretical background. To avoid unnecessary repetition, the current review is mainly devoted to studies that were published in the last 20 years. Likewise, the theoretical background of Nagy (1995) is briefly extended.

This chapter reviews clay dissolution kinetics. The theoretical background of clay dissolution kinetics is part of the kinetic theory that has been developed to describe and quantify mineral dissolution, particularly silicate minerals. Considering this, the different clay dissolution rate laws and interpretations of the clay dissolution mechanisms that have been proposed over the last four decades contribute to enlarge the current knowledge on the kinetic theory of mineral dissolution.

Clay minerals compiled in this chapter belong to the following two groups: the 1:1 clay minerals (**kaolinite** (kaolinite)) and the 2:1 clay minerals (**dioctahedral smectite** (montmorillonite, beidellite, and nontronite); **trioctahedral smectite** (hectorite and saponite); **dioctahedral mica** (illite and muscovite) and **trioctahedral mica** (biotite and phlogopite); **vermiculite** (vermiculite); and **chlorite** (chlorite)) (Bergaya et al., 2011).

Environmental variables such as pH, temperature, catalysts/inhibitors, solution saturation state, and ionic strength affect clay minerals dissolution kinetics. Thus, in this chapter, the role of each of these variables is examined. Although adsorption is very important for understanding kinetics of surface reaction (i.e., adsorption of H^+, in particular of edges of particles, is the first step of the overall dissolution process), it is not included in the review.

In Appendix are reported the chemical composition and structural formula of the clay minerals used in the reviewed studies (Table A.1) and the list of experimental conditions and obtained experimental results (Table A.2).

4.2 THEORETICAL BACKGROUND: CLAY MINERAL DISSOLUTION KINETICS

Dissolution rate can be written by considering a general law, as that proposed by Lasaga (1998):

$$Rate = k_0 \cdot A_{\min} \cdot e^{\frac{-E_a}{RT}} \cdot a_{\mathrm{H}}^{n_{\mathrm{H}}} \cdot \prod_i a_i^{n_i} \cdot g(I) \cdot f(\Delta G_r) \quad (4.1)$$

where k_0 is a constant, $A_{\min}$ is the reactive surface area of the mineral, E_a is the apparent activation energy of the overall reaction, R is the gas constant, T is the temperature (K), a_{H} is the activity of protons in the solution, n_{H} is the order of the reaction with respect to H^+ and the term involving the activities of other aqueous species, a_i, incorporate other possible catalytic or inhibitor effects on the overall rate. The $g(I)$ term indicates a possible dependence of the rate on the ionic strength (I) in addition to that entering through the activities of specific ions. The $f(\Delta G_r)$ term accounts for the variation from equilibrium ($\Delta G_r = 0$; $0 \leq f(\Delta G_r) \leq 1$). The formulation of Eqn (4.1) is very useful because it relates the reaction rate to activities of ions in solution that may be obtained directly from the solution chemistry. However, silicate dissolution is a surface process and therefore, it is more appropriate to express the dependence of the rate on the concentrations (or activities) of ions adsorbed on the surface rather than on the bulk activities in solution. Hence, Eqn (4.1) can be written in terms of surface concentrations ($C_{S,i}$)

$$Rate = k_0 \cdot A_{\min} \cdot e^{\frac{-E_a}{RT}} \cdot C_{s,\mathrm{H}}^{n_{\mathrm{H},ads}} \cdot \prod_i C_{s,i}^{n_{i,ads}} \cdot g(I) \cdot f(\Delta G_r) \quad (4.2)$$

where the coefficients $n_{H,ads}$ and $n_{i,ads}$ are the reaction orders with respect to the adsorbed surface species. Given the above and based on Eqns (4.1) and (4.2), there is a diversity of rate laws that account for the dissolution kinetics of the different clay minerals, as it is described along the chapter.

Over the last forty years, several mechanistic models have been built to account for the effect of deviation from equilibrium ($f(\Delta G_r)$) on mineral dissolution kinetics. Among others, these mechanisms include: (1) the most widely extended one was derived from transition state theory (TST; Eyring, 1935) which was developed for elementary reactions and may be applied to overall reactions governing the dissolution process and relates the variation in solution composition with mineral dissolution and precipitation rates (Lasaga and Kirkpatrik, 1981; Aagard and Helgeson, 1982; Oelkers et al., 1994; Lasaga, 1998; references therein; Oelkers, 2001). TST may be applied only in cases in which defects such as dislocations do not control the dissolution kinetics (Lasaga, 1998); (2) At the start of the twenty-first century, a different conceptual model took off. Lasaga and Lüttge (2001) proposed a model of mineral dissolution that directly links the surface reactivity of minerals with their dissolution kinetics; (3) In a third approach, dissolution of some aluminosilicates can be understood through the same mechanistic theory of nucleation developed for mineral growth (Dove et al., 2005).

Lasaga and Lüttge's (2001) model predicts that dissolution occurs via the formation of etch pits and the generation of dissolution stepwaves at the outskirts of the pit walls. If the undersaturation of the solution is high enough, i.e., $\Delta G < \Delta G_{crit}$ (critical free energy), etch pits not only open up but they are the source for trains of steps that travel across the surface. By moving away from the defect site and lowering the silicate surface, the step propagation eventually controls the bulk dissolution rate. Several studies using vertical scanning interferometry (VSI) have demonstrated this mechanism and suggest that etch pits in combination with step movement control the dissolution process of several minerals independent of their degree of structural complexity (e.g., Lüttge and Arvidson, 2010; Kurganskaya and Lüttge, 2013).

Dove et al. (2005) showed that mineral dissolution can be understood through the same mechanistic theory of nucleation developed for mineral growth. The derived dislocation model can predict variation of the normal aluminosilicate dissolution rate (R_n) with degree of undersaturation (σ). It is argued that two-dimensional nucleation dominates rates measured at higher temperatures of 150–200°C. In contrast, the 80°C data for kaolinite is predicted by the dislocation defect model. Nonetheless, this model has not been further developed for clay minerals other than kaolinite.

Implementation of Lasaga and Lüttge's (2001) model gains terrain as real-time microscopic observations of reacted mineral surfaces offer new insights into the dissolution mechanisms (Kurganskaya et al., 2012). Also, the validity of the TST-based rate laws to be sufficient to explain mineral dissolution kinetics as a function of distance from equilibrium has been questioned (Lüttge, 2006), mainly based on the fact that TST allows a relation between the forward and reverse rate constants but only of an elementary reaction. The adequacy of simple surface adsorption models to explain the complexity of the three-dimensional lattice and its participation in the dissolution process is also argued. Lüttge (2006) showed that the use of a stochastic approach that treats crystal dissolution as a many-body problem has led to the development of the stepwave model and some in-depth exploration of the dissolution mechanism.

As it is extensively shown in this chapter, most of the descriptions and interpretations of the clay dissolution kinetics are, at the moment, TST-based. Therefore, TST is further discussed below in Section 4.2.4. The terms integrated in Eqns (4.1) and (4.2), other than the $f(\Delta G_r)$, account for the effect of different environmental variables on clay dissolution rates. A brief description of these terms with their derived expressions is given below.

In order to model the effects of pH, temperature, catalysis/inhibitors, ionic strength on clay mineral dissolution rates, it is important to separate between direct and indirect effects of the environmental variables involved. By direct effect we mean an effect related to surface processes and therefore, one that can be used to understand the reaction mechanism (Cama et al., 2002). As changes in most environmental variables affect also the degree of saturation, their direct effect on the rate should be measured in experiments that are carried out under far-from-equilibrium conditions where $f(\Delta G_r) = 1$.

4.2.1 pH Dependence

The effect of pH on the dissolution rates of oxides and silicates, and hence clay minerals, has been exhaustively examined (Bandstra and Brantley, 2008). For the pH-dissolution rate dependence, in the absence of catalysts/inhibitors, the dissolution rate has been classically interpreted by three reaction mechanisms (e.g., Huertas et al., 1999): proton-promoted, hydroxyl-promoted and water-promoted, that dominate the reaction rate under acidic, alkaline, or near-neutral conditions, respectively:

$$Rate = k_{H^+} \cdot a_{H^+}^{n_{H^+}} + k_W + k_{OH^-} \cdot a_{OH^-}^{n_{OH^-}} \tag{4.3}$$

where a_{H^+} is the proton activity; a_{OH^-} is the hydroxyl activity; k_{H^+}, k_W, and k_{OH^-} are the rate constants under acidic, neutral, and basic conditions, respectively, and n_{H^+} and n_{OH^+} are the corresponding orders of reaction. The dissolution rates of most aluminosilicates as a function of pH exhibit a U-shape (Figure 4.1(a)). In some studies the pH dependence of dissolution rate is modeled using a surface complexation model (e.g., Huertas et al., 1998, 1999):

$$Rate = k_H \cdot C_{S,H} + k_W + k_{OH} \cdot C_{S,OH} \tag{4.4}$$

where k_i is the rate coefficient of reaction mechanism i, and $C_{s,i}$ is the mineral surface concentration of adsorbed species i.

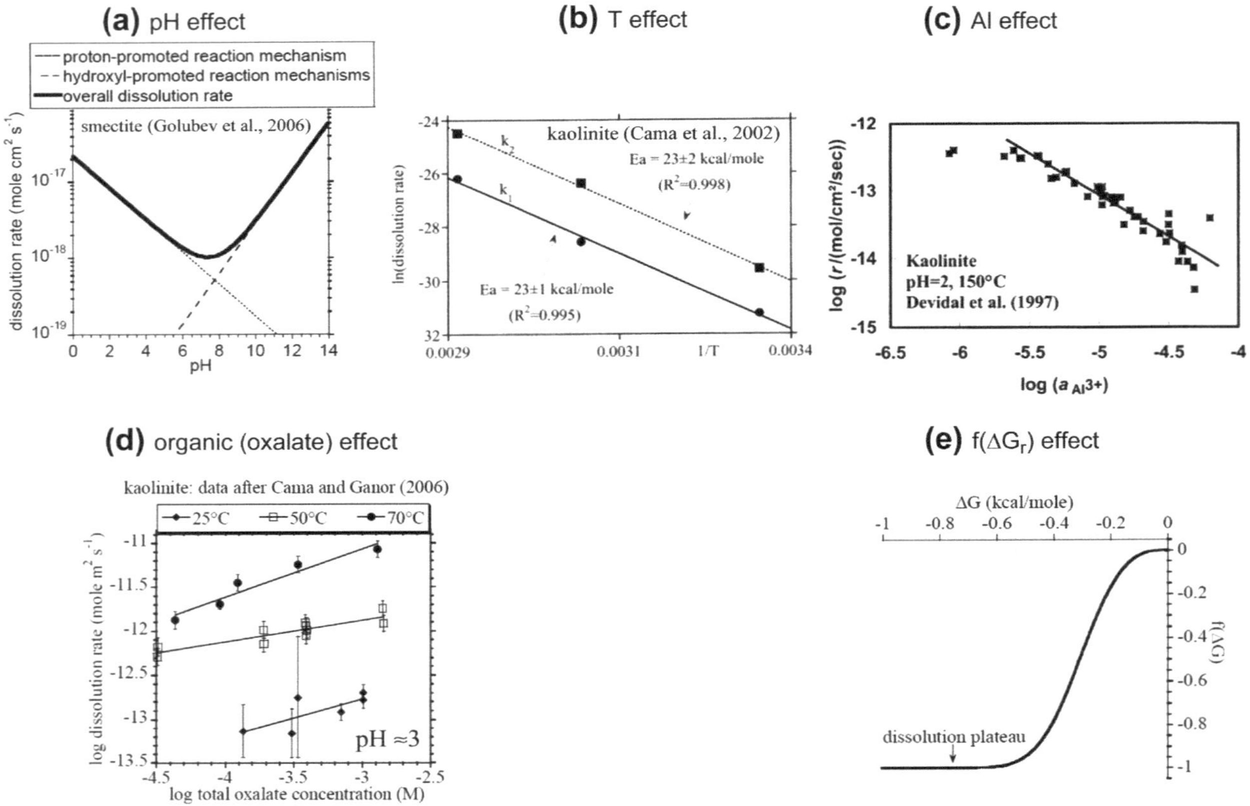

FIGURE 4.1 Mineral dissolution rates as a function of: (a) pH, (b) 1/T, (c) $a_{Al^{3+}}$, (d) oxalate concentration, and (e) $f(\Delta G_r)$.

4.2.2 T Dependence

The temperature dependence of the dissolution rate generally follows the Arrhenius law

$$Rate = A \cdot e^{-E_a/RT} \tag{4.5}$$

where A is the preexponential factor, E_a is the apparent activation energy of the overall reaction, R is the gas constant, and T is the temperature (K). In order to obtain the apparent activation energy experimentally, it is common to conduct several experiments at different temperatures while all the other experimental variables (e.g., pH) are held constant (Figure 4.1(b)). A critical assumption in calculating activation energies using this method is that the preexponential factor, A, is the same in all experiments. This preexponential factor includes other effects of the experimental variables on the rate. As mineral dissolution is a surface process, the pH effect on dissolution rate is controlled by protons adsorbed on the surface (Eqn. (4.2)) and not directly by the pH. Therefore, conducting experiments under the same pH is not enough to keep the pre-exponential factor constant, and one should keep the surface concentration constant (Cama et al., 2002).

Under far-from-equilibrium conditions, and keeping all parameters except pH and temperature constant, Eqn (4.1) becomes

$$Rate = k \cdot e^{-E_a/RT} \cdot a_{H^+}^{n_{H^+}} \tag{4.6}$$

where k is a coefficient that is constant under these conditions.

4.2.3 Catalytic/Inhibitor Effect

4.2.3.1 Al Effect

Oelkers et al. (1994) showed strong Al inhibition of kaolinite dissolution reaction under very-far-from-equilibrium conditions at 150°C and pH = 2. The reaction order of this inhibition was −1 (i.e., rate ∞ $1/a_{Al}$). The authors proposed that Al-silicate dissolution is controlled by a Si-rich precursor complex formed by the exchange of Al for protons at the surface. According to this mechanism, dissolution rate of certain aluminosilicates dissolving at low pH is expressed by

$$Rate = k_1 \cdot \left[\frac{\left(\frac{a_{H^+}^3}{a_{Al^{3+}}}\right)^n \prod_i a_{A_i}^{v_i}}{1 + K_1 \cdot \left(\frac{a_{H^+}^3}{a_{Al^{3+}}}\right)^n \prod_i a_{A_i}^{v_i}} \right] \tag{4.7}$$

where k_1 and K_1 are rate coefficient and equilibrium constant, respectively, n is a stoichiometric coefficient, A_i, stands for the ith aqueous species involved in the formation of the precursor complex, v_i represents stoichiometric coefficients, which may be either positive or negative, and a_{H^+} and $a_{Al^{3+}}$ are the solution activities of $H^+$and Al^{3+}, respectively (Figure 4.1(c)).

4.2.3.2 Organic Matter Effect

A thorough review of the effect of organic matter (OM) on dissolution kinetics of minerals may be found in Ganor et al. (2009). Due to differences in the surface properties of different minerals, there is a significant variability in their tendencies to adsorb OM. Among the naturally abundant minerals, the most effective adsorbents are clay minerals and oxides (Ganor et al., 2009). Although both inhibition and enhancement of clay mineral dissolution rates were observed, the latter effect is much more common. Low molar mass organic acids (e.g., oxalate, malonate, citrate, salicylate, *o*-phthalate, and gluconate) through ligand-promoted dissolution at mineral surfaces and complexation of metal cations in solution have been shown to enhance dissolution of kaolinite and montmorillonite (Chin and Mills, 1991; Wieland and Stumm, 1992; Vandevivere et al., 1994; Ganor and Lasaga, 1994; Cama and Ganor, 2006; Golubev et al., 2006). Golubev et al. (2006) found that acetate, gluconate, alginate, glucosamine, and glucuronate may also inhibit dissolution of montmorillonite. The empirical rate laws that account for the effect of OM on silicate dissolution are of the form

$$Rate = k_{OM} \cdot C_{OM}^{n} \tag{4.8}$$

where k_{OM} is the rate coefficient and C_{OM} is the OM concentration, and n is the reaction order with respect to OM concentration (e.g., Figure 4.1(d); Cama and Ganor, 2006).

4.2.4 Effect of Deviation from Equilibrium (ΔG_r Effect)

TST has provided the basis to describe the form of the $f(\Delta G_r)$ term in Eqns (4.1) and (4.2) (Nagy, 1995; Oelkers, 1996; Lasaga, 1998; Brantley, 2008; Ganor et al., 2009; Schott et al., 2009, 2012). In the last twenty years, the experimental observations lead to fully nonlinear rate laws, i.e., rate laws in which the rate is not a linear function of the Gibbs free energy even very close to equilibrium, in contrast to the TST prediction for elementary reactions. Therefore, the rate dependence on deviation from equilibrium was described using functions such as (Nagy and Lasaga, 1992):

$$f(\Delta G_r) = 1 - \exp\left(m \cdot \left(\frac{\Delta G_r}{RT}\right)^n\right) \tag{4.9}$$

The shape of the curve $f(\Delta G_r)$ function represented in Eqn (4.9) (Figure 4.1(e)) is only controlled by the mathematical formulas and the parameters m and n. It is important to note that, Eqn (4.9) is not based on a first principle theory, and therefore there is no physical meaning to the values of the coefficients m and n (Cama et al., 2000). A particular form of Eqn (4.9) given by

$$f(\Delta G_r) = 1 - \exp\left(\frac{\Delta G_r}{\sigma RT}\right) \tag{4.10}$$

where σ is known as the Temkin's average stoichiometric number, which is equal to the ratio of the rate of destruction of the activated complex to the overall reaction rate, has been used to describe the dissolution of silicate minerals (Aagard and Helgeson, 1982; Schott et al., 2009, 2012).

4.2.5 Surface Area and Surface Reactivity

Reactive surface area is one of the key parameters for studying the kinetics of mineral dissolution. The determination of the reactive surface area is not trivial in clay minerals as they possess both internal and external surface areas, and in which many cases the dissolution is controlled by the chemical attack on the edge surface. The common practice in experimental kinetics is to normalize the dissolution rate to the surface area measured by the BET method. As the BET method measures the total surface area, it was suggested that it is not a valid proxy for the reactive surface area of clays (Metz et al., 2005b).

4.2.6 Ionic Strength Effect

TST predicts that the rate constant depends on the activity coefficients of aqueous species (Lasaga, 1998). For not too high concentrated solutions (e.g., $I = 0.1$ M), the logarithm of the rate constant of an elementary reaction involving ions depends on the square root of the ionic strength of the solution

$$\log k_+ = \log k_0 - \frac{A_{DH}(z_A + z_B)^2\sqrt{I}}{1 + \bar{a}B_{DH}\sqrt{I}} \tag{4.11}$$

where k_+ is the rate constant, k_0 is a reference constant, z_i is the charge of the ion, I is the ionic strength of the aqueous solution, $\bar{a}$ is the effective diameter of the ion, and A_{DH} and B_{DH} are constants (Debye–Hückel formula).

4.3 EXPERIMENTAL METHODOLOGY

Most of the knowledge on mineral dissolution kinetics acquired over the last fifty years comes from the ability to accurately measure the variation in aqueous chemistry as a function of time when a mineral (usually as powder) and a solution are put together to react. This approach has been used to produce adequate data to formulate empirically derived dissolution rate laws that account for the effects of pH, temperature, aqueous Al, organic acids, solution saturation state, and ionic strength on clay mineral dissolution rates. In the last fifteen years, additional micro- and nanoscale data has come out from the improved capacity of new techniques such as atomic force microscopy (AFM), interferometry techniques (VSI and PSI (phase shifting interferometry)), and laser confocal microscopy with differential interference contrast microscopy (LCM-DIM) to study the mineral—water interface (Brantley, 2003; references therein; Lüttge and Arvidson, 2008; Cappelli et al., 2013; Satoh et al., 2013). Integration of both strategies is leading to a remarkable level of understanding of the dissolution kinetics.

4.3.1 Aqueous Chemistry: Chemical Reactors

Two types of chemical reactors have been used to measure mineral dissolution rates (Brantley and Conrad, 2008): batch reactors and flow-through reactors, which include different forms (stirred and nonstirred flow-through reactors, fluidized bed reactors, thin-film continuous flow reactors, column and plug-flow reactors (Malmström and Banwart, 1997; Brantley, 2003 and references therein). An important difference among the clay mineral kinetic studies is in the basic design of the dissolution experiments. The dissolution rate in a batch experiment is obtained by measuring the concentration of a component j at different times during the experiment and using the expression

$$Rate \cdot v_j = -\frac{dC_j}{dt} \cdot \frac{V}{A} \tag{4.12}$$

where C_j is the concentration of component j, v_j is the stoichiometric coefficient of j in the dissolution reaction, A is the total reactive surface area of the mineral, and V is the volume of the solution in the reactor. When the change in concentration is linear with time during the entire stage of dissolution, the dissolution rate may be readily obtained by linear regression. However, under changing surface and solution conditions, a batch experiment will yield only the integrated kinetics over the time period of the experiment. In a flow-through experiment, the dissolution rate is calculated at steady state under constant kinetic conditions (Nagy et al., 1991)

$$Rate \cdot v_j = -\frac{q}{A} \cdot \left(C_{j,out} - C_{j,inp}\right) \tag{4.13}$$

where $C_{j,inp}$ and $C_{j,out}$ are the concentrations of component j in the input and the output solutions, and q is the fluid volume flux through the system. The advantages of the batch experiment over the flow-through are its simplicity and the ability to model the change in rate over changing experimental conditions. However, there are some important advantages of the flow-through experiment that make it a more accurate, better-controlled method, so that the results are simpler to interpret.

The most important advantage of the flow-through experiment is that the dissolution rate is determined at steady state when the concentrations of all the components involved in the reaction are constant, including the concentrations of reactants, products, and all possible catalysts and inhibitors that are included in the experiment. For instance, this enables the determination of the dissolution rate at a constant pH, without requiring a buffer that may influence the reaction rate. In order to achieve constant pH in a batch experiment, either a buffer or a pH-stat must be used. Both possibilities complicate the system and increase the number of variables. In addition, flow-through experiments are ideal for determining the dissolution rate as a function of saturation state because, at steady state, ΔG_r is known and constant. In a batch experiment, the concentration and, as a result, the ΔG_r value continually changes as it approaches zero (equilibrium).

4.3.2 Solid Sample Surface Topography: AFM, VSI, PSI, and LCM-DIM

The advance in microscopy techniques that are used to characterize the evolution of clay mineral surfaces while dissolving represents a firm step to delve deeper into dissolution mechanisms.

AFM is useful to quantify surface microtopography on the order of the unit cell. Bickmore et al. (1999) developed three methods that allowed for the imaging of kaolinite, Na^+-montmorillonite, phlogopite, and vermiculite in aqueous solutions with AFM. Regarding smectite, Bosbach et al. (2000) and Bickmore et al. (2001) investigated the dissolution of hectorite and nontronite under acidic conditions at room temperature. Under highly alkaline conditions, Yokoyama et al. (2005) studied montmorillonite dissolution by in situ AFM ($pH = 13.3$ and $I = 0.3$ M), and Kuwahara (2006) carried out tapping mode and contact mode atomic force microscopy analyses to investigate smectite dissolution at 25°C. Recently, Satoh et al. (2013) have conducted ex situ measurements of previously reacted samples of compacted montmorillonite at 70°C in 0.3 M NaOH solution (pH 12.1 at 70°C) using AFM to quantify the ratio of edge surface area to total surface area. Regarding mica dissolution, Aldushin et al. (2004a,b, 2006) used hydrothermal AFM (HAFM) to study in situ the mechanism and kinetics of alteration of apophyllite (sheet silicate with structural similarity to micas) and phlogopite, as well as their basal plane reactivities at temperatures from 20 to 140°C. Kuwahara (2008) investigated the dissolution behavior of muscovite under alkaline conditions at 25–50°C using in situ AFM with an air/fluid heater system. Kurganskaya et al. (2012) performed AFM experiments involving muscovite to evaluate crystallographic controls on mica dissolution in buffered basic solution (pH 9.4; Na_2BO_4 and NaOH) and high temperature (155°C). Rufe and Hochella (1999) studied the alteration of phlogopite in HCl-solutions (pH 2) and distilled water (pH 5.7) with AFM in situ experiments at room temperature. Chlorite dissolution was investigated by Brandt et al. (2003) who performed microscopic experiments on chlorite basal surfaces.

Using the VSI technique, it is possible to quantitatively map the variations in clay mineral surface topography at the subnanometer- and micrometer-scale resolutions (vertical and lateral resolution, respectively). Kurganskaya et al. (2012) carried out ex situ VSI measurements of reacted muscovite at pH 9.4, 155°C, and far from equilibrium, obtaining rates of normal surface retreat and local rates at deep etch pits. Satoh et al. (2013) carried out in situ measurements to quantify montmorillonite dissolution rates at a compaction pressure ranging from 0.04 to 10 MPa and temperature of 70°C in 0.3 M NaOH solution (pH 12.1 at 70°C) using VSI and an auto-compaction cell.

The LCM-DIM technique allows in situ surface exploration with vertical nanometric accuracy at different temperatures (Cappelli et al., 2013). Although unambiguous quantitative information on the vertical surface topography is unavailable with it, horizontal quantification is possible, and measurements of step edge retreat with time can be performed. This type of observation benefits from the advantage that broad surface regions can be scanned for long time periods and allows the visualization of the mechanisms on clay minerals' surface areas with variable roughness and topography.

4.4 KAOLINITE

The overall dissolution reaction of kaolinite under acidic and basic pH conditions can be respectively expressed as

$$Al_2Si_2O_5(OH)_4 + 6H^+ \leftrightarrow 2Al^{3+} + 2H_4SiO_4 + H_2O$$

and

$$Al_2Si_2O_5(OH)_4 + 5H_2O + 2OH^- \leftrightarrow 2Al(OH)_4^- + 2H_4SiO_4$$

Several studies (e.g., Nagy et al., 1991; Ganor et al., 1995; Devidal et al., 1997) showed that steady-state kaolinite dissolution was stoichiometric with respect to Al/Si ratio both in acid and alkaline solutions. To assess the possible initial incongruency of kaolinite dissolution reported by Carroll and Walther (1990), Devidal et al. (1997) performed X-ray photoelectron spectroscopy (XPS) surface composition analyses. The XPS kaolinite surface composition examined after acid and basic dissolution was found to be depleted in Al relative to the initial mineral, suggesting an initial preferential Al release. These data, together with the stoichiometric nature of steady-state dissolution, imply that the slightly depleted Al surface persists, following its initial development, throughout the duration of the experiments. This initial Al preferential release may be related to the formation of a rate controlling precursor complex.

Also, the pH of the solutions used to pretreat the starting kaolinite samples may change the relative concentration of Al and Si sites on the surface toward the stoichiometric ratio (Walther, 1996).

4.4.1 pH and T Dependence

Ganor et al. (1995) evaluated the combined effect of pH and temperature on far-from-equilibrium kaolinite dissolution rates (temperatures of 25, 50, and 80°C and in the pH range 2–4.2). Far from equilibrium conditions, Eqn (4.6) became

$$Rate = 10^{-5.7\pm0.7} \cdot e^{-7.0\pm1.1/RT} \cdot a_{H^+}^{0.40\pm0.14} \tag{4.14}$$

The resulting apparent activation energy was 7.0 ± 1.1 kcal mol^{-1} and the reaction order was 0.40 ± 0.14. This pH dependence is similar to the pH dependence of the proton density on the kaolinite surface $\chi_{ads,\,H^+}^{ns} = \alpha a_{H^+}^{0.31}$ (Wieland and Stumm, 1992) where n_S is the reaction order with respect to H^+ on the surface. It follows that the rate of dissolution is proportional to $\chi_{ads,\,H^+}^{ns}$ and n_S is 1.3 ± 0.5 (i.e., 0.40/0.31). This linear dependence of the kaolinite dissolution rate on adsorbed concentrations of protons in acidic solution was explained by a sequential reaction mechanism involving the serial rupture of Al-O-Si linkages in the presence of an adsorbed H^+ ion.

Bauer and Berger (1998) studied kaolinite dissolution using batch experiments in high molar (0.1–4 M) KOH solutions at 35°C and 80°C. The kaolinite dissolution was characterized by a linear release of silica and Al as a function of the log of time. This relationship can be explained by a reaction affinity effect which is controlled by the octahedral layer dissolution. Under far-from-equilibrium conditions, the rates were proportional to $a_{OH^-}^{0.56\pm0.12}$ at 35°C and to $a_{OH^-}^{0.81\pm0.12}$ at 80°C. The activation energy of kaolinite dissolution increased from 33 ± 8 kJ mol^{-1} in 0.1 M KOH solutions to 51 ± 8 kJ mol^{-1} in 3M.

Huertas et al. (1999) postulated a general rate law to account for the kaolinite dissolution rate-pH dependence at 25°C based on the coordination theory (Stumm, 1992). The kinetic data were combined with surface speciation data by Huertas et al. (1998). It was possible to express the dissolution rate using a simple power function of the concentration of the surface sites active in various pH ranges.

$$R = 10^{-8.25}\left[>Al_2OH_2^+\right] + 10^{-10.82}\left[>AlOH_2^+\right]^{0.5}$$
$$+10^{-9.1}\left[>Al_2OH + >AlOH + >SiOH\right] + 10^{3.78}\left[>Al_2O^- + >AlO^-\right]^3 \tag{4.15}$$

This model based on the proposed surface speciation of kaolinite (Figure 4.2) shows that the dissolution kinetics of this mineral is controlled by aluminum surface complexes, without direct contribution of the $>SiO^-$ sites, stressing the important role played by the gibbsite basal plane (octahedral sheet) and edges. Under acidic conditions the dissolution mechanism is governed by the presence of the strong ($>Al_2OH_2^+$) and weak ($>AlOH_2^+$) acid Al sites.

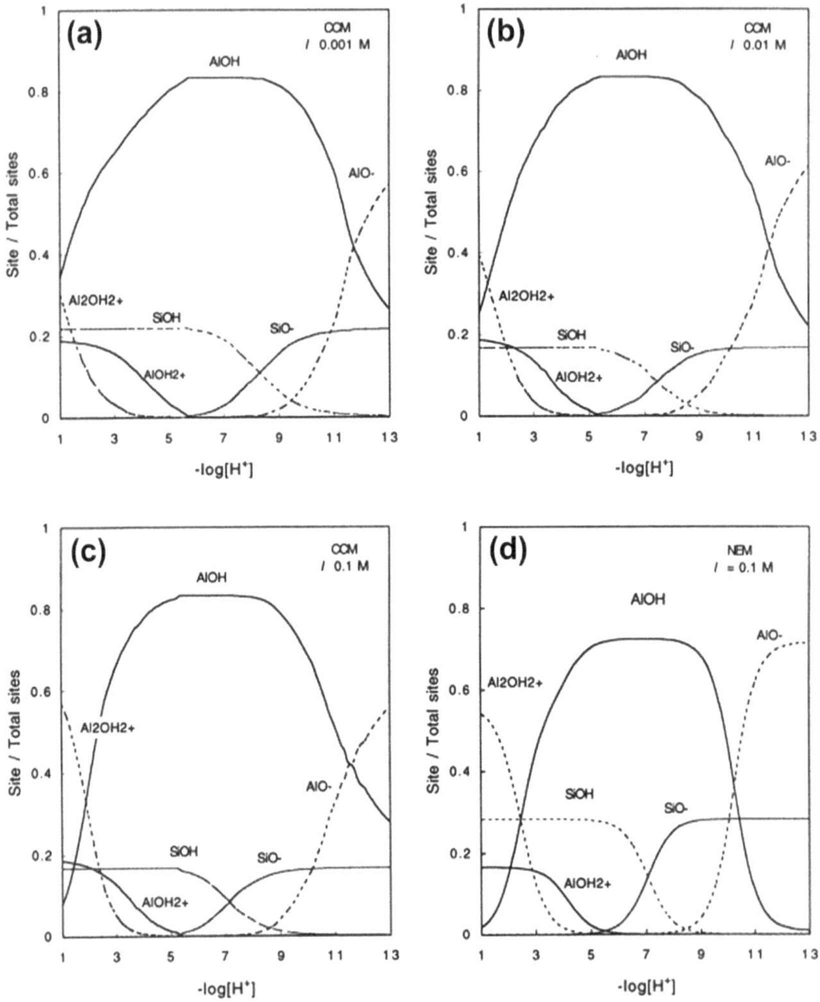

FIGURE 4.2 Predicted distribution of the surface species of kaolinite as a function of pH for various ionic strengths according to the constant capacitance model (CCM) (a–c). The results obtained with the nonelectrostatic model (NEM) for *I* 0.1 M (d) are shown for comparison. Site density is expressed in terms of molar fraction of the total surface density. *Modified from Huertas et al. (1998).*

Following the mechanism that was proposed by Huertas et al. (1999) for the proton-promoted reaction of kaolinite at 25°C, Cama et al. (2002) proposed a mechanism for the combined effect of pH and temperature on kaolinite dissolution rate, under far from equilibrium. The first of the two independent proton-promoted reaction paths controls the overall dissolution rate at pH $\geq$ 2.5, whereas the second path controls it below pH 0.5. Between pH 0.5 and 2.5 the two reaction paths influence the rate. Using this model the effects of pH and temperature on the overall dissolution rate of kaolinite under acidic condition can be described by

$$\begin{aligned} Rate = {} & 2\cdot 10^{2}\cdot e^{-22/RT}\cdot\frac{2\cdot 10^{-10}\cdot e^{19/RT}\cdot a_{H^+}}{1+2\cdot 10^{-10}\cdot e^{19/RT}\cdot a_{H^+}} \\ & + 5\cdot 10^{7}\cdot e^{-28/RT}\frac{1.4\cdot 10^{-7}\cdot e^{10/RT}\cdot a_{H^+}}{1+1.4\cdot 10^{-7}\cdot e^{10/RT}\cdot a_{H^+}} \end{aligned} \tag{4.16}$$

where R is the gas constant, T is the temperature (K), and a_{H^+} is the activity of protons in solution. The results of the proposed model show that by proper separation of variables, the kaolinite dissolution rate may be described by a rate law using activation energies that are independent of pH.

4.4.2 Effect of Deviation from Equilibrium (ΔG_r Effect)

The dependence of the dissolution rate on the deviation from equilibrium, $f(\Delta G_r)$, may take a variety of forms (Lasaga, 1998). Thus, an empirical $f(\Delta G_r)$ function for the dissolution of kaolinite at 80°C and pH = 3 was derived by Nagy et al. (1991) to be

$$f(\Delta G_r) = -\left(1 - e^{\left(\frac{\Delta G_r}{RT}\right)^{0.9\pm 0.02}}\right) \tag{4.17}$$

Therefore, at constant temperature and surface area, the dissolution reaction rate will be independent of ΔG_r far from equilibrium (large negative ΔG_r). This region in rate versus ΔG_r space where the function $f(\Delta G_r)$ is flat (Figure 4.1(e)) is termed the dissolution plateau (Lasaga, 1998).

Devidal et al. (1997) determined steady-state dissolution and precipitation rates of kaolinite as a function of chemical affinity ($A = -\Delta G_r$, the Gibbs free energy of the overall reaction) and aqueous Si and Al concentrations using a mixed flow reactor at 150°C, 40 bars, and a pH of 2, 6.8, and 7.8, and proposed the following rate law:

$$Rate = k_+\left[\frac{\frac{K^* a_{H^+}^3}{a_{Al^{3+}}}}{1+\frac{K^* a_{H^+}^3}{a_{Al^{3+}}}}\right]\left[1-(A)^{0.5}\right] \tag{4.18}$$

where a_{H^+} and $a_{Al^{3+}}$ represent the proton and Al^{3+} activities, respectively, and $K^* = 3.83$ and $k_+ = 4.2\times 10^{-13}$ mol cm^{-2} s^{-1}. A more complex rate law which includes the $Al(OH)_4^-$ and H_4SiO_4 activities was postulated to describe kaolinite dissolution and precipitation rates in alkaline solutions. For chemical affinities (A) ranging between 2 and 10 kcal mol^{-1}, measured dissolution rates are an inverse function of aqueous Al concentration at acidic pH, and of aqueous Al and silica concentrations at acidic and alkaline pH, respectively. As equilibrium is approached ($-2 < A < 2$ kcal mol^{-1}), these rates become increasingly controlled by chemical affinity. The variation of dissolution/crystallization rates with chemical affinity and aqueous Al and Si concentrations are described using a coupled TST/Langmuir adsorption model by adopting the concept that reaction rate is controlled by the decomposition of a silica-rich/aluminum-deficient precursor complex. This complex is formed by the exchange of three hydrogen ions for one aluminum atom at the kaolinite surface. At alkaline conditions, this reaction competes with precursor complex condensation and/or aqueous silica adsorption forming unreactive siloxane groups.

Yang and Steefel (2008) proposed a different $f(\Delta G_r)$, Eqn (4.10) to that of Nagy et al. (1991) and Devidal et al. (1997) to relate the kaolinite dissolution rate at 22°C and the reaction affinity by a TST rate formulation with a Temkin coefficient of 2

$$f(\Delta G_r) = \left(1 - \exp\left(\frac{-\Delta G_r}{2RT}\right)\right) \tag{4.19}$$

yielding

$$R_{diss}\ \left(\text{mol m}^{-2}\ \text{s}^{-1}\right) = 1.15 \times 10^{-13} \cdot \left(1 - \exp\left(\frac{-\Delta G_r}{2RT}\right)\right) \tag{4.20}$$

This implies that the activated complexes involved in the rate-limiting step for kaolinite dissolution involved a single Al or Si atom. In contrast to Devidal et al. (1997), dissolution with input solutions having high Al concentration showed no evidence for significant Al inhibition under the experimental conditions of their study. Yang and Steefel (2008) argued that extending the stepwave dissolution model presented by Lasaga and Luttge (2001, 2003) from minerals having relative well-developed framework crystal structures (e.g., calcite, feldspar, or quartz) to layered silicates is problematic, since dissolution may occur primarily along the edges of the layers. In addition, an attempt to interpret the kaolinite kinetic data collected at 22°C (low temperature) with the Dove et al. (2005) proposed models based on crystal growth theory was made. Neither produced a satisfactory fit, suggesting that direct extrapolation of these rate models to lower temperature may be unjustified.

4.4.3 Catalytic/Inhibition Effect

Dissolution of kaolinite at $pH = 2$ is specifically inhibited by Al at 150°C (Oelkers et al., 1994; Devidal et al., 1997). These authors demonstrated that this aluminum inhibition is independent of the degree of saturation and occurs under far-from-equilibrium conditions, and therefore is not an effect of deviation from equilibrium. The reaction order of this inhibition was −1, and thus $Rate = k \cdot a_{Al^{3+}}^{-1}$. This rate law is not supported by a mechanistic model.

Ganor and Lasaga (1998) presented mechanistic models describing the possible effect of inhibition on mineral dissolution rate. Under acidic conditions the kaolinite dissolution reaction is proton-promoted, and the Al inhibition may be a result of Al adsorption near or at the same site in which the protons are adsorbed. The predictions of the proposed competition model and the independent adsorption model of Ganor and Lasaga (1998) given by

$$Rate = 0.021 \cdot \frac{9.7 \cdot 10^{-9} \cdot a_{H^+}}{1 + 9.7 \cdot 10^{-9} \cdot a_{H^+} + 620000 \cdot c_{Al}} \quad (4.21)$$

and

$$Rate = 0.02 \cdot \left(1 - 1 \cdot \frac{4 \cdot 10^5 \cdot c_{Al}}{1 + 4 \cdot 10^5 \cdot c_{Al}}\right) \times \frac{7.6 \cdot 10^{-9} \cdot a_{H^+}}{1 + 7.6 \cdot 10^{-9} \cdot a_{OH^-}} \quad (4.22)$$

where a_{H^+} and c_{Al} are the proton activity and the total aluminum concentration proved to be valid on the examined kaolinite experimental data by Oelkers et al. (1994). Ganor et al. (1995) argued that according to the agreement of their data with results of previous studies with very different Al concentrations (Carroll and Walther, 1990; Wieland and Stumm, 1992), there are reasons to believe that the dissolution rate of kaolinite at low temperatures (<80°C) is not influenced by Al inhibition similar to the inhibition that was observed by Oelkers et al. (1994) at 150°C. However, Ward and Brady (1998) showed that Al appears to exchange 1:3 for protons at less than neutral pH and 1:4 with OH^- at high pH.

4.4.4 Ionic Strength Effect

Devidal et al. (1997) investigated the ionic strength effect on the far-from-equilibrium dissolution rates at 150°C and pH = 7.8 in NH_4OH and NH_4OH/NH_4Cl solutions. Kaolinite dissolution rates were strongly dependent on ionic strength. The rates measured in the dissolution plateau region were increased by a factor of ~5 (from 7.6×10^{-13} to 4×10^{-12} mol cm^{-2} s^{-1}) when the ionic strength was increased from 2×10^{-4} to 2.5×10^{-2} M.

4.4.5 Surface Area Effect

Reactive surface area of kaolinite has been investigated exhaustively. Xie and Walther (1992) presented an analysis of the surface site density of kaolinite from surface potentiometric titration data, indicating that the surface area of kaolinite and possibly other clay minerals available for forming surface complexes and sites for cation detachment is significantly larger than that measured by standard gas adsorption methods. Hence, it is believed that H^+ and OH^- can penetrate between the layers of the clay minerals which is not possible for the gas, Kr, used in the BET analysis.

Metz and Ganor (2001) carried out flow-through experiments at pHs 2–4, different stirring speeds, and temperatures of 25°C, 50°C, and 70°C to examine kaolinite dissolution rate. Although kaolinite dissolution is surface-controlled and not diffusion-controlled at this low-temperature range, the calculated dissolution rate was enhanced by stirring. The stirring effect increased by decreasing temperature and by increasing pH. Very fine particles formed as a result of spalling-off or abrasion of the kaolinite, having a very high ratio of reactive surface area to specific surface area and therefore the dissolution rate increased with stirring. As a result, kinetic factors obtained under stirred conditions will be smaller than those obtained under nonstirred conditions. The authors recommended the standard use of non-stirred reaction vessels for kinetic experiments of mineral dissolution and precipitation, at least for slow reactions that are surface-controlled.

4.5 SMECTITE

The dissolution kinetics of dioctahedral smectites (montmorillonite, beidellite and nontronite) and trioctahedral smectites (hectorite and saponite) has been extensively studied. The overall dissolution reactions of any smectite at acidic conditions may be expressed as (Metz et al., 2005a)

$$
\begin{aligned}
&(K_aNa_bCa_cMg_d)\left[Mg_xFe_yAl_z\right]^{VI}[Al_wSi_{8-w}]^{IV}O_{20}(OH)_4\cdot(H_2O)_n \\
&\quad + (12+4w)H^+ + (8-4w)H_2O) \\
&\quad \leftrightarrow aK^+ + bNa^+ + cCa^{2+} + (d+x)Mg^{2+} + yFe^{3+} \\
&\quad + (z+w)Al^{3+} + (8-w)H_4SiO_4 + nH_2O
\end{aligned}
$$

where the values of m, n, a, b, c, d, w, x, y, and z of each sample are derived from its structural formula.

4.5.1 pH and T Dependence

4.5.1.1 Acid pH

Zysset and Schindler (1996) studied the kinetics of the proton-promoted dissolution of K^+-montmorillonite at 25°C by batch experiments in the pH range of 1–5 in solutions of constant KCl concentrations of 0.03, 0.10, and 1.0 M, respectively. The dissolution rates R_{Si} and R_{Al} (mol g^{-1} h^{-1}) were obtained from the observed increase with time of both dissolved Si (IV) and Al(III). R_{Si} was found to be linearly dependent on $[H^+]$ (i.e., $R_{Si} = k\cdot[H^+]$), and adsorption of H^+ was interpreted in terms of the surface complexation model. Proton uptake by the solid phase was formally attributed to H^+-K^+ exchange and to protonation of edge surface aluminol groups. In the range $1 < pH < 5$ the dissolution of K^+-montmorillonite is nearly congruent. Deviation from congruency was observed at low KCl concentration, being attributed to adsorption of dissolved Al (III) by cation exchanger sites. The contribution of basal planes was negligible. The dissolution process was preponderantly located at particle edges.

Amram and Ganor (2005) proposed a rate law describing the combined effect of pH (1–4.5) and temperature (25, 50, and 70°C) on smectite dissolution rate, under far-from-equilibrium conditions using nonstirred flow-through reactors. Smectite dissolution rates were obtained based on the release of silicon and aluminum at steady state. Low Al/Si ratios, obtained in experiments conducted at $pH \geq 4$, are explained by precipitation of gibbsite and/or diaspore. Dissolution rate increased with temperature and decreased with increasing pH. Dissolution rates of experiments in which $\Delta G_r \leq -21\ \text{kcal mol}^{-1}$, were not affected by deviation from equilibrium. A simple model based on that proposed by Cama et al. (2002) was used to describe the combined effect of pH and temperature on smectite dissolution rate. In the model of Amram and Ganor (2005) the dissolution rate was linearly proportional to the concentration of adsorbed protons on the mineral surface, and proton adsorption was described using a Langmuir adsorption isotherm. The resulting rate law was

$$Rate = 220 \cdot e^{-17460/RT} \cdot \frac{3 \cdot 10^{-6} \cdot e^{10700/RT} \cdot a_{H^+}}{1 + 3 \cdot 10^{-6} \cdot e^{10700/RT} \cdot a_{H^+}} \tag{4.23}$$

where a_{H^+} is the activity of protons in solution. According to the model, the dependence of dissolution rate on temperature is affected by the activation energy ($17 \pm 2\ \text{kcal mol}^{-1}$) and the adsorption enthalpy ($-11 \pm 2\ \text{kcal mol}^{-1}$). The results of this study neither supported nor contradicted the suggestion of Zysset and Schindler (1996) that the active site that controls the rate under acidic conditions is Al-O-Si or Al-OH-Al group.

Metz et al. (2005a) studied the dissolution stoichiometry of smectite-rich bentonites SAz-1, STx-1, and SWy-1 at 50°C and pH 2 and 3 using flow-through reactors. These rich bentonite samples contained considerable amounts of silica phases (quartz, cristobalite, and/or amorphous silica). As a result, the molar Al/Si ratios of the bulk samples are significantly lower than those of the pure smectite. The change in the release rate of the different elements during the dissolution of the smectite samples was initially non-stoichiometric. During this nonstoichiometric stage that lasted from few hundreds to few thousand hours, there was an ion exchange reaction between the solution and the interlayer sites, and a significant release of excess Si due to the fast dissolution of accessory silica. The initial dissolution rate of the smectite itself was slow, being this attributed to coating or cementation of the smectite aggregates by amorphous silica. When the release of Si from smectite dominated that of accessory silica, the elemental release becomes stoichiometric and the Al/Si ratio in solution approached a constant value.

Golubev et al. (2006) obtained dissolution rates of Na–Montmorillonite (Wyoming) SWy-2 smectite at pH from 1 to 12 at 25°C using mixed-flow reactors equipped with interior dialysis compartment. The Si-based rates decreased with increasing pH at $1 \leq pH \leq 8$ with a slope close to -0.2 and increased at $9 \leq \text{pH} \leq 12$ with a slope of ≈ 0.3. In contrast, nonstoichiometric Mg release weakly depended on pH at $1 \leq \text{pH} \leq 12$ and decreased with increasing pH. The empirical expression describing Si-release rates obtained at $I = 0.01$ M was given by

$$Rate\left(\text{mol cm}^{-2}\,\text{s}^{-1}\right) = 2.2 \cdot 10^{-17} \cdot a_{\text{H}^+}^{0.21} + 1.0 \cdot 10^{-20} + 6 \cdot 10^{-17} \cdot a_{\text{OH}^-}^{0.33} \quad (4.24)$$

It was observed an invariability of Ca release versus pH which implied that interlayer Ca^{2+} cation exchange was pH independent at $2 \leq \text{pH} \leq 8$.

Likewise, Rozalén et al. (2008, 2009) studied the effect of pH on the kinetics of smectite (K^+-montmorillonite) dissolution over the pH range of 1−13.5 (KNO_3 solutions) at 25, 50, and 70°C in batch and stirred flow-through reactors. The dissolution rates were obtained based on the release of Si and Al at steady state under far-from-equilibrium conditions. Dissolution was nonstoichiometric between pH 5 and 10, due to adsorption/precipitation of Al. Sample pretreatment and the nature of the interlayer cation did not affect the steady-state dissolution rate or stoichiometry of cation release. The rate dependences on pH at $I = 0.01$ M and 25, 50, and 70°C were described by

$$R_{25^\circ\text{C}}\left(\text{mol m}^{-2}\,\text{s}^{-1}\right) = 10^{-12.30} \cdot a_{\text{H}^+}^{0.40} + 10^{-14.37} + 10^{-13.5} \cdot a_{\text{OH}^-}^{0.27} \quad (4.25)$$

$$R_{50^\circ\text{C}}\left(\text{mol m}^{-2}\,\text{s}^{-1}\right) = 10^{-11.04} \cdot a_{\text{H}^+}^{0.59} + 10^{-14.20} + 10^{-11.98} \cdot a_{\text{OH}^-}^{0.36} \quad (4.26)$$

$$R_{70^\circ\text{C}}\left(\text{mol m}^{-2}\,\text{s}^{-1}\right) = 10^{-10.30} \cdot a_{\text{H}^+}^{0.75} + 10^{-14.00} + 10^{-11.15} \cdot a_{\text{OH}^-}^{0.42} \quad (4.27)$$

At high and low pH dissolution rates become more pH-dependent with increasing temperature. At acid pH the dissolution rate of montmorillonite is probably controlled by positively charged aluminol sites ($>AlOH_2^+$), and cation exchange reactions can only contribute to the dissolution at very acid pH. At basic pH dissolution rates may be controlled by negatively charged aluminol groups ($>AlO^-$). It was observed that ionic strength has a minor effect on surface charge.

4.5.1.2 Basic pH

Bauer and Berger (1998) studied the dissolution of two industrial smectites (one was pure montmorillonite and the other had a significant component of beidellitic charge) using batch experiments in high molar (0.1−4 M) KOH solutions at 35°C and 80°C. The two smectites dissolved independently of the aqueous silica or Al concentrations. At far-from-equilibrium conditions, the rates were proportional to $a_{\text{OH}^-}^{0.15\pm0.06}$ at 35°C and 80°C. The activation energy of smectite dissolution appeared to be pH-independent (52 ± 4 kJ mol^{-1}). The authors showed that the rate-limiting step for smectite dissolution was dissolution of the tetrahedral sheet, in contrast to dissolution of the octahedral sheet for kaolinite dissolution.

Nakayama et al. (2004) obtained dissolution rates of montmorillonite in compacted sand–bentonite mixture specimens under highly alkaline, simulated groundwater conditions. Under the employed experimental conditions of pH 13–14 and temperatures of 90–170°C, dissolution of montmorillonite was described by linear dependence on time

$$\mathrm{W}_{(t)} = \mathrm{W}_{(0)} - R_A t, \tag{4.28}$$

where $W_{(t)}$ denotes the density of montmorillonite (milligram of montmorillonite per meter cube of sand–bentonite mixture), $W_{(0)}$ the initial density, R_A (Mg m^{-3} s^{-1}) the rate of density decrease, and t(s) the time after the contact with the simulated groundwater. R_A is a function of pH and temperature (K), and expressed as $R_A = (0.013 \pm 0.007)\exp[-(3.7 \pm 0.2) \times 10^4/\mathrm{RT}]$ for the "pH 14.0" simulated groundwater, where R is the gas constant.

Ueta et al. (2013) investigated the effect of alkaline pH on the dissolution rate of montmorillonite aggregates at 23°C using an enhanced PSI technique combined with an internal refraction interferometry method. Molecular resolution was provided during the optical observation of the dissolution phenomena in real time and in situ. Normal dissolution velocities as low as 1×10^{-4} to 1×10^{-3} nm s^{-1} were obtained. The molar dissolution fluxes of montmorillonite obtained gave considerably faster dissolution rates in comparison to those obtained in previous investigations by solution analysis methods. The pH dependence of montmorillonite dissolution rate was qualitatively in good agreement with those reported in previous investigations.

4.5.2 Effect of Deviation from Equilibrium (ΔG_r Effect)

Cama et al. (2000) observed stoichiometric dissolution of smectite at 80°C and pH 8.8. Pretreatment of the smectite surfaces was necessary to obtain reliable and stoichiometric kinetic results. The dissolution rate of the sample reflected the dissolution rate of the montmorillonitic sheets. Under these experimental conditions smectite dissolution rate was not inhibited by aluminum. The dissolution rate of smectite decreased as a function of silicon concentration. This observation may be explained both by the effect of deviation from equilibrium on dissolution rate and by silicon inhibition. The former, may be expressed as

$$\frac{Rate}{A_{\min}}\left(\mathrm{mol\ m^{-2}\ s^{-1}}\right) = k \cdot \left(1 - \exp\left(m \cdot \left(\frac{\Delta G_r}{RT}\right)^n\right)\right) \tag{4.29}$$

where the coefficients k, m, and n were calculated from a nonlinear regression using least squares ($k = -8.1 \times 10^{-12}$ mol m^{-2} s^{-1}, $m = -6 \times 10^{-10}$, and $n = 6$). Cama et al. (2000) emphasized that different combinations of the three coefficients yield other curves that adequately describe the experimental data. The possible effect of silicon inhibition may be expressed as

$$\frac{Rate}{A_{\min}}\left(\text{mol m}^{-2}\text{ s}^{-1}\right) = \left(\frac{3.7\times10^{-17}}{c_{\text{Si}}}\right) \tag{4.30}$$

The authors claimed that the obtained data set could not be used to differentiate between these two possible reaction mechanisms. Marty et al. (2011) investigated the effect of pH and Gibbs energy on the dissolution rate of a synthetic Na^+-montmorillonite using flow-through experiments at 25 and 80°C and pH 7 and 9. The dissolution reaction took place stoichiometrically at 80°C, whereas at 25°C preferential release of Mg over Si and Al was observed. The TEM-EDX analyses (transmission electronic microscopy with quantitative chemical analysis) of the solid phase after the experiment at 25°C showed the presence of newly formed Si-rich phases, which accounted for the Si deficit. Depletion of Si concentration was attributed to incongruent clay mineral dissolution with the formation of detached Si tetrahedral sheets (i.e., alteration product), whereas the Al behavior remains uncertain (e.g., possible incorporation into Al-rich phases). Based on the Mg release at steady state, and normalizing the rates to the initial edge surface area (11.2 $m^2\ g^{-1}$), a dissolution rate law for the dissolution reaction as a function of ΔG_r at 25°C and pH $\approx$ 9, similar to that used in Cama et al. (2000) was proposed

$$Rate\left(\text{mol m}^{-2}\text{ s}^{-1}\right) = 3.82\times10^{-12}\left[1-\exp\left(-8.23\times10^{-10}\left|\frac{\Delta G_r}{RT}\right|^{5.47}\right)\right] \tag{4.31}$$

The adjustable parameters are not unique and different expressions could be used. This nonlinear rate law is similar to other nonlinear expressions that are used to express the effect of solution saturation state on clay mineral dissolution rates. Figure 4.3 shows the variation of the dissolution rates as function of Gibbs free energy ($f(\Delta G_r$-rate dependence) obtained by Cama et al. (2000) and Marty et al. (2011). The dissolution plateau was not unambiguously reached even at $\Delta G_r \approx -100$ kJ mol^{-1}. It is hard to obtain rates at lower ΔG_r values because the output concentrations are well below detection limit (Rozalén et al., 2008). Hence, the obtainment of far-from-equilibrium rates (dissolution plateau) at near neutral pH is experimentally limited.

4.5.3 Catalytic/Inhibition Effect

Al^{3+} inhibits the K^+-montmorillonite dissolution reaction at pH $\leq$ 3 (Zysset and Schidnler, 1996). It was suggested that this inhibition originated from adsorption of Al^{3+} on particle edge surface sites.

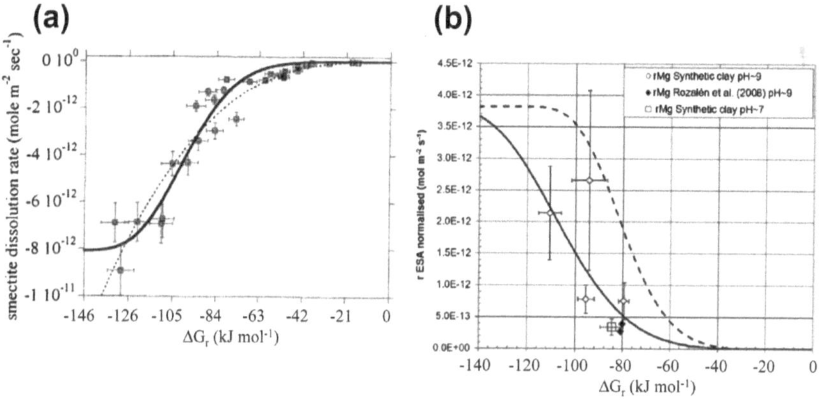

FIGURE 4.3 Effect of the degree of saturation on smectite dissolution rate: (a) the solid and the dashed lines are two possible fittings of the $f(\Delta G_r)$ function to the BET-normalized dissolution rates at pH ≈ 8.8 and 80°C (*Modified from Cama et al., 2000*); (b) Variation of edge surface area (ESA) dissolution rates as function of $f(\Delta G_r)$ at pH ≈ 9 and 25°C. Black curve fits experimental data as considered by Marty et al. (2011) and dashed curve uses deviation from equilibrium proposed by Cama et al. (2000). *Modified from Marty et al. (2011).*

Metz (2001) examined the effects of silicon and aluminum on smectite dissolution rate at 50°C: under far-from-equilibrium conditions ($\Delta G_r \leq -30$ kcal mol^{-1}) smectite dissolution rate was independent both of Al concentration (3 < [Al] < 16 μM at pH 3 and 21 < [Al] <139 μM at pH 2) and of Si concentration (9 < [Si] < 41 μM at pH 3 and 56 < [Si] < 329 μM at pH 2). In contrast to the observations of Furrer et al. (1993) and Zysset and Schindler (1996), Metz (2001) found that far-from-equilibrium smectite dissolution rate is independent of both Al and Si concentrations. Likewise, Amram and Ganor (2005) suggested that the changes in far-from-equilibrium dissolution rates were not significantly influenced by the variability in Al and Si concentration. It is important to note that the present study ranges of Al (0.05−843 μM) and Si (2.8−2387 μM) concentrations were significantly larger than the ranges which were thoroughly studied by Metz (2001).

Van Rompey et al. (2002) studied the effect of acid type (such as HCl, HNO_3, and H_2SO_4 with $0 \leq pH \leq 0.86$) and concentration on the dissolution rate of hectorite using stirred batch experiments. As expected, the rate of dissolution increased with increasing acid concentration. From the halftimes of dissolution, the dissolution curves, and the XRD data, it could be concluded that the dissolution rate of hectorite decreased in the order $H_2SO_4 > HNO_3 \geq HCl$ at the same molar concentration.

Golubev et al. (2006) studied the effect of dissolved carbonate (0.5−10 mM), phosphate (10^{-5} to 0.03 M), and nine organic ligands (acetate, oxalate, citrate, EDTA, alginate, glucuronic acid, 3,4-dihydroxybenzoic acid, gluconate, and glucosamine) on the dissolution rates of SWy-2 smectite at circumneutral pH and 25°C. The Si-release-based dissolution was promoted by the addition of the following ligands ranked by decreasing effectiveness: EDTA > 3,4-DHBA > citrate ≥ oxalate. In contrast, phosphate, glucuronate, glucosamine, gluconate, alginate, and acetate inhibit the dissolution reaction, while HCO_3^- and CO_3^{2-} exhibited no effect on dissolution rate. Non-stoichiometric, non-steady-state Mg release was very weakly affected by the presence of ligands. It was proposed that the effect of ligands on smectite dissolution could be modeled conceptually as surface adsorption assuming the overall dissolution rate is the sum of all the different parallel dissolution reactions promoted at active centers (presumably, Mg–O–Si and/or Al–O–Si bonds) by various reactants/ligands which compete for available surface sites (Stumm, 1992). Accordingly, the following expression was used to match the experimental results

$$Rate = k_{\mathrm{Si}}^{\#}\cdot\left(1-\frac{K_{S-L}^{*}\cdot\left[L^{n-}\right]}{1+K_{S-L}^{*}\cdot\left[L^{n-}\right]}\right)+k_{L}^{\#}\cdot\frac{K_{S-L}^{*}\cdot\left[L^{n-}\right]}{1+K_{S-L}^{*}\cdot\left[L^{n-}\right]} \qquad (4.32)$$

where $k_i^{\#} = k_i \cdot S_T$ with $i = \mathrm{Si}$, L, and k_L are the empirical kinetic constant pertinent to each ligand. The solid curves (Figure 4.4) computed with Eqn (4.32) closely match the experimental data for a very broad range of a ligand's aqueous concentration, demonstrating its validity. It was concluded that the consequence of the relatively weak effect of bacterial cell components and microbial exometabolites on smectite reactivity in aqueous solutions was that the impact of plants and biota on mineral chemical weathering in soils may be weaker than generally argued for clay minerals. As a result, atmospheric CO_2 consumption on the land, due to direct dissolution of soil clay minerals and Ca, Mg-associated transport of HCO_3^-, is unlikely to be strongly influenced by organic ligands present in soil solutions.

4.5.4 Ionic Strength Effect

Amram and Ganor (2005) assessed the effects of ionic strength and Na and NO_3 concentrations on the determination of the observed pH dependence of smectite dissolution rate. The ionic strength ranged from 0.000032 M (pH 4.5) to 0.1 M (pH 1) in some experiments. In the rest of the experiments the ionic strength and NO_3^- concentration was maintained at about 0.32 M by adding suitable amounts of $NaNO_3$ into the input solutions. However, owing to the experimental noise, it was hard either to prove that the small differences observed between the experiments represented a real effect of the addition of salt, nor to rule out this possibility.

4.5.5 Surface Area Effect

Metz et al. (2005b) examined the usability of the BET surface area and the AFM edge surface area as proxies for the reactive surface area of dissolving smectite. The dissolution rate of smectite is considered to be controlled by the

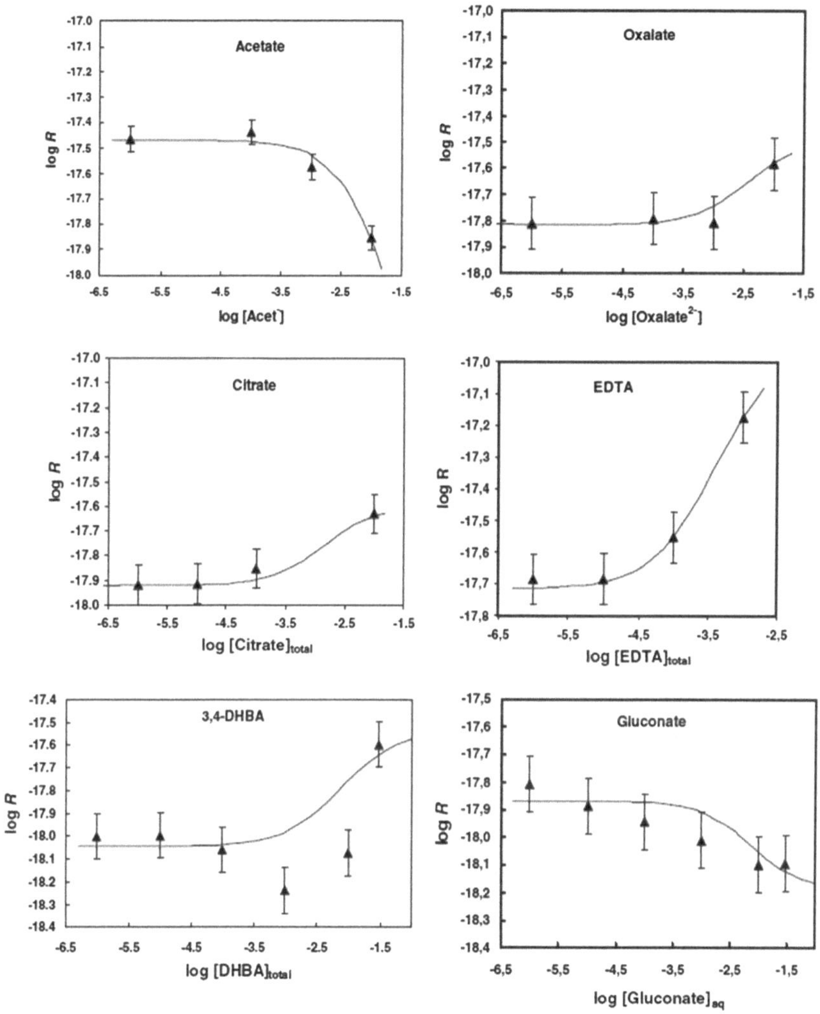

FIGURE 4.4 Effect of added ligands on smectite Si-based steady-state dissolution rates at circumneutral pH. Triangles represent Si-release-based rates and the solid lines represent the fit according to the model by Golubev et al. (2006). *Modified from Golubev et al. (2006).*

chemical attack on the edge surface, and hence the rate should be normalized to its edge surface area. It may be normalized to the external surface area if the external surface area is proportional to the edge surface area. Since a smectite is very microporous, its BET surface area is not an accurate estimate of the external surface area (see Chapter 1, in this Volume). In addition, exchange reactions between interlayer cations and protons, which may change the interlayer expansion, affect the extent by which gases penetrate into the micropores, and as a result influence the BET surface area of the smectite.

To use the BET surface area as a proxy for the reactive surface area of smectite, one should assume that the reactive surface area is proportional to the external surface area. The specific external surface area of a layered particle depends linearly on the height reciprocal. In contrast, the specific edge surface area is independent of the particle's height and depends linearly on the sum of the reciprocals of the length of the axes. Therefore, Metz et al. (2005b) argued that there is no reason to expect a correlation between the total and the edge surface areas, and hence, the BET surface area of smectite might not serve as a proxy for its reactive surface area. This conclusion may be extrapolated to clay minerals which form layered particles. Although in several studies, the dissolution rate of smectite was normalized by the sample mass (Zysset and Schindler, 1996; Amram and Ganor, 2005), in general, the usage of the mass as a normalization factor cannot be justified. However, in the absence of a better proxy, it is better to normalize the dissolution rate of smectite by its mass than normalizing it using the BET surface area or not normalizing it at all. Metz et al. (2005b) suggested to use AFM measurements of the specific edge surface area as an alternative proxy for the reactive surface area of smectite.

Yokoyama et al. (2005) obtained dissolution rates from the edge surface area by investigating the montmorillonite dissolution under highly alkaline conditions (pH $= 13.3$; $I = 0.3$ M) at 30, 50, and 70°C. The AFM observations indicated that dissolution occurred at edge surfaces. Although etch pits formed on the basal surface of a few particles, they did not appear to be the dominant dissolution sites. This suggests that the reduction in particle volume was due mainly to dissolution along edge surfaces. Therefore, the edge surface area should be used to calculate the dissolution rate for montmorillonite under alkaline conditions. Dissolution rates of individual particles with different morphologies estimated by AFM were similar to rates estimated from bulk dissolution experiments.

By means of in situ AFM experiments at 25°C and alkaline conditions, Kuwahara (2006) showed that smectite particles dissolved via the retreat of the edge surfaces, and the basal surface was unreactive within the experimental duration. Moreover, the anisotropic dissolution behavior was also observed along the edge surface area, that is, the retreat of dissolution fronts along the {110} faces was much faster than that along {010} faces. This difference can be explained by smectite dissolution under alkaline conditions being controlled by OH^- attack, which is catalyzed by protonated Al-OH groups, on the bridging O atoms of both Al-O-Si sites, which locate only on the {110} surfaces, and Al_2-O-Si sites on the {010} and {110} surfaces. The bridging O atoms of Al_2-O-Si sites on the {010} surfaces are buried more deeply in the structure than those of Al-O-Si sites on the {110} surfaces, which would be more difficult to attack on the {010} surfaces.

To assess the effect of several organic acids with concentrations that ranged from 0.001 to 0.01 M on smectite dissolution rate at circumneutral pH, Golubev et al. (2006) performed analysis of reacted solid products of Sw-2 smectite using XRD, FT-IR, and XPS. No major change in the smectite structure, surface chemical composition or specific surface area as a function of pH, ligand concentration, and duration of experiments was observed.

Rozalén et al. (2008) showed that at 25°C, the obtained dissolution rates of smectite and other clay minerals (illite, micas, and chlorites) became similar when normalized to the mass of solid or edge surface area. The effect of pH on the dissolution rate is likewise similar. Grybos et al. (2010) investigated how the changes in clay mineral surfaces (basal vs edge) as a result of a change in particle size influence the extent of dissolution using size-selected colloidal particles of two nontronites. In this case, acid dissolution involves jointly the leaching of octahedral iron from the edge faces and the dissolution of tetrahedral iron from the basal faces. Lack of correlation between the extent of Fe^{3+} dissolution and the increase in edge surfaces with a decrease of particle size implies that reactive surface sites are evenly distributed on the basal planes, and therefore the dissolution of nontronites is significantly influenced by basal surfaces. Marty et al. (2011) showed that in reacted powdered samples of synthetic smectite at pH $\approx$ 9 and 80°C, dissolution features such as etch pits on the basal planes were not observed with direct AFM examination, suggesting that smectite dissolution proceeded via edge dissolution. Hence, in this study the rates were normalized to the external surface area.

By means of ex situ AFM and in situ VSI, Satoh et al. (2013) quantified the montmorillonite dissolution rates at a compaction pressure ranging from 0.04 to 10 MPa and temperature of 70°C in 0.3 M NaOH solution (pH 12.1 at 70°C). The observed dissolution rates were 1–2 orders of magnitude slower than those expected for dispersed montmorillonite under the same conditions. This rate reduction can be explained by limited water flow in the nanopores in which the dissolution of surrounding grains takes place. The effective total surface area of montmorillonite as a function of pressure, $A_{min}(P)$ may be able to predict the dissolution rate of compacted montmorillonite during compaction due to the variable exposed surface area. The effective total surface area of montmorillonite as a function of pressure, $A_{min}(P)$ can be empirically formulated as follows:

$$A_{min}(P) = \left(1.0 + 0.64628 \cdot P^{0.1527}\right) \cdot A_{min} \quad (4.33)$$

where A_{min} is the initial edge surface area before compaction.

4.6 MICAS

4.6.1 Illite

Köhler et al. (2003) studied illite dissolution over a wide range of pH (1.4–12.4) at 5, 20, 25, and 50°C using closed system reactors. The preferential initial Al release at acidic conditions was consistent with the observation of rapid Al exchange reactions from clay mineral surfaces in natural near-surface environments (Sposito, 1989). The authors proposed a dissolution rate law to account for the variation in illite dissolution rates as a function of pH and temperature

$$\frac{Rate}{\text{mol m}^{-2}\text{ s}^{-1}} = 2.2 \times 10^{-4} \exp\left(\frac{-E_{AH^+}}{RT}\right) a_{H^+}^{0.6} + 2.5 \times 10^{-13} \exp\left(\frac{-E_{AH_2O}}{RT}\right) a_{H^+}^{0.6} + 0.27 \exp\left(\frac{-E_{AOH^-}}{RT}\right) a_{OH^-}^{0.6} \tag{4.34}$$

where E_{AH^+}, E_{AH_2O}, and E_{AOH^-} equaled 46, 14, and 67 kJ mol^{-1}, respectively. It is inferred that the activation energy decreases with increasing pH at acidic conditions, minimizes at near to neutral pH, and increases with increasing pH at alkaline conditions. This E_A-pH variation is similar to that reported by Carroll and Walther (1990) for kaolinite dissolution. It was also concluded that since the rates obtained were similar to other clay minerals (e.g., kaolinite, montmorillonite, and smectite; Figure 4.5), the rate limiting step could be identical for the dissolution of a wide variety of clay minerals. Based on the observations and interpretation of illite dissolution, Köhler et al. (2005) showed that clay mineral dissolution rates decrease continuously with elapsed time, both in the laboratory and in the field, due to the preferred dissolution of reactive edge surfaces. Therefore, it was proposed that clay mineral dissolution does not attain steady state in

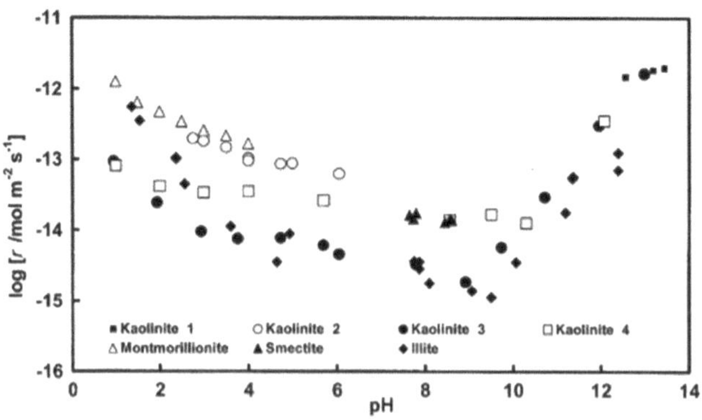

FIGURE 4.5 Comparison of apparent illite dissolution rates obtained by Köhler et al. (2003) with those of other clay minerals. All silica dissolution rates have been normalized to 10 oxygen atoms. Closed system experiments having durations in excess of 1 month are represented by filled symbols; those of shorted duration are denoted by open symbols. The sources and clay mineral identity of each data set are given below: illite du Puy early stage (Köhler et al., 2003); kaolinite 1 (Bauer and Berger, 1998); kaolinite 2 (Wieland and Stumm, 1992); montmorillonite (Zysset and Schindler, 1996); smectite (Huertas et al., 2001); kaolinite 3 (Huertas et al., 1999); and kaolinite 4 (Carroll and Walther, 1990). *Modified from Köhler et al. (2003).*

nature, but rather their dissolution rates decrease continuously during their dissolution.

Bibi et al. (2011) used flow-through reactors to study illite dissolution rate as a function of pH (1.0–4.25; H_2SO_4) and two ionic strengths (0.01 and 0.25 M maintained using NaCl solution). Like in Köhler et al. (2003), preferential release of Al^{3+} and K^+ over Si during the initial stage of experiments was observed. With increasing time the release rates of Si and Al became stoichiometric except for the experiment at pH >4 where possible adsorption of dissolved Al^{3+} led to a very low Al/Si ratio. An elevated release of Al^{3+} and K^+ occurred at the higher ionic strength, however, no such effect was observed on Si release rates. The K^+ release is controlled by H^+ exchange for K^+ at very low pH, whereas at pH >1, K release is driven by Na^+ for K^+ exchange reaction. A dissolution rate law showing the dependence of illite dissolution rate on proton concentration ($R_i = [H^+]^n$) in the acid-sulfate solutions was derived from the steady state dissolution rates, becoming useful in predicting the impact of illite dissolution in saline acid-sulfate environments. The fractional reaction orders (n) calculated from the linear regression of plot of log R_{Si} against pH at the higher and lower ionic strength were 0.32 and 0.36, respectively; corresponding values obtained from R_{Al} were 0.32 and 0.44, respectively. The dissolution rate of illite was therefore mainly controlled by solution pH and no effect of ionic strength was observed on the dissolution rates. The similarity in R_{Si} and R_{Al} in the majority of the experiments suggested stoichiometric dissolution of illite.

4.6.2 Muscovite

Kalinowski and Schweda (1996) studied the kinetics of muscovite dissolution in acid pH (1–4) and room temperature. Muscovite dissolution was close to stoichiometric, in contrast to the reported nonstoichiometric dissolution of biotite and phlogopite, suggesting a limited diffusion of H^+ into the solid. The pH-rate dependence was described as $R = 1.7 \times 10^{-12} \cdot a_{H^+}^{0.14}$. The lower reaction order for muscovite compared to that of biotite and phlogopite may indicate the existence of differences in reaction mechanisms between the dioctahedral and trioctahedral micas. In fact, a nonstoichiometric decationization reaction obtained with the trioctahedral micas implies protonation of internal surfaces or reaction sites to balance the excess negative charge of the residual structure.

Oelkers et al. (2008) obtained steady-state muscovite dissolution rates at higher temperature (from 60 to 201°C) and broader pH range (from 1 to 10.3) as a function of reactive solution K, Si, and Al concentration. The variation of rates with reactive solution composition depends on the solution pH. At pH $\leq$ 7 rates were found to decrease significantly with increasing reactive fluid Al activity but be independent of aqueous SiO_2 activity. In contrast at basic pH muscovite dissolution rates depend on both reactive fluid Al and Si activity.

In accordance to these findings, two rate laws were proposed: at pH ≤7 from 60 to 175°C:

$$R\left(\text{mol cm}^{-2}\ \text{s}^{-1}\right) = \left(10^{-6.53} - \exp\left(\frac{-58.2\ \text{kJ mol}^{-1}}{RT}\right)\right)\left(\frac{a^3_{\text{H}^+}}{a_{\text{Al}^{3+}}}\right)^{0.50} \tag{4.35}$$

and at pH range from 8.5 to 10.5 and 150°C:

$$R\left(\text{mol cm}^{-2}\ \text{s}^{-1}\right) = \left(10^{-9.195} - \exp\left(\frac{-89.1\ \text{kJ mol}^{-1}}{RT}\right)\right)\left(\frac{a^3_{\text{H}^+}}{a_{\text{Al}^{3+}}}\right)^{0.50}\left(a_{\text{SiO}_2}\right)^{-1} \tag{4.36}$$

Oelkers et al. (2008) argued that the contrasting behaviors suggest a change in dissolution mechanism with pH. At acidic pH rates appear to be controlled by the breaking of tetrahedral Si—O bonds after adjoining tetrahedral Al have been removed by proton exchange reaction (known as digestion reaction). At basic pH rates may be controlled by the breaking of octahedral Al—O bonds after adjoining tetrahedral Al and Si have been removed from the muscovite structure. Moreover, it was reported that any aqueous anion which tends to complex Si or Al in solution (such as organic acids at mildly acidic conditions) will increase muscovite dissolution rates.

Kuwahara (2008) investigated the dissolution behavior of muscovite under alkaline conditions at 25—50°C using in situ AFM. The muscovite dissolution took place only at the edge surfaces that are less than a few percent of the total surface area, while the basal surfaces were completely unreactive. The edge surface area-normalized dissolution rates derived from this AFM study were consistent with the dissolution rates renormalized to the estimated edge surface area of the earlier studies. A model dissolution rate equation, which simultaneously includes the effect of pH and temperature, was deduced

$$\log r_T = 0.23\text{pH} - 13.26 - \left[(6700\text{pH} - 23900)\left(\frac{1}{T} - \frac{1}{298}\right) \Big/ 2.303RT\right] \tag{4.37}$$

where r_T is in mol m^{-2} s^{-1} and T ranges from 25 to 80°C. The mechanism proposed to account for the muscovite dissolution can be applied to montmorillonite dissolution under alkaline conditions: OH^- anions would attack the bridging O bonds of both Al-O-Si and Al_2-O-Si sites existing on the edge surfaces, catalyzed by protonated Al-OH groups (Kuwahara, 2006). This mechanism that would dominate at higher pH and higher temperature is consistent with the above model where the dissolution rates of muscovite become increasingly sensitive to pH at elevated temperature.

Kurganskaya et al. (2012) performed VSI and AFM experiments with muscovite at pH 9.4 and 155°C to evaluate crystallographic controls on mica dissolution. The minimal global rate of normal surface retreat observed in VSI data was 1.42×10^{-10} mol m^{-2} s^{-1} while the local rate, observed at deep etch pits, reached 416×10^{-10} mol m^{-2} s^{-1}. Complementary AFM data clearly showed crystallographic control of mica dissolution, both in terms of step advance and the geometric influence of interlayer rotation (stacking periodicity). These observations indicate that basal/edge surface area ratios are highly variable and change continuously over the progression of the reaction, thus obviating their utility as characteristic parameters defining mica reactivity. The observations show that dissolution mechanism of mica may not be fundamentally different from minerals such as feldspars, calcite, etc., in which point and line defect centers serve as the foci of step movement. In muscovite, these defects serve as the centers of tetrahedral-octahedral-tetrahedral (TOT) step propagation; bulk removal of material then occurs by step movement at etch pit walls (Figure 4.6). Local variation in the frequency of step coalescence, partly a function of local defect density, gives rise to a complex, three-dimensional surface, whose roughness may in turn be annihilated by the arrival of steps from distant defects. The cumulative effect of these processes leads to the

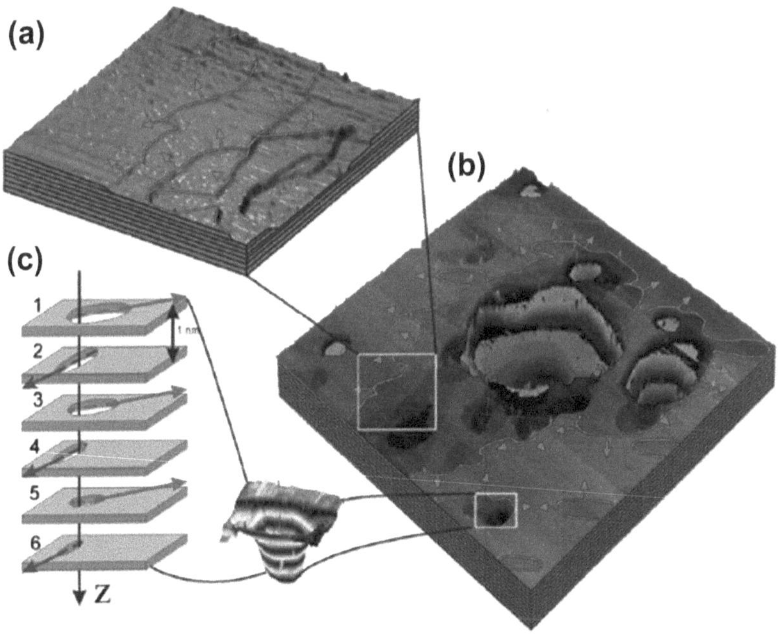

FIGURE 4.6 Schematic illustration of stepwave formation on muscovite {001} surface. (a) Consecutive steps moving outward from a dislocation center. (b) Steps emanating from two major etch pits (linear defects) and surrounding monolayer pits (point defects) form a single stepwave of complex morphology. (c) Multilayer etch pit formation, showing layer-by-layer dissolution in two principal directions. The TOT layers are shown as gray parallel blocks enumerated from top to bottom). *Modified from Kurganskaya et al. (2012).*

global removal of material from the surface. These observations are consistent with the stepwave model of Lasaga and Luttge (2001).

4.6.3 Biotite

Kalinowski and Schweda (1996) studied the dissolution kinetics of biotite, phlogopite, and muscovite in acid solution ($1 < pH < 4$) and room temperature using reactors where clay mineral dispersions were separated from eluant solutions with dialysis membranes. Rates were normalized to total surface area and may have little relevance for micas, since reactive sites probably are concentrated on edge surfaces which comprise a small fraction of total surface. It was observed that surface area changed during the experiments, partly due to delamination along basal surfaces, which affected the calculation of rate constants and the pH-dependence of rates. During biotite dissolution an initial loss of interlayer K^+ occurred. This is in accordance with a recent observation of biotite near-surface, which in acidic conditions is comprised exclusively of a remnant Si, O, and H framework (Bray et al., 2014). For biotite, the pH-rate dependence was described as $R = 3.2 \times 10^{-10} \cdot a_{H^+}^{0.61}$ in units of mol ($O_{20}(OH)_4$ formula units) $m^{-2}\,s^{-1}$. In addition, high ratios of dissolved Fe^{2+}/Fe^{3+} implied that biotite Fe^{2+} was oxidized by previously released Fe^{3+}. Malmström et al. (1996) performed batch and flow-through experiments in the near-neutral pH region ($7 < pH < 8.5$) and 25°C. The initial dissolution of biotite in the near-neutral pH region, under $N_2(g)$ atmosphere was highly nonstoichiometric, being K^+ release faster than Fe^{3+} release. The rates of Mg, Al, and Si release were between those for K^+ and Fe^{3+} and approach that of Fe^{3+} with time. Malmström and Banwart (1997) studied biotite dissolution in both acid and basic solutions (pH from 2 to 10) using thin-film continuous flow reactors at 25°C. While the release of interlayer K was relatively fast, becoming diffusion-controlled within a few days, the release rates of framework ions (Mg, Al, Fe, and Si) were much slower and reach an apparent steady state. The stoichiometry and rate of dissolution varied greatly with pH as shown in the empirical rate law proposed

$$Rate\ \left(\text{mol m}^{-2}\,\text{h}^{-1}\right) = k_H\left[H^+\right]^m + k_0 + k_{OH}\left[H^+\right]^n \qquad (4.38)$$

where the k_i values vary with the ith element (Si, Fe, Mg, and Al) as $-4.45 \le k_H \le -5.31$, $k_0 = -7.31$ (for Si), and $-10.57 \le k_{OH} \le -15.36$. Characterization of the reacted solid phase indicated that biotite was transformed to vermiculite during the weathering process. This alteration process corresponds to changes in composition of the original inter-, octahedral-, and tetrahedral-sheets while retaining the layer structure. The transformation may take place either in only the outermost part of the mineral (altered layer) or in the entire mineral (bulk). In the experiments of Malmström and Banwart (1997), where the altered layer extended through most of the particle, the difference in the calculated stoichiometry between these conceptually different cases was small. The calculated octahedral and tetrahedral stoichiometries of the formed vermiculite at pH 6.8 were close to those of the unweathered biotite.

Taylor et al. (2000) performed dissolution experiments at around pH 3 using column experiments. Cation release rates indicate that different sites within the sheets of the clay mineral structure reacted at different rates. Interlayer cations were generally released faster than octahedral cations, which reacted slightly faster than the tetrahedral cation, and interlayer potassium release was considered to be controlled by diffusion with a moving boundary condition. The log value of the overall dissolution rate (mol m^{-2} s^{-1}) was determined to be −11.6.

Murakami et al. (2003) conducted batch experiments of fresh and altered biotites in granite at 150°C of up to 56 days to elucidate certain aspects of the weathering process. Observations revealed that in the early stage, biotite dissolution proceeds from the edges of particles inward and secondary minerals such as Fe oxide are precipitated mostly at the edges, with only a few secondary minerals found on the basal surfaces. A higher Mg content in biotite facilitates the formation of vermiculite, at least in the early stage. Vermiculitization by layer-by-layer transformation does not occur for biotite with less than 0.4 Mg per $O_{10}(OH)_2$, but occurs for that with more than 0.8 Mg in the early stage. Because vermiculitization occurs even when a solution is not supersaturated with respect to smectite (and probably to vermiculite), vermiculitization is related to a local microenvironment rather than to the bulk solution. Because vermiculite dissolves at a much slower rate than biotite (Kirkwood and Nesbitt, 1991; Kalinowski and Schweda, 1996, 2007), Mg-rich biotite dissolves at a slower rate than Fe-rich biotite.

Ganor et al. (2005) compared the dissolution rates of biotite/chlorite from a bulk granite to the dissolution rates of the same mineral in mineral-rich fractions that were separated from the granite sample. Powder samples were placed in flow-through reactors at pH 1 and 25°C. Based on the observed release rates of the major elements, it was suggested that the biotite/chlorite-rich fraction dissolved noncongruently under near-equilibrium conditions. In contrast, the biotite and chlorite within the bulk granite sample dissolved congruently under far-from-equilibrium conditions. These differences result from variations in the degree of saturation of the solutions with respect to both the dissolving biotite/chlorite and to the nontronite, which probably was precipitating during the dissolution of the biotite and chlorite-rich fraction.

Samson et al. (2005) studied biotite dissolution under conditions of high pH (8−14) and high concentrations of aluminum (up to 0.055 M Al^{3+}), sodium (up to 7 M Na^{+}), and nitrate (up to 6 M NO_3^-) using continuously stirred flow-through reactors at 22−25°C. Initial dissolution transients (intervals of rapid release rates that decay to slower steady-state rates) showed fast preferential release of K^{+} followed by near-stoichiometric release of Si, Al, and

Mg, and slower release of Fe^{3+}. K^+ release was a function of the availability of Na^+ for exchange and accessibility to the biotite interlayers. Alteration to vermiculite was enhanced in sodium-enriched solutions. The pH dependence of quasi-steady-state release rates of Si, Al, Mg, Fe, and K exhibited the following pattern: $8 > 10 > 11 \approx 12 < 13 < 14$. At high pH, precipitation of secondary phases (e.g., phyllosilicates (sepiolite and kerolite) and zeolites (sodalite and analcime)) likely reduced the actual dissolution rates by passivating the biotite surface. Hodson (2006a,b) obtained biotite dissolution rates at pH 3 and 20°C using powder samples of different grain sizes in flow-through reactors to elucidate on the suitability of normalizing dissolution rates to initial and final BET surface area, geometric surface area, mass and geometric edge surface area. Initial BET surface area was a good normalizing term as before any dissolution occurs it will be proportional to reactive site density. Simple modeling indicated that biotite grain edges dissolved 71−132 times faster than basal surfaces.

Using AFM, McMaster et al. (2008) and Haward et al. (2011) suggested that etch pit nucleation at the biotite basal plane controls the biotite dissolution in the presence of oxalic acid. In the absence of oxalate, however, formation of discrete etch pits at room temperature was not observed. Pachana et al. (2012) observed stair-shaped etch pits on biotite and muscovite during hydrothermal (200°C) alteration in acidic solutions (pH 1.1−5.7). Cappelli et al. (2013) performed a systematic, in situ, and real-time exploration of broad regions of the reacting biotite (001) cleavage surface (in the order of tens of thousands square micrometers) in long-term flow-through experiments at pH 1 (T = 11.5−70°C) using LCM-DIM. Dissolution was controlled by the retreat of the step edges that followed an approximately normal direction to the terrace limits. Swelling of basal surface preceded the contraction and breaking of the layers (Figure 4.7). Mica dissolution is thought to be dominated by an edge attack mechanism that produces selective leaching of interlayer (K^+, Na^+) and octahedral (Mg^{2+}, Fe^{2+}, Fe^{3+}, and Al^{3+}) cations through the {hk0} surfaces and forms an altered silica-enriched rim (Kaviratna and Pinnavaia, 1994; Turpault and Trotignon, 1994; Rufe and Hochella, 1999; Aldushin et al., 2006). Fast exchange of interlayer potassium by protons and loss of octahedral cations lead to biotite layers curling and peeling-off (Turpault and Trotignon, 1994).

Dissolution from existing defects on the basal plane was not observed during biotite dissolution. Formation of basal etch pits did not occur during the reaction time of the experiments (up to six days) in the temperature range of (Cappelli et al., 2013) 11.5−70°C. Based on measured horizontal retreat rates at different temperature, the activation energy value was calculated (61.5 kJ mol^{-1}).

Bray et al. (2014) derived the chemical composition and charge of the biotite near surface in contact with NaCl bearing aqueous solutions at 25°C from pH 1 to 12 by means of zeta potential measurements and potentiometric

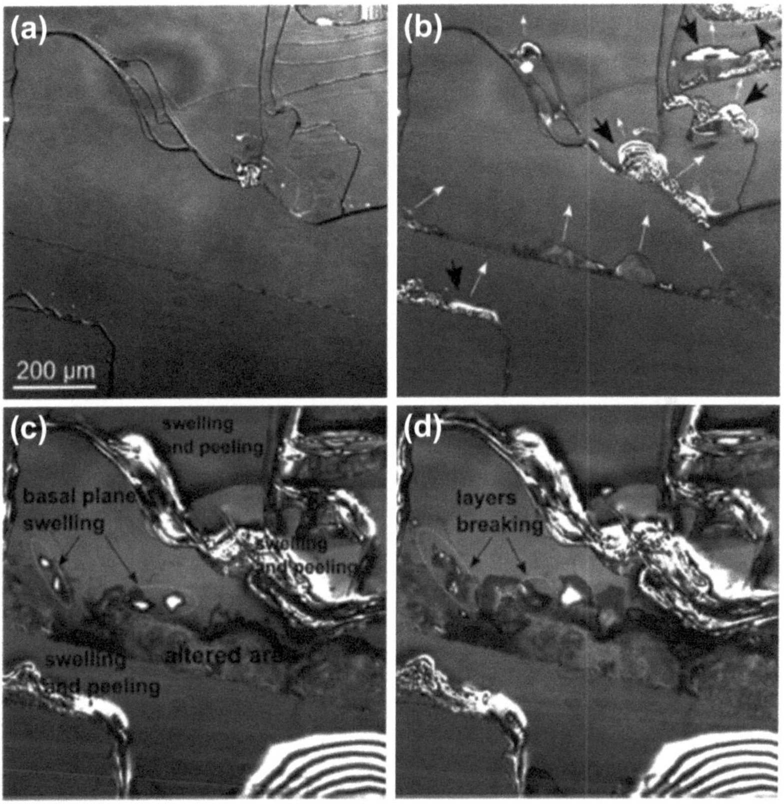

FIGURE 4.7 Laser confocal microscopy with differential interference contrast microscopy (LCM-DIM) images of biotite (001) cleavage surface: pristine surface (a) and after reaction at pH 1 and 11.5°C for 21 h (b), ~5 days (c), and ~6 days (d). Note the presence of high step edges (dark outlines). Dissolution started irregularly along the edges (white arrows in (b) indicate dissolution direction). Large black arrows indicate overlighted areas (regions with PMT saturation) where roughening and peeling of layers occur. *Modified from Cappelli et al. (2013).*

titrations. It was shown that metal release reactions are critical in controlling biotite dissolution rates. As the pH of minimum biotite dissolution rate (pH ≈ 7) differs from both the isoelectric point (3.02) and immersion pH (9.66), the proposed biotite dissolution rates were not directly related to proton consumption at the surface. In contrast, the biotite surface reactivity appears to be dominated by noncharge conservative metal release, which alters the surface composition of biotite as a function of pH. Such metal release involves the breaking of metal–oxygen bonds, both at the biotite surface and at a varying depth within the mineral, with a shift toward a relatively negatively charged biotite near-surface. The breaking of such bonds and the charge imbalance would weaken the biotite structure near the fluid interface, facilitating the mineral's eventual dissolution.

4.6.4 Phlogopite

Kalinowski and Schweda (1996) showed that the pH dependence of phlogopite dissolution rate could be described as $R = 3.4 \times 10^{-11} \cdot a_{H^+}^{0.40}$ in units of mol ($O_{20}(OH)_4$ formula units) $m^{-2}\,s^{-1}$. Like biotite, phlogopite dissolution was preceded by preferential loss of interlayer K^+. Taylor et al. (2000) performed experiments at around pH 3 with powdered sample using column experiments. As it occurred with biotite, different sites within the clay mineral sheets reacted at different rates. Interlayer cations were generally released more rapidly than octahedral cations, and interlayer potassium release was considered to be controlled by diffusion with a moving boundary condition. Cation release during phlogopite dissolution was also much less stoichiometric than during biotite weathering. The log value of the overall dissolution rate (mol $m^{-2}\,s^{-1}$) was determined to be -11.2.

Rufe and Hochella (1999) performed long-term in situ AFM imaging of phlogopite in contact with HCl solution (pH 2) providing evidence of the reactivity of phlogopite {001} surfaces. Results revealed a change in the phlogopite layer volume induced by alteration in HCl solutions that was driven by depolymerization and repolymerization reactions occurring during leaching layer formation. After the initial leaching of octahedral cations, the hydrolysis of Si-O-Si and Si-O-Al in the tetrahedral sheet allows the layer to expand. The consecutive cross-linking of neighboring Si-OH groups to re-form Si-O-Si (repolymerization) might cause layer contraction by the expulsion of water.

Aldushin et al. (2006) investigated the dissolution of the basal surface of phlogopite with acidic solutions from 20 to 140°C by HAFM. In highly acidic solutions (pH 1.5−2) and high temperature (100−140°C), phlogopite dissolution occurred via nucleation of etch pits on the pristine surface, implying that the basal surface plays an important role in the dissolution processes, at least at elevated temperatures. The formation of an altered layer was also detected at these conditions. Due to the low reactivity of phlogopite at room temperature, there is little information about a change in dissolution mechanism. Nevertheless, the kinetic data show that the phlogopite dissolution rate is clearly influenced by the reactivity of the basal surface and that the absolute amount of released material has comparable contributions from both basal surfaces and edge surfaces.

Shao et al. (2011) investigated the effects of four organic ligands—oxalate, malonate, acetate, and propionate—on phlogopite dissolution and on changes in its surface morphology under geological carbon sequestration conditions (95°C and 102 atm). The presence of organic ligands was found to influence phlogopite dissolution: the rate of phlogopite dissolution followed the order: oxalate > malonate > acetate ≈ propionate, which appeared to be the same sequence as the stability of the complexation with mineral surface metals.

4.7 VERMICULITE

Kalinowski and Schweda (2007) studied the dissolution kinetics of a tri-octahedral vermiculite that contains phlogopite and interstratifications of 10 and 14Å layers in HCl solution within the pH range 2–6.5 at 22°C by placing powdered samples in dialysis reactors. High initial cation release rates reflected both rapid selective leaching of the exchangeable cations and fast dissolution of 2:1 layers. Initial protonation of the exchange sites balanced charge losses due to leaching of interlayer Mg^{2+} and K^{+}. At times when 10% of the respective sample was dissolved, $\log R = \log k - 0.44$ pH. The unreacted vermiculite contained phlogopite and interstratifications of 10 and 14 Å layers. In the experiment with highest rates at pH 2, a regularly interstratified hydromica appeared to be the most weathering resistant phase. It was argued that comparison between rates measured at different pH values on an equal reaction time basis is irrelevant, as the changes of the sample composition are most rapid at low pH.

The vermiculite dissolution reaction under acidic pH and constant temperature (pH 2.7 and 50°C) led rapidly to Al-hydroxy-interlayer vermiculite which has lower cation exchange capacity compared to the initial mineral (Mareschal et al., 2009). In stirred flow-through experiments run for 672 h, vermiculite was found to dissolve nonstoichiometrically for 500 h. Thereafter, the release rate for Si, Mg, and Al became stoichiometric, whereas Fe sustained nonstoichiometric release throughout the whole experiment. Both Al and Fe precipitated in the interlayer space. Microscopic observations and measurements showed that dissolution mainly occurred at the expense of basal surfaces, which is less common in studies of clay mineral dissolution.

4.8 CHLORITE

Malmström et al. (1996) performed batch and flow-through experiments in the near-neutral pH region ($7 < \text{pH} < 8.5$) and 25°C and observed that the initial release rates of the different elements were nearly stoichiometric ($\approx 9.3 \times 10^{-9}$ mol m^{-2} h^{-1}) and similar to the iron release rate from biotite.

Rochelle et al. (1996) obtained dissolution rates at basic pH using powdered Fe-rich chlorite samples ($\log R$ (mol m^{-2} s^{-1}) $= -12.46$, -12.00, and -11.22 at pH of 7.7, 9.9, and 12.5, respectively). Apparent progressive incongruency with pH was attributed to formation of secondary Fe and Al minerals.

Shake flask experiments were used by Hamer et al. (2003) to obtain dissolution rates in acid pH (1.5–4.5). Incongruent dissolution was observed above pH 3.5, and progressive increase in dissolution in the order nitric (0.97) ≈ hydrochloric (1.00) ≈ acetic (1.01) < sulfuric (1.19) < citric (2.70) < oxalic acid (3.27) for the pH range studied was reported.

Brandt et al. (2003) studied the dissolution of chlorite with intermediate Fe-content. Macroscopic and microscopic results were obtained via mixed flow experiments and AFM, respectively. BET surface area normalized steady-state dissolution rates at 25°C for pH 2–5 vary between 10^{-12} and 10^{-13} mol m^{-2} s^{-1}. The order of the dissolution reaction with respect to protons was calculated to be about 0.29. For pH 2–4, chlorite was found to dissolve nonstoichiometrically, with a preferred release of the octahedrally coordinated cations. The additional release of octahedrally coordinated cations may be due to the transformation of chlorite to interstratified chlorite/vermiculite from the grain edges inward. Dissolution of the basal chlorite surfaces was controlled by a defect (crack or compositional inhomogeneities) dissolution mechanism. The formation of nanoscale etch pits within flat terraces terminated by the TOT layer was observed in situ. Molecular steps which originated at these surface defects were the most reactive surface sites during acid dissolution. The macroscopic and microscopic chlorite dissolution rates could be linked via the reactive surface area as identified by AFM. The reactive surface area with respect to dissolution consisted of only 0.2% of the BET-surface area. A dissolution rate of 2.5×10^{-9} mol m^{-2} s^{-1} was calculated from macroscopic and microscopic dissolution experiments at pH 2, when normalized to the reactive surface area.

Gustafsson and Puigdomenech (2003) used thin film continuous flow-through to obtain the dissolution rates of a relatively low Fe chlorite in the pH region 4–12 at 25°C (Figure 4.8). Lowson et al. (2005) obtained far from equilibrium, quasi-steady-state dissolution rates of an iron rich chlorite for the pH range 3–10.5 and at 25°C using flow-through cells. The empirical rate law proposed allows the rate to be expressed as a function of H^+

$$\log R = -\left[9.79 \cdot a_{H^+}^{0.49\pm0.07} + 13.00 + 16.79/a_{H^+}^{0.43\pm0.2}\right] \qquad (4.39)$$

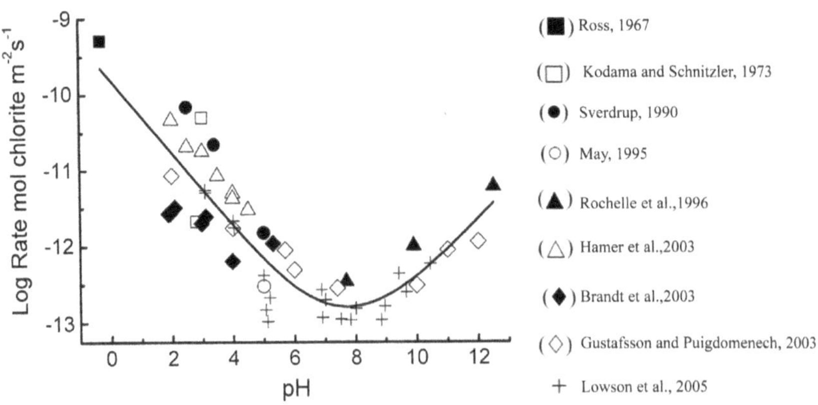

FIGURE 4.8 Chlorite dissolution rates as a function of pH at 25°C. *Modified from Lowson et al. (2005).*

where R stands for the dissolution rate in mol m^{-2} s^{-1} and a_i refers to the activity of the subscripted aqueous species (Figure 4.8). Additionally, when the data for the Al^{3+} concentration at the outlet were incorporated into the analysis, it was found that the rate could also be described by the relationship

$$R = 10^{-10.46}\left(\frac{a_{H^+}^3}{a_{Al^{3+}}}\right)^{0.27} \tag{4.40}$$

Lowson et al. (2007) enlarged the experimental conditions and obtained dissolution rates over a pH range of 3−10 and temperature range 25−95°C. While the rate-determining step is within the silica dissolution regime, the rate is a function of the H^+ and Al^{3+} concentrations and the dominant aqueous Al species. Individual rates may be described by a generic rate law applicable across the spectrum of Al species

$$r_n = k_n\left(\frac{K_n}{\beta_{Al_pL_q^{(3p-zq)}}}\left(\frac{a_{H^+}^{3p}}{a_{Al^{3+}}^p}\right)\Big/\left(1+\frac{K_n}{\beta_{Al_pL_q^{(3p-zq)}}}\left(\frac{a_{H^+}^{3p}}{a_{Al^{3+}}^p}\right)\right)\right)^m \tag{4.41}$$

where r_n is the rate subscripted for the nth Al species, k is the rate constant of the rate controlling step, K is the surface exchange constant, β is the solution stability constant subscripted for the Al species, a is the species activity subscripted for species and raised to the power of the stoichiometry, p and q are stoichiometric coefficients, z is the ligand charge, and m is the fractional coefficient for the precursor of the rate-defining step. The observed rate is the sum of the individual rates. When the observed rate is dominated by the formation of a single aluminum species, and in the absence of strong complexing agents such as oxalate, the proposed rate is proportional to $(a_{H^+}^3/a_{Al^{3+}})^m$ where m is the fractional coefficient defining the numerical excess of clusters of partially hydrolyzed silica tetrahedra with respect to the number of precursor complexes on the surface and may be experimentally obtained. Activation energy values of 94.3 and 40.5 kJ mol^{-1} were calculated for the acid and basic pH, respectively.

In the context of CO_2 geological storage, Smith et al. (2013) studied the dissolution kinetics of a Mg-rich chlorite under far-from-equilibrium conditions using a mixed-flow reactor over temperatures of 100−275°C at pH values of 3.0−5.7 in a background solution matrix of 0.05 m NaCl. Over this temperature range, near-stoichiometric release of magnesium with respect to silica for both $CO_{2(aq)}$ and CO_2-free experiments was observed. In contrast, iron and aluminum were released in highly nonstoichiometric quantities. Far from equilibrium, chlorite dissolution rate behavior (normalized by surface area) from 100 to 275°C, pH 3.3−5.7, and $CO_{2(aq)}$ from 0 to 0.5 m can be described with an H^+-catalyzed dissolution mechanism

$$rate = k_{acid,298K}\cdot\exp\left[\left(\frac{-E_{acid}}{R}\right)\cdot\left(\frac{1}{T}-\frac{1}{298K}\right)\right]a_{H^+}^{0.49} \tag{4.42}$$

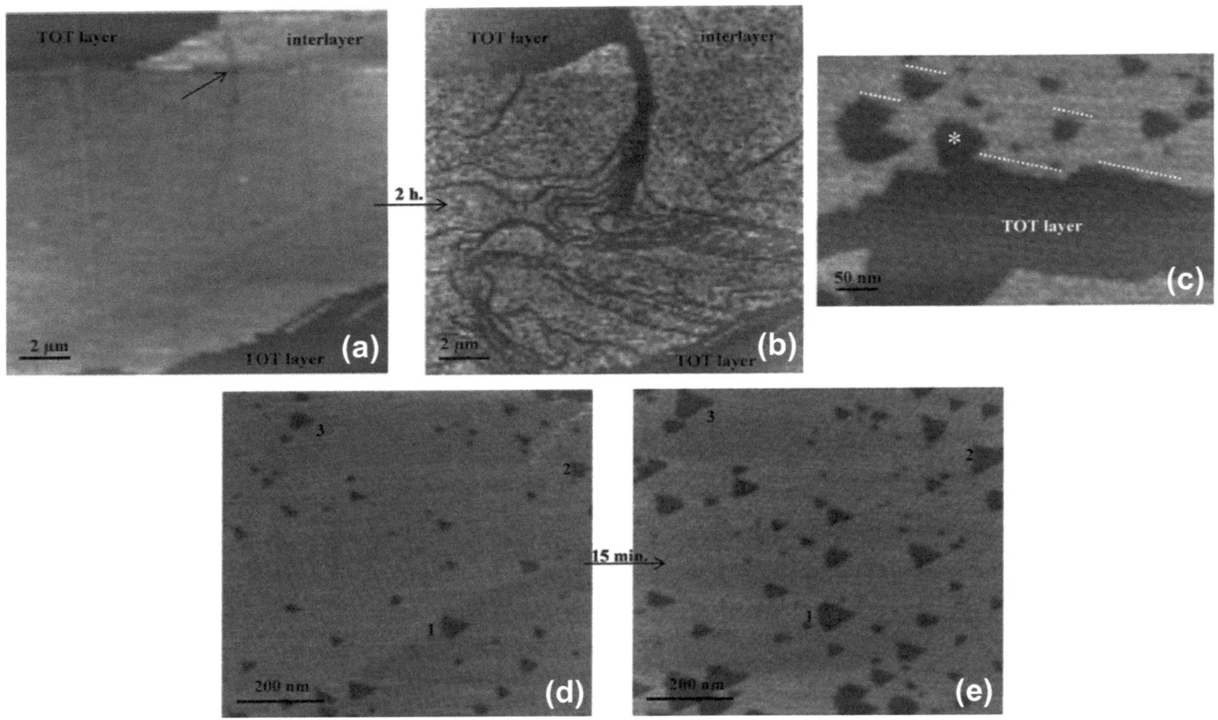

FIGURE 4.9 The interlayer surface dissolves through dissolution channels (b, c) and etch pits (c, d, e). Following 2 h in contact with water, the initial formation of the main channel is indicated by an arrow in a. After further 2 h, channels extend in the interlayer surface (b, c). In c, a higher magnification of a channel segment is displayed, where the asterisk indicates an etch pit that is coalescing with the channel. Channel edges are parallel to etch pits edges, which have a mono-oriented triangular shape (c, d, e). The area in a has been captured about halfway of the overall dissolution reaction. Some of the expanding etch pits are indicated by numbers in d and e. Z ranges: 0–10 nm (a, b), 0–2 nm (c), and 0–5 nm (d, e). *Modified from Gazzè et al. (2014).*

where $k_{acid,298\ K}$ and E_{acid} represent the 25°C kinetic rate constant and activation energy, respectively, for the acid dissolution mechanism ($k_{acid} = 10^{-9.91}$ mol m^{-2} s^{-1} and $E_{acid} = 25.1$ kJ mol^{-1}) and a_{H^+} denotes the proton activity. It was concluded that the resulting rate equation is dependent on temperature and pH, but independent of pCO_2 for the lack of significant differences in dissolution rate attributable to elevated CO_2 beyond decreases in solution pH.

Gazzè et al. (2014) characterized chlorite structure and dissolution in the near-pristine state by AFM imaging of surfaces of the chlorite in pure water at near-neutral pH. Chlorite reactivity was investigated with a special focus on the interlayer component. Upon cleavage, the TOT layer showed atomically flat terraces and steps, while the interlayer presented strips and voids. In pure water, dissolution channels and equilateral, mono-oriented triangular etch pits formed in the interlayer and led to progressively increased solubilization. Dissolution channels were proposed to originate from structural defects, while a conceptual model was discussed to explain the presence of triangular etch pits (Figure 4.9). In this model, their formation is driven by the different reactivity of the two octahedral configurations along the etch pits.

4.9 SUMMARY AND CONCLUSIONS

From this review on clay dissolution kinetics the conclusions are

1. Dissolution stoichiometry
   Early fast release rates of interlayer cations from clay minerals, induced by ion exchange reactions between the solution and the interlayer sites, have been observed in several studies, yielding to early nonstoichiometric dissolution of clay minerals, which is pH dependent. In most cases steady-state dissolution was stoichiometric. Nonstoichiometric dissolution under steady state condition is mostly the result of precipitation of secondary phases.
2. Clay dissolution rate—pH dependence
   Similarly to other silicates, the dissolution rates of clay minerals slow down with increasing pH in the acidic range, become constant at an intermediate pH, and increase with increasing pH in the basic range. This pattern of change with pH in the dissolution rate of oxides and silicates was classically interpreted by three reaction mechanisms: proton-promoted, hydroxyl-promoted, and water-promoted, that dominate the reaction rate under acidic, basic, or near-neutral conditions, respectively (White and Brantley, 1995). The fractional reaction orders calculated from the linear regression of plot of log R_{Si} and/or R_{Al} against pH are between 0.32 and 0.60 (illite), 0.40 and 0.75 (montmorillonite). The proton-dependent reaction order for trioctahedral micas (phlogopite and biotite) ranged between 0.25 and 0.61 compared to 0.14–0.20 for dioctahedral mica

(muscovite). The fractional reaction orders for pH are between 0.44 and 0.45 (vermiculite) and 0.29 and 0.49 (chlorite).

3. Clay dissolution rate—$f(\Delta G_r)$ dependence
The functional dependence of the dissolution rate on the deviation from equilibrium expressed as Gibbs free energy of reaction, $f(\Delta G_r)$, is essential to any attempt to apply experimental kinetic data to natural processes such as mineral weathering and soil degradation or to processes in nuclear waste repositories in which dissolution/precipitation rates of clay minerals are decisive. Up to date, the $f(\Delta G_r)$ functionality has been experimentally tested at different conditions only for dissolution of kaolinite and montmorillonite in the case of clay minerals. The nonlinear curves that account for the $f(\Delta G_r)$-rate dependency of clay minerals fall in the category of nonlinear expressions that are used to quantify the solution saturation state effect on aluminosilicate dissolution rates. The equations predict the slow dissolution rate decrease close to equilibrium and the sharp decrease in dissolution rate before the dissolution plateau is approached. However, arisen question marks concerning the use of TST-based expressions for the mineral dissolution (Luttge, 2006; Lüttge et al., 2013) should lead us to develop a more consistent theory of crystal dissolution. Within this context, dissolution/precipitation of clay minerals needs to be further investigated by exploring the reacted surfaces by means of suitable analytical techniques (e.g., AFM and VSI).

4. Effect of catalytic/inhibitor on clay mineral dissolution rate
Dissolution of kaolinite at pH = 2 and 150°C is specifically inhibited by Al. Addition of Al^{3+} into solution was found to inhibit the K^+-montmorillonite dissolution reaction at pH $\leq$3 and 25°C but not at 50°C ($\approx 10^{-6}$ M < [Al] < $\approx$ 10^{-4} M). Ganor and Lasaga (1998) presented a mechanistic model describing the effect of inhibition on mineral dissolution rate. The surface reaction mechanism consists of fast adsorption of Al^{3+} on the mineral surface (crystal edge surface sites) followed by a slow hydrolysis step. Golubev et al. (2006) demonstrated that very high concentrations (0.001−0.01 M) of organic ligands, whether they are originated from OM enzymatic degradation or bacterial metabolic activity are necessary to appreciably affect smectite dissolution. Hence, the effect of natural organics on the weathering rate of smectite is expected to be weak.

5. Ionic strength effect on the clay mineral dissolution rate
Experimental works to study the effect of ionic strength on kaolinite, montmorillonite, and illite show that the kaolinite dissolution rates were dependent on ionic strength. The dissolution rates of the other two clay minerals were mainly controlled by solution pH, and no effect of ionic strength was observed on the dissolution rates. In the case of illite dissolution, an increase in the release rates of Al^{3+} and K^+ occurred by increasing the ionic strength from 0.01 M to 0.25 M, however, no such effect was observed on Si release rates.

6. Reactive surface area of clay minerals

It is widely recognized that dissolution and precipitation of clay minerals (and phyllosilicates) occur preferentially at the edge, i.e., the (hk0) surfaces. Lack of correlation between the total and the edge surface area of smectite makes the smectite BET surface area an ineffective proxy for its reactive surface area. This conclusion may be extrapolated to all clay minerals which form layered particles. Hence, it is suggested to use AFM measurements of the specific edge surface area as an alternative proxy for the reactive surface area of smectite and the other clay minerals. However, the recent VSI and AFM experiments with muscovite by Kurganskaya et al. (2012) indicate that basal/edge surface area ratios are highly variable and change continuously over the progression of the reaction, thus obviating their utility as characteristic parameters defining mica reactivity.

APPENDIX

TABLE A.1 Composition of the clay Minerals Used in the Reviewed Studies

Kaolinite
Kaolinite St Austell (Bauer and Berger, 1998) $Al_2Si_{1.99}Ca_{0.0025}K_{0.0016}O_5(OH)_4$
KGa1 kaolinite (Huertas et al., 1998) $Al_2Si_2O_5(OH)_4$
KGa2 kaolinite (Metz and Ganor, 2001) $Al_2Si_2O_5(OH)_4$ $Al_{3.66}Fe^{III}_{0.07}Ti_{0.16}Si_4O_{10}(OH)_8$
Smectite
K^+-montmorillonite (Zysset and Schindler, 1996) $K_{0.318}[Al_{1.509}Fe_{0.205}Mg_{0.283}]_{10}[Si_{3.975}Al_{0.025}](OH)_2$
Smectite Ceca (Wyoming-Na) (Bauer and Berger, 1998) $(Si_{7.98}Al_{0.02})^{IV}(Al_{3.105}Fe^{3+}_{0.3128}Fe^{2+}_{0.0443}Mg_{0.5109})^{IV}O_{22}(Na^+_{0.4475}K^+_{0.012}Ca^{2+}_{0.0984})^{CE}$
Smectite Ibeco-Ca (Bauer and Berger, 1998) $(Si_{7.65}Al_{0.35})^{IV}(Al_{2.8}Fe^{3+}_{0.47}Fe^{2+}_{0.06}Mg_{0.7})^{IV}O_{22}(K^+_{0.04}Ca^{2+}_{0.5})^{CE}$
SHCa-1 hectorite (Bosbach et al., 2000) $(K_{0.05}Na_{0.42}Mg_{0.56})[Mg_{4.60}Li_{1.39}Ti_{0.01}][Si_{7.75}Al_{0.17}Fe(III)_{0.05}]O_{20}(OH)_4$
Smectite Cabo de Gata (Cama et al., 2000) $K_{0.19}Na_{0.51}Ca_{0.195}Mg_{0.08}(Al_{2.56}Fe_{0.42}Mg_{1.02})(Si_{7.77}Al_{0.23})O_{20}(OH)_4$
Smectite Serrata de Níjar (Huertas et al., 2001) $Ca_{0.18}Mg_{0.13}Na_{0.20}K_{0.20}[Al_{2.83}Fe_{0.37}Mg_{0.89}][Si_{7.58}Al_{0.42}]O_{20}(OH)_4$

Continued

TABLE A.1 Composition of the clay Minerals Used in the Reviewed Studies—cont'd

Not pure smectite SAz1 (Amram and Ganor, 2005; Metz et al., 2005a) $K_{0.02}Na_{0.05}Ca_{0.41}Mg_{0.18}[Al_{2.77}Fe_{0.17}Mg_{1.11}][Si_{7.70}Al_{0.30}]O_{20}(OH)_4$
Smectite STx-1 (Si-corrected) (Metz et al., 2005a) $K_{0.03}Na_{0.11}Ca_{0.33}Mg_{0.10}[Al_{3.06}Fe_{0.16}Mg_{0.83}][Si_{7.68}Al_{0.32}]O_{20}(OH)_4$
Smectite SWy-1 (Si-corrected) (Metz et al., 2005a) $Na_{0.77}Ca_{0.08}[Al_{3.00}Fe_{0.40}Mg_{0.59}][Si_{7.68}Al_{0.32}]O_{20}(OH)_4$
Na^+-montmorillonite SWy-2 (Golubev et al., 2006) $Ca_{0.06}Na_{0.56}[Al_{3.08}Fe(III)_{0.38}Mg_{0.54}][Si_{7.93}Al_{0.07}]O_{20}(OH)_4$
Kunipia P smectite (Kuwahara, 2006) $Na_{0.78}K_{0.02}Ca_{0.12}(Al_{3.02}Mg_{0.66}Fe_{0.18}Ti_{0.02})(Si_{7.74}Al_{0.26})O_{20}(OH)_4$
K-smectite Cabo de Gata (Rozalén et al., 2008) $K_{0.44}(Al_{1.27}Fe^{3+}_{0.22}Mg_{0.56})(Si_{3.95}Al_{0.05})(OH)_2$ $Na_{0.68}(Si_8)(Al_{3.32}Mg_{0.68})O_{20}(OH)_{3.8}F_{0.2}$
Saz-1 smectite (Metz et al., 2005a) $K_{0.02}Na_{0.05}Ca_{0.41}Mg_{0.18}[Mg_{1.11}Fe_{0.17}Al_{2.77}][Si_{7.70}Al_{0.30}]O_{20}(OH)_4$
NAu1 nontronite (Grybos et al., 2010) $[Si_{6.98}Al_{0.95}Fe_{0.07}][Mg_{0.04}Fe_{3.61}Al_{0.36}]O_{20}(OH)_4Na_{1.05}$
Synthetic montmorillonite (Marty et al., 2011) $Na_{0.68}[Si_8][Mg_{0.68}Al_{3.32}]O_{20}(OH)_{3.8}F_{0.2}$
Mica
Illite
Illite de Puy (Kohler et al., 2003)
$(Ca_{0.01}K_{0.53}Na_{0.01}X^{I}_{0.12})(Al_{1.27}Mg_{0.44}Fe^{3+}_{0.36})(Si_{3.55}Al_{0.45})O_{10}(OH)_2$
X = 93% Na, 2% Ca and 3% K
Na-saturated Silver Hill illite (Bibi et al., 2011) $K_{1.38}Na_{0.05}(Al_{2.87}Mg_{0.46}Fe^{3+}_{0.39}Fe^{2+}_{0.28}Ti_{0.07}[Si_{7.02}Al_{0.98}]O_{20}(OH)_4$
Biotite
Biotite Moen (Kalinowski and Schweda, 1996) $Ca_{0.02}Na_{0.04}K_{1.82}(Al_{0.33}Fe^{3+}_{0.82}Fe^{2+}_{1.71}Mn_{0.01}Mg_{2.70})(Si_{5.80}Al_{2.20})O_{20}(OH)_{3.97}F_{0.03}$
Biotite LK5210 (Malmström et al., 1996) $K_{0.80}(Al_{0.44}Fe^{3+}_{0.22}Fe^{2+}_{1.10}Mg_{0.84})(Si_{3.18}Al_{0.82})O_{10}(OH)_2$
Biotite LK5210 (Malmström and Banwart, 1997) $K_{0.79}(Al_{0.58}Fe^{3+}_{019}Fe^{2+}_{1.03}Mg_{0.83})(Si_{3.17}Al_{0.67}Fe^{3+}_{0.16})O_{10}(OH)_2$
Bancroft biotite (Samson et al., 2005) $K_{0.939}Na_{0.055}Ca_{0.001}(Mn_{0.031}Fe^{3+}_{0.058}Fe^{2+}_{1.058}Mg_{1.574}Ti_{0.183})(Si_{2.966}Al_{0.989})$ $O_{10}(OH_{1.601}F_{0.390}Cl_{0.009})$

TABLE A.1 Composition of the clay Minerals Used in the Reviewed Studies—cont'd

$K_{1.007}Na_{0.016}Ca_{0.000}(Mn_{0.056}Fe^{3+}_{0.059}Fe^{2+}_{1.075}Mg_{1.609}Ti_{0.121})(Si_{2.981}Al_{0.991})$ $O_{10}(OH_{1.709}F_{0.283}Cl_{0.008})$
Biotite Grasåsen (Bray et al., 2014) $K_{0.913}Na_{0.011}Ca_{0.0002}(Mn_{0.004}Cr^{2+}_{0.0004}Fe^{2+}_{1.098}Mg_{1.436})(Si_{2.832}Al_{1.312}Ti^{4+}_{0.184})O_{10}(OH)_2$
Phlogopite
Phlogopite Burgess (Kalinowski and Schweda, 1996) $Ca_{0.01}Na_{0.09}K_{1.92}(Al_{0.12}Fe^{3+}_{0.13}Fe^{2+}_{0.10}Mn_{0.005}Mg_{5.62})(Si_{5.77}Al_{2.23})O_{20}(OH)_{3.61}F_{0.39}$
Phlogopite (Shao et al., 2011) $KMg_{2.87}Si_{3.07}Al_{1.23}O_{10}(F,OH)_2$
Muscovite
Muscovite Tuftane (Kalinowski and Schweda, 1996) $Ca_{0.03}Na_{0.15}K_{1.82}(Al_{3.39}Fe^{3+}_{0.40}Fe^{2+}_{0.15}Mn_{0.04}Mg_{0.04})(Si_{6.18}Al_{1.82})O_{20}(OH)_{3.91}F_{0.09}$
Muscovite Ishikawa-Chou (Kuwahara, 2008) $Na_{0.18}K_{1.77}(Al_{3.70}Fe_{0.42}Mg_{0.07})(Si_{5.95}Al_{2.05})O_{20}(OH)_4$
Muscovite Madagascar pegmatite (Oelkers et al., 2008) $(Na_{0.09}K_{0.86})Fe_{0.05}Si_{3.05}Al_{2.92}O_{10}(OH_{1.95}F_{0.06})$
Vermiculite
Vermiculite Chile (Kalinowski and Schweda, 2007) $Ca_{0.01}Na_{0.01}K_{1.05}Mg_{0.35}(Fe^{3+}_{0.23}Fe^{2+}_{0.06}Mn_{0.003}Mg_{5.66}(Si_{5.93}Al_{1.77}Fe^{3+}_{0.30})O_{20}(OH)_4$
Vermiculite Andalusia (Mareschal et al., 2009) $Na_{0.80}[Ti_{0.02}Al_{0.17}Mn_{0.01}Mg_{2.54}][Si_{2.75}Al_{1.25}]O_{10}(OH)_2$
Chlorite
Chlorite (Rochelle et al., 1996) $(Al_{1.31}Fe_{1.97}Mg_{2.68})[Si_{2.78}Al_{1.23}O_{20}](OH)_8$
Chlorite 740665 (Malmström et al., 1996) $(Fe^{3+}_{0.1}Fe^{2+}_{0.1}Al_{0.7}Mg_{4.9})(Si_{3.5}Al_{0.5}O_{20})(OH)_8$
Fe^{2+}-chlorite (ripidolite) (Hamer et al., 2003) $(Al_{1.259}Fe^{2+}_{1.95}Al_{0.7}Mg_{2.80})(Si_{2.75}Al_{1.25}O_{10})(OH)_8$
CCa-2 Flagstaff Hill (Brandt et al., 2003) $(Fe^{3+}_{0.94}Fe^{2+}_{3.02}Al_{2.48}Mg_{5.54})[Si_{5.33}Al_{2.66}O_{20}](OH)_{16}$
Maln India (Brandt et al., 2003) $(Fe^{3+}_{0.93}Fe^{2+}_{2.27}Al_{2.48}Mg_{6.33})[Si_{5.33}Al_{2.66}O_{20}](OH)_{16}$
Sietronics Ltd chlorite (Lowson et al., 2005) $(Fe^{3+}_{0.07}Fe^{2+}_{1.90}Al_{0.97}Mg_{2.76})[Si_{2.48}Al_{1.52}O_{10}](OH)_8$
Chlorite (Gazzè et al., 2014) $(Mg_{4.90}Fe_{0.31}Ti_{0.001}Mn_{0.005}Al_{0.74})[Si_{3.34}Al_{0.66}]O_{10}(OH)_8$

TABLE A.2 Experimental Conditions, Reaction Orders with Respect to a_{H^+} and a_{OH^-} and Apparent Activation Energy Values Obtained in the Reviewed Studies

Mineral	pH Range	Reaction Order a_{H^+}; a_{OH^-}	T (°C)	E_{app} (kJ mol^{-1})	Experiment Type	Reference
Kaolinite	11.47–13.92	0.56 ± 0.12 (35°C)	35–80	33 ± 8 (0.1 M)[a]	Batch reactor	Bauer and Berger (1998)
		0.81 ± 0.12 (80°C)		51 ± 8 (3 M)[a]		
Kaolinite	1–13	0.55; 0.75	25	–	Batch reactor	Huertas et al. (1999)
Kaolinite	0.5–4.5	–	25–70	96.2	Nonstirred flow-through reactor	Cama et al. (2002)
Smectite	11.47–13.92	0.15 ± 0.06 (35–80°C)	35–80	52 ± 4 (3 M)[a]	Batch reactor	Bauer and Berger (1998)
Montmorillonite	7.6–8.5	0.34	20–60	–	Semibatch reactor	Huertas et al. (2001)
Na^+-montmorillonite	1–12	0.21; 0.33	25	–	Mixed-flow reactor	Golubev et al. (2006)
K^+-montmorillonite	1–6; 8–14	0.4; 0.27	25	–	Batch and stirred flow-through reactor	Rozalén et al. (2009)
K^+-montmorillonite	1–6; 8–14	0.59; 0.36	50	–	Batch and stirred flow-through reactor	Rozalén et al. (2009)

K^+-montmorillonite	1–6; 8–14	0.75; 0.42	70	–	Batch and stirred flow-through reactor	Rozalén et al. (2009)
Illite	1.3–6.9; 8.6–12.4	0.6; 0.6	5–50	46; 67	Batch reactor	Köhler et al. (2003)
Illite	1.0–4.25	0.32–0.36	25	–	Flow-through reactor	Bibi et al. (2011)
Muscovite	1–4	0.14–0.20	25	–	Dialysis cell	Kalinowski and Schweda (1996)
Muscovite	1–7; 8.5–10.5	–	60–201	58.2; 89.1	Mixed-flow reactor	Oelkers et al. (2008)
Biotite	1–4	0.61	25	–	Dialysis cell	Kalinowski and Schweda (1996)
Biotite	11–14	0.25–0.27	22–25	–	Stirred flow-through reactor	Samson et al. (2005)
Phlogopite	1–4	0.4	25	–	Dialysis cell	Kalinowski and Schweda (1996)
Vermiculite	2–6	0.44–0.45	22	–	Dialysis cell	Kalinowski and Schweda (2007)
Chlorite	2–5	0.29	25	–	AFM	Brandt et al. (2003)
Chlorite	3–10.5	0.49; 0.43	25	–	Flow-through cell	Lowson et al. (2005)
Chlorite	3–10.5	–	25–95	94.3; 40.5	Flow-through cell	Lowson et al. (2007)

AFM, atomic force microscopy
[a]*indicates KOH solution concentration.*

REFERENCES

Aagaard, P., Helgeson, H.C., 1982. Thermodynamic and kinetic constraints on reaction rates among minerals and aqueous solutions. I. Theoretical considerations. Am. J. Sci. 282, 237–285.

Aldushin, K., Jordan, G., Rammensee, W., Schmahl, W.W., Becker, H.-W., 2004a. Apophyllite (001) surface alteration in aqueous solutions studied by HAFM. Geochim. Cosmochim. Acta 68, 217–226.

Aldushin, K., Jordan, G., Fechtelkord, M., Schmahl, W.W., Becker, H.-W., Rammensee, W., 2004b. On the mechanism of apophyllite alteration in aqueous solutions. A combined AFM, XPS and MAS NMR study. Clays Clay Miner. 52, 432–442.

Aldushin, K., Jordan, G., Schmahl, W.W., 2006. Basal plane reactivity of phyllosilicates studied in situ by hydrothermal atomic force microscopy (HAFM). Geochim. Cosmochim. Acta 70, 4380–4391.

Amram, K., Ganor, J., 2005. The combined effect of pH and temperature on smectite dissolution rate under acidic conditions. Geochim. Cosmochim. Acta 69, 2535–2546.

Bandstra, J.Z., Brantley, S.L., 2008. Data fitting techniques with applications to mineral dissolution kinetics. In: Brantley, S.L., White, A.F. (Eds.), Kinetics of Water-rock Interaction. Springer, pp. 211–257.

Bauer, A., Berger, G., 1998. Kaolinite and smectite dissolution rate in high molar KOH solutions at 35 ° and 80°C. Appl. Geochem. 13, 905–916.

Bergaya, F., Theng, B.K., Lagaly, G. (Eds.), 2011. Handbook of clay science, vol. 1. Elsevier.

Bibi, I., Singh, B., Silvester, E., 2011. Dissolution of illite in saline–acidic solutions at 25°C. Geochim. Cosmochim. Acta 75, 3237–3249.

Bickmore, B.R., Hochella Jr, M.F., Bosbach, D., Charlet, L., 1999. Methods for performing atomic force microscopy imaging of clay minerals in aqueous solutions. Clays Clay Min. 47, 573–581.

Bickmore, B.R., Bosbach, D., Hochella Jr, M.F., Charlet, L., Rufe, E., 2001. In situ atomic force microscopy study of hectorite and nontronite dissolution: Implications for phyllosilicate edge surface structures and dissolution mechanisms. Amer. Min. 86, 411–423.

Bosbach, D., Charlet, L., Bickmore, B., Hochella Jr, M.F., 2000. The dissolution of hectorite: In-situ, real-time observations using atomic force microscopy. Amer. Mineral. 85, 1209–1216.

Brandt, F., Bosbach, D., Krawczyk-Barsch, E., Arnold, T., Bernhard, G., 2003. Chlorite dissolution in the acid pH-range: a combined microscopic and macroscopic approach. Geochim. Cosmochim. Acta 67, 1451–1461.

Brantley, S.L., 2003. Reactions kinetics of primary rock-forming minerals under ambient conditions. In: Treatise of Geochemistry, Vol. 5. Elsevier Ltd, pp. 73–117.

Brantley, S.L., 2008. Kinetics of mineral dissolution. In: Brantley, S.L., White, A.F. (Eds.), Kinetics of Water-rock Interaction. Springer, pp. 151–210.

Brantley, S.L., Conrad, C.F., 2008. Analysis of rates of geochemical reactions. In: Brantley, S.L., White, A.F. (Eds.), Kinetics of Water-rock Interaction. Springer, pp. 1–37.

Bray, A.W., Benning, L.G., Bonneville, S., Oelkers, E.H., 2014. Biotite surface chemistry as a function of aqueous fluid composition. Geochim. Cosmochim. Acta 128, 58–70.

Cama, J., Ganor, J., Ayora, C., Lasaga, C.A., 2000. Smectite dissolution kinetics at 80°C and pH 8.8. Geochim. Cosmochim. Acta 64, 2701–2717.

Cama, J., Metz, V., Ganor, J., 2002. The effect of pH and temperature on kaolinite dissolution rate under acidic conditions. Geochim. Cosmochim. Acta 66, 3913–3926.

Cama, J., Ganor, J., 2006. The effects of organic acids on the dissolution of silicate minerals: a case study of oxalate catalysis of kaolinite dissolution. Geochim. Cosmochim. Acta 70, 2191–2209.

Cappelli, Ch, Van Driessche, A., Cama, J., Huertas, F.J., 2013. In situ observation of biotite dissolution at pH 1 using advanced optical microscopy. Cryst. Growth Des. 13, 2880–2886.

Carroll, S.A., Walther, J.V., 1990. Kaolinite dissolution at 25, 60, and 80°C. Amer. J. Sci. 290, 797–810.

Chin, P.-K.F., Mills, G.L., 1991. Kinetics and mechanisms of kaolinite dissolution: effects of organic ligands. Chem. Geol. 90, 307–317.

Devidal, J.L., Schott, J., Dandurand, J.L., 1997. An experimental study of kaolinite dissolution and precipitation kinetics as a function of chemical affinity and solution composition at 150°C, 40 bars, and pH 2, 6.8, and 7.8. Geochim. Cosmochim. Acta 61, 5165–5186.

Dove, P.M., Han, N., De Yoreo, J.J., 2005. Mechanisms of classical crystal growth theory explain quartz and silicate dissolution behavior. Proc. Natl. Acad. Sci. 102, 15357–15362.

Eyring, H., 1935. The activated complex in chemical reactions. J. Chem. Phys. 3, 107–115.

Furrer, G., Zysset, M., Schindler, P.W., 1993. Weathering kinetics of montmorillonite: Investigations in batch and mixed-flow reactors. In: Manning, D.A.C., Hall, P.L., Hughes, C.R. (Eds.), Geochemistry of clay-pore fluid interaction, vol. 4. Chapman & Hall, pp. 243–262.

Ganor, J., Lasaga, A.C., 1994. The effects of oxalic acid on kaolinite dissolution rate. Min. Mag. A 58, 315–316.

Ganor, J., Mogollón, J.L., Lasaga, A.C., 1995. The effect of pH on kaolinite dissolution rates and on activation energy. Geochim. Cosmochim. Acta 59, 1037–1052.

Ganor, J., Lasaga, A.C., 1998. Simple mechanistic models for inhibition of a dissolution reaction. Geochim. Cosmochim. Acta 62, 1295–1306.

Ganor, J., Roueff, E., Erel, Y., Blum, J.D., 2005. The dissolution kinetics of a granite and its minerals-implications for comparison between laboratory and field dissolution rates. Geochim. Cosmochim. Acta 69, 607–621.

Ganor, J., Reznick, I.J., Rosenberg, Y.O., 2009. Organics in water-rock interactions. In: Thermodynamics and Kinetics of Water-rock Interaction. Reviews in Mineralogy, vol. 70, pp. 259–369.

Gazzè, S.A., Stack, A.G., Ragnarsdottir, V.K., McMaster, T.J., 2014. Chlorite topography and dissolution of the interlayer studied with atomic force microscopy. Am. Min. 99, 128–138.

Golubev, S.V., Bauer, A., Pokrovsky, O.S., 2006. Effect of pH and organic ligands on the kinetics of smectite dissolution at 25°C. Geochim. Cosmochim. Acta 70, 4436–4451.

Grybos, M., Michot, L.J., Skiba, M., Billard, P., Mustin, C., 2010. Dissolution of anisotropic colloidal mineral particles: evidence for basal surface reactivity of nontronite. J. Coll. Inter. Sci. 343, 433–438.

Gustafsson, Å.B., Puigdomenech, I., 2003. The effect of pH on chlorite dissolution rates at 25°C. Mat. Res. Symp. Proc. 757, 649–655.

Hamer, M., Graham, R.C., Amrhein, C., Bozhilov, K.N., 2003. Dissolution of ripidolite (Mg, Fe-Chlorite) in organic and inorganic acid solutions. Soil Sci. Soc. Am. J. 67, 654–661.

Haward, S.J., Smits, M.M., Ragnarsdóttir, K.V., Leake, J.R., Banwart, S.A., McMaster, T.J., 2011. Geochim. Cosmochim. Acta 75, 6870–6881.

Hodson, M.E., 2006a. Searching for the perfect surface area normalizing term —a comparison of BET surface area-, geometric surface area- and mass-normalized dissolution rates of anorthite and biotite. J. Geochem. Expl. 88, 288–291.

Hodson, M.E., 2006b. Does reactive surface area depend on grain size? Results from pH 3, 25°C far-from-equilibrium flow-through dissolution experiments on anorthite and biotite. Geochim. Cosmochim. Acta 70, 1655–1667.

Huertas, J.F., Chou, L., Wollast, R., 1998. Mechanism of kaolinite dissolution at room temperature and pressure: part 1. Surface speciation. Geochim. Cosmochim. Acta 62, 417–431.

Huertas, J.F., Chou, L., Wollast, R., 1999. Mechanism of kaolinite dissolution at room temperature and pressure. Part II: kinetic study. Geochim. Cosmochim. Acta 63, 3261–3275.

Huertas, F.J., Caballero, E., Jimenez de Cisneros, C., Huertas, F., Linares, J., 2001. Kinetics of montmorillonite dissolution in granitic solutions. Appl. Geochem. 16, 397–407.

Kalinowski, B.E., Schweda, P., 1996. Kinetics of muscovite, phlogopite and biotite dissolution and alteration at pH 1 to 4, room temperature. Geochim. Cosmochim. Acta 60, 367–385.

Kalinowski, B.E., Schweda, P., 2007. Rates and nonstoichiometry of vermiculite dissolution at 22°C. Geoderma 142, 197–209.

Kaviratna, H., Pinnavaia, T.J., 1994. Acid hydrolysis of octahedral Mg (super 2+) sites in 2:1 layered silicates; an assessment of edge attack and gallery access mechanisms. Clays Clay Miner. 42, 717–723.

Kirkwood, D.E., Nesbitt, H.W., 1991. Formation and evolution of soils from an acidified watershed: plastic Lake, Ontario, Canada. Geochim. Cosmochim. Acta 55, 1295–1308.

Köhler, S.J., Dufaud, F., Oelkers, E.H., 2003. An experimental study of illite dissolution kinetics as a function of pH from 1.4 to 12.4 and temperature from 5 to 50°C. Geochim. Cosmochim. Acta 67, 3583–3594.

Köhler, S.J., Bosbach, D., Oelkers, E.H., 2005. Do clay mineral dissolution rates reach steady state? Geochim. Cosmochim. Acta 69, 1997–2006.

Kurganskaya, I., Arvidson, R.S., Fischer, C., Lüttge, A., 2012. Does the stepwave model predict mica dissolution kinetics? Geochim. Cosmochim. Acta 97, 120–130.

Kurganskaya, I., Lüttge, A., 2013. A comprehensive stochastic model of phyllosilicate dissolution: structure and kinematics of etch pits formed on muscovite basal face. Geochim. Cosmochim. Acta 120, 545–560.

Kuwahara, Y., 2006. In-situ AFM study of smectite dissolution under alkaline conditions at room temperature. Am. Mineral. 91, 1142–1149.

Kuwahara, Y., 2008. In situ observations of muscovite dissolution under alkaline conditions at 25–50°C by AFM with an air/fluid heated system. Am. Miner. 93, 1028–1033.

Lasaga, A.C., Kirkpatrick, R.J., 1981. Kinetics of geochemical processes. Rev. Mineralogical Soc. America 3 (8), 1–408.

Lasaga, A.C., 1998. Kinetic Theory in the Earth Sciences. Princeton University Press.

Lasaga, A.C., Lüttge, A., 2001. Variation of crystal dissolution rate based on a dissolution stepwave model. Science 291, 2400–2404.

Lasaga, A.C., Luttge, A., 2003. A model for crystal dissolution. Eur. J. Mineral. 15, 603–615.

Lowson, R.T., Comarmond, M.-C.J., Rajaratnam, G., Brown, P.L., 2005. The kinetics of the dissolution of chlorite as a function of pH and at 25°C. Geochim. Cosmochim. Acta 69, 1687–1699.

Lowson, R.T., Brown, P.L., Comarmond, M.-C.J., Rajaratnam, G., 2007. The kinetics of chlorite dissolution. Geochim. Cosmochim. Acta 71, 1431–1447.

Lüttge, A., 2006. Crystal dissolution kinetics and Gibbs free energy. J. Electron. Spectrosc. Relat. Phenom. 150, 248–259.

Lüttge, A., Arvidson, R.S., 2008. The mineral-water interface. In: Brantley, S.L., White, A.F. (Eds.), Kinetics of Water-rock Interaction. Springer, pp. 73–107.

Lüttge, A., Arvidson, R.S., 2010. Reactions at surfaces: a new approach integrating interferometry and kinetic simulations. J. Am. Ceram. Soc. 93, 3519–3530.

Lüttge, A., Arvidson, R.S., Fischer, C., 2013. A stochastic treatment of crystal dissolution kinetics. Elements 9, 183–188.

Malmström, M., Banwart, S., Lewenhagen, J., Duro, L., Bruno, J., 1996. The dissolution of biotite and chlorite at 25°C in the near-neutral pH region. J. Cont. Hydro 21, 201–213.

Malmström, M., Banwart, S., 1997. Biotite dissolution at 25°C: the pH dependence of dissolution rate and stoichiometry. Geochim. Cosmochim. Acta 61, 2779–2799.

Mareschal, L., Ranger, J., Turpaul, M.P., 2009. Stoichiometry of a dissolution reaction of a trioctahedral vermiculite at pH 2.7. Geochim. Cosmochim. Acta 73, 307–319.

Marty, N., Cama, J., Sato, T., Chino, D., Villiéras, F., Razafitianamaharavo, A., Brendlé, J., Giffaut, E., Soler, J.M., Gaucher, E.C., Tournassat, C., 2011. Dissolution kinetics of synthetic Na-smectite. An integrated experimental approach. Geochim. Cosmochim. Acta 75, 5849–5864.

McMaster, T.J., Smits, M.M., Haward, S.J., Leake, J.R., Banwart, S., Ragnarsdottir, K.V., 2008. Mineral. Mag. 72, 115–120.

Metz, V., 2001. Dissolution Kinetics of Smectite and Kaolinite. Ph.D. Thesis, Ben-Gurion University of the Negev.

Metz, V., Ganor, J., 2001. Stirring effect on kaolinite dissolution rate. Geochim. Cosmochim. Acta 65, 3475–3490.

Metz, V., Amram, K., Ganor, J., 2005a. Stoichiometry of smectite dissolution reaction. Geochim. Cosmochim. Acta 69, 1755–1772.

Metz, V., Raanan, H., Pieper, H., Bosbach, D., Ganor, J., 2005b. Towards the establishment of a reliable proxy for the reactive surface area of smectite. Geochim. Cosmochim. Acta 69, 2581–2591.

Murakami, T., Utsunomiya, S., Yokoyama, T., Kasama, T., 2003. Biotite dissolution processes and mechanisms in the laboratory and in nature: early stage weathering environment and vermiculitization. Am. Min. 88, 377–386.

Nagy, K.L., Blum, A.E., Lasaga, A.C., 1991. Dissolution and precipitation kinetics of kaolinite at 80°C and pH 3: the dependence on solution saturation state. Amer. J. Sci. 291, 649–686.

Nagy, K., Lasaga, A., 1992. Dissolution and precipitation kinetics of gibbsite at 80°C and pH 3: the dependence on solution saturation state. Geochim. Cosmochim. Acta 56, 3093–3111.

Nagy, K.L., 1995. Dissolution and precipitation kinetics of sheet silicates. In: Chemical Weathering Rates of Silicate Minerals. Reviews in Mineralogy, vol. 31, pp. 173–233.

Nakayama, S., Sakamoto, Y., Tamaguchi, T., Akai, M., Tanaka, T., Sato, T., Lida, Y., 2004. Dissolution of montmorillonite in compacted bentonite by highly alkaline aqueous solutions and diffusivity of hydroxide ions. Appl. Clay Sci. 27, 53–65.

Oelkers, E.H., Schott, J., Devidal, J.L., 1994. The effect of aluminum, pH, and chemical affinity on the rates of aluminosilicate dissolution reactions. Geochim. Cosmochim. Acta 58, 2011–2024.

Oelkers, E.H., 1996. Physical and chemical properties of rocks and fluids for chemical mass transport calculations. In: Reactive Transport in Porous Media. Rev. Min, 34, pp. 131–191.

Oelkers, E.H., 2001. General kinetic description of multioxide silicate mineral and glass dissolution. Geochim. Cosmochim. Acta 65, 3703–3719.

Oelkers, E.H., Schott, J., Gauthier, J.M., Herrero-Roncal, T., 2008. An experimental study of the dissolution mechanism and rates of muscovite. Geochim. Cosmochim. Acta 72, 4948–4961.

Pachana, K., Zuddas, P., Censi, P., 2012. Appl. Geochem. 27, 1738–1744.

Rochelle, C.A., Bateman, K., MacGregor, R., Pearce, J.M., Savage, D., Wetton, P.D., 1996. Experimental determination of chlorite dissolution rates. In: Murakami, T., Ewing, R.C. (Eds.), Materials Research Society Symposium, No 353. Scientific Basis for Nuclear Waste Management XVIII, Kyoto 1994, pp. 149–156.

Rozalén, M.L., Huertas, F.J., Brady, P.V., Cama, J., García-Palma, S., Linares, J., 2008. Experimental study of the effect of pH on the kinetics of montmorillonite dissolution at 25°C. Geochim. Cosmochim. Acta 72, 4224–4253.

Rozalén, M.L., Brady, P.V., Huertas, F.J., 2009. Surface chemistry of K-montmorillonite: Ionic strength, temperature dependence and dissolution kinetics. J. Coll. Inter. Sci. 333, 474–484.

Rufe, E., Hochella, M.F., 1999. Quantitative assessment of reactive surface area of phlogopite during acid dissolution. Science 285, 874–876.

Samson, S.D., Nagy, K.L., Cotton III, W.B., 2005. Transient and quasi-steady-state dissolution of biotite at 22–25°C in high pH, sodium, nitrate, and aluminate solutions. Geochim. Cosmochim. Acta 69, 399–413.

Satoh, H., Ishii, T., Owada, H., 2013. Dissolution of compacted montmorillonite at hyperalkaline pH and 70°C: in situ VSI and ex situ AFM measurements. Clay Minerals 48, 285–294.

Schott, J., Pokrovsky, O.S., Oelkers, E.H., 2009. The link between mineral dissolution/precipitation kinetics and solution chemistry. In: Thermodynamics and Kinetics of Water-rock Interaction. Reviews in Mineralogy, vol. 70, pp. 207–258.

Schott, J., Oelkers, E.H., Bénézeth, P., Goddéris, Y., François, L., 2012. Can accurate kinetic laws be created to describe chemical weathering? C. R. Geosci. 344, 568–585.

Shao, H., Ray, J.R., Jun, Y.-S., 2011. Effects of organic ligands on supercritical CO_2-induced phlogopite dissolution and secondary mineral formation. Chem. Geol. 290, 121–132.

Smith, M.M., Wolery, T.J., Carroll, S.A., 2013. Kinetics of chlorite dissolution at elevated temperatures and CO_2 conditions. Chem. Geol. 347, 1–8.

Sposito, G., 1989. The Chemistry of Soils. Oxford University Press.

Stumm, W., 1992. Chemistry of the Solid-Water Interface: Processes at the Mineral–Water and Particle–Water Interface in Natural Systems. J. Wiley & Sons.

Taylor, A.S., Blum, J.D., Lasaga, A.C., MacInnis, I.N., 2000. Kinetics of dissolution and Sr release during biotite and phlogopite weathering. Geochim. Cosmochim. Acta 64, 1191–1208.

Turpault, M.P., Trotignon, L., 1994. The dissolution of biotite single crystals in dilute HNO3 at 24°C: evidence of an anisotropic corrosion process of micas in acidic solutions. Geochim. Cosmochim. Acta 58, 2761–2775.

Ueta, S., Satoh, H., Kato, H., Ueda, A., Tsukamoto, K., 2013. A novel technique of in situ phase-shift interferometry applied for faint dissolution of bulky montmorillonite in alkaline solution. J. Nucl. Sci. Tech. 50, 731–741.

Vandevivere, P., Welch, S.A., Ullman, W.J., Kirchman, D.L., 1994. Enhanced dissolution of silicate minerals by bacteria at near-neutral pH. Microb. Ecol. 27, 241–251.

Van Rompaey, K., Van Ranst, E., De Coninck, F., Vindevogel, N., 2002. Dissolution characteristics of hectorite in inorganic acids. App. Clay Sci. 21, 241–256.

Walther, J.V., 1996. Relation between rates of aluminosilicate mineral dissolution, pH, temperature, and surface charge. Am. J. Sci. 296, 693–728.

Ward, D.B., Brady, P.V., 1998. Effect of Al and organic acids on the surface chemistry of kaolinite. Clays Clay Miner. 46, 453–465.

White, A.F., Brantley, S.L., 1995. Chemical weathering rates of silicate minerals. In: Ribbe, P.H. (Ed.), Reviews in Mineralogy, vol. 31. Mineralogical Society of America, Washington, p. 583.

Wieland, E., Stumm, W., 1992. Dissolution kinetics of kaolinite in acidic aqueous solutions at 25°C. Geochim. Cosmochim. Acta 56, 3339–3355.

Xie, Z., Walther, J.V., 1992. Incongruent dissolution and surface area of kaolinite. Geochim. Cosmochim. Acta 56, 3357–3363.
Yang, L., Steefel, C.I., 2008. Kaolinite dissolution and precipitation kinetics at 22°C and pH 4. Geochim. Cosmochim. Acta 72, 99–116.
Yokoyama, S., Kuroda, M., Sato, T., 2005. Atomic force microscopy study of montmorillonite dissolution under highly alkaline conditions. Clays Clay Miner. 53, 147–154.
Zysset, M., Schindler, P.W., 1996. The proton promoted dissolution kinetics of K-montmorillonite. Geochim. Cosmochim. Acta 60, 921–931.

Chapter 5

Stability of Clay Barriers Under Chemical Perturbations

Olivier Bildstein[a] **and Francis Claret**[b]
[a] *Atomic Energy and Alternative Energies Commission, Nuclear Energy Division, Cadarache, Saint Paul-lez-Durance, France;* [b] *Water, Environment and Ecotechnology Division, French Geological Survey (BRGM), Orléans, France*

Chapter Outline

5.1 INTRODUCTION

Clay barriers are key components in deep geological storage applications. The efficiency of such systems relies on the confinement properties of the natural or engineered clay barriers: low permeability, low diffusivity, high retention

Developments in Clay Science, Vol. 6C. http://dx.doi.org/10.1016/B978-0-08-100027-4.00005-X

and swelling capacity. In this context, having confidence that these properties will persist over the long term, say thousands of years for CO_2 storage, to hundreds of thousands of years for radioactive waste disposal, is essential. Natural systems have demonstrated that such durability is indeed attainable for very long periods of time as attested by the existence of efficient clay caprocks retaining oil, hydrocarbon gas, and CO_2 gas in reservoirs, as well as host-rocks in billion-year-old natural ore deposits (e.g., Cigar Lake, Canada).

The phenomena associated with complex chemical evolution will be described essentially in two systems, CO_2 storage and radioactive waste disposal (Figure 5.1). However, other systems for which these results are pertinent will also be mentioned (e.g., permeable reactive barriers using zero-valent iron). The reactivity of clay barriers is intimately linked to the nature and properties of their constituent minerals as well as their transport and retention properties. Starting from the initial physicochemical conditions (pH and redox potential, aqueous species concentrations) different types of perturbations are identified in the near field of a drift in the repository, around the casing of an injection/production well, or in a permeable reactive barrier. These perturbations may be caused by a unique aggressive agent such as supercritical CO_2, or by the interactions between clay and different materials such as concrete, steel, and CO_2 close to wellbores in CO_2 storage, concrete and bitumen in medium-level radioactive waste disposal, or glass and steel in high-level waste disposal.

In all the systems investigated, clay barriers react due to changes in the initial physicochemical conditions or to the introduction of "foreign" materials. In this chapter, the response of clay barriers to three types of perturbation is examined: (1) perturbation due to processes such as oxidation, desaturation, microbiological reactions, and interactions with drilling fluids; (2) perturbation linked to the interactions between clay and allochthonous "engineered" solid materials (iron, steel, concrete, glass, bitumen, etc.), and (3) perturbation by different gases (CO_2, H_2, etc.) introduced into deep geological environments.

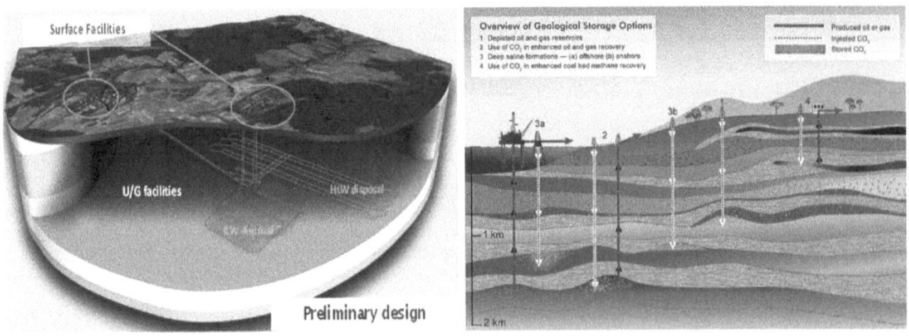

FIGURE 5.1 Examples of industrial applications using clay barrier as host-rock or caprock: left, high-level radioactive waste disposal (Andra); right, deep geological CO_2 storage (Metz et al., 2005).

The first type of responses typically occurs during the excavation, drilling, and operation phases in underground facilities or wellbores. Although the interactions of drilling fluids with clay are a great challenge to the oil industry, this topic will not be addressed here; only oxidation and desaturation processes will be considered in this chapter. Indeed, water-based drilling fluids are increasingly being used for oil and gas exploration instead of oil-based or synthetic-based fluids because they are suitable for environmental reasons. However, clay mineral hydration and swelling may lead to significantly increased oil-well construction costs (Anderson et al., 2010). Moreover, in the case of shale gas development, interactions between the clay matrix and a high volume of hydraulic fracturing fluids, necessary for resource exploration, might lead to a potential risk to water resources (e.g., Vengosh et al., 2014).

The impact of the different perturbations on clay barriers will be described in terms of dissolution of primary minerals and precipitation of secondary minerals, as well as modifications of clay mineral properties (especially cation exchange capacity (CEC), cation content, and swelling ability) and transport properties through modifications in porosity, permeability, and tortuosity.

The challenging scientific approach used to tackle the problem of predicting long-term clay barriers behavior will be emphasized by showing that the results from experiments conducted in the laboratory and in underground research laboratories (URL), from natural/archaeological analogs (MacKinley and Alexander, 1992, 1993; Smellie et al., 1997), and from explanatory/predictive modeling complement each other. Indeed, numerical calculations are one of the mainstays of the environmental sciences (Miller et al., 2010), used as a bridge between current process knowledge and predictive capabilities. This integrated approach is necessary to solve the complexity of the multispace and temporal-scale issues arising both from the experimental and modeling methods. The space scale covers the nanometer to kilometer range, ranging, for example, from surface aqueous complexation in the interlayer space of the clay mineral to the size of a nuclear waste repository or a well drilled for hydrocarbon exploration. In addition, experiments integrate different scales from laboratory to natural analogs via URL (Savage, 2011). The temporal-scale ranges from picoseconds to millions of years, from phenomena occurring at the molecular scale to the times targeted by performance assessment.

5.2 PERTURBING THE PHYSICOCHEMICAL CONDITIONS IN THE SUBSURFACE: DESATURATION AND OXIDATION

Deep clay-rocks foreseen for nuclear waste disposal are reducing environments (see Chapter 3, in this volume). In these sedimentary formations some iron-bearing minerals exist. Iron may be present (1) in the structure of clay minerals (Fe(II) and Fe(III), the latter being predominant (Stucki, 2013)); (2) in the structure of pyrite or siderite (Fe(II)), (3) adsorbed on the edge surfaces of the clay mineral, or (4) in an exchangeable form in the interlayer space (Hadi et al., 2013; Didier et al., 2014). Oxidation will occur in the anaerobic host-rock during construction (excavation, drilling operations) and operations (gallery ventilation) in a geological repository. Under these conditions the prevailing reducing condition will be perturbed, redox-sensitive minerals will react, and this may affect the hydromechanical host-rock properties (Schmitz et al., 2007).

This phenomenon is well known and has been observed for example in mining environments where pyrite reacts under oxidative conditions leading to acid-mine drainage. Likewise, sulfate concentration increase has been observed in clay-rock pore water (De Craen et al., 2004, 2008) after exposure to air. However, in the case of clay-rock, the pH buffer capacity is much higher. Strong pH decrease is unlikely, because such a decrease will be compensated by calcite dissolution or amphoteric clay layer edge sites reactivity. With calcite dissolution, Ca^{2+} is released into the pore water and triggers ion exchange reactions that increase Na^+, K^+, and Mg^{2+} concentrations (De Craen et al., 2004). In addition, the high Ca^{2+} and SO_4^{2-} concentrations induce gypsum precipitation (Charpentier et al., 2001). This could also happen during clay-rock desaturation likely to occur as a result of gallery ventilation (Lerouge et al., 2014). Jarosite precipitation is also reported concomitantly with the absence of calcite (De Craen et al., 2008), which is in agreement with the fact that jarosite is an indicator of acidic conditions (Elwood Madden et al., 2012). Other sulfate minerals phases such as celestite, bassanite, and natrojarosite have been observed (Charpentier et al., 2004; Vinsot et al., 2014) as well as precipitation of iron hydroxides. Clay-rocks also contain organic matter (Courdouan et al., 2007; Deniau et al., 2008; Schäfer et al., 2009) that, when oxidized, releases oxygen-functionalized compounds (e.g., carboxyl groups); these compounds might be mobilized by water and participate to further reactions (Faure et al., 1999; Faure and Peiffert, 2007; Blanchart et al., 2012).

Apart from the above-mentioned mineral dissolution and precipitation and organic matter reactivity, mineralogy itself seems only weakly affected. Detailed analysis of Toarcian shale samples from the Tournemire site (France) revealed minor differences between the illite–smectite (I-Sm) mixed-layered mineral composition of preserved and oxidized samples, the latter being slightly enriched in Sm layer (Charpentier et al., 2004). The results rely on X-ray diffraction (XRD) pattern deconvolution, and more advanced analytical identification such as multispecimen methods (Lanson et al., 2009; Sakharov and Lanson, 2013) might be used to unambiguously confirm this point. Electron energy-loss spectroscopy measurements indicate an increase in the Fe(III)/total Fe ratio of I-Sm particles. Though relevant mineral pathways induced by oxidation of clay-rock are well established, some uncertainties remain concerning both the identification of all the species involved in the oxygen reduction and the reaction kinetics. An in situ experiment was designed by ANDRA to tackle these issues (Vinsot et al., 2012).

The experimental setup is based on gas circulation in a borehole and seepage water chemistry being monitored as a function of time. The in situ experiment is still running and the comparison between experimental and modeling data will help understanding the different mechanisms related to oxidation phenomena as has already been done for pristine pore water.

Also relevant for safety analysis and long-term clay-rock evolution, De Craen et al. (2008) reported that interaction time seems to have a limited impact on Boom clay since observations performed on a drift excavated in 1987 and a connecting gallery excavated in 2002 indicate the same behavior, from the mineralogical and pore water composition point of view. Mineralogical changes have been observed within a distance of 4.5 cm around the lining concrete/clay-rock interface, whereas pore water composition was reported to be modified within 1 m of the clay-rock when compared with the pristine pore water. On another clay-rock formation (Toarcian shale) and on a longer timescale (100 years), the impact of oxidation also appears to be limited and localized at the clay-rock surface (Charpentier et al., 2004). For Opalinus clay, constraints on oxidation phenomena and processes have been derived from studies at two localities: the 140-year-old Hauenstein railway tunnel and the 6-year-old Mont-Terri tunnel (Mäder and Mazurek, 1997). Associated with the excavation-disturbed zone and the fracture networks, brownish oxidation zones extending 3–15 mm into matrix clay-rock have been identified at Hauenstein. In all these studies, even after a rather long period of time ($\sim$100 years), the oxidation front is located not more than a few centimeters from the surface exposed to the atmospheric air, following the geometry of the excavation-disturbed zone or fracture network. This demonstrates that these are a good path for oxygen transfer. Recently, Vinsot et al. (2014) reported on a comprehensive study on oxidized features that have been observed on 115 boreholes cored in the URL (Meuse Haute Marne). Observations were made on cores sampled from a few days earlier to 6.5 years prior with some samples drilled parallel or perpendicular to the horizontal major stress field. At a macroscopic scale, three main oxidizing features were observed: (1) oxidized sedimentary elements, mainly bioturbation filled by pyrite and sometimes fossils, marked by a rust-brown color, (2) oxidized patina, thin layers of iron oxides and hydroxides, identified by a rust color observed on fracture walls, and (3) white gypsum spots. Their locations depend on the fracture network geometry, which itself depends on the orientation of the drift in relation to the orientation of the in situ stress field (Armand et al., 2014). Associated with the excavation-induced fracture pattern, two zones are distinguished: traction and a shear zone in which the hydraulic conductivity is greater than and similar to that of the pristine zone respectively. With increasing time, it seems that oxygen diffusion and interaction with the clay-rocks starts in the traction zone and, after few years ($\sim$2 years), the shear zone is then invaded. It is worth noting that oxygen did not reach the limit of the excavation-induced fracture zone as, after six years' oxidation, features have been observed at up to 1.8 m, whereas this zone can extend as far as 4.5 m. In addition to oxygen diffusion, drilling is also associated with desaturation and water evaporation, leading to increased salt concentration in the pore water (Zheng et al., 2008; Vinsot et al., 2013). Even though changes in the clay-rock porosity caused by oxidation and the associated mineral dissolution/precipitation may seem weak (Zheng et al., 2008), the pore water that will seep into the drifts and the gallery (e.g., after closure of the repository) will interact with the oxidation products and the salt inherited by water evaporation. This saltier water will interact first with the repository materials. The effect of the oxic transient on repository material like cement-based materials or carbon steel is not addressed in this chapter, since discussion will appear selectively in the next sections, but clearly a complex oxidizing/reducing front will develop (De Windt et al., 2014).

5.3 INTRODUCING ALLOCHTHONOUS SOLID MATERIALS IN THE GEOLOGICAL ENVIRONMENT

5.3.1 Concrete/Clay Interactions

Although cement and calcium silicate hydrates (CSH, an important component of concrete) share two essential properties with Sm, namely, a layered structure and electrically charged surface (Grangeon et al., 2013a,b; Van Damme and Pellenq, 2013), the chemical interactions and compatibility of cement-based materials with clay-rocks have been widely studied in recent decades. Indeed, the pH of pore water in either clay-rocks or bentonite (Bent) is in the range 7–8 (see Chapter 3, in this volume), whereas the progressive degradation of cement materials leads to a pH in the cement-pore fluids ranging over time from 13.5 to 10 (Vieillard and Rassineux, 1992). Early calculations discussed by Gaucher and Blanc (2006) and Savage et al. (2007), based on mass balance assumptions, lead to the estimate of approximately 0.2–1 m^3 of bentonite will be needed to buffer 1 m^3 of concrete. In fact, things are much more complex, and as described later, the spatial extension of the alkaline plume is much more limited. However, recalling this early calculation is interesting because it probably explains why scientific communities put a lot of effort into examining the impact of an alkaline plume on clay barriers in the context of deep geological disposal. Thanks to these efforts, a better understanding of this complex interface has emerged at least from the chemical and mineralogical point of view. These efforts also contributed to the development of low-alkaline concrete (Bach et al., 2012; Lothenbach et al., 2012; Dauzères et al., 2014) in order to reduce the pH gradient at clay barrier/concrete interfaces.

Comprehensive reviews (Gaucher and Blanc, 2006; Savage et al., 2007) and data summary tables (Dauzères et al., 2010) on clay mineral concrete interactions already exist. The focus will therefore be put on the challenges highlighted in these reviews.

5.3.1.1 Batch Experiments: Interaction between Cement-Pore Water and Clay Mineral

Because of their ease and potential for covering a large range of experimental conditions, many experiments were conducted in batches, (1) on pure clay minerals, bentonite, clay-rocks, and clay mineral fraction of clay-rocks (e.g., Claret et al., 2002); (2) in a wide temperature range (20–300°C); (3) in a pH range of 9.5–13.5 that is representative of changes in chemical cement-pore fluids changes over time; (4) for reaction times that can reach roughly 2 years but generally around 1–2 months; and (5) at last but not least, with a great variety of liquid to solid ratios. Even though batch experiments present some drawbacks (Gaboreau et al., 2012), their analysis can give insight into Sm reactivity. Smectites are the most studied minerals because they can be found both in sedimentary formations targeted for hosting repositories (Claret et al., 2004) and in bentonite investigated as geotechnical barriers and backfill materials (Dohrmann et al., 2013). In addition to the presence of accessory minerals like carbonate and gypsum that play a role, Sm reactivity depends on its composition and layer-charge localization (Kaufhold and Dohrmann, 2009, 2010, 2011; Fernandez et al., 2014). The general trend of montmorillonite (Mt, a subgroup of smectite) alterations is summarized by Gaucher and Blanc (2006): first ion exchange occurs followed by a beidellization or an illitization, next stage being the neoformation of secondary phases like zeolites, CSH, and aluminum-substituted calcium silicate hydrate (C-A-S-H). As the so-called "early" cement-pore fluids might contain a high amount of potassium (Anderson et al., 1989), many experiments focus on the influence of potassium on Sm stability. As already stated above, illitization via mixed-layered mineral formation is often reported, in line with the pioneering work of Eberl et al. (1993) and Bauer and Velde (1999). It is worth noting that this illitization process can be overestimated if the XRD patterns are not examined carefully (Kaufhold and Dohrmann, 2009, 2010; Ferrage et al., 2011). Indeed most often the results are based solely on the ethylene glycol solvation test performed after the reaction of the clay mineral fraction with KOH, and, in doing so, the collapsed layers are wrongly interpreted as illite. As in a real storage situation, both the temperature and the liquid to solid ratio will be even less favorable for illitization, the preeminent mechanism is probably the modification of the exchangeable cation population that can be also identified in in situ samples (Gaboreau et al., 2012).

5.3.1.2 From Batch Experiment to Cement-Based Materials/ Clay-Rock Interface

Even though clay mineral fraction reactivity has been widely studied, it should be reminded that in addition to the clay mineral fraction complexity itself (sedimentary formations often contain kaolinite (Kaol) and chlorite (Chl) in addition to mixed-layered minerals with various illite content, (see for example Chermak, 1993; Claret et al., 2004; Honty et al., 2010, 2012), other rock-forming minerals are often observed such as carbonates, quartz, and pyrite. Thanks to this mineral assemblage, the partial pressure of CO_2 in sedimentary formations can be more than 10–100 times greater than atmospheric pressure of CO_2 (Gaucher et al., 2009), inducing thus a strong buffer capacity. Concrete also cannot simply be mimicked by a hyperalkaline fluid or a fluid at equilibrium with portlandite. Its composition will depend on its formulation but among other minerals it will often contain calcium hydroxide (CH, portlandite), calcium silica hydroxides having different Ca/Si (C/S) ratios, and aluminate, calcium, and sulfate bearing phases like ettringite and calcium monosulfato-aluminate hydrate (CmSAH) (Van Damme and Pellenq, 2013). In addition, as reported by Dauzères et al. (2010) and Gaboreau et al. (2012), until recently little attention was paid to the cement alteration itself.

Some laboratory experiments try to reconcile the necessity of looking at the reactivity of the clay-rock-forming minerals in contact to a cement-based material by performing experiments that put into contact discs of clay and cement materials (Fernandez et al., 2006; Dauzères et al., 2010). The experiments of Dauzères et al. (2010) are probably more realistic in terms of temperature and transport regime chosen. They clearly demonstrate the alteration of both cement and clay-rock adjacent to the interface with the carbonation of the interface, portlandite dissolution and a C/S decrease in the CSH phases, and ettringite precipitation. More disputable is the Sm to illite transformation described by the authors (see the discussion on illitization above). This mineralogical transformation does not seem to induce porosity clogging, whereas this has been observed for in situ samples (Gaboreau et al., 2011, 2012). This apparent difference between lab and in situ observations can be linked to a different contact time and also the difficulty in accurately reproducing in situ parameters such as pCO_2, pH, and Eh and the exact composition of the pore water in the laboratory. Some in situ experiments already exist (Read et al., 2001; Tinseau et al., 2006). In the HADES URL (in Mol, Belgium) ordinary Portland cement has interacted with Boom clay over a period of 18 months at 85°C (Read et al., 2001). The altered zone across the interface is narrow (100–250 μm) and in addition to a porosity increase in the zone of portlandite dissolution, a narrow Mg-Al-Si rich band in the clay close to the contact is reported. The analyses conducted indicate the formation of a diphasic (Mg-aluminate hydroxide and Mg-silicate hydroxide) gel with low crystallinity and compositions similar to hydrotalcite and sepiolite (Sep), respectively. Such a complex zonation with Mg enrichment adjacent to the interface has also been described by Jenni et al. (2014) and this is also correlated to the nature of the concrete, namely, ordinary Portland cement versus low-pH cement. This reactivity difference between the two cements is also supported by leaching experiments (Dauzères et al., 2014). Even if this Mg phase found at the interface has not yet been clearly identified (Is it M-S-H, M-CSH, Sep, or something else?), all the in situ experiments clearly indicate portlandite dissolution, decreased C/S ratio for the CSH, carbonation at the interface, and modification of the cation population within the clay mineral interlayer spaces. Associated with these mineralogical changes, the porosity is also modified. Based on autoradiography measurements, Gaboreau et al. (2012) showed clogging porosity in the clay-rock while the porosity increases in the cement in some cases; they also clearly depict a more complex picture and at least one heterogeneous process that depends on conditions experienced by the samples (e.g., saturated versus nonsaturated conditions, the interface geometry, the existence of a fissure network). This heterogeneity was also described by Jenni et al. (2014). In addition to consistent mineralogical paragenesis, the described alteration zones are in the micrometer range. One may wonder if the small size of the altered zone is linked to the interaction time (<15 years) in these experiments. On this aspect industrial and natural analogs are very useful because in addition to validating reaction pathways, how the altered zone extends can be evaluated for long timescales.

The following section emphasizes industrial and natural analogs that can give insight on interaction distance rather than an exhaustive description (which can be found in the review performed by Savage, 2011). In the 125-year-old Tournemire railroad tunnel, the Toarcian clay-rock has been in contact with the tunnel masonry (siliceous lime) approximately 70 m over the Cernon fault (Tinseau et al., 2006). The observed mineral pathways depend both on water flow rate and on dry or wet conditions. In any case the reported mineralogical modifications (dolomite neoformation and leaching of Chl and Kaol occur) are limited to few centimeters. One of the most famous analogs regarding clay mineral concrete interaction is the Maqarin analog in the north of Jordan (Khoury et al., 1985; Alexander et al., 1992; Khoury et al., 1992). In this area, hyperalkaline groundwater is the product of low-temperature leaching of an assemblage of natural cement minerals produced as a result of high-temperature/low-pressure metamorphism of marls (i.e., clay biomicrites) and limestones. On the hydraulic downstream of the cement zone, alkaline groundwater circulated through fractures within the biomicrite clay. On the edge of the fractures, calcite, Kaol, silica, low amounts of illite, albite, and organic matter dissolved. Within the fractures, different opening/clogging stages may have occurred, leading to a complex mineralogical pathway. Aragonite precipitated first, followed by a solid solution of ettringite–thaumasite and gypsum. Fracture activation precipitated jennite and tobermorite. In addition, some zeolites corresponding to the last stage of mineralization were locally observed. Yugawaralite ($CaAl_2Si_6O_{16}\cdot 4H_2O$) and mordenite ($(Ca,Na_2,K_2)Al_2Si_{10}O_{24}\cdot 7H_2O$) were observed for the water oversaturated with quartz; laumontite ($Ca(AlSi_2O_6)_2\cdot 4H_2O$) and epistilbite ($CaAl_2Si_6O_{16}\cdot 5H_2O$) for the water at equilibrium with quartz; chabazite ($(Ca,Na_2,K_2,Mg)Al_2Si_4O_{12}\cdot 6H_2O$) for the water undersaturated with quartz. The alteration front within the clay-rock matrix had an extension of 1–4 mm congruent to thaumasite and ettringite precipitation. In total, for a 100000-year interaction period, the perturbation front can be evaluated at 40 mm around the fracture.

5.3.1.3 Reactive Transport Modeling: A Tool to Both Describe Experiments and Predict Clay Mineral Concrete Interactions on a Long Timescale

One question addressed to the scientific community is how clay mineral concrete interfaces will behave over long term in a safety context. Even though the way to link the scientific data to safety analysis is still a matter of debate (Grambow and Bretesché, 2014; Grambow et al., 2014), reactive transport modeling has been widely used to describe the mineralogical changes occurring at clay mineral concrete interface and to predict the behavior of this interface on a long timescale, i.e., 100000 years (Steefel and Lichtner, 1994, 1998; Savage et al., 2002; Soler, 2003; De Windt et al., 2004; Gaucher et al., 2004; Vieillard et al., 2004; Trotignon et al., 2006, 2007; De Windt et al., 2008; Marty et al., 2009; Trotignon et al., 2009; Savage et al., 2010a; Kosakowski and Berner, 2013; Shao et al., 2013; Liu et al., 2014). A simplified analytical model has been proposed (Neretnieks, 2014), useful from a performance assessment point of view, however, it cannot reproduce the complexity of the different phenomena occurring at the interface, and using reactive transport modeling is more reliable. Various reactive transport codes exist (Steefel et al., 2014) and their accuracy, robustness, completeness, and numerical stability to describe multicomponent reactive transport across a clay-rock/cement interface has been successfully benchmarked (Marty et al., 2015). It is worth noting that over the last decade, simulations have been performed using modeling strategies of growing complexity. With increasing code capability, and among other things, more minerals have been taken into account, mineral dissolution–precipitation has been considered both at local equilibrium and then kinetically controlled, and special discretization has been refined. At the same time, thermodynamic databases for clay mineral and concrete phases have been improved (Blanc et al., 2012; Giffaut et al., 2014). Therefore in order to evaluate the impact of different modeling assumptions, Marty et al. (2014) have performed calculations with a consistent set of data and input parameters arranged with increasing orders of complexity. This standardized approach allows for proper comparison of numerical results and shows that modeled reaction pathways appear to be independent from the modeling assumptions chosen. Another important finding is that the geochemical transformations remain located very close to the clay mineral cement interface, in agreement with previously described findings.

Based on the existing experiments and modeling, it can be concluded that the extension of the mineralogical change will be spatially limited (within the centimeter range) across the clay mineral concrete interface. Finally, transient state influence (oxidation, desaturation, hydrogen production) on clay concrete interfaces has seldom been studied, at least from the experimental point of view, and some efforts from scientists would probably be necessary on this aspect too.

5.3.2 Steel Corrosion in Clay

In the context of deep geological disposal of nuclear waste in clay environments, the physicochemical conditions encountered during most of the lifetime of the repository are circumneutral pH and reducing redox potential, with temperatures ranging from 20°C to 90°C. The transport conditions in the vicinity of the iron-clay interface are diffusive considering the low permeability of compacted bentonite or clay-rock.

In terms of chemical behavior, carbon steel is unstable in the presence of water. When corrosion of metallic iron or steel occurs in reduced conditions, aqueous iron is released in solution and hydrogen is produced as a result of the dissociation of water and the reduction of protons, leading to a pH increase and a decrease in the redox potential according to the following overall reaction: $Fe(s) + 2H_2O \rightarrow Fe^{2+} + H_2 + 2OH^-$. Aqueous iron precipitates at the surface of the metallic component to form a dense corrosion layer, usually iron oxides and hydroxides. When corrosion occurs in the geological environment, iron may also be partly transferred further along to react, diffuse into the clay environment, and be retained on the surface of clay minerals. In the first internal corrosion layer, an additional amount of hydrogen may be produced if the corrosion products contain iron in ferric form, such as magnetite (potentially resulting from the transformation of initial metastable ferrous hydroxide): $3Fe^{2+} + 6OH^- \rightarrow 3Fe(OH)_2(s) \rightarrow Fe_3O_4(s) + 2H_2O + H_2$. The thickness of this layer depends on the geochemical environment and the ability to evacuate the reaction products (e.g., King, 2009; King et al., 2014b). In the outer corrosion layer, other corrosion products may form by incorporating chemical elements such as carbonates, silicates, and sulfides, provided by the dissolution of primary minerals from the clay barrier, and therefore potentially driving further mineral transformations. During the course of iron-clay interactions, this mineral paragenesis can evolve along with the pH and Eh parameters. The specific effect of the hydrogen, on the other hand, will be considered in Section 5.4.1.

A sensitivity analysis shows that the corrosion rate and the chemical reactivity of hydrogen gas are of paramount importance in determining the extent of the perturbations in the system (Bildstein et al., 2012). The corrosion rate is strongly linked to the nature of the minerals constituting a passive layer (iron oxides/hydroxides precipitate at the higher corrosion rates, iron carbonates and silicates at lower rates) and its thickness, and it also influences the paragenesis of secondary phases. A wide range of values for the initial corrosion rate can be found in the literature, depending on the experimental and environmental conditions, up to about 100 μm/year (see data review in

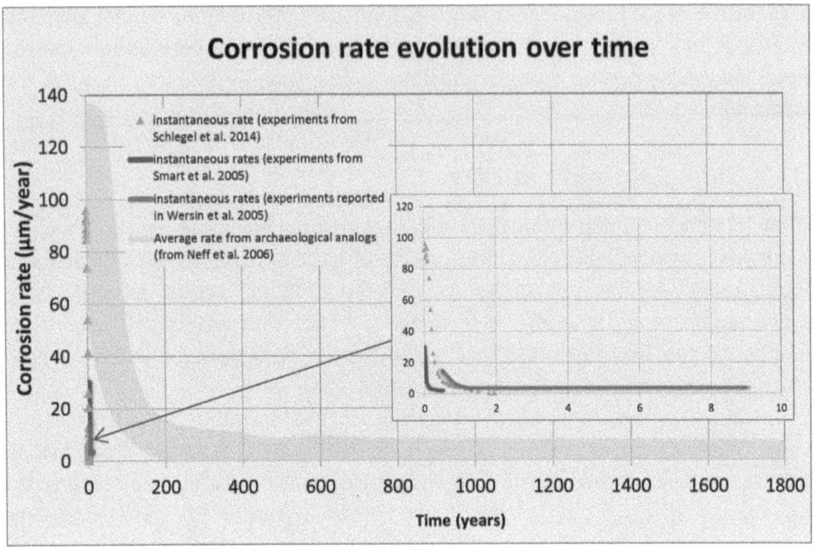

FIGURE 5.2 Instantaneous corrosion rate of metallic iron and steel measured in laboratory and in situ experiments, and average corrosion rate estimated from archaeological artifacts.

Neff et al., 2006; Féron et al., 2008; Wersin et al., 2008; King, 2009). Experiments show that the corrosion rate is strongly dependent on temperature and iron/clay mineral ratio, and that the system's "confinement" character (powder vs compacted or massive materials) plays a role in the long-term rate (several months): values <3–4 µm/year for batch experiments (from 25 to 90°C) down to 0.1 µm/year for longer-term integrated experiments (Figure 5.2). This trend is confirmed by average corrosion rates estimated from archaeological analogs (Neff et al., 2006) and by most in situ experiments.

Recent reviews of experimental work emphasizing mineral paragenesis and clay mineral alteration show a rather convergent set of observations (Wersin et al., 2008; King, 2009; Mosser-Ruck et al., 2010; Savage, 2012). Mineral characterization from iron-clay mineral interaction systems in a wide range of conditions show that magnetite (sometimes associated with maghemite γ-Fe_2O_3) is commonly observed adhering to the iron surface in most of the occurrences (experiments and archaeological artifacts) either as (1) a submicrometric internal corrosion layer (Lantenois et al., 2005; Charpentier et al., 2006; Smart et al., 2006; Carlson et al., 2007; Martin et al., 2008; Schlegel et al., 2008, 2010; De Combarieu et al., 2011; Schlegel et al., 2014) or (2) a thinner, sometimes discontinuous, layer (~10–100 nm) as observed using scanning transmission X-ray microscopy (Michelin et al., 2012). This internal corrosion layer is thought to be at the origin of the passivation effect controlling the corrosion rate. In some experiments, amorphous iron hydroxide $Fe(OH)_2$ or iron oxide are observed in place of magnetite in a 6-month time frame at 50°C (replaced by magnetite after 2 years in Milodowski et al., 2009) but also in long-term experiments (10 years at 80°C; Ishidera et al., 2008).

External corrosion products precipitating at the surface of magnetite include iron carbonates (Ca incorporating siderite $(Fe,Ca)CO_3$ and chukanovite $Fe_2(OH)_2CO_3$) and iron silicates (Fe-serpentine-type minerals, greenalite $Fe_3Si_2O_5(OH)_4$, cronstedtite $Fe_4SiO_5(OH)_4$) (Wilson et al., 2006a; Martin et al., 2008; Perronnet et al., 2008; Schlegel et al., 2008, 2010; De Combarieu et al., 2011). The actual mineral paragenesis results from the competition between the mineral dissolution and precipitation kinetics, the local geochemical conditions (pH, Eh), and the transport of reaction products. In some archaeological artifacts, siderite (enriched in Ca^{2+}) and chukanovite are present at the contact with the iron surface and a thin discontinuous magnetite layer is displaced toward the interface with clay mineral (Neff et al., 2005; Saheb et al., 2009, 2010). This feature can be interpreted as relics of an early oxic stage.

In a large number of experiments, the pH at the interface (which is difficult to measure in integrated experiments), or the pH of solutions in batch experiments, is slightly alkaline as a result of the corrosion process, with values increasing up to about 9–10 (Wilson et al., 2006b; De Combarieu et al., 2007: Perronnet et al., 2007; Ishidera et al., 2008; Perronnet et al., 2008). This pH perturbation is responsible for the presence of a transformed clay mineral layer showing systematic dissolution of primary clay minerals: in experiments with clay-rocks, illite, and I-Sm (Jodin-Caumon et al., 2012) tend to dissolve preferentially along with the occasional destabilization of quartz and dolomite (Bourdelle et al., 2014). In experiments involving bentonite, the primary Sm is strongly altered (De Combarieu, 2007; Perronnet et al., 2007; Mosser-Ruck et al., 2010). Along with the influx of iron, the system evolves with the precipitation of either Fe-serpentine (Lantenois et al., 2005; Perronnet et al., 2008; Schlegel et al., 2008, 2010; De Combarieu et al., 2011; Jodin-Caumon et al., 2012; Bourdelle et al., 2014) or iron-rich Sm (Guilllaume et al., 2004; Charpentier et al., 2006; Wilson et al., 2006a).

Recent experiments confirm these results, but they also highlight the differences in the mineral paragenesis between batch and integrated experiments. In the batch experiments described by Bourdelle et al. (2014), a 0.1 ratio of iron and Callovo-Oxfordian clay-rock (COx), both in the form of powder, reacted for 3 months in an agitated vessel at 90°C. The total corrosion of iron occurred at circumneutral pH without precipitation of magnetite. A Fe-serpentine mineral was observed as the main corrosion product resulting from the total dissolution of the interstratified I-Sm mineral and partial dissolution of illite and quartz. Similar results were obtained in the same type of experiments in a 90–40°C thermal gradient (Pignatelli et al., 2014), but with magnetite precipitating over the whole temperature range. The iron-rich secondary phyllosilicate minerals were identified as greenalite and cronstedtite. These results are to be compared with those obtained in the series of integrated

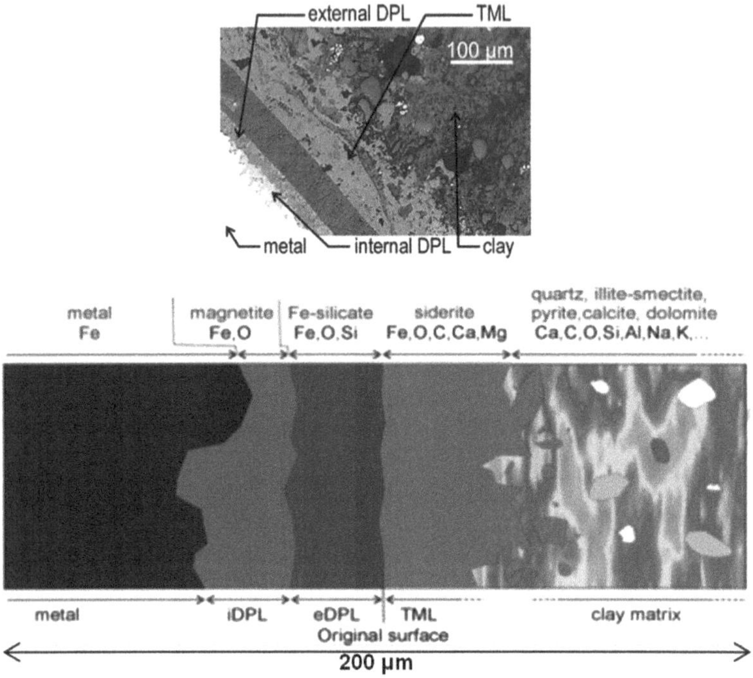

FIGURE 5.3 Electron microscopy in backscattered electron mode and corrosion products identified in the internal and external dense product layer (DPL) and in the transformed clay matrix layer (TML) after 2 years of iron corrosion in contact with clay-rock at 90°C (Schlegel et al., 2008, 2014).

experiments called "Arcorr" (for "corrosion in argilites") involving the corrosion of an iron rod and a "microcontainer" in massive COx clay-rock for 2 years at 90°C (Figure 5.3). In these experiments, a layer of magnetite was identified at the contact with iron, along with a layer of Fe-rich phyllosilicates and chukanovite (for the microcontainer) in the external corrosion layer, and Ca-enriched siderite in the micrometric-scale transformed clay-rock (Martin et al., 2008; Schlegel et al., 2008, 2010, 2014).

Alteration of the clay material is usually associated with the pH perturbation and, to a lesser extent, to the migration of aqueous iron. The alteration zone extends to distance from 0.1 to 2 mm in integrated experiments in the average time frame of a year (Smart et al., 2006; Carlson et al., 2007; Ishidera et al., 2008; Schlegel et al., 2008; Milodowski et al., 2009) and in in situ experiments (after 6 years in Gaudin et al., 2013), and up to 3 mm in several hundred years old archaeological analogs (Neff et al., 2005, 2006). Within this zone, physicochemical alteration is observed which may ultimately affect properties such as permeability, diffusivity, and CEC. In batch experiments with FoCa7 Bent at 80°C, Perronnet et al. (2008) measured a decrease of up to 50% of the CEC as the iron/clay mineral ratio increased. In integrated experiments however, the results were more diverse: no significant change in mineralogy and diffusion properties (Xia et al., 2005; Ishidera et al., 2008); no Mt transformation, CEC decrease, and permeability increase of up to two orders of magnitude (Carlson et al., 2007); no significant changes in mineralogy, no change in CEC but slight change in swelling pressure suggesting that the original Mt was transformed into iron-rich dioctahedral Sm (Gaudin et al., 2013; Milodowski et al., 2009).

The effect of the transition between oxic and anoxic conditions has also been investigated very recently in laboratory experiments (Ishidera et al., 2008; Jeannin et al., 2010, 2011; El Hajj et al., 2013), in situ experiments (Gaudin et al., 2013), and archaeological analogs (Saheb et al., 2014). Interestingly, the mineral phases precipitating during the transient oxic phase tend to be destabilized and to form the corrosion products and secondary minerals similar to those observed in the anoxic phase (magnetite, ferrous carbonates, and 7Å Fe-rich phyllosilicates).

Reactive transport modeling has also been intensively used to investigate the iron-clay interactions using Bent or clay-rock for the barrier and including processes such as kinetics for mineral dissolution/precipitation, ion exchange, and variable porosity. In the absence of a descriptive geochemical model for corrosion in reactive transport codes, authors usually assume a constant corrosion rate in the modeling studies (based on an average value for the specified temperature) or take into account the decreasing corrosion rate by using a solubility limit for iron (Marty et al., 2010). Other modeling work fits experimental data with an analytical curve (Hunter et al., 2007; Wersin et al., 2008; Bildstein et al., 2015). A new direction has been taken lately with models calculating the corrosion rate as limited by the diffusion of reactants and products of the reaction through the dense corrosion product (magnetite) layer (Peña et al., 2008; Bildstein et al., 2015).

The modeling results overall converge to predict a pH increase up to 8–9 (Montes et al., 2005; Hunter et al., 2007; Lu et al., 2011; Wersin and Birgersson, 2014), and even higher up to 10–11 at the iron-clay interface (Bildstein et al., 2006; Samper et al., 2008; Wersin et al., 2008; Marty et al., 2010; Savage et al., 2010a). The lower pH values are usually associated with high precipitation rates (or local equilibrium assumption) or low corrosion rates for the temperature considered (≤ 1 μm/year). Farther into the clay material, simulations show a rapid decrease in pH irrespective of the hypotheses concerning the kinetics of mineral reactions or the presence of ion exchange in the simulation. The corrosion products predicted by the models are dominated by magnetite, along with Fe-carbonates (Ca-siderite, chukanovite) and Fe-silicates (greenalite, cronstedtite, berthierine) in the dense layer. In some cases, the precipitation of magnetite is only transient, usually due to competition with iron silicates, especially when high reactive surface areas and/or precipitation rate for secondary minerals are used (De Combarieu et al., 2007; Savage et al., 2010a; Ngo et al., 2014).

In all the simulations performed in the waste disposal conditions, the spatial extent of the perturbation in the clay barrier remains limited (about 10 cm/10000 years). This thin alteration zone is characterized by the dissolution of the primary clay minerals (Mt, illite, and I-Sm, in different proportions depending on the clay mineral and the simulation conditions), and also in some conditions, of quartz and carbonates. Precipitation of iron carbonates is also predicted (siderite or Ca-rich siderite, but chukanovite is difficult to obtain). Fe-bearing secondary minerals also precipitate as a result of the supply of silica and aluminum: Fe-saponite, Fe-Mt, and nontronite. Fe-bearing Chl is also predicted in some simulations (Marty et al., 2010), representing the longer term end product of the transformation (these minerals are observed only in high temperature experiments and introduced in the simulations as the product of transient Fe-rich Sm ripening process).

Porosity reduction (clogging) is predicted in the corrosion product layer in long-term calculations, but it is usually accompanied by porosity increase in the clay barrier (Montes et al., 2005; Bildstein et al., 2006; Samper et al., 2008; Marty et al., 2010), so the net balance of pore volume is not easy to interpret. Moreover, the petrophysical and transport properties, and the mechanical durability of these alteration zones are not well known.

In an effort to refine the understanding of the corrosion process and provide a coupling between the iron corrosion and the clay mineral alteration process, phenomenological electrochemical models for corrosion have also been developed. This type of model is usually based on the existence of an oxide layer controlling the corrosion rate: a thin magnetite layer forms at the interface with iron (internal layer) and dissolves at the contact with an external layer or clay, thus maintaining a nonzero layer thickness and propagating toward the internal part of the steel component (e.g., Bataillon et al., 2010; King et al., 2014b). These mechanisms are also invoked to explain the mineral paragenesis in the experiments (e.g., Martin et al., 2008; Schlegel et al., 2008, 2010) and are supported by ^{18}O diffusion experiments in the dense corrosion layer (Vega et al., 2005, 2007), and by impedance measurements in the same conditions (Jeannin et al., 2011). The coupling of this type of models with reactive transport models in clay environments does however remain a challenge for long-term simulations due to the small time step required to solve the electrochemical equations.

5.4 CHEMICAL PERTURBATIONS DUE TO ALLOCHTHONOUS GAS

Gas transport may be fast but is also controlled by changes in the clay mineral properties especially for CO_2 injected in geological storage and H_2 produced in nuclear waste repositories (cf. Chapter 8). According to gases' chemical reactivity rates, the alteration can therefore be both very localized, e.g., around the casing of an injection well, or affects a large region at the interface between the reservoir and the caprock or along the migration path in the heterogeneities (technological gaps) in the host-rock.

5.4.1 H_2 Injection/Production

The presence of dihydrogen gas (H_2) mainly results from injection into subsurface reservoirs for storage purposes and from in situ production by chemical reactions (metal corrosion) or radiolytic reactions (water radiolysis, bitumen self-irradiation, etc.).

When H_2 is produced in situ, a gas phase may form if solubility is exceeded, depending on the local physicochemical conditions (temperature, pressure) and the amount of H_2 released in solution. These conditions are generally reached in the vicinity of deep geological nuclear repository and in permeable reactive barriers since H_2 solubility is quite low and the corrosion rate imposes H_2 production that surpasses evacuation through aqueous diffusion: overpressure of 15−45 bars is attainable with an average corrosion rate of 1−10 μm/year in repository conditions (Talandier et al., 2006; Xu et al., 2008; Senger et al., 2011). In these conditions, H_2 is present in excess in the geochemical system.

The reactivity of H_2 and its effect on the redox potential are however still under debate. It is admitted that when H_2 is produced in situ, for instance from the corrosion of iron or steel in reduced conditions, the redox potential locally drops dramatically (to about −500 mV Sakamaki et al., 2014). In this case, the redox potential value lies on the H_2/H_2O stability line in the Eh/pH Pourbaix diagram. In the absence of corrosion processes (once H_2 has been accumulated, or been injected in solution), H_2 may oxidize but another active redox couple must be present. However, aqueous species do not readily react with H_2 in deep geological conditions, e.g., sulfates, nitrates, and carbonates; some reactions need to be catalyzed either by a solid surface or by microorganisms. Effective abiotic reduction of sulfates is observed only at high temperature (>250°C), the reaction being too slow at lower temperature (Truche et al., 2009) unless it is catalyzed by sulfate-reducing bacteria (e.g., Libert et al., 2014). The abiotic reduction of nitrates using H_2 has been documented in the conditions of deep geological repositories (90°C), showing that catalysis by a metallic surface was required (Truche et al., 2013a). This reaction increases the pH and produces ammonium cations (NH_4^+), which have a strong affinity for clay mineral surfaces, and are therefore capable of shifting the cation exchange content, potentially at the expense of radionuclides (e.g., Missana et al., 2004). Both these reactions also produce an alkaline pH perturbation in clay minerals (Bildstein et al., 2012).

The most direct impact of H_2 on the clay barriers that has been confirmed by laboratory experiments is the dissolution of primary minerals due to the reduction of structural chemical elements, either directly or as microbiologically catalyzed processes. For instance, clay minerals can be destabilized as a result of the reduction of structural ferric cations: this reaction has been observed both in abiotic (in unsaturated conditions; Didier et al., 2014) and in biotic conditions (Esnault et al., 2013). Another reaction involves pyrite which is transformed into pyrrhotite through a dissolution/precipitation mechanism, releasing sulfides in solution (Truche et al., 2010, 2013b). The effect of pyrite transformation on the clay barrier stability is however very limited (a slight decrease of pH is observed) (Truche et al., 2013b). The dissolution of corrosion products such as magnetite also can result from biotic reduction of Fe(III) (Esnault et al., 2011; Kerber-Schütz et al., 2014), leading to an increased corrosion rate, and being a source of additional iron that is released in the system. The quantification of the amount of dissolved minerals is however difficult to establish and to extrapolate to compacted or plug core systems; the long-term effect of these processes on clay barrier stability, especially those involving microorganisms, needs therefore to be consolidated at the field scale. Survival and potential role of microorganisms may be hindered by physicochemical conditions and the accommodation in the pore structure in the clay barrier (Stroes-Gascoyne et al., 2011). However, gaps at the interface between the different materials in the storage facility or close to wellbores, as well as fractures created by the excavation or drilling operations are thought to be potential locations for microbial activity (Libert et al., 2014). Microorganisms have been shown to be active and their influence on H_2 consumption was confirmed through sulfate and nitrate reduction (Libert et al., 2011, 2014) as well as for ferric iron reduction (Kerber-Schütz et al., 2014).

Note that a side effect of H_2 gas production is also the decrease of water content in clay barrier. The decrease in water flow (due to capillary effects) is responsible for reduced permeable reactive barrier efficiency. Unsaturated conditions in clay barrier may also limit the chemical reactivity (Lassin et al., 2005). The potential effect of gas overpressure is addressed in Chapter 7 in this volume.

5.4.2 Injection of CO_2

Caprock integrity in the context of CO_2 sequestration has received increasing attention in the last decade (e.g., Johnson et al., 2005; Gaus et al., 2005; Fleury et al., 2010; Kaszuba et al., 2013; Song and Zhang, 2013). The injection of supercritical CO_2 (SC-CO_2) into deep geological reservoirs, and the subsequent dissolution of large amounts of CO_2 in the formation water (due to its high solubility in water), induce strong acidic perturbation in the system (pH values of 3.5–4.5 under the conditions expected for CO_2 sequestration). This acidic perturbation is a cause of concern for the confinement properties because of the high reactivity of acidified brines, especially once the buoyant fluid has reached the top of the reservoir and entered the clay caprock (directly in the form of SC-CO_2 or dissolved in the pore water). The first primary minerals to dissolve are carbonates, as they have high reaction kinetics. In a second step, less-reactive clay minerals (illite, Chl, feldspars) may be destabilized in the longer term. Once the pH has been buffered by the earlier reactions, other secondary carbonate and clay minerals also precipitate (e.g., Hellevang et al., 2013; Kampman et al., 2014). This in turn may affect clay barrier porosity and permeability properties.

These processes have been confirmed by experiments on interactions between CO_2 and clay-rocks: siderite, "corroded" magnesite, and analcime precipitate in CO_2-shale interactions at 200°C (Kaszuba et al., 2005); in experiments at 80–150°C with clay and carbonate-rich rock, primary carbonate minerals dissolves and mixed Fe-Ca-Mg carbonate precipitates, and Kaol and Sm dissolve in I-Sm while Fe-Mg-Sm (saponite?) precipitates (Crédoz et al., 2009); in shales reacting at temperatures from 80 to 250°C, ankerite, calcite, Chl, and illite dissolve and secondary calcite and Sm precipitate (Alemu et al., 2011); in the same conditions, K-feldspar dissolves and pore-bridging I-Sm precipitates (Liu et al., 2012).

Illitization of primary clay minerals has also been confirmed by dedicated batch experiments looking at individual clay minerals in interactions with acidified CO_2-rich brine; for instance, I-Sm minerals (obtained by purifying the caprock clay material from the Chinle Formation, Colorado, USA) were illitized in the presence of K-feldspar impurities at 80°C after three months reaction (with identification of secondary individual illite crystals; Crédoz et al., 2011). Smectite was also "illitized" in experiments with caprocks and brine (partial transformation into a K-rich beidellite after 3 months reaction at 80°C; Crédoz et al., 2009). Interestingly, experiments with carbonate-rich caprock from the Paris Basin (France) reacting with dry SC-CO_2, simulating the direct ingress of an SC-CO_2 phase into the caprock, also revealed illitized Sm along with precipitated Kaol (Kohler et al., 2009).

Caprock mineralogical alteration induced by CO_2 migration causes pore-texture changes and creates preferential pathways with increased porosity, permeability, and capillary properties ($\times 10$ in permeability and $+45\%$ in the entry pressure value; Wollenweber et al., 2010). It also increases diffusion coefficients (Busch et al., 2008; Wollenweber et al., 2010; Berthe et al., 2011). Differences between carbonate-rich and clay caprock behavior can be identified since these two mineral families have different reactivity. For instance, the fast dissolution of large amounts of calcite increased permeability in a carbonate-rich caprock (Ellis et al., 2010). This can also be coupled with reactivation of cracks due to increased pore pressure (Angeli et al., 2009). By contrast, global permeability decreases were observed in a clay caprock due to the accumulation of clay mineral particles coating the walls of fractures (Noiriel et al., 2007).

Studies on natural CO_2 reservoirs offer limited data for CO_2/caprock interactions because they are essentially focused on reservoir properties, potential leakage, and trace elements immobilized by acidic perturbation

(Liu et al., 2012; Kampman et al., 2014). Nevertheless, insights can be gained into the long-term behavior of CO_2-rich environments from some natural analogs: an interesting example is given by the siltstone caprock in Green River (Nevada, USA) where the diffusion of large amounts of CO_2 over a distance of 10 cm/100000 year has resulted in the dissolution of dolomite cement and iron oxide grain coatings as well as precipitation of carbonate minerals in fractures (e.g., Kampman et al., 2014).

These processes have only recently been examined using reactive transport modeling, trying to simulate clay barrier integrity over long periods of time, starting with the pioneering work of Johnson et al. (2005) and Gaus et al. (2005). In these simulations, changing fronts over distances of up to 1 m/ 1000 years shows dissolution of primary silicates (illite, albite) and precipitation of chalcedony, calcite, siderite and Kaol. Simulations with sealed and fractured caprocks show reduced porosity in homogeneous caprock through excess calcite precipitation compared with dissolution (Gherardi et al., 2007). Another series of calculations in homogeneous and fractured caprocks shows that the mineral alteration occurs over a maximal distance of ~0.1–1 m/ 10000 years, with increased porosity (+20–25%) depending on whether or not pore water invades fractures in the caprock (Bildstein et al., 2010). This study also shows that if the pore water is already pH-buffered in the reservoir (high carbonate mineral content and/or residence time) then the impact on the caprock is much lower than in cases where aggressive pore water (pH ~ 3.5) directly enters the caprock. Note that in some studies, the porosity of the caprock even decreases as a result of anhydrite precipitation in sulfate-rich shale pore water (Tian et al., 2014).

Clay minerals also demonstrate high CO_2 adsorption capacity (Busch et al., 2008), which can cause additional swelling if it occurs in the interlayer space. It can close preexisting fractures and decrease permeability (Ilton et al., 2012; De Jong et al., 2014). Fractures may be reactivated as a result of coupled chemical alteration, pressure buildup, and thermal stress (e.g., Rutqvist and Tsang, 2002; Shukla et al., 2010; Vilarrasa et al. 2014).

Finally, impurities coinjected with CO_2 as a result of CO_2 capture processes (mainly hydrogen sulfide, methane, nitrogen, nitrous oxide, sulfur dioxide, and O_2) may also have a chemical effect on clay barrier through oxidation (see Section 5.2) and further acidification of brines, but very few people have studied this effect. Experiments conducted with shales and SC-CO_2 containing a few percent of O_2 show oxidative dissolution of pyrite and precipitation of ferric iron oxide as well as mobilization of uranium; brine acidification induced calcite and dolomite dissolution and triggered gypsum precipitation (Jung et al., 2013). Modeling work is more abundant but has focused on interactions with the reservoir and indicates that the effect of H_2S and SO_2 on brine acidification should be limited upon reaching the caprock, due to solubility in SC-CO_2 and/or pH buffering of carbonate minerals (e.g., Ellis et al., 2010).

5.5 CONCLUSION: WHAT IS KNOWN AND WHAT NEEDS TO BE IMPROVED

A large number of studies have been conducted over the past decade to better understand the mechanisms of clay mineral alteration and to assess the durability of the properties of clay host-rock or caprock with respect to chemical perturbations. The overall objective was twofold: (1) to assess the intensity and the extent of the chemical perturbation, and (2) to determine the long-term effect on the stability of the clay barrier. For this purpose, a large amount of data has been acquired concerning the interactions of clay barriers with different types of solid, liquid, and gaseous materials that will be introduced into natural and engineered underground storage facilities. The mechanisms of clay mineral alteration are much better understood now, from laboratory and in situ experiments. However, these remain on small spatial and temporal scales compared to the industrial application scales. Archaeological and natural analogs have, to some degree, allowed us to extend our vision beyond the limits of experiments and helped to improve the robustness of long-term numerical simulations and therefore the predictability of the physicochemical behavior of such systems. Using the established knowledge, modeling work performed on the long-term evolution of storage systems, in a wide range of conditions, shows no significant detrimental effect on clay barrier: most barrier properties alteration is predicted over very small distances compared to the thickness of the barriers considered (a few decimeters in waste disposal and a few meters in CO_2 storage systems on a timescale of 10000 years). These predictions often result from considering that clay barriers have homogeneous properties.

The knowledge that has been acquired also raises new questions and, as with all extrapolation exercises, calls for deeper and refined understanding: heterogeneities and space/timescale upscaling are identified as the most important challenges to overcome.

Upscaling the experimental results to the field scale remains a difficult task. Reconciling the results from experiments using grinded or powdered materials with those using compacted clay mineral or plug core and steel coupons cannot be resolved solely by considering simple hypotheses for reactive surface areas: pore structure effects, heterogeneities and access to water must be taken into account (Peters, 2009; Landrot et al., 2012; Noiriel et al., 2012; Tian et al., 2014). Predicting the long-term behavior of this type of systems is also still hindered by the uncertainty in the data for the kinetics of the processes involved, especially mineral precipitation (i.e., nucleation and growth; Fritz and Noguéra, 2009; Savage et al., 2010b; Kampman et al., 2014). The key may lie in understanding the differences between the kinetic rate/reactive surface area determined in batch experiments, in integrated/in situ experiments, and in the field (e.g., Maher et al., 2006). The question of the long-term evolution of processes such as corrosion is still open even though the average rate can be somewhat bracketed on a 1000-year timescale by archaeological analogs (e.g., Neff et al., 2006; King, 2009).

The redox buffering capacity of clay minerals also remains debatable. Though the pH buffering capacity of the clay barriers (both in acidic and alkaline conditions) is now abundantly evidenced, the question of how redox potential is controlled in bentonite or clay-rock, in reduced conditions, is still subject to question. This is mainly due to the fact that the presence of readily active redox couples has not been clearly demonstrated: the sulfate reduction rate is low (considered to require biotic catalysis with H_2 or organic matter) and the amount of ferric iron in clay minerals is also low. In radioactive waste disposal, during the corrosion process, the ingress of large amounts of H_2 produces "stable," strongly reduced conditions (e.g., Duro et al., 2014). By contrast, if H_2 has escaped out of the system after completion of the corrosion phase (mainly through aqueous diffusion), the clay barrier may only be weakly buffered with respect to additional redox perturbations. This is a particularly important issue that needs to be resolved for radionuclide migration in the postcorrosion phase.

Both investigations of laboratory experiments, aged sample interfaces retrieved from URL, and numerical modeling clearly indicated localized but significant porosity modifications (clogging or opening) induced by the geochemical reactions. These porosity changes are poorly understood in terms of mechanical behavior and also regarding changes to the transport properties at the interface. Numerical codes are able to treat the latter point (Xie et al., 2015) and some coupling between reactive transport codes and mechanical codes exist (e.g., Rutqvist and Tsang, 2002; King et al., 2014a), however this remains a challenge from both the numerical and experimental point of view.

REFERENCES

Alemu, B.L., Aagaard, P., Munz, I.A., Skurtveit, E., 2011. Caprock interaction with CO_2: a laboratory study of reactivity of shale with supercritical CO_2 and brine mixtures at 250°C and 110 bars. Appl. Geochem. 26, 1975–1989.

Alexander, W.R., Dayal, R., Eagleson, K., Eikenberg, J., Hamilton, E., Linklater, C.M., McKinley, I.G., Tweed, C.J., 1992. A natural analogue of high pH cement pore waters from the Maqarin area of northern Jordan. II: results of predictive geochemical calculations. J. Geochem. Explor 46, 133–146.

Anderson, K., Allard, B., Bengtsson, M., Magnusson, B., 1989. Chemical composition of cement pore waters. Cement Concrete Res. 19, 327–332.

Anderson, R.L., Ratcliffe, I., Greenwell, H.C., Williams, P.A., Cliffe, S., Coveney, P.V., 2010. Clay swelling – a challenge in the oilfield. Earth-Sci Rev. 98, 201–216.

Angeli, M., Soldal, M., Skurtveit, E., Aker, E., 2009. Experimental percolation of supercritical CO_2 through a caprock. Energy Procedia 1, 3351–3358.

Armand, G., Leveau, F., Nussbaum, C., de La Vaissiere, R., Noiret, A., Jaeggi, D., Landrein, P., Righini, C., 2014. Geometry and properties of the excavation-induced fractures at the Meuse/Haute-Marne URL drifts. Rock Mech. Rock Eng. 47, 21–41.

Bach, T.T.H., Coumes, C.C.D., Pochard, I., Mercier, C., Revel, B., Nonat, A., 2012. Influence of temperature on the hydration products of low pH cements. Cement Concrete Res. 42, 805–817.

Bataillon, C., Bouchon, F., Chainais-Hillairet, C., Desgranges, C., Hoarau, E., Martin, F., Perrin, S., Tupin, M., Talandier, J., 2010. Corrosion modelling of iron based alloy in nuclear waste repository. Electrochim. Acta 55 (15), 4451–4467.

Bauer, A., Velde, B., 1999. Smectite transformation in high molar KOH solutions. Clay Miner. 34, 259–273.

Berthe, G., Savoye, S., Wittebroodt, C., Michelot, J.-L., 2011. Changes in containment properties of clay-rock caprocks induced by dissolved CO_2 seepage. Energy Procedia 4 (2011), 5314–5319.

Bildstein, O., Trotignon, L., Perronnet, M., Jullien, M., 2006. Modelling iron-clay interactions in deep geological disposal conditions. Phys. Chem. Earth 31 (10–14), 618–625.

Bildstein, O., Kervévan, C., Lagneau, V., Delaplace, P., Credoz, A., Audigane, P., Perfetti, E., Jullien, M., 2010. Integrative modeling of caprock integrity in the context of CO_2 storage: evolution of transport and geochemical properties and impact on performance and safety assessment. Oil Gas Sci. Technol. 65 (3), 481–502.

Bildstein, O., Lartigue, J.E., Pointeau, I., Cochepin, B., Munier, M., Michau, N., 2012. In: Proceedings of International Meeting "Clays in Natural and Engineered Barriers for Radioactive Waste Confinement", October 22–25, Montpellier, France, pp. 84–85.

Bildstein, O., Lartigue, J.E., Schlegel, M.L., Bataillon, C., Cochepin, B., Munier, I., Michau, N., 2015. Gaining insight into corrosion rate and secondary mineral paragenesis from numerical simulations of integrated iron-clay-rock experiments. In: Proceedings of "Clays in Natural and Engineered Barriers for Radioactive Waste Confinement", March 23–26, Brussels, Belgium.

Blanc, P., Lassin, A., Piantone, P., Azaroual, M., Jacquemet, N., Fabbri, A., Gaucher, E.C., 2012. Thermoddem: a geochemical database focused on low temperature water/rock interactions and waste materials. Appl. Geochem. 27, 2107–2116.

Blanchart, P., Faure, P., De Craen, M., Bruggeman, C., Michels, R., 2012. Experimental investigation on the role of kerogen and clay minerals in the formation of bitumen during the oxidation of Boom Clay. Fuel 97, 344–351.

Bourdelle, F., Truche, L., Pignatelli, I., Mosser-Ruck, R., Lorgeoux, C., Roszypal, C., Michau, N., 2014. Iron-clay interactions under hydrothermal conditions: impact of iron specific surface area on reaction pathway revealed by mineralogy and a continuous monitoring of pH, solution and gas compositions. Chem. Geol. 381, 194–205.

Busch, A., Alles, S., Gensterblum, Y., Prinz, D., Dewhurst, D.N., Raven, M.D., Stanjek, H., Krooss, B.M., 2008. Carbon dioxide storage potential of shales. Int. J. Greenh. Gas Con. 2, 297–308.

Carlson, L., Karnland, O., Oversby, V.M., Rance, A.P., Smart, N.R., Snellman, M., Vähänen, M., Werme, L.O., 2007. Experimental studies of the interactions between anaerobically corroding iron and bentonite. Phys. Chem. Earth 32, 334–345.

Charpentier, D., Cathelineau, M., Mosser-Ruck, R., Bruno, G., 2001. Mineralogical evolution of argillites in dehydrated-oxidised zones: the example of the argillitic walls from Tournemire tunnel. Cr. Acad. Sci. Ii A 332, 601–607.

Charpentier, D., Mosser-Ruck, R., Cathelineau, M., Guillaume, D., 2004. Oxidation of mudstone in a tunnel (Tournemire, France): consequences for the mineralogy and crystal chemistry of clay minerals. Clay Miner. 39, 135–149.

Charpentier, D., Devineau, K., Mosser-Ruck, R., Cathelineau, M., Villiéras, F., 2006. Bentonite-iron interactions under alkaline condition: an experimental approach. Appl. Clay Sci. 32, 1–13.

Chermak, J.A., 1993. Low temperature experimental investigation of the effect of high pH KOH solutions on the Opalinus shale, Switzerland. Clays Clay Miner. 41, 365–372.

Claret, F., Bauer, A., Schafer, T., Griffault, L., Lanson, B., 2002. Experimental investigation of the interaction of clays with high-pH solutions: a case study from the Callovo-Oxfordian formation, Meuse-Haute Marne underground laboratory (France). Clays Clay Miner. 50, 633–646.

Claret, F., Sakharov, B.A., Drits, A., Velde, B., Meunier, A., Griffault, L., Lanson, B., 2004. Clay minerals in the Meuse-Haute Marne underground laboratory (France): possible influence of organic matter on clay mineral evolution. Clays Clay Miner. 52, 515–532.

Courdouan, A., Christl, I., Meylan, S., Wersin, P., Kretzschmar, R., 2007. Characterization of dissolved organic matter in anoxic rock extracts and in situ pore water of the Opalinus Clay. Appl. Geochem. 22, 2926–2939.

Crédoz, A., Bildstein, O., Jullien, M., Raynal, J., Pétronin, J.-C., Lillo, M., Pozo, C., Geniaut, G., 2009. Experimental and modeling study of geochemical reactivity between clayey caprocks and CO_2 in geological storage conditions. Energy Procedia 1, 3445–3452.

Crédoz, A., Bildstein, O., Jullien, M., Raynal, J., Trotignon, L., Pokrovsky, O., 2011. Mixed-layer illite-smectite reactivity in acidified solutions: implications for clayey caprock stability in CO(2) geological storage. Appl. Clay Sci. 53, 402–408.

Dauzères, A., Le Bescop, P., Sardini, P., Coumes, C.C.D., 2010. Physico-chemical investigation of clayey/cement-based materials interaction in the context of geological waste disposal: experimental approach and results. Cement Concrete Res. 40, 1327–1340.

Dauzères, A., Le Bescop, P., Cau-Dit-Coumes, C., Brunet, F., Bourbon, X., Timonen, J., Voutilainen, M., Chomat, L., Sardini, P., 2014. On the physico-chemical evolution of low-pH and CEM I cement pastes interacting with Callovo-Oxfordian pore water under its in situ CO_2 partial pressure. Cement Concrete Res. 58, 76–88.

De Combarieu, G., 2007. Altération du verre de confinement de déchets type R7T7 en condition de stockage géologique. Ph.D. Thesis. Paris 11.

De Combarieu, G., Barboux, P., Minet, Y., 2007. Iron corrosion in Callovo-Oxfordian argilite: from experiments to thermodynamic/kinetic modelling. Phys. Chem. Earth 32, 346–358.

De Combarieu, G., Schlegel, M.L., Neff, D., Foy, E., Vantelon, D., Barboux, P., Gin, S., 2011. Glass-iron-clay interactions in a radioactive waste geological disposal: an integrated laboratory-scale experiment. Appl. Geochem. 26, 65–79.

De Craen, M., Van Geet, M., Wang, L., Put, M., 2004. High sulphate concentrations in squeezed Boom Clay pore water: evidence of oxidation of clay cores. Phys. Chem. Earth, Parts A/B/C 29, 91–103.

De Craen, M., Geet, M.V., Honty, M., Weetjens, E., Sillen, X., 2008. Extent of oxidation in Boom Clay as a result of excavation and ventilation of the HADES URF: experimental and modelling assessments. Phys. Chem. Earth, Parts A/B/C 33 (Suppl. 1), S350–S362.

De Jong, S.M., Spiers, C.J., Busch, A., 2014. Development of swelling strain in smectite clays through exposure to carbon dioxide. Int. J. Greenh. Gas Con. 24 (2014), 149–161.

De Windt, L., Pellegrini, D., van der Lee, J., 2004. Coupled modeling of cement/clay-rock interactions and radionuclide migration. J. Contam. Hydrol 68, 165–182.

De Windt, L., Marsal, F., Tinseau, E., Pellegrini, D., 2008. Reactive transport modeling of geochemical interactions at a concrete/argillite interface, Tournemire site (France). Phys. Chem. Earth 33, S295–S305.

De Windt, L., Marsal, F., Corvisier, J., Pellegrini, D., 2014. Modeling of oxygen gas diffusion and consumption during the oxic transient in a disposal cell of radioactive waste. Appl. Geochem. 41, 115–127.

Deniau, I., Devol-Brown, I., Derenne, S., Behar, F., Largeau, C., 2008. Comparison of the bulk geochemical features and thermal reactivity of kerogens from Mol (Boom Clay), Bure (Callovo–Oxfordian argillite) and Tournemire (Toarcian shales) underground research laboratories. Sci. Total Environ. 389, 475–485.

Didier, M., Géhin, A., Grenèche, J.-M., Charlet, L., Giffaut, E., 2014. Method development for evaluating the redox state of Callovo-Oxfordian clayrock and synthetic montmorillonite for nuclear waste management. Appl. Geochem. http://dx.doi.org/10.1016/j.apgeochem.2014.06.015.

Dohrmann, R., Kaufhold, S., Lundqvist, B., 2013. Chapter 5.4-the role of clays for safe storage of nuclear waste. In: Bergaya, F., Lagaly, G. (Eds.), Developments in Clay Science. Elsevier, pp. 677–710.

Duro, L., Domènech, C., Grivé, M., Roman-Ross, G., Bruno, J., Källström, K., 2014. Assessment of the evolution of the redox conditions in a low and intermediate level nuclear waste repository (SFR1, Sweden). Appl. Geoch. http://dx.doi.org/10.1016/j.apgeochem.2014.04.015.

Eberl, D.D., Velde, B., McCormick, T., 1993. Synthesis of illite-smectite from smectite at earth surface temperatures and high pH. Clay Miner. 28, 49–60.

El Hajj, H., Abdelouas, A., El Mendili, Y., Karakurt, G., Grambow, B., Martin, C., 2013. Corrosion of carbon steel under sequential aerobic–anaerobic environmental conditions. Corros. Sci. 76, 432–440.

Ellis, B.R., Crandell, L.E., Peters, C.A., 2010. Limitations for brine acidification due to CO_2 co-injection in geologic carbon sequestration. Int. J. Greenh. Gas Con. 4 (3), 575–582.

Elwood Madden, M.E., Madden, A.S., Rimstidt, J.D., Zahrai, S., Kendall, M.R., Miller, M.A., 2012. Jarosite dissolution rates and nanoscale mineralogy. Geochim. Cosmochim. Acta 91, 306–321.

Esnault, L., Jullien, M., Mustin, C., Bildstein, O., Libert, M., 2011. Metallic corrosion processes reactivation sustained by iron-reducing bacteria: implication on long-term stability of protective layers. Phys. Chem. Earth 36, 1624–1629.

Esnault, L., Libert, M., Bildstein, O., Mustin, C., Marsal, F., Jullien, M., 2013. Impact of iron-reducing bacteria on the properties of argillites in the context of radioactive waste geological disposal. Appl. Clay Sci. 83–84, 42–49.

Faure, P., Landais, P., Griffault, L., 1999. Behavior of organic matter from Callovian shales during low-temperature air oxidation. Fuel 78, 1515–1525.

Faure, P., Peiffert, C., 2007. Lixiviation of argillaceous rocks: impact of oxidation degree states on organic compound mobilization. Fuel 86, 426–433.

Fernandez, R., Cuevas, J., Sanchez, L., de la Villa, R.V., Leguey, S., 2006. Reactivity of the cement-bentonite interface with alkaline solutions using transport cells. Appl. Geochem. 21, 977–992.

Fernandez, R., Ruiz, A.I., Cuevas, J., 2014. The role of smectite composition on the hyperalkaline alteration of bentonite. Appl. Clay Sci. 95, 83–94.

Féron, D., Crusset, D., Gras, J.M., 2008. Corrosion issues in nuclear waste disposal. J. Nucl. Mat. 379 (2008), 16–23.

Ferrage, E., Vidal, O., Mosser-Ruck, R., Cathelineau, M., Cuadros, J., 2011. A reinvestigation of smectite illitization in experimental hydrothermal conditions: results from X-ray diffraction and transmission electron microscopy. Am. Mineral. 96, 207–223.

Fleury, M., Pironon, J., Le Nindre, Y.M., Bildstein, O., Berne, P., Lagneau, V., Broseta, D., Pichery, T., Fillacier, S., Lescanne, M., Vidal, O., 2010. Evaluating sealing efficiency of caprocks for CO_2 storage: an overview of the Géocarbone-Integrité program and results. Oil Gas Sci. Technol. 65 (3), 435–444.

Fritz, B., Noguéra, C., 2009. Mineral precipitation kinetics. Rev. Mineral. Geochem. 70, 371–410.

Gaboreau, S., Prêt, D., Tinseau, E., Claret, F., Pellegrini, D., Stammose, D., 2011. 15 years of in situ cement–argillite interaction from Tournemire URL: characterisation of the multi-scale spatial heterogeneities of pore space evolution. Appl. Geochem. 26, 2159–2171.

Gaboreau, S., Lerouge, C., Dewonck, S., Linard, Y., Bourbon, X., Fialips, C.I., Mazurier, A., Pret, D., Borschneck, D., Montouillout, V., Gaucher, E.C., Claret, F., 2012. In situ interaction of cement paste and shotcrete with clay-rocks in a deep disposal context. Am. J. Sci. 312, 314–356.

Gaucher, E.C., Blanc, P., Matray, J.M., Michau, N., 2004. Modeling diffusion of an alkaline plume in a clay barrier. Appl. Geochem. 19, 1505–1515.

Gaucher, E.C., Blanc, P., 2006. Cement/clay interactions – a review: experiments, natural analogues, and modeling. Waste Manage. 26, 776–788.

Gaucher, E.C., Tournassat, C., Pearson, F.J., Blanc, P., Crouzet, C., Lerouge, C., Altmann, S., 2009. A robust model for pore water chemistry of clayrock. Geochim. Cosmochim. Acta 73, 6470–6487.

Gaudin, A., Bartier, D., Truche, L., Tinseau, E., Foct, F., Dyja, V., Maillet, A., Beaufort, D., 2013. First corrosion stages in Tournemire clay-rock/steel interaction: in situ experiment and modelling approach. Appl. Clay Sci. 83–84, 457–468.

Gaus, I., Azaroual, M., Czernichowski-Lauriol, I., 2005. Reactive transport modelling of the impact of CO_2 injection on the clayey cap rock at Sleipner (North Sea). Chem. Geol. 217, 319–337.

Gherardi, F., Xu, T., Pruess, K., 2007. Numerical modeling of self-limiting and self-enhancing caprock alteration induced by CO_2 storage in a depleted gas reservoir. Chem. Geol. 244, 103–129.

Giffaut, E., Grivé, M., Blanc, P., Vieillard, P., Colàs, E., Gailhanou, H., Gaboreau, S., Marty, N., Madé, B., Duro, L., 2014. Andra thermodynamic database for performance assessment: ThermoChimie. Appl. Geochem. 49, 225–236.

Grambow, B., Bretesché, S., 2014. Geological disposal of nuclear waste: II. From laboratory data to the safety analysis – Addressing societal concerns. Appl. Geochem. 49, 247–258.

Grambow, B., Landesman, C., Ribet, S., 2014. Nuclear waste disposal: I. Laboratory simulation of repository properties. Appl. Geochem. 49, 237–246.

Grangeon, S., Claret, F., Lerouge, C., Warmont, F., Sato, T., Anraku, S., Numako, C., Linard, Y., Lanson, B., 2013a. On the nature of structural disorder in calcium silicate hydrates with a calcium/silicon ratio similar to tobermorite. Cement Concrete Res. 52, 31–37.

Grangeon, S., Claret, F., Linard, Y., Chiaberge, C., 2013b. X-ray diffraction: a powerful tool to probe and understand the structure of nanocrystalline calcium silicate hydrates. Acta Crystallogr. B B69, 465–473.

Guillaume, D., Neaman, A., Cathelineau, M., Mosser-Ruck, R., Peiffert, C., Abdelmoula, M., Dubessy, J., Villiéras, F., Michau, N., 2004. Experimental study of the transformation of smectite at 80°C and 300°C in the presence of Fe oxides. Clay Miner. 39, 1734.

Hadi, J., Tournassat, C., Ignatiadis, I., Greneche, J.M., Charlet, L., 2013. Modelling CEC variations versus structural iron reduction levels in dioctahedral smectites. Existing approaches, new data and model refinements. J. Colloid Interf. Sci. 407, 397–409.

Hellevang, H., Pham, V.T., Aagaard, P., 2013. Kinetic modelling of CO_2–water–rock interactions. Int. J. Greenh. Gas Con. 15, 3–15.

Honty, M., De Craen, M., Wang, L., Madejová, J., Czímerová, A., Pentrák, M., Stríček, I., Van Geet, M., 2010. The effect of high pH alkaline solutions on the mineral stability of the Boom Clay – batch experiments at 60°C. Appl. Geochem. 25, 825–840.

Honty, M., Wang, L., Osacký, M., Uhlík, P., Czímerová, A., Madejová, J., 2012. Experimental interactions of the Opalinus Clay and Boom Clay with various repository relevant solutions at 90 °C under closed conditions. Appl. Clay Sci. 59–60, 50–63.

Hunter, F.M.I., Bate, F., Heath, T.G., Hoch, A.R., 2007. Geochemical Investigation of Iron Transport into Bentonite as Steel Corrodes. SKB Technical Report TR-07-09.

Ilton, E.S., Todd Schaef, H., Qafoku, O., Rosso, K.M., Felmy, A.R., 2012. In situ X-ray diffraction study of Na^+ saturated montmorillonite exposed to variably wet super critical CO_2. Env. Sci. Tech. 46, 4241–4248.

Ishidera, T., Ueno, K., Kurosawa, S., Suyama, T., 2008. Investigation of montmorillonite alteration and form of iron corrosion products in compacted bentonite in contact with carbon steel for ten years. Phys. Chem. Earth 33, 269–275.

Jeannin, M., Calonnec, D., Sabot, R., Refait, P., 2010. Role of a clay sediment deposit on the corrosion of carbon steel in 0.5 mol L^{-1} NaCl solutions. Corros. Sci. 52, 2026–2034.

Jeannin, M., Calonnec, D., Sabot, R., Refait, P., 2011. Role of a clay sediment deposit on the passivity of carbon steel in 0.1 mol dm^{-3} $NaHCO_3$ solutions. Electrochim. Acta 56, 1466–1475.

Jenni, A., Mäder, U., Lerouge, C., Gaboreau, S., Schwyn, B., 2014. In situ interaction between different concretes and Opalinus Clay. Phys. Chem. Earth, 70–71, 71–83.

Jodin-Caumon, M.-C., Mosser-Ruck, R., Randi, A., Pierron, O., Cathelineau, M., Michau, N., 2012. Mineralogical evolution of a clay-rock after reaction with iron under thermal gradient. Clays Clay Miner. 60, 443–455.

Johnson, J.W., Nitao, J.J., Morris, J.P., 2005. Reactive transport modeling of cap rock integrity during natural and engineered CO_2 storage. In: Thomas, D.C., Benson, S.M. (Eds.), Carbon Dioxide Capture for Storage in Deep Geologic Formations, vol. 2, pp. 787–813.

Jung, H.B., Um, W., Cantrell, K.J., 2013. Effect of oxygen co-injected with carbon dioxide on Gothic shale caprock–CO_2–brine interaction during geologic carbon sequestration. Chem. Geol. 354 (2013), 1–14.

Kampmann, N., Bickle, M., Wigley, M., Dubacq, B., 2014. Fluid flow and CO_2–fluid–mineral interactions during CO_2-storage in sedimentary basins. Chem. Geol. 369, 22–50.

Kaszuba, J.P., Janecky, D.R., Snow, M.G., 2005. Experimental evaluation of mixed fluid reactions between supercritical carbon dioxide and NaCl brine: relevance to the integrity of a geologic carbon repository. Chem. Geol. 217, 277–293.

Kaszuba, J.P., Yardley, B., Andreani, M., 2013. Experimental perspectives of mineral dissolution and precipitation due to carbon dioxide-water-rock interactions. Rev. Mineral. Geochem. 77, 153–188.

Kaufhold, S., Dohrmann, R., 2009. Stability of bentonites in salt solutions: I. Sodium chloride. Appl. Clay Sci. 45, 171–177.

Kaufhold, S., Dohrmann, R., 2010. Stability of bentonites in salt solutions: II. Potassium chloride solution — initial step of illitization? Appl. Clay Sci. 49, 98–107.

Kaufhold, S., Dohrmann, R., 2011. Stability of bentonites in salt solutions: III. Calcium hydroxide. Appl. Clay Sci. 51, 300–307.

Kerber-Schütz, M., Moreira, R., Bildstein, O., Lartigue, J.E., Schlegel, M.L., Tribollet, B., Vivier, V., Libert, M., 2014. Combined geochemical and electrochemical methodology to quantify corrosion of carbon steel by bacterial activity. Bioelectrochemistry 97, 61–68.

King, F., 2009. Corrosion of carbon steel under anaerobic conditions. Nagra Technical Report NTB 08-12. Nagra, Wettingen, Switzerland.

King, F., Burt, D., Ganeshalingam, J., Gardner, P., Sanderson, D., Watson, S., Padovani, C., 2014a. Coupled analysis of mechanical- and corrosion-related degradation of carbon steel spent fuel container. Corros. Eng. Sci. Technol. 49 (6), 442–449.

King, F., Kolar, M., Keech, P.G., 2014b. Simulations of long-term anaerobic corrosion of carbon steel containers in Canadian deep geological repository. Corros. Eng. Sci. Technol. 49 (6), 455–459.

Khoury, H.N., Salameh, E., Abdul-Jaber, Q., 1985. Characteristics of an unusual highly alkaline water from the Maqarin area, northern Jordan. J. Hydrol. 81, 79–91.

Khoury, H.N., Salameh, E., Clark, I.D., Fritz, P., Bajjali, W., Milodowski, A.E., Cave, M.R., Alexander, W.R., 1992. A natural analogue of high pH cement pore waters from the Maqarin area of northern Jordan. I: introduction to the site. J. Geochem. Explor 46, 117–132.

Kohler, E., Parra, T., Vidal, O., 2009. Clayey caprock behavior in H_2O–CO_2 media at low pressure and temperature conditions: an experimental approach. Clays and Clay Miner. 57, 616–637.

Kosakowski, G., Berner, U., 2013. The evolution of clay rock/cement interfaces in a cementitious repository for low- and intermediate level radioactive waste. Phys. Chem. Earth, Parts A/B/C 64, 65–86.

Landrot, G., Ajo-Franklin, J., Yang, L., Cabrini, S., Steefel, C.I., 2012. Measurement of accessible reactive surface area in a sandstone, with application to CO_2 mineralization. Chem. Geol. 318–319, 113–125.

Lanson, B., Sakharov, B.A., Claret, F., Drits, V.A., 2009. Diagenetic smectite-to-illite transition in clay-rich sediments: a reappraisal of X-Ray diffraction results using the multi-specimen method. Am. J. Sci. 309, 476–516.

Lantenois, S., Lanson, B., Muller, F., Bauer, A., Jullien, M., Plancon, A., 2005. Experimental study of smectite interaction with metal Fe at low temperature: 1. Smectite destabilization. Clays Clay Miner. 53, 597–612.

Lassin, A., Azaroual, M., Mercury, L., 2005. Geochemistry of unsaturated soil systems: aqueous speciation and solubility of minerals and gases in capillary solutions. Geochim. Cosmochim. Acta 69 (22), 5187–5201.

Lerouge, C., Claret, F., Boyer, B., Borschneck, D., Gaboreau, S., Grangeon, S., Linard, Y., 2014. Constraints from sulfur isotopes on the origin of gypsum at concrete/clay-rock interfaces. Phys. Chem. Earth, 70–71, 84–85.

Libert, M., Esnault, L., Jullien, M., Bildstein, O., 2011. Molecular hydrogen: an abundant energy source for bacterial activity in nuclear waste repositories. Phys. Chem. Earth 36, 1616–1623.

Libert, M., Kerber-Schütz, M., Esnault, L., Féron, D., Bildstein, O., 2014. Impact of microbial activity on the radioactive waste disposal: long term prediction of biocorrosion processes. Bioelectrochemistry 97, 162–168.

Liu, F., Lu, P., Griffith, C., Hedges, S.W., Soong, Y., Hellevang, H., Zhu, C., 2012. CO_2-brine-caprock interaction: reactivity experiments on Eau Claire Shale and a review of relevant literature. Int. J. Greenh. Gas Con. 7, 153–167.

Liu, S., Jacques, D., Govaerts, J., Wang, L., 2014. Conceptual model analysis of interaction at a concrete–Boom Clay interface. Phys. Chem. Earth, Parts A/B/C 70–71, 150–159.

Lothenbach, B., Le Saout, G., Ben Haha, M., Figi, R., Wieland, E., 2012. Hydration of a low-alkali CEM III/B–SiO_2 cement (LAC). Cement Concrete Res. 42, 410–423.

Lu, C., Samper, J., Fritz, B., Clement, A., Montenegro, L., 2011. Interactions of corrosion products and bentonite: an extended multicomponent reactive transport model. Phys. Chem. Earth 36, 1661–1668.

MacKinley, I.G., Alexander, W.R., 1992. A review of the use of natural analogues to test performance assessment models of a cementitious near field. Waste Manage. 12, 253–259.

MacKinley, I.G., Alexander, W.R., 1993. Assessment of radionuclide retardation: uses and abuses of natural analogue studies. J. Contam. Hydrol. 13, 249–259.

Mäder, U.K., Mazurek, M., 1997. Oxidation phenomena and processes in Opalinus Clay: evidence from the excavation-disturbed zones in Hauenstein and Mt. Terri Tunnels and Siblingen open clay pit. MRS Online Proc. Lib. 506.

Maher, K., Steefel, C.I., DePaolo, D., Viani, B.E., 2006. The mineral dissolution rate conundrum: insights from reactive transport modeling of U isotopes and pore fluid chemistry in marine sediments. Geochim. Cosmochim. Acta 70, 337–363.

Martin, F.A., Bataillon, C., Schlegel, M.L., 2008. Corrosion of iron and low alloyed steel within a water saturated brick of clay under anaerobic deep geological disposal conditions: an integrated experiment. J. Nucl. Mater. 379, 80–90.

Marty, N.C.M., Tournassat, C., Burnol, A., Giffaut, E., Gaucher, E.C., 2009. Influence of reaction kinetics and mesh refinement on the numerical modelling of concrete/clay interactions. J. Hydrol. 364, 58–72.

Marty, N.C.M., Fritz, B., Clément, A., Michau, N., 2010. Modelling the long term alteration of the engineered bentonite barrier in an underground radioactive waste repository. Appl. Clay Sci. 47, 82–90.

Marty, N.C.M., Munier, I., Gaucher, E., Tournassat, C., Gaboreau, S., Vong, C., Giffaut, E., Cochepin, B., Claret, F., 2014. Simulation of cement/clay interactions: feedback on the increasing complexity of modelling strategies. Transport Porous Med. 104, 385–405.

Marty, N.C.M., Bildstein, O., Blanc, P., Claret, F., Cochepin, B., Gaucher, E.C., Jacques, D., Lartigues, J.-E., Liu, S., Mayer, K.U., Meussen, J.C.L., Munier, I., Pointeau, I., Su, D., Steefel, C.I., 2015. Benchmarks for multicomponent reactive transport across a cement/clay interface. Computat. Geosci. http://dx.doi.org/10.1007/s10596-014-9463-6.

Metz, B., Davidson, O., de Coninck, H.C., Loos, M., Meyer, L.A., 2005. IPCC Special Report on Carbon Dioxide Capture and Storage, Prepared by Working Group III of the Intergovernmental Panel on Climate Change. Cambridge University Press, Cambridge, United Kingdom and New York, NY, USA, 442 p.

Michelin, A., Drouet, E., Foy, E., Dynes, J.J., Neff, D., Dillmann, P., 2012. Investigation at the nanometre scale on the corrosion mechanisms of archaeological ferrous artefacts by STXM. J. Anal. At. Spectrom. 28, 59–66.

Miller, A.W., Rodriguez, D.R., Honeyman, B.D., 2010. Upscaling sorption/desorption processes in reactive transport models to describe metal/radionuclide transport: a critical review. Environ. Sci. Technol. 44, 7996–8007.

Milodowski, A.E., Cave, M.R., Kemp, S.J., Taylor, H., Green, K., Williams, C.L., Shaw, R.A., Gowing, C.J.B., Eatherington, N.D., 2009. Mineralogical Investigations of the Interaction between Iron Corrosion Products and Bentonite from the NF-PRO Experiments (Phase 2). SKB Report TR-09-03.

Missana, T., Garcia-Gutiérrez, M., Alonso, U., 2004. Kinetics and irreversibility of cesium and uranium sorption onto bentonite colloids in a deep granitic environment. Appl. Clay Sci. 26, 137–150.

Montes, H.G., Marty, N., Fritz, B., Clement, A., Michau, N., 2005. Modelling of long-term diffusion reaction in a bentonite barrier for radioactive waste confinement. Appl. Clay Sci. 30, 181–198.

Mosser-Ruck, R., Cathelineau, M., Guillaume, D., Charpentier, D., Rousset, D., Barres, O., Michau, N., 2010. Effects of temperature, pH, and iron/clay and liquid/clay ratios on experimental conversion of dioctahedral smectite to berthierine, chlorite, vermiculite, or saponite. Clays Clay Miner. 58, 280–291.

Neff, D., Dillmann, P., Bellot-Gurlet, L., Beranger, G., 2005. Corrosion of iron archaeological artefacts in soil: characterisation of the corrosion system. Corros. Sci. 47, 515–553.

Neff, D., Dillmann, P., Descostes, M., Beranger, G., 2006. Corrosion of iron archaeological artefacts in soil: estimation of the average corrosion rates involving analytical techniques and thermodynamic calculations. Corros. Sci. 48, 2947–2970.

Neretnieks, I., 2014. Development of a simple model for the simultaneous degradation of concrete and clay in contact. Appl. Geochem. 43, 101–113.

Ngo, V.V., Delalande, M., Clément, A., Michau, N., Fritz, B., 2014. Coupled transport-reaction modeling of the long-term interaction between iron, bentonite and Callovo-Oxfordian claystone in radioactive waste confinement systems. Appl. Clay Sci. 101, 430–443.

Noiriel, C., Madé, B., Gouze, P., 2007. Impact of coating development on the hydraulic and transport properties in argillaceous limestone fracture. Water Resour. Res. 43.

Noiriel, C., Steefel, C.I., Yang, L., Ajo-Franklin, A., 2012. Upscaling calcite precipitation rates from pore to continuum scale. Chem. Geol. 318–319, 60–74.

Peña, J., Torres, E., Turrero, M.J., Escribano, A., Martín, P.L., 2008. Kinetic modelling of the attenuation of carbon steel canister corrosion due to diffusive transport through corrosion product layers. Corros. Sci. 50, 2197–2204.

Perronnet, M., Villieras, F., Jullien, M., Razafitianamaharavo, A., Raynal, J., Bonnin, D., 2007. A link between the energetic heterogeneities of the edge faces of smectites. Geochim. Cosmochim. Acta 71, 1463–1479.

Perronnet, M., Jullien, M., Villiéras, F., Raynal, J., Bonnin, D., Bruno, G., 2008. Evidence of a critical content in Fe(0) on FoCa7 bentonite reactivity at 80 °C. Appl. Clay Sci. 38, 187–202.

Peters, C., 2009. Accessibilities of reactive minerals in consolidated sedimentary rock: an imaging study of three sandstones. Chem. Geol. 265, 198–208.

Pignatelli, I., Mugnaioli, E., Mosser-Ruck, R., Barres, O., Kolb, U., Michau, N., 2014. A multi-technique, micrometer- to atomic-scale description of a synthetic analogue of chukanovite, Fe-2(CO3)(OH)(2). Eur. J. Mineral. 26 (2), 221–229.

Read, D., Glasser, F.P., Ayora, C., Guardiola, M.T., Sneyers, A., 2001. Mineralogical and microstructural changes accompanying the interaction of Boom Clay with ordinary Portland cement. Adv. Cem Res. 13, 175–183.

Rutqvist, J., Tsang, C.F., 2002. A study of caprock hydromechanical changes associated with CO_2-injection into a brine formation. Environ. Geol. 42, 296–305.

Sakamaki, K., Kataoka, M., Maeda, T., Iida, Y., Kamoshida, M., Yamaguchi, T., Tanaka, T., 2014. Experimental verification of models assessing Eh evolution induced by corrosion of carbon steel overpack. Corros. Eng. Sci. Tech. 49 (6), 450–454.

Saheb, M., Neff, D., Dillmann, P., Matthiesen, H., Foy, E., Bellot-Gurlet, L., 2009. Multisecular corrosion behaviour of low carbon steel in anoxic soils: characterisation of corrosion system on archaeological artefacts. Mater. Corros. 60, 99–105.

Saheb, M., Descostes, M., Neff, D., Matthiesen, H., Michelin, A., Dillmann, P., 2010. Iron corrosion in an anoxic soil: comparison between thermodynamic modelling and ferrous archaeological artefacts characterised along with the local in situ geochemical conditions. Appl. Geochem. 25, 1937–1948.

Saheb, M., Gallien, J.-P., Descostes, M., Raimbault, L., Perez, A., Neff, D., Marsal, F., Pellegrini, D., Dillmann, P., 2014. Influence of an aerated/anoxic transient phase on the long-term corrosion of iron. Corros. Sci. 86, 71–80.

Sakharov, B.A., Lanson, B., 2013. Chapter 2.3-X-ray identification of mixed-layer structures: modelling of diffraction effects. In: Bergaya, F., Lagaly, G. (Eds.), Developments in Clay Science. Elsevier, pp. 51–135.

Samper, J., Lu, C., Montenegro, L., 2008. Reactive transport model of interactions of corrosion products and bentonite. Phys. Chem. Earth 33, 306–316.

Savage, D., Noy, D., Mihara, M., 2002. Modelling the interaction of bentonite with hyperalkaline fluids. Appl. Geochem. 17, 207–223.

Savage, D., Walker, C., Arthur, R., Rochelle, C., Oda, C., Takase, H., 2007. Alteration of bentonite by hyperalkaline fluids: a review of the role of secondary minerals. Phys. Chem. Earth 32, 287–297.

Savage, D., Benbow, S., Watson, C., Takase, H., Ono, K., Oda, C., Honda, A., 2010a. Natural systems evidence for the alteration of clay under alkaline conditions: an example from Searles Lake, California. Appl. Clay Sci. 47, 72–81.

Savage, D., Watson, C., Benbow, S., Wilson, J., 2010b. Modelling iron-bentonite interactions. Appl. Clay Sci. 47, 91–98.

Savage, D., 2011. A review of analogues of alkaline alteration with regard to long-term barrier performance. Mineral. Mag. 75, 2401–2418.

Savage, D., 2012. Prospects for Coupled Modelling. Finnish Radiation and Nuclear Safety Authority report STUK-TR 13, 44 p.

Schäfer, T., Michel, P., Claret, F., Beetz, T., Wirick, S., Jacobsen, C., 2009. Radiation sensitivity of natural organic matter: Clay mineral association effects in the Callovo-Oxfordian argillite. J. Electron Spectrosc. Relat. Phenom. 170, 49–56.

Schlegel, M.L., Bataillon, C., Benhamida, K., Blanc, C., Menut, D., Lacour, J.-L., 2008. Metal corrosion and argillite transformation at the water-saturated, high temperature iron-clay interface: a microscopic scale study. Appl. Geochem. 23, 2619–2633.

Schlegel, M.L., Bataillon, C., Blanc, C., Prêt, D., Foy, E., 2010. Anodic activation of iron corrosion in clay media under water-saturated conditions at 90 °C: characterization of the corrosion interface. Environ. Sci. Technol. 44, 1503–1508.

Schlegel, M.L., Bataillon, C., Brucker, F., Blanc, C., Prêt, D., Foy, E., Chorro, M., 2014. Corrosion of metal iron in contact with anoxic clay at 90°C: characterization of the corrosion products after two years of interaction. Appl. Geochem. http://dx.doi.org/10.1016/j.apgeochem.2014.09.002.

Schmitz, R.M., Schroeder, C., Charlier, R., 2007. A general approach to assess hydro-mechanical changes of natural clay barriers due to physico-chemical interactions with radwaste in deep disposal sites. Phys. Chem. Earth, Parts A/B/C 32, 922–928.

Senger, R., Ewing, J., Zhang, K., Avis, J., Marschall, P.M., Gaus, I., 2011. Modeling approaches for investigating gas migration from a deep low/intermediate level waste repository (Switzerland). Transp. Porous Med. 90, 113–133.

Shao, H., Kosakowski, G., Berner, U., Kulik, D.A., Mäder, U., Kolditz, O., 2013. Reactive transport modeling of the clogging process at Maqarin natural analogue site. Phys. Chem. Earth, Parts A/B/C 64, 21–31.

Shukla, R., Ranjith, P., Haque, A., Choi, X., 2010. A review of studies on CO_2 sequestration and caprock integrity. Fuel 89, 2651–2664.

Smart, N.R., Rance, A.P., Carlson, L., Werme, L.O., 2006. Further studies of the anaerobic corrosion of steel in bentonite. In: MRS Proceedings, 932, 32.1. http://dx.doi.org/10.1557/PROC-932-32.1.

Smellie, J.A.T., Karlsson, F., Alexander, W.R., 1997. Natural analogue studies: present status and performance assessment implications. J. Contam. Hydrol. 26, 3–17.

Soler, J.M., 2003. Reactive transport modeling of the interaction between a high-pH plume and a fractured marl: the case of Wellenberg. Appl. Geochem. 18, 1555–1571.

Song, J., Zhang, D., 2013. Comprehensive review of caprock-sealing mechanisms for geologic carbon sequestration. Environ. Sci. Technol. 47 (1), 9–22.

Steefel, C.I., Lichtner, P.C., 1994. Diffusion and reaction in rock matrix bordering a hyperalkaline fluid-filled fracture. Geochim. Cosmochim. Acta 58, 3595–3612.

Steefel, C.I., Lichtner, P.C., 1998. Multicomponent reactive transport in discrete fractures: I. Controls on reaction front geometry. J. Hydrol. 209, 186–199.

Steefel, C.I., Appelo, C.A.J., Arora, B., Jacques, D., Kalbacher, T., Kolditz, O., Lagneau, V., Lichtner, P.C., Mayer, K.U., Meeussen, J.C.L., Molins, S., Moulton, D., Shao, H., Šimůnek, J., Spycher, N., Yabusaki, S.B., Yeh, G.T., 2014. Reactive transport codes for subsurface environmental simulation. Computat. Geosci. 1–34.

Stroes-Gascoyne, S., Sergeant, C., Schippers, A., Hamon, C., Nèble, S., Vesvres, M.-H., Barsotti, V., Poulain, S., Le Marrec, C., 2011. Biogeochemical processes in a clay formation in situ experiment: Part D–microbial analyses–synthesis of results. Appl. Geochem. 26, 980–989.

Stucki, J.W., 2013. Chapter 11-properties and behaviour of iron in Clay minerals. In: Faïza, B., Gerhard, L. (Eds.), Developments in Clay Science. Elsevier, pp. 559–611.

Talandier, J., Mayer, G., Croisé, J., 2006. Simulations of the hydrogen migration out of intermediate-level radioactive waste disposal drifts using TOUGH2. In: TOUGH Symp. LBNL, Berkeley, pp. 1–8.

Tian, H., Xu, T., Wang, F., Patil, V.V., Sun, Y., Yua, G., 2014. A numerical study of mineral alteration and self-sealing efficiency of a caprock for CO_2 geological storage. Acta Geotech. 9, 87–100.

Tinseau, E., Bartier, D., Hassouta, L., Devol-Brown, I., Stammose, D., 2006. Mineralogical characterization of the Tournemire argillite after in situ interaction with concretes. Waste Manage. 26, 789–800.

Trotignon, L., Peycelon, H., Bourbon, X., 2006. Comparison of performance of concrete barriers in a clayey geological medium. Phys. Chem. Earth 31, 610–617.

Trotignon, L., Devallois, V., Peycelon, H., Tiffreau, C., Bourbon, X., 2007. Predicting the long term durability of concrete engineered barriers in a geological repository for radioactive waste. Phys. Chem. Earth 32, 259–274.

Trotignon, L., Thouvenot, P., Munier, I., Cochepin, B., Piault, E., Treille, E., Bourbon, X., Mimid, S., 2009. Numerical simulation of atmospheric carbonation of concrete components in a deep geological radwaste disposal during operating period using Toughreact. In: Proceedings, Tough Symposium 2009, Lawrence Berkeley National Laboratory, California, September 14–16, 2009.

Truche, L., Berger, G., Destrigneville, C., Pages, A., Guillaume, D., Giffaut, E., Jacquot, E., 2009. Experimental reduction of aqueous sulphate by hydrogen under hydrothermal conditions: implication for the nuclear waste storage. Geochim. Cosmochim. Acta 73, 4829–4835.

Truche, L., Berger, G., Destrigneville, C., Guillaume, D., Giffaut, E., 2010. Kinetics of pyrite to pyrrhotite reduction by hydrogen in calcite buffered solutions between 90 and 180 °C: implications for nuclear waste disposal. Geochim. Cosmochim. Acta 74, 2894–2914.

Truche, L., Berger, G., Albrecht, A., Domergue, L., 2013a. Engineered materials as potential geocatalysts in deep geological nuclear waste repositories: a case study of the stainless steel catalytic effect on nitrate reduction by hydrogen. Appl. Geochem. 35, 279–288.

Truche, L., Jodin-Caumon, M.C., Lerouge, C., Berger, G., Mosser-Ruck, R., Giffaut, E., Michau, N., 2013b. Sulphide mineral reactions in clay-rich rock induced by high hydrogen pressure. Application to disturbed or natural settings up to 250°C and 30 bar. Chem. Geol. 351, 217–228.

Van Damme, H., Pellenq, R.J.M., 2013. Chapter 14.3-cement hydrates. In: Bergaya, F., Lagaly, G. (Eds.), Developments in Clay Science. Elsevier, pp. 801–817.

Vega, E., Berger, P., Dillmann, P., 2005. A study of transport phenomena in the corrosion products of ferrous archaeological artefacts using ^{18}O tracing and nuclear microprobe analysis. Nucl. Instr. Meth. B 240, 554–558.

Vega, E., Dillmann, P., Berger, P., Fluzin, P., 2007. Species transport in the corrosion products of ferrous archaeological analogues: contribution to the modelling of long-term iron corrosion mechanisms. In: Dillmann, P., Béranger, G., Piccardo, P., Matthiessen, H. (Eds.), Corrosion of Metallic Heritage Artefacts: Investigation Conservation and Prediction for Long Term Behaviour. Woodhead Publishing, Cambridge, pp. 92–108.

Vengosh, A., Jackson, R.B., Warner, N., Darrah, T.H., Kondash, A., 2014. A critical review of the risks to water resources from unconventional shale Gas development and hydraulic fracturing in the United States. Environ. Sci. Technol. 48, 8334–8348.

Vieillard, P., Rassineux, F., 1992. Thermodynamic and geochemical modelling of the alteration of two cement matrices. Appl. Geochem. 7 (Suppl. 1), 125–136.

Vieillard, P., Ramirez, S., Bouchet, A., Cassagnabere, A., Meunier, A., Jacquot, E., 2004. Alteration of the Callow-Oxfordian clay from Meuse-Haute Marne underground laboratory (France) by alkaline solution: II. Modelling of mineral reactions. Appl. Geochem. 19, 1699–1709.

Vilarrasa, V., Olivella, S., Carrera, J., Rutqvist, J., 2014. Long term impacts of cold CO_2 injection on the caprock integrity. Int. J. Greenh. Gas Con. 24 (2014), 1–13.

Vinsot, A., Claret, F., Lundy, M., Wechner, S., 2012. In Situ Measurement of the Rate of Oxygen Consumption by the Callovo-Oxfordian Argillaceous Rock, Clays in Natural and Engineered Barriers for Radioactive Waste Confinement. France, Montpellier.

Vinsot, A., Linard, Y., Lundy, M., Necib, S., Wechner, S., 2013. Insights on desaturation processes based on the chemistry of seepage water from boreholes in the Callovo-Oxfordian argillaceous rock. Procedia Earth Planetary Sci. 7, 871–874.

Vinsot, A., Leveau, F., Bouchet, A., Arnould, A., 2014. Oxidation Front and Oxygen Transfer in the Fractured Zone Surrounding the Meuse/Haute-Marne URL Drifts in the Callovian–oxfordian Argillaceous Rock. Geological Society. Special Publications 400, London.

Wersin, P., Birgersson, M., Olsson, S., Karnland, O., Snellman, M., 2008. Impact of Corrosion-Derived Iron on the Bentonite Buffer within the KBS-3H Disposal Concept: The Olkiluoto Site as Case Study. SKB Report R-08-34, 61 p.

Wersin, P., Birgersson, M., 2014. Reactive Transport Modelling of Iron–Bentonite Interaction within the KBS-3H Disposal Concept: The Olkiluoto Site as a Case Study. Geological Society, London. http://dx.doi.org/10.1144/SP400.24. Special Publications, 400.

Wilson, J., Cressey, G., Cressey, B., Cuadros, J., Ragnarsdottir, K.V., Savage, D., Shibata, M., 2006a. The effect of iron on montmorillonite stability. (II) experimental investigation. Geochim. Cosmochim. Acta 70, 323–336.

Wilson, J., Savage, D., Cuadros, J., Shibata, M., Ragnarsdottir, K.V., 2006b. The effect of iron on montmorillonite stability. (I) background and thermodynamic considerations. Geochim. Cosmochim. Acta 70, 306–322.

Wollenweber, J., Alles, S., Busch, A., Krooss, B.M., Stanjek, H., Littke, R., 2010. Experimental investigation of the CO_2 sealing efficiency of caprocks. Int. J. Greenh. Gas Con. 4, 231–241.

Xia, X., Idemitsu, K., Arima, T., Inagi, Y., Ishidera, T., Kurosawa, S., Iijima, K., Sato, H., 2005. Corrosion of carbon steel in compacted bentonite and its effect on neptunium diffusion under reducing condition. Appl. Clay Sci. 28, 89–100.

Xie, M., Mayer, K.U., Claret, F., Alt-Epping, P., Jacques, D., Steefel, C.I., Chiaberge, C., Simunek, J., 2015. Implementation and evaluation of permeability-porosity and tortuosity-porosity relationships linked to mineral dissolution-precipitation. Computat. Geosci. http://dx.doi.org/10.1007/s10596-014-9458-3.

Xu, T., Senger, R., Finsterle, S., 2008. Corrosion-induced gas generation in a nuclear waste repository: reactive geochemistry and multiphase flow effects. Appl. Geochem. 23 (12), 3423–3433.

Zheng, L., Samper, J., Montenegro, L., Mayor, J.C., 2008. Multiphase flow and multicomponent reactive transport model of the ventilation experiment in Opalinus Clay. Phys. Chem. Earth, Parts A/B/C 33 (Suppl. 1), S186–S195.

Chapter 6

Self-Diffusion of Water and Ions in Clay Barriers

Ian C. Bourg[a,b] and Christophe Tournassat[b,c]
[a] *Department of Civil and Environmental Engineering, Princeton University, Princeton, NJ, USA;*
[b] *Earth Sciences Division, Lawrence Berkeley National Laboratory, Berkeley, CA, USA;*
[c] *Water, Environment and Ecotechnology Division, French Geological Survey (BRGM), Orléans, France*

Chapter Outline

6.1 INTRODUCTION

A well-known characteristic of clay-rocks and engineered bentonite barriers is their very low hydraulic permeability k. The existing database indicates that k ranges from $\sim 10^{-21}$ to 10^{-17} m^2 with little sensitivity, at least in some cases, to the presence of fractures or faults (Neuzil, 1994; Oscarson et al., 1996; Bock et al., 2010; Barbour et al., 2012). In porous media with such low

Developments in Clay Science, Vol. 6C. http://dx.doi.org/10.1016/B978-0-08-100027-4.00006-1

k values, at hydraulic pressure gradients representative of subsurface conditions, molecular diffusion is an important mass transfer mechanism on length scales up to hundreds of meters and millions of years, with implications in the isolation of landfills and contaminated sites (Johnson et al., 1989; Jo et al., 2006; Lange et al., 2009; De Soto et al., 2012), the persistence of dense nonaqueous-phase liquids (DNAPL) in contaminated aquifers (Parker et al., 2004; Chapman and Parker, 2005), the distribution of inert tracers in clay formations (Patriarche et al., 2004b; Lavastre et al., 2005; Mazurek et al., 2011), and the geologic storage of CO_2 (Ketzer et al., 2005; Gherardi et al., 2007) and high-level radioactive waste (HLRW) (Altmann et al., 2012).

Because of the important role of molecular diffusion in clay materials, extensive experimental efforts have aimed to characterize the associated diffusion coefficients. Characterization of these diffusion coefficients is nontrivial, because of the need to surmount significant experimental challenges (Glaus et al., 2007) and theoretical complexities associated with the fine-grained nature and negative surface charge of clay minerals (Berend et al., 1995; Revil et al., 2005; Leroy et al., 2006; Gajo and Loret, 2007). The present chapter summarizes the experimental database on diffusion in clay-rocks and engineered clay barriers and briefly discusses its implications regarding the fundamental phenomena associated with diffusion in these materials.

For the sake of brevity, the present chapter focuses exclusively on the macroscopic scale self-diffusion of water tracers (HTO, HDO), halides (I^-, Br^-, Cl^-), and alkali and alkaline earth metals (Na^+, Cs^+, Sr^{2+}). This choice is motivated by the abundance of experimental data that exist for these species and by the relative simplicity of their aqueous geochemistry. As evidenced below, this simplicity is instrumental in identifying the key trends associated with water and ion diffusion in clay materials. The diffusion of more complex or less extensively studied species such as oxyanions (Aertsens et al., 2003; Jansson and Eriksen, 2004; Sato and Miyamoto, 2004), actinides (Tachi et al., 2010; Joseph et al., 2013), lanthanides (Seida et al., 2010), trace metals (Lange et al., 2009), dissolved gases (Higashihara et al., 2004, 2005), and organic molecules (Chapman and Parker, 2005; Martens et al., 2010) is not addressed here despite its interest in a range of areas. Diffusion in unsaturated conditions—an important phenomenon in soils (Calvet, 1973; Nye, 1979) and, potentially, in HLRW repositories in certain conditions (Savoye et al., 2010)—is not discussed. Emergent phenomena that arise from couplings between the mass fluxes of different species during salt diffusion or advection in clay barriers are examined in Chapter 8 (in this Volume). The scale-dependence of diffusion in clay materials is examined in more detail in Chapter 11 (in this Volume).

6.2 MACROSCOPIC SCALE DIFFUSION COEFFICIENTS: DEFINITION AND MEASUREMENT

6.2.1 Fickian Expressions

Diffusion of dilute conservative solutes in porous media can be described using an effective diffusion coefficient D_e ($m^2\,s^{-1}$) defined by the following Fickian expression (written here in one dimension for simplicity):

$$N = -D_e \frac{\partial C_b}{\partial x} \tag{6.1}$$

In Eqn (6.1), N is the solute mass flux density in the x direction and C_b is the volumetric concentration of the species of interest in free (bulk liquid) pore water. In clay, bulk liquid water may not always exist, because the interfacial region of cation adsorption and anion exclusion on clay mineral surfaces—the electrical double layer (EDL)—can fill the entire pore space (Birgersson and Karnland, 2009; Tachi and Yotsuji, 2014; see also Chapter 1, in this Volume). Nevertheless, Eqn (6.1) can be applied by defining a fictitious bulk liquid water reservoir in equilibrium with the pore fluid at coordinate x, where C_b can be evaluated using an adsorption model.

Conservation of mass at any location in the porous medium can be expressed as follows:

$$\frac{\partial C_t}{\partial t} = -\frac{\partial N}{\partial x} \tag{6.2}$$

where C_t is the total concentration of the species of interest (in solution or adsorbed on the clay mineral surfaces) per volume of porous medium. If adsorption is linear and reaches equilibrium instantaneously, the relationship between C_t and C_b can be expressed using a "rock capacity factor" α invariant with C_b:

$$\alpha = \frac{C_t}{C_b} \tag{6.3}$$

where $\alpha = \phi$, the dimensionless porosity, in the case of inert tracers. Combination of Eqns (6.1)–(6.3) yields the relation:

$$\alpha \frac{\partial C_b}{\partial t} = \frac{\partial}{\partial x}\left(D_e \frac{\partial C_b}{\partial x}\right) \tag{6.4}$$

A second macroscopic scale diffusion coefficient, the apparent diffusion coefficient D_a, is defined by the expression:

$$\frac{\partial C_t}{\partial t} = \frac{\partial}{\partial x}\left(D_a \frac{\partial C_t}{\partial x}\right) \tag{6.5}$$

If α is invariant in space, Eqns (6.4) and (6.5) show that D_e and D_a are related by a factor equal to α:

$$D_e = \alpha D_a \tag{6.6}$$

Alternative terminologies and notations for the macroscopic diffusion coefficients are sometimes used (Nye, 1979; Shackelford, 1991; Hendry et al., 2009; Shackelford and Moore, 2013). For example, the ratio α/ϕ is often called the retardation factor (Chapman and Parker, 2005).

6.2.2 Experimental Techniques

Experimental data on D_e or D_a in clay samples are most frequently measured on length scales of centimeters in compacted powders or intact rock samples. Commonly used experimental methodologies are summarized below (Shackelford, 1991). Depending on the setup, diffusion experiments may yield either D_a or D_e or both.

In *closed cell* (CC) experiments, the porous medium is saturated with water, doped with a known spatial distribution of the species of interest, and enclosed in a diffusion cell. At a time t after the beginning of the experiment, the porous medium is examined to determine the new spatial distribution of the species of interest. A frequently used methodology consists in placing a uniform planar source of the species of interest at $x = x_0$ (Figure 6.1(a)). The concentration profile at time t, then, has the form (Sato et al., 1992; Kozaki et al., 1998):

$$C_t(x,t) = \frac{M}{2\sqrt{\pi D_a t}} e^{\frac{-(x-x_0)^2}{4D_a t}} \tag{6.7}$$

where M is the total amount of the diffusing species per unit area of the planar source. The CC method is compelling because of its experimental simplicity and because it avoids experimental challenges (described in the next section) that arise when the porous medium is placed in contact with a bulk liquid water reservoir. The main disadvantage of this method is that it only yields information on D_a.

In *through-diffusion* (TD) experiments, the porous medium is placed between two bulk liquid water reservoirs, one of which is doped with the species of interest. The water reservoirs are frequently replenished in order to impose constant boundary conditions, and the flux of the species of interest reaching the low concentration reservoir is monitored as a function of time (Figure 6.1(b)). In the ideal case where the filters and tubing have no impact on solute mass fluxes, the long-time limit of the cumulative mass flux density of the species of interest reaching the low concentration reservoir should follow the expression (Shackelford, 1991; Van Loon et al., 2003):

$$\lim_{t\to\infty} \int_0^t N(x = L) = \frac{D_e C_0}{L} t - \frac{\alpha L C_0}{6} \tag{6.8}$$

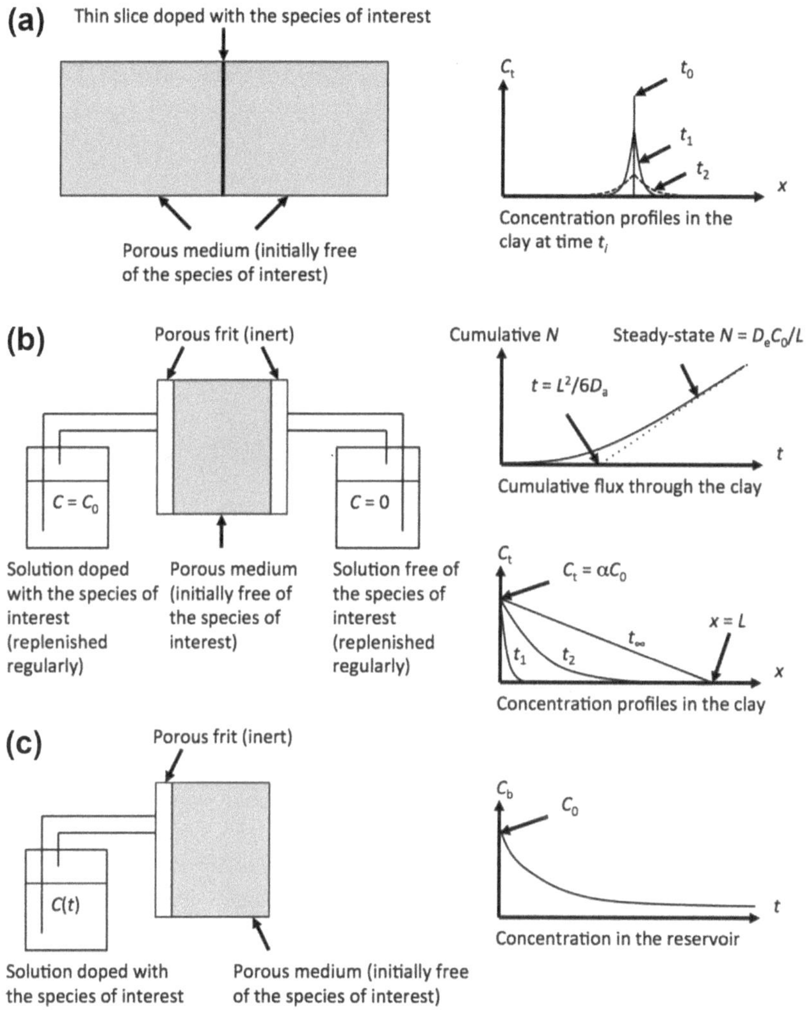

FIGURE 6.1 Schematic description of the three methodologies most commonly used to obtain macroscopic scale self-diffusion coefficients in clay materials: (a) closed cell, (b) through-diffusion, (c) in-diffusion.

where L is the thickness of the clay sample and C_0 is the concentration of the species of interest in the high concentration reservoir. A key advantage of this method is that it allows a direct evaluation of both D_e and α, hence also D_a in conditions where Eqn (6.6) applies. In addition, artifacts associated with end effects and filter-plates can be accounted for by determining the concentration profile of the species of interest in the porous medium at the end of the experiment. The main disadvantage of the TD method is that establishing steady state conditions requires long durations in the case of strongly adsorbed species (Shackelford, 1991; Alonso et al., 2009). Variants of the TD methodology are sometimes used in which concentration is allowed to vary in one or both reservoirs (Glaus et al., 2007), which makes the experiments somewhat less arduous.

In *in-diffusion* (ID) experiments, the porous medium is placed in contact with a single water reservoir doped with a concentration $C_r(0)$ of the species of interest. In the simplest case, C_r then varies freely with time as the solute diffuses into the porous medium (Figure 6.1(c)). In the ideal case where the filters and tubing have no impact on solute mass fluxes, C_r decreases with time according to the relation (Shackelford, 1991; Cormenzana et al., 2003):

$$\frac{C_r(t)}{C_r(0)} = \exp\left(\frac{\alpha D_e}{H_r^2}t\right)\operatorname{erfc}\left(\sqrt{\frac{\alpha D_e}{H_r^2}t}\right) \tag{6.9}$$

where H_r is the volume of the reservoir per cross-section area of the porous medium in the direction normal to diffusion. Key advantages of the ID methodology are that it requires relatively short experimental durations (Shackelford, 1991) and can be applied in situ in clay-rock formations (Wersin et al., 2004, 2008). Furthermore, the ID methodology, like the TD methodology, allows a direct evaluation of both D_e and α if the concentration profile in the porous medium is determined at the end of the experiment. The main weakness of this type of experiment is that experimental artifacts at the boundary between the bulk water reservoir and the clay sample (discussed in the next section) can strongly influence experimental results (Palut et al., 2003).

Two alternative methodologies have contributed to the database on α and D_a discussed in the present chapter. The first consists in electromigration experiments whereby a clay sample enriched with a uniform planar source of the species of interest is exposed to an electric field E in the direction normal to the planar source (Maes et al., 1999; Higashihara et al., 2004; Tanaka et al., 2011). Information on D_a is obtained by carrying out such experiments at several values of E and, then, extrapolating either the electromigration velocity or the dispersion coefficient to $E = 0$. This extrapolation is necessary, because the electric field induces both ion electromigration and electroosmotic flow of water (Chapter 8, in this Volume).

The second alternative methodology consists in using nuclear magnetic resonance (NMR) to quantify the translational motion of "tagged" water molecules (more precisely, protons) on timescales of $\sim 10^{-1}$ s using the pulsed gradient spin echo (PGSE) technique. The PGSE ^1H NMR technique can accurately measure D_a in clay samples at dry bulk densities up to about 0.6 kg dm^{-3} (Nakashima, 2001, 2003a, 2004).

Finally, molecular diffusion in individual clay nanopores on timescales of picoseconds to tens of nanoseconds has been extensively examined using quasi-elastic neutron scattering (QENS) experiments (Cebula et al., 1981; Marry et al., 2008b; González Sánchez et al., 2009) and molecular dynamics (MD) simulations (Chang et al., 1995; Greathouse et al., 2000; Marry and Turq, 2003; Kosakowski et al., 2008; Marry et al., 2008a; Bourg and Sposito, 2010, 2011a; Holmboe and Bourg, 2014). The QENS technique probes the motions of water H atoms, but the interpretation of experimental data can be challenging because diffusion at the pore scale involves strong anisotropy, heterogeneity, and non-Fickian diffusion (Marry et al., 2008b). MD simulations provide an atomistic-level view of molecular diffusion on the same length and timescales, but they can be sensitive to the choice of simulation methodology (Bourg and Sposito, 2010; Ferrage et al., 2011; Marry et al., 2011; Holmboe and Bourg, 2014). When applied jointly, these two techniques provide a detailed view of the dynamics of water and ions in clay nanopores (Malikova et al., 2005; Michot et al., 2007; Holmboe and Bourg, 2014) as discussed in Chapter 11 (in this Volume).

6.2.3 Experimental Challenges

In experimental setups in which a compacted clay sample is placed in contact with one or several bulk water reservoirs, the use of porous filter-plates (required to control clay mineral swelling) introduces a significant artifact, because the filters act as barriers to diffusion (Suzuki et al., 2003; Glaus et al., 2007). As noted above, this difficulty is obviated in CC experiments, where filters are absent. In other types of experiments (TD, ID), the effect of the filters can be accounted for using analytical (Descostes et al., 2008; Yaroshchuk et al., 2009) or numerical calculations (Jansson and Eriksen, 2004; Glaus et al., 2007; Wersin et al., 2008; Appelo et al., 2010). Nevertheless, the uncertainty of solute D_e values in the filters inevitably influences the precision of reported data (Jansson and Eriksen, 2004; Glaus et al., 2007, 2010; González-Sánchez et al., 2008; Yaroshchuk et al., 2009).

Another important experimental difficulty is the existence of "end effects" (alteration of sample surfaces) associated with sample preparation and storage (in the case of clay-rocks) or with the migration of clay particles into the filter-plates (in the case of compacted clay powders). These end effects can cause the porosity of clay samples to increase significantly near their extremities (Glaus et al., 2010, 2011). This type of artifact most significantly influences the results of ID experiments (Van Loon and Jakob, 2005). In situations where end effects and mass transfer through the porous filters play important roles, the precision of calculated diffusion parameters can be significantly increased by measuring the concentration profile of the species of interest in the porous medium at the end of the experiment (Glaus et al., 2011).

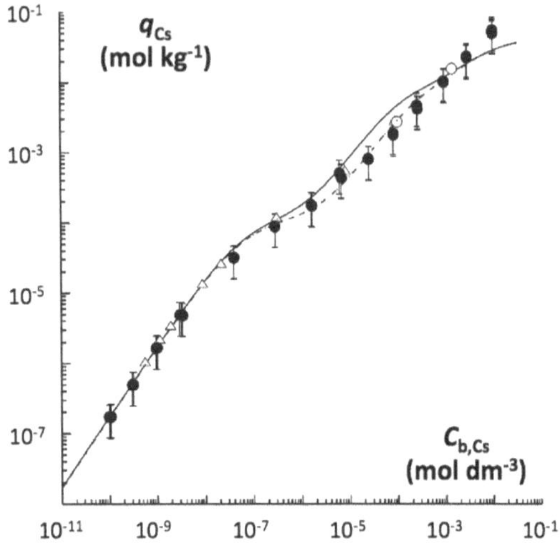

FIGURE 6.2 Adsorption isotherm of Cs on Opalinus clay-rock samples (q is the concentration of adsorbed Cs in moles per kg of solid). Experimental data were obtained with intact (open symbols) or crushed Opalinus clay (closed symbols). The lines show predictions obtained with two different adsorption models. *Figure reprinted with permission from Van Loon et al. (2009).*

Interpretations of CC, TD, and ID diffusion data almost invariably rely on Eqn (6.3), which assumes that adsorption is linear and rapid relative to the timescale of diffusion. These assumptions are rarely examined and may not always be valid (Moridis, 1999). For example, in the important case of Cs^+ adsorption on micaceous minerals (illite, mica), Cs^+ adsorbs preferentially on high affinity edge surface sites (so-called "frayed edge sites") and, more weakly, on basal surface sites (Comans et al., 1991; Poinssot et al., 1999; Steefel et al., 2003). The presence of different adsorption sites results in Cs^+ adsorption increasing nonlinearly with Cs^+ concentration in solution (Figure 6.2). The consideration in Eqn (6.2) of a nonlinearity in the adsorption process yields the following alternative to Eqn (6.4):

$$\left(\alpha + C_b\frac{\partial \alpha}{\partial C_b}\right)\frac{\partial C_b}{\partial t} = \frac{\partial}{\partial x}\left(D_e\frac{\partial C_b}{\partial x}\right) \qquad (6.10)$$

where the derivative of the retention term, $\partial\alpha/\partial C_b$, must be taken into account in order to accurately interpret experimental results (Appelo et al., 2010).

One potential approach to improving the accuracy of measured D_a and D_e values consists in using batch adsorption experiments to constrain the value of α. An important question, then, is whether α values obtained from batch experiments at low solid-water ratios are relevant to adsorption in porous media at higher solid-water ratios. In the important case of Cs^+ diffusion in clay barriers, several studies have suggested that α values measured in batch experiments may not be relevant to diffusion in porous media (Tsai et al., 2001; Jakob et al., 2009; Shackelford and Moore, 2013). In at least one clay-rock, however, direct measurements of Cs^+ adsorption in intact and crushed rock yielded identical results (Figure 6.2; Van Loon et al., 2009; Chen et al., 2014), and α values obtained by adsorption on crushed rock and by diffusion in intact rock could be reconciled by using Eqn (6.10) to account for the nonlinearity of the adsorption isotherm (Appelo et al., 2010). In the case of bentonite, adsorption experiments indicate that the K_d value for Cs^+ adsorption is independent of the solid-water ratio at low degrees of compaction (Montavon et al., 2006) but increases with the solid-water ratio in highly compacted clay (Van Loon and Glaus, 2008), a change in behavior that may reflect the higher affinity of Cs^+ for clay mineral surfaces at lower levels of interlayer hydration (Laird and Shang, 1997; Bourg and Sposito, 2011b). In short, the use of adsorption parameters obtained for dispersed clay to describe diffusion in compacted systems requires careful consideration of the species, material, and conditions of interest.

In the case of very slowly diffusing species, specially designed methodologies may be required to characterize concentration profiles with characteristic thicknesses on the order of tens to hundreds of micrometers. Abrasive peeling coupled with $\alpha/\beta/\gamma$-spectroscopy allows measurements of concentration profiles at the millimeter scale (Van Loon and Eikenberg, 2005). A laser-induced breakdown spectroscopy microprobe (μLIBS) technique has been used to characterize concentration profiles at submillimeter scales (Savoye et al., 2013). Nondestructive analytical techniques such as X-ray radiography (XRR) (Nakashima, 2002, 2003b; Cavé et al., 2009; Loomer et al., 2013) and Rutherford backscattering spectrometry (RBS) (Alonso et al., 2009) have been used to probe concentration profiles on even finer length scales ($\sim$25 nm in the case of RBS). The XRR and RBS techniques, however, are limited to heavy elements (atomic number > 56 in the case of RBS; somewhat lighter elements in the case of XRR); they involve significant experimental challenges associated with the calibration and interpretation of experimental signals (Alonso et al., 2009; Loomer et al., 2013); and they require large concentrations of the diffusing species (Nakashima, 2003b; Loomer et al., 2013).

6.3 CONCEPTUAL MODELS OF D_a AND D_e

In macroporous media, D_e depends only on porosity, a geometric factor G ($\geq$1) that describes the influence of pore-network geometry (i.e., the orientation, shape, and connectivity of pores), and the self-diffusion coefficient D_0 of the species of interest in bulk liquid water (Dykhuizen and Casey, 1989; González Sánchez et al., 2008):

$$D_e = (\phi/G) \times D_0 \tag{6.11}$$

Theoretical calculations predict $G = 3$ for a well-connected network of straight cylindrical pores (Dykhuizen and Casey, 1989; Kato et al., 1995). In clay materials, however, Eqn (6.11) overestimates the diffusion of anions and underestimates the diffusion of cations if the G value is fitted to water diffusion results. This discrepancy obtains from the significant fraction of pore water that is located in the EDL, the interfacial water region where cation adsorption and anion exclusion screen the negative clay surface charge density (Altmann et al., 2012; Tachi and Yotsuji, 2014). In engineered bentonite barriers and clay-rocks considered for use in HLRW storage, most of the pore fluid is located in pores narrower than 10 nm (Holmboe et al., 2012; Keller et al., 2013). For comparison, the characteristic thickness of the EDL is on the order of 2 to 6 nm for 1:1 electrolytes with concentrations of 0.01 to 0.1 mol L^{-1} according to the well-known Gouy–Chapman theory (Sposito, 2004, Chapter 1; in this Volume). Because of this similarity between pore width and EDL thickness, much of the pore water in smectite-rich materials has properties distinct from those of bulk liquid water (Sato, 2008; Montavon et al., 2009; Mazurek et al., 2011) and these materials exhibit strong couplings between their aqueous chemistry, mechanics, and transport properties (Laird and Shang, 1997; Jo et al., 2006; Gajo and Loret, 2007).

Despite the extensive database on diffusion in clay-rocks and bentonite, a widely accepted conceptual model of diffusion in these systems has yet to emerge. A first approach consists in modifying Eqn (6.11) to account for the effect of pore-scale couplings in a semiempirical manner. In this approach, the diffusion of uncharged species, anions, and cations can be modeled with the equations:

$$D_{\mathrm{e,uncharged}} = (\phi/G) \times q_{\mathrm{nano}} D_0 \tag{6.12}$$

$$D_{\mathrm{e,anion}} = (\phi_{\mathrm{e}}/G) \times D_0 \tag{6.13}$$

$$D_{\mathrm{e,cation}} = (\phi/G) \times D_0 + (\rho_{\mathrm{b}} K_{\mathrm{d}}) \times D_{\mathrm{s}} \tag{6.14}$$

where $q_{\mathrm{nano}} \leq 1$ is a parameter that accounts for the slower diffusion dynamics of water and solutes in the vicinity of clay surfaces than in bulk liquid water (Kemper et al., 1964; Li and Gregory, 1974; Bourg et al., 2006; González Sánchez et al., 2009; Holmboe and Bourg, 2014), $\phi_{\mathrm{e}} \leq \phi$ is an effective "anion-accessible" porosity that accounts for anion exclusion in the EDL (van Schaik and Kemper, 1966; Muurinen et al., 1989; Mazurek et al., 2011; Shackelford and Moore, 2013), $K_{\mathrm{d}} = q/C_b$ is a linear adsorption coefficient, ρ_b is the dry bulk density of the porous medium, and D_{s} is a surface diffusion coefficient that accounts for the mobility of adsorbed cations (Jenny and Overstreet, 1939; van Schaik et al., 1966; Nye, 1979; Molera and Eriksen, 2002; Jakob et al., 2009; Gimmi and Kosakowski, 2011). Equations (6.12)–(6.14) provide convenient phenomenological descriptions of D_{e}, but their predictive value is limited unless they can be coupled to an accurate model of water and ion distribution and dynamics near individual clay mineral particle surfaces, i.e., a model of q_{nano}, ϕ_{e}, K_{d}, and D_{s}.

During the last decade, several fundamental models of adsorption and diffusion in individual slit-shaped nanopores have been developed and used to predict water and ion D_e and D_a values as well as clay swelling mechanics and semipermeable membrane properties. At present, these models disagree significantly on their treatment of the microstructure of clayey media—the pore space being treated as consisting of either a single type of pore (Leroy et al., 2006; Appelo and Wersin, 2007; Birgersson and Karnland, 2009; Jougnot et al., 2009; Appelo et al., 2010; Tachi and Yotsuji, 2014; Tachi et al., 2014) or a mixture of interlayer nanopores and larger pores (Wersin et al., 2004; Bourg et al., 2006, 2007, 2008; Bourg and Sposito, 2010; Churakov and Gimmi, 2011; Tournassat and Appelo, 2011; Wang et al., 2013)—as well as in their treatment of adsorption and diffusion in the EDL.

Two significant challenges limit current efforts to constrain pore-scale models of D_a and D_e. Firstly, the structure of the EDL and the dynamics of water and ions near clay—water interfaces have never been directly observed: existing models are based on theoretical calculations (see Chapter 1, in this Volume) or atomistic simulations (Kosakowski et al., 2008; Marry et al., 2008a; Tournassat et al., 2009; Bourg and Sposito, 2011a; Holmboe and Bourg, 2014). These methods, however, sometimes disagree significantly on important features such as the fraction of the clay mineral surface charge that is screened in the Stern layer (by surface complexes) versus the diffuse ion swarm (by long-range electrostatic interactions). Secondly, the microstructure of water-saturated clay material is not precisely known at the microscopic (pore network) scale despite advances in X-ray diffraction (Liu et al., 2003a; Wilson et al., 2004; Ferrage et al., 2005; Holmboe et al., 2010, 2012), X-ray computed tomography (Liu et al., 2003b; Tomioka et al., 2010; Keller et al., 2013), small angle neutron scattering (Swift et al., 2014), NMR (Montavon et al., 2009), and electron microscopy (Hicher et al., 2000; Melkior et al., 2009), because none of these techniques can probe the full range of length scales characteristic of clay layers and their mesoscale assemblages (from 10^{-9} to 10^{-6} m) in conditions that exist in compacted, water-saturated swelling clays.

Despite these difficulties, important progress has been made in the last decade regarding the structure of the EDL, the dynamics of water on clay mineral surfaces, and the microstructure of bentonite and clay-rocks. The following sections reexamines the existing database on D_a and α values (recalling that $D_e = \alpha D_a$) in engineered clay barriers and clay-rocks in light of these advances and discusses the implications of the database regarding the fundamental phenomena that influence diffusion in clay materials.

6.4 SUMMARY OF MEASURED D_a AND D_e VALUES

6.4.1 Parameters that Influence Diffusion in Clay Barriers

A large fraction of the database on diffusion in clay barriers was derived using compacted water-saturated smectite, bentonite, or smectite-quartz mixtures.

Diffusion data obtained in bentonite and smectite-quartz mixtures can be directly compared with data obtained in pure smectite if the compaction of the porous medium is expressed as the effective dry bulk density of the clay matrix, $\rho_{b,clay}$, defined by the relation (Liu et al., 2003b; Sato and Suzuki, 2003):

$$\rho_{b,clay} = \frac{f_{clay}\rho_b}{1 - \frac{(1-f_{clay})\rho_b}{\rho_{other}}} \tag{6.15}$$

In Eqn (6.15), f_{clay} is the mass fraction of clay minerals in the solid, ρ_b is the dry bulk density of the porous medium, and ρ_{other} is the average density of nonclay minerals in the solid (Liu et al., 2003a). In order to directly compare data on engineered and natural clay barriers, diffusion coefficients in clay-rocks are reported as a function of $\rho_{b,clay}$ for rocks of known mineralogy. In applying Eqn (6.15), different definitions are possible as to what constitutes the clay matrix. The type of clay mineral (e.g., smectite vs illite) and the details of the clay mineral structural charge (e.g., whether it arises from isomorphic substitutions in the octahedral or tetrahedral sheets) also may influence the diffusion coefficients D_a and D_e, as suggested by the evidence that cations have a greater tendency to form inner-sphere surface complexes on tetrahedral charge sites (Chang et al., 1995).

In addition to $\rho_{b,clay}$, other parameters that influence diffusion in clay materials are temperature T (K) (González Sánchez et al., 2009), the direction of the diffusion measurement (Sato and Suzuki, 2003; Mazurek et al., 2011), ionic strength I (mol dm^{-3}) (Glaus et al., 2007), the type of exchangeable cation (Choi and Oscarson, 1996; González Sánchez et al., 2008; Melkior et al., 2009; Kozaki et al., 2010), and the species of interest. The T-dependence of α has not been extensively studied. The T-dependence of D_a can be expressed as an activation energy of diffusion E_a (kJ mol^{-1}) based on the Arrhenius relation (González Sánchez et al., 2009):

$$D_a \propto e^{-Ea/RT} \tag{6.16}$$

where R is the ideal gas constant and T is absolute temperature. The anisotropy of the diffusion coefficient tensor has been evaluated almost exclusively in the case of water tracers (HTO, HDO). Results are reported here as an anisotropy factor $A_{p/n} = D_{a,p}/D_{a,n} = D_{e,p}/D_{e,n}$, where the subscripts p and n indicate directions parallel or normal to compaction (in bentonite) or sedimentation (in clay-rocks). The background ionic strength I is that of the solution that was equilibrated with the medium. The impact of the type of exchangeable cation is discussed briefly in the case of Na^+- versus Ca^{2+}-exchanged clay minerals (very few data are available on the impact of other exchangeable cations) and quantified by the ratio $E_{Ca/Na} = D_{a,Ca}/D_{a,Na}$, where the subscripts Na and Ca indicate Na^+- and Ca^{2+}-exchanged clay minerals.

In the case of species that have a pH- or redox-dependent speciation, D_a, D_e, and α may vary with pH and redox potential. This type of behavior is not expected for the species considered here. Diffusion coefficients also may depend on pH in the case of species that adsorb on clay mineral edge surfaces, as these surfaces have a pH-dependent surface charge and reactivity (Bourg et al., 2007; Tournassat et al., 2013). Wang and Liu (2004) and Wang et al. (2005) found that the D_a and α values of Sr^{2+} in bentonite are essentially invariant with pH in the range of 5–9; in the case of Cs^+, however, α increased and D_a decreased with increasing pH, perhaps reflecting a pH-dependent adsorption of Cs^+ on edge surface sites.

6.4.2 Compilation of Diffusion Data

Experimental data on the D_a, D_e, and α values of water (H_2O, HDO, HTO), alkali and alkaline earth metals (Na^+, Cs^+, Sr^{2+}), and halides (Cl^-, I^-, Br^-) in bentonite and clay-rocks were compiled from the literature for use in the present section. Studies that focused on very low solid-water ratios (Nakashima et al., 1999), that used high concentrations of the diffusing species (Nakashima, 2002, 2003b; Loomer et al., 2013), or that studied the codiffusion of multiple species (De Soto et al., 2012) were excluded, as were data obtained with bentonites other than Na^+-, Ca^{2+}-, or Na^+/Ca^{2+}-exchanged bentonite (Manjanna et al., 2009; Melkior et al., 2009) and data obtained with clay-rocks that showed evidence of significant alteration (Cavé et al., 2009) or for which the clay mineralogy was not fully characterized (Cole et al., 2000; Hendry et al., 2009; Barbour et al., 2012). The studies included in the database are listed in Table 6.1 (for bentonite) and Table 6.2 (for clay-rocks).

Dry bulk densities of bentonite samples were converted to $\rho_{b,clay}$ values using Eqn (6.15) with $\rho_{other} = 2.7 \pm 0.1$ kg dm^{-3} (Sato and Suzuki, 2003; Appelo et al., 2010). In cases where ρ_b was measured by drying at 100°C, the smectite fraction of the bentonite samples was assumed to include 2 ± 1 mass% of adsorbed water (Bourg, 2004). In the case of clay-rocks, reported ϕ and f_{clay} values were used to calculate ρ_b (using a solid grain density $\rho_s = 2.7 \pm 0.1$ kg dm^{-3}) and $\rho_{b,clay}$ (using Eqn (6.15)). Equation (6.6) was used to convert D_e values to D_a values. Equation (6.16) was applied to extrapolate D_a values to 298 K using measured values of the activation energy of diffusion in bulk liquid water: $E_a = 18.3$ kJ mol^{-1} for water (Mills, 1973) and 17.4, 16.2, 20.1, and 18.4 kJ mol^{-1} for Na^+, Cs^+, Sr^{2+}, and Cl^-, respectively (Robinson and Stokes, 1959). The E_a value of Cl^- was used for all anions. Relative apparent diffusion coefficients D_a/D_0 were calculated using $D_0 = 2.30$, 2.27, and 2.24 10^{-9} m^2 s^{-1} for H_2O, HDO, and HTO (Mills, 1973) and 1.33, 2.07, 0.794, 2.03, 2.01, and 2.00 10^{-9} m^2 s^{-1} for Na^+, Cs^+, Sr^{2+}, Cl^-, Br^-, and I^- at 298 K, respectively (Li and Gregory, 1974).

TABLE 6.1 Macroscopic Scale Experimental Studies of Self-Diffusion in Water-Saturated Bentonite

Solid[a]	Species of Interest	I (mol dm^{-3}) and Main Cation[b]	T (K)[c]	$\rho_{b,clay}$ (kg dm^{-3})	Result[d]	Technique[e]	References
Bentonite (MX)	Cl^-, Na^+, Cs^+, Sr^{2+}	0.001–0.5 (Na)	n.s.	1.0–1.5	D_a, α	TD + f + p	Muurinen and Penttilä-Hiltunen (1987), Muurinen et al. (1989, 1990)
Smectite (KP)	HTO, I^-, Cs^+, Sr^{2+}	0 (Na)	n.s.	0.2–2.0	D_a	CC	Sato et al. (1992)
Bentonite (MX)	Cl^-, Cs^+, Sr^{2+}	0 (Na)	n.s.	0.8–1.7	D_a	CC	Kim et al. (1993)
Bentonite (MX)	I^-	0 (Na)	n.s.	1.4	D_a	TD	Madsen and Kahr (1993)
Bentonite (KG)	HTO, Cs^+	0 (Na)	n.s.	0.4–1.5	D_a	CC	Sato et al. (1993)
Bentonite (KG)	HTO	0 (Na)	n.s.	0.3–1.5	D_e, α	TD	Kato et al. (1995)
Bentonite (AV)	HTO, I^-, Sr^{2+}	0 (Na or Ca)	296 ± 2	1.1	D_e, α	TD	Choi and Oscarson (1996)
Bentonite (AV)	I^-, Cs^+	0.26 (Na/Ca) SGW	296 ± 2	1.1	D_e, α	TD	Oscarson et al. (1996)
Bentonite (MX)	I^-, Na^+, Sr^{2+}	0.02–0.2 SGW	298	1.5	D_e, α	TD + f	Eriksen and Jansson (1996), Eriksen et al. (1999)

Smectite (KP)	HTO, Cl^-, Na^+, Cs^+, Sr^{2+}	0 (Na)	278–323 ± 0.1	0.7–1.8	D_a, E_a	CC	Kozaki et al. (1996, 1997, 1998, 1999b, 2001, 2005)
Smectite (KP)	HTO, Na^+, Cs^+, Ca^{2+}	0 (Na/Ca)	288–323 ± 0.1	1.0, 1.8	D_a, E_a	CC	Kozaki et al. (1999a, 2005, 2010)
Smectite (KP)	HTO	0 (Na)	278–323 ± 1.5	1.0–2.0	D_a, E_a	CC	Nakazawa et al. (1999)
Smectite (KP, SA, VK, SW), bentonite (KG)	H_2O	0 (Na)	284–343	0–0.67	D_a, E_a	NMR	Nakashima (2001, 2003a, 2004), Nakashima and Mitsumori (2005)
Bentonite (MX)	Cl^-, I^-, Na^+, Cs^+, Sr^{2+}	0.01–0.5 (Na)	n.s.	0.3–1.5	D_e, α	TD + f or TD + f + p	Molera and Eriksen (2002), Molera et al. (2003)
Smectite (KP)-sand mixtures	Na^+, Sr^{2+}	0 (Na)	278–323	1.0–1.7	D_a, E_a	CC	Liu et al. (2003a,b)
Smectite (KP), bentonite (KG)	HTO	0.35–1.1 (Na) SGW	298	0.57–1.46	$\boldsymbol{D_a}$	TD + f + p	Sato and Suzuki (2003)
Bentonite (FX)	HTO, Cl^-	0 (n.s.)	n.s.	0.93–1.57	D_e, α		García-Gutiérrez et al. (2004)
Bentonite (MX)	I^-	SGW (Äspö ground water)	288	1.54	D_a, α		Jansson and Eriksen (2004)
Smectite (KP)	HDO	0 (Na)	298–333	0.9, 1.35	$\boldsymbol{D_a}$, E_a	TD	Suzuki et al. (2004)

Continued

TABLE 6.1 Macroscopic Scale Experimental Studies of Self-Diffusion in Water-Saturated Bentonite—cont'd

Solid[a]	Species of Interest	I (mol dm^{-3}) and Main Cation[b]	T (K)[c]	$\rho_{b,clay}$ (kg dm^{-3})	Result[d]	Technique[e]	References
Bentonite (MX)	Sr^{2+}, Cs^+	0.1 (Na)[f]	n.s.	0.8	D_e, α	ID + p	Wang and Liu (2004), Wang et al. (2005)
Bentonite (VK)	Cl^-	0.01–1.0 (Na)	298 ± 1	1.04–1.64	D_e, α	TD + p	Van Loon et al. (2007)
Smectite (MI)	HTO	0.01–1.0 (Na or Ca)	273–333 ± 1	1.90 ± 0.05	D_e, α, E_a	TD + f	González Sánchez et al. (2008)
Smectite (KP)	Na^+	0.05–0.5 (Na)	288–323 ± 0.1	0.98	D_a, E_a	CC	Kozaki et al. (2008)
Bentonite (MX)	HTO, Na^+	0.01 or 0.1 (Na or Ca); 0.1 (Na/Ca/Mg) SGW	296	1.2	D_e, α	TD + f	Melkior et al. (2009)
Smectite (MI)	HTO, Cl^-, Na^+, Sr^{2+}	0.1–2.0 (Na)	298 ± 1	1.95	D_e, α	TD + f or TD + f + p	Glaus et al. (2007, 2010, 2011)
Smectite (KP)	HTO, I^-, Na^+, Cs^+	0.01–0.5 (Na)	295.5 ± 2.5	0.8	D_e, D_a, α	TD + p or CC	Tachi and Yotsuji (2014)

[a]*Smectite: KP: Kunipia-F montmorillonite; SW: SWy-2 montmorillonite; SA: Saponite from the Japan Clay Science Society (standard sample JCSS3501); Bentonite: KG: Kunigel-V1 bentonite (47.5 ± 1.5 mass% smectite); VK: Volclay KWK bentonite (64 mass% smectite, 7% chlorite/smectite; uncertainty on the total clay content was assumed equal to ±5 mass%); MX: MX-80 bentonite (72.5 ± 7.5 mass% smectite); AV: Avonlea bentonite (80 mass% smectite, 10 mass% illite; uncertainty on the total clay content was assumed equal to ±5 mass%); FX: FEBEX bentonite (93 ± 2 mass% smectite).*
[b]*Cation concentration in moles of charge (0 M indicates that the clay was saturated with pure water). The main cation is indicated in parentheses (Na/Ca indicates a mixed Na/Ca-exchanged clay). The abbreviation SGW indicates the use of a synthetic ground-water.*
[c]*If not specified (n.s.), T was assumed equal to 288 ± 5 K.*
[d]*Bold font indicates D values that were measured parallel and normal to compaction. All other D values were measured parallel to compaction.*
[e]*CC = closed cell; TD = through-diffusion; ID = in-diffusion; NMR = nuclear magnetic resonance; f = data analysis accounting for the impact of filter-plates; p = data analysis supported by concentration profile of the species of interest in the porous medium.*
[f]*Experiments were carried out at a range of pH values (3–12).*

TABLE 6.2 Macroscopic Scale Experimental Studies of Self-Diffusion in Water-Saturated Clay-Rocks

Solid[a]	Species of Interest	Aqueous Chemistry ($mol\ dm^{-3}$)[b]	T (K)	Result	Technique	References
OPA-MT	HTO	SGW	n.s.	D_e	ID + p	Palut et al. (2003)
OPA-MT[c]	HTO, Br^-, I^-, Na^+, Cs^+, Sr^{2+}, Co^{2+}	SGW	289 ± 3	D_e, α	ID	Wersin et al. (2004, 2008)
OPA-MT	HTO, I^-, Cl^-	SGW	n.s.	D_e, α	ID + f	Wersin et al. (2004)
OPA-MT, OPA-BK	HTO, I^-, Cl^-	SGW	296 ± 2	D_e or $\mathbf{D_e}$, α	TD, OD	Van Loon et al. (2004, 2005b)
OPA-MT, OPA-BK	HTO, Na^+, Cl^-	SGW	273–343 ± 0.1	D_e, α, E_a	TD	Van Loon et al. (2005a)
COX	Na^+, Cs^+	SGW	296	D_e, α	TD + f	Melkior et al. (2007)
OPA-MT	Cs^+	SGW	296 ± 2	D_e	TD + f	Jakob et al. (2009)
COX	HTO, Cl^-, I^-	SGW	294 ± 1	D_e, α	TD + f	Descostes et al. (2008)
TOA	HTO, Cl^-, I^-	SGW, SGW + 0.1 M NaCl	n.r.	D_e, α	TD	Wittebroodt et al. (2012)
OPA-MT	HTO, Cl^-, Br^-, I^-, Na^+ Cs^+, Sr^{2+}	SGW	296 ± 2	D_e, α	TD + f	Appelo et al. (2010)
TOA	HTO	SGW	296 ± 1	$\mathbf{D_e}$	TD	Motellier et al. (2007)

Continued

TABLE 6.2 Macroscopic Scale Experimental Studies of Self-Diffusion in Water-Saturated Clay-Rocks—cont'd

Solid[a]	Species of Interest	Aqueous Chemistry (mol dm^{-3})[b]	T (K)	Result	Technique	References
COX	HTO, Cl^-, Na^+, Cs^+	SGW	294, 353 ± 1	D_e, α, E_a	TD	Savoye et al. (2011)
WAK	HTO, I^-, Cs^+	0.1 × SGW, SGW, 3 × SGW	295.5 ± 2.5	D_e, α	TD	Tachi et al. (2009, 2011)
BOO-MO	HTO, I^-	SGW	298	D_a, α	TD	Aertsens et al. (2003)
BOO-MO	HTO, I^-, Sr^{2+}	SGW	n.s.	D_a	EM	Maes et al. (1999)
TOA	Cl^-	SGW	n.s.	D_a, α	OD	Patriarche et al. (2004a)

[a]*OPA-MT: Opalinus clay at Mont Terri ($\phi = 0.18 \pm 0.05$; $f_{clay} = 0.6$); OPA-BK: Opalinus clay at Benken ($\phi = 0.12 \pm 0.01$; $f_{clay} = 0.6$); COX: Callovo-Oxfordian clay at Bure ($\phi = 0.17 \pm 0.04$; $f_{clay} = 0.4$); TOA: Toarcian clay at Tournemire ($\phi = 0.10 \pm 0.01$; $f_{clay} = 0.5$); WAK: Wakkanai mudstone ($\phi = 0.38 \pm 0.04$; $f_{clay} = 0.7$); BOO-MO: Boom clay at Mol ($\phi = 0.37 \pm 0.02$; $f_{clay} = 0.5$).*

[b]*Chemistry of the bulk solutions that were placed in contact with the solid. SGW denotes the use of synthetic ground water representative of in situ conditions: Opalinus clay at Mont Terri roughly 0.23 M NaCl + 0.04 M $CaCl_2$-$MgCl_2$ ($I = 0.39$ M); Opalinus clay at Benken roughly 0.15 M NaCl + 0.01 M $CaCl_2$-$MgCl_2$ ($I = 0.2$ M); Callovo-Oxfordian clay roughly 0.04 M NaCl + 0.02 M $CaCl_2$-$MgCl_2$ ($I = 0.1$ M); Toarcian clay roughly 0.01 M NaCl; Wakkanai mudstone roughly 0.31 M NaCl + 0.06 M $NaHCO_3$ ($I = 0.24$ M); Boom clay roughly 0.015 M $NaHCO_3$ ($I = 0.02$ M).*

[c]*In situ experiments carried out in underground research laboratories (other experiments were carried out in the laboratory).*

6.4.3 Water Diffusion

Experimental data on the α and D_a values of water (H_2O, HDO, HTO) are compiled in Figures 6.3 and 6.4. Reported α values consistently indicate that $\alpha = \phi$, i.e., water diffusion occurs through the entire pore space of the medium, with no inaccessible or dead-end pores (Aertsens et al., 2003; García-Gutiérrez et al., 2004; Hendry et al., 2009). This is evidenced by the plot of measured α values and predicted ϕ values as a function of ρ_b in Figure 6.3.

Experimental data on the D_a/D_0 values of water as a function of $\rho_{b,clay}$ are shown in Figure 6.4(a) (in the case of bentonite) and Figure 6.4(b) (in the case of clay-rocks). In both figures, data on diffusion in the direction normal versus parallel to compaction (in bentonite) or sedimentation (in clay-rocks) are highlighted. The results compiled in Figure 6.4(a) and (b) indicate that D_a decreases significantly with $\rho_{b,clay}$, particularly in the direction parallel to compaction or sedimentation. At identical values of $\rho_{b,clay}$, water tends to diffuse more slowly in bentonite than in clay-rocks. This difference is consistent with the clay mineral fraction in bentonite consisting almost entirely of smectite, whereas the clay mineral fraction in clay-rocks tends to consist of illite or illite-smectite mixtures (hence, the fraction of pore water located in direct contact with clay mineral surfaces must be greater in bentonite than in clay-rocks at the same value of $\rho_{b,clay}$). According to Eqn (6.15) and the relationship between porosity and dry bulk density ($\phi = 1 - \rho_b/\rho_s$), any model that attempts to describe the D_a or D_e values of water based solely on porosity—including the well-known model $D_e/D_0 = \phi^m$ (Suzuki et al., 2004; Leroy et al., 2006; Descostes et al., 2008; Jougnot et al., 2009; Glaus et al., 2010; Mazurek et al., 2011) and other models (Boudreau, 1996; Boudreau and Meysman, 2006)—would erroneously predict that water diffuses more rapidly in pure smectite than in clay-rocks for the same value of $\rho_{b,clay}$.

Pore-scale studies of water diffusion in bentonite indicate that the dependence of D_a on $\rho_{b,clay}$ at effective clay dry bulk densities greater than about 0.2 kg/dm^3 reflects two phenomena. The first phenomenon is the slower

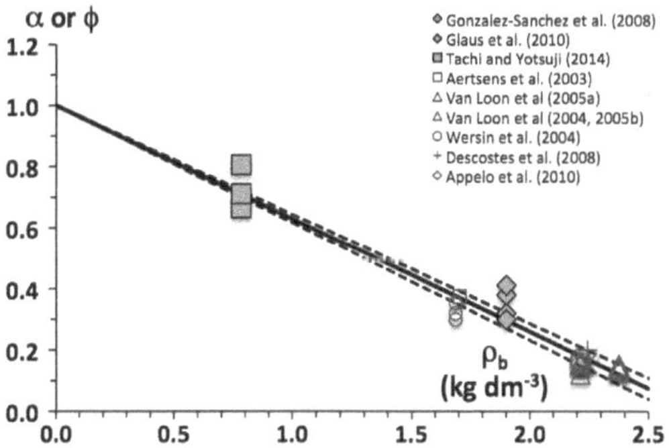

FIGURE 6.3 Experimental data on α versus ρ_b for water diffusion in bentonite (filled red symbols (gray in print versions)) and clay-rocks (open blue symbols (dark gray in print versions)). The solid line (with confidence interval shown as dashed lines) shows the relationship between ϕ and ρ_b calculated if $\rho_s = 2.7 \pm 0.1$ kg dm^{-3}.

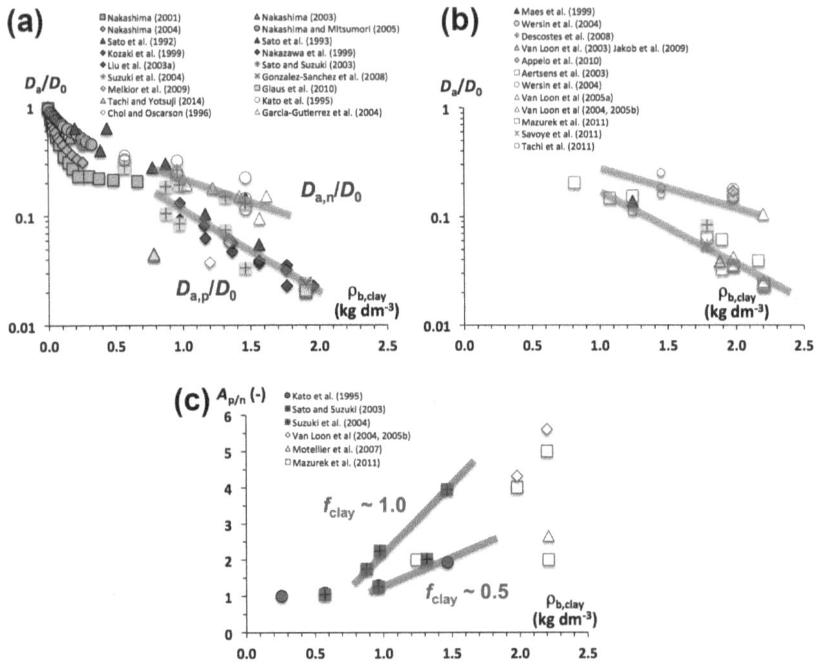

FIGURE 6.4 Experimental data on D_a versus $\rho_{b,clay}$ for water diffusion in clay materials: (a) D_a/D_0 in bentonite (filled symbols: *best available data obtained by standard measurement techniques* (green (darkest gray in print versions)) *or by NMR* (yellow (dark gray in print versions)); pale blue symbols (light gray in print versions): *data obtained with no sampling of concentration profiles in the clay*; open orange symbols (white in print versions): *data obtained from TD or ID methods without accounting for the influence of filter-plates*); (b) D_a/D_0 in clay-rocks; (c) anisotropy factor $A_{p/n}$ in bentonite (filled red symbols (dark gray in print versions)) and clay-rocks (open blue symbols (white in print versions)). In Figures 6.4(a) and (b), shaded regions highlight the trends obtained in the directions parallel versus normal to compaction or sedimentation. In Figure 6.4(c), shaded regions highlight the trends obtained using bentonite with ~50 or 100% clay content.

diffusion of water molecules located within about one nanometer from the surface (Marry et al., 2008a; Bourg and Sposito, 2011a). The importance of this "surface water" increases with $\rho_{b,clay}$, causing the mean principal value of the apparent diffusion coefficient tensor ($D_{a,mp} = 1/3\ D_{a,p} + 2/3\ D_{a,n}$) to decrease with $\rho_{b,clay}$ (Bourg and Sposito, 2010; Holmboe and Bourg, 2014). At $\rho_{b,clay} = 1.5\ \mathrm{kg\ dm^{-3}}$ (where most water molecules in compacted smectite are in direct contact with a clay mineral surface at the molecular scale), experimental data yield $D_{a,mp}/D_0 \sim 0.1$. Therefore, surface water is clearly not "immobile" or "frozen" as assumed in certain models (Moridis, 1999; Nakashima, 2003a).

The second phenomenon that determines the dependence of D_a on $\rho_{b,clay}$ is that the apparent diffusion coefficient tensor becomes increasingly anisotropic as $\rho_{b,clay}$ increases, as reflected in the anisotropy factor $A_{p/n} > 1$. This anisotropy reflects the tendency of clay particles to become increasingly arranged with their basal surfaces oriented normal to the direction of loading (Hicher et al., 2000). Experimental data compiled in Figure 6.4(c) confirm that $A_{p/n}$ increases with $\rho_{b,clay}$, particularly at $\rho_{b,clay} > 0.7$ kg dm^{-3}. In the case of bentonite, the anisotropy factor $A_{p/n}$ tends to increase with clay content, perhaps indicating that large mineral grains disrupt the ability of externally applied stresses to reorient clay particles (Sato and Suzuki, 2003; Liu et al., 2003b; Devineau et al., 2006; Robinet et al., 2012). The data shown in Figure 6.4(c) for shales (obtained with rocks where $f_{clay} = 0.5-0.6$) are roughly consistent with the data for bentonites with $f_{clay} \sim 0.5$, suggesting that the relationship between $A_{p/n}$ and $\rho_{b,clay}$ is broadly similar in bentonite and clay-rocks despite important differences in clay mineralogy (smectite vs illite) and compaction mechanism (compaction of dry powder vs consolidation of water-saturated sediments). Data obtained with species other than water suggest that $A_{p/n}$ does not depend strongly on the species of interest (Van Loon et al., 2004).

The impact of other variables on water D_a values remains scarcely examined. The database on the impact of the type of exchangeable cation (as quantified by $E_{Ca/Na}$) suggests that water may diffuse more rapidly in Ca^{2+}- than in Na^+-bentonite, particularly at high values of $\rho_{b,clay}$ (Choi and Oscarson, 1996; González Sánchez et al., 2008), perhaps because Ca^{2+}-smectite particles tend to consist of a greater number of stacked clay mineral layers and, therefore, may retain more meso- or macro-porosity than Na^+-smectite (Bérend et al., 1995; Holmboe et al., 2012). However, experimental data on diffusion in films of well-oriented smectite lamellae suggest that the mobility of water in individual nanopores is lower in Ca^{2+}- than in Na^+-smectite (Kemper et al., 1964), which should at least partly counterbalance the effect of stacking differences. Measured D_a/D_0 values have little dependence on temperature (Nakazawa et al., 1999; González Sánchez et al., 2008; Holmboe and Bourg, 2014) and salinity (González Sánchez et al., 2008).

6.4.4 Anion Diffusion

Experimental data on the diffusion of halide anions (Cl^-, I^-, Br^-) are compiled in Figure 6.5 and 6.6. The fraction of pore space that is accessible to anions (i.e., the retardation factor α/ϕ) is shown in Figure 6.5(a) for bentonite equilibrated with 0.1 mol L^{-1} NaCl solutions (filled symbols) and for clay-rocks equilibrated with synthetic groundwater thought to be in equilibrium with the rock (open symbols). The results shown in Figure 6.5(a) indicate that α/ϕ in bentonite decreases significantly with $\rho_{b,clay}$ ($\alpha/\phi = 0.08 \pm 0.04$ at $\rho_{b,clay} > 1.5$ kg dm^{-3}). In clay-rocks, $\alpha/\phi = 0.46 \pm 0.05$ with little sensitivity to $\rho_{b,clay}$. At an ionic strength of 0.1 mol L^{-1} in a 1:1 electrolyte, Gouy–Chapman theory indicates that the characteristic thickness of the anion exclusion region on negatively charged clay mineral surfaces is on the order of 2 nm (two Debye lengths; see Chapter 1, in this volume). Therefore, the data in Figure 6.5(a) suggest that roughly 50% of the

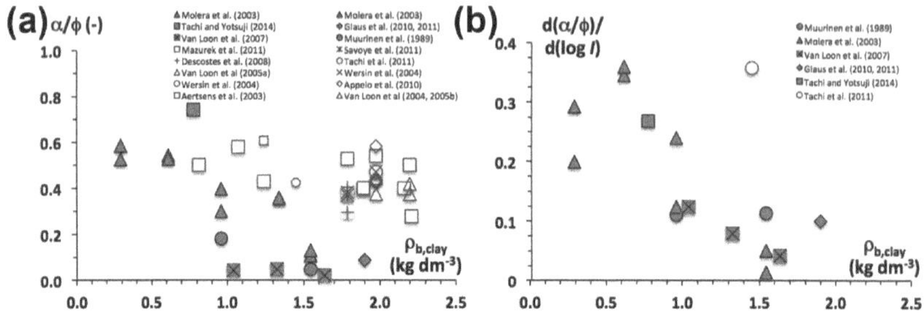

FIGURE 6.5 Experimental data on α/ϕ for halide anions: (a) α/ϕ as a function of $\rho_{b,clay}$ in bentonite equilibrated with a 0.1 mol L^{-1} NaCl solution (filled red symbols (gray in print versions)) and in clay-rocks equilibrated with synthetic groundwater (open blue symbols (dark gray in print versions)); (b) $d(\alpha/\phi)/d(\log I)$ as a function of $\rho_{b,clay}$ in bentonite and in clay-rocks.

pore space in shales, and almost all of the pore space in bentonite, is located within about 2 nm from a clay mineral surface. The difference between bentonite and clay-rocks is consistent with the higher proportion of swelling clay minerals (smectite) in bentonite than in clay-rocks. The influence of ionic strength on α/ϕ is shown by the data on $d(\log(\alpha/\phi))/d(\log I)$ versus $\rho_{b,clay}$ in Figure 6.5(b). The fraction of the pore space that is accessible to anions increases with salinity in agreement with the predictions of the Gouy–Chapman theory (Sposito, 2004).

Experimental data on D_a/D_0 for anions in bentonite and shales are reported in Figure 6.6. Data on D_a/D_0 in bentonite (available only in the direction parallel to compaction) show a strong dependence on $\rho_{b,clay}$. Data on the ratio $(D_a/D_0)_{anion}/(D_a/D_0)_{water}$, compiled in Figure 6.6(b), show that the $\rho_{b,clay}$-dependence of D_a/D_0 is essentially identical for anions and water up to $\rho_{b,clay} = 1.0$ kg dm^{-3}. These results indicate that anions diffuse as fast as water in the anion-accessible pore space of bentonite and clay-rocks, as already demonstrated long ago (van Schaik and Kemper, 1966). At $\rho_{b,clay} > 1.0$ kg dm^{-3}, the ratio $(D_a/D_0)_{anion}/(D_a/D_0)_{water}$ decreases significantly to an average value of 0.65 ± 0.12 in clay-rocks and 0.26 ± 0.10 in bentonite at $\rho_{b,clay} > 1.5$ kg dm^{-3}.

Remarkably, the D_a/D_0 values of halide anions have a significant salinity-dependence with a value that peaks near $\rho_{b,clay} = 1.0$ kg dm^{-3}. This salinity-dependence of D_a/D_0 suggests that anions diffuse not only in bulk-liquid-like water, but also in water near the clay mineral surfaces that is influenced by the (salinity-dependent) electrostatic influence of the clay mineral surfaces. The effect is relatively small (D_a decreases by a factor of ~2 as salinity increases from 0.01 to 1.0 M NaCl at $\rho_{b,clay} = 1.0$ kg/dm^3) but it is inconsistent with models that treat the pore width (Tachi and Yotsuji, 2014; Tachi et al., 2014) or electrostatic potential (Leroy et al., 2006; Birgersson and Karnland, 2009; Jougnot et al., 2009) as spatially invariant, a finding consistent with X-ray diffraction studies showing the coexistence of multiple swelling states in hydrated smectite (Ferrage et al., 2005; Holmboe et al., 2012). Experimental data

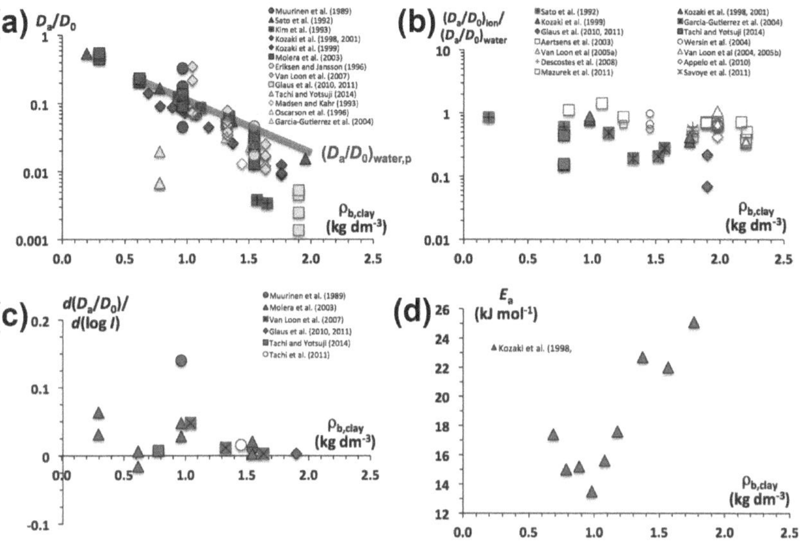

FIGURE 6.6 Experimental data on D_a for halide anions: (a) D_a/D_0 as a function of $\rho_{b,clay}$ in bentonite (filled green symbols (dark gray in print versions): *best available data*; pale blue symbols: *data obtained with no sampling of concentration profiles in the clay*; open orange symbols: *data obtained from TD or ID methods without accounting for the influence of filter-plates*); (b) $(D_a/D_0)_{anion}/(D_a/D_0)_{water}$ as a function of $\rho_{b,clay}$ in bentonite (filled red symbols (gray in print versions)) and in clay-rocks (open blue symbols (light gray in print versions)); (c) $d(D_a/D_0)/d(\log I)$ as a function of $\rho_{b,clay}$ in bentonite and in clay-rocks; (d) E_a as a function of $\rho_{b,clay}$ in bentonite.

on the activation energy of diffusion indicate that the T-dependence of D_a in bentonite is significantly greater than in bulk liquid water at $\rho_{b,clay} >$ 1.3 kg dm^{-3}, the same condition where the meso- and macro-porosity of bentonite essentially disappears according to Figure 6.5(a). The anisotropy of D_a and the impact of the type of exchangeable cation remain relatively scarcely examined.

6.4.5 Cation Diffusion

Experimental data on the diffusion of alkali and alkaline earth metals (Na^+, Cs^+, Sr^{2+}) in clay materials are compiled in Figures 6.7–6.10. Data on α versus $\rho_{b,clay}$ are shown in Figure 6.7, with each figure corresponding to a different ion. The dashed lines in Figure 6.7 are calculated based on the assumption that cation adsorption can be described with a linear adsorption coefficient $K_d = q/C_b$, which yields:

$$\alpha = \phi + \rho_b K_d \tag{6.17}$$

According to Eqn (6.17), the α values of cations should increase with ρ_b (and with $\rho_{b,clay}$, if f_{clay} is fixed) as observed experimentally in Figure 6.7 in the case

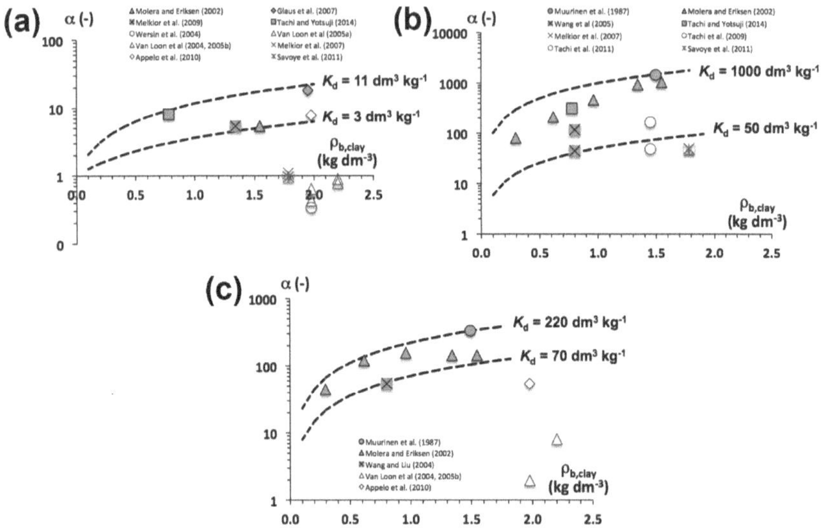

FIGURE 6.7 Experimental data on α as a function of $\rho_{b,clay}$ for (a) Na^+, (b) Cs^+, and (c) Sr^{2+} in bentonite equilibrated with a 0.1 mol L^{-1} NaCl solution (filled red symbols (gray in print versions)) and in clay-rocks equilibrated with synthetic groundwater (open blue symbols (light gray in print versions)).

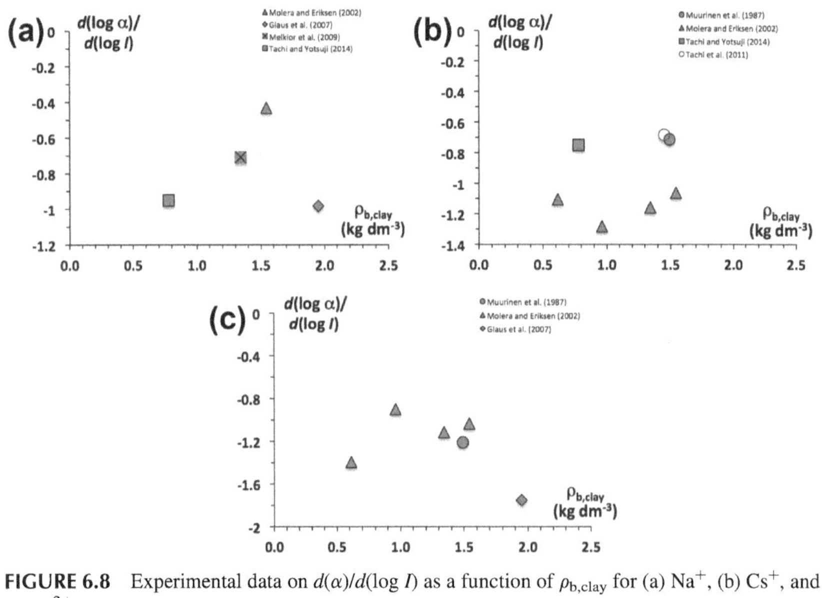

FIGURE 6.8 Experimental data on $d(\alpha)/d(\log I)$ as a function of $\rho_{b,clay}$ for (a) Na^+, (b) Cs^+, and (c) Sr^{2+} in bentonite (filled red symbols (gray in print versions)) and in clay-rocks (open blue symbol (light gray in print versions)).

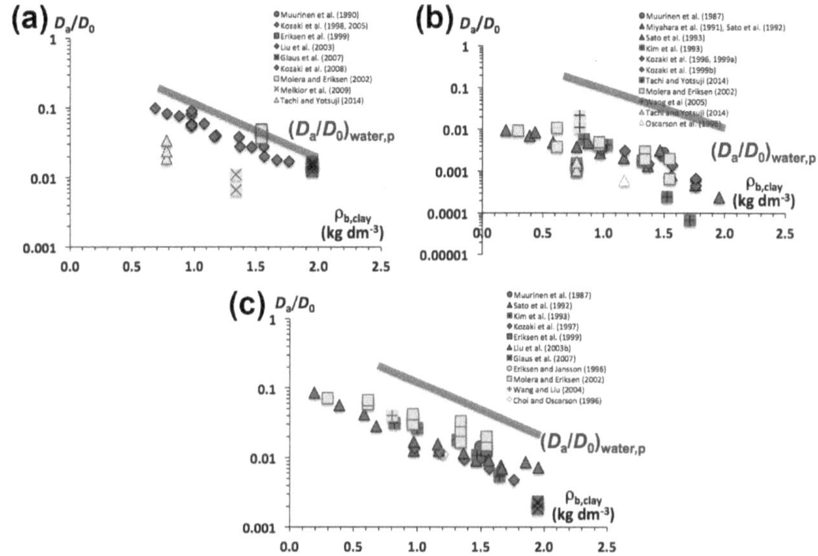

FIGURE 6.9 Experimental data on D_a/D_0 as a function of $\rho_{b,clay}$ in bentonite for (a) Na^+, (b) Cs^+, and (c) Sr^{2+} (filled green symbols (dark gray in print versions): *best available data*; pale blue symbols (gray in print versions): *data obtained with no sampling of concentration profiles in the clay*; open orange symbols (light gray in print versions): *data obtained from TD or ID methods without accounting for the influence of filter-plates*).

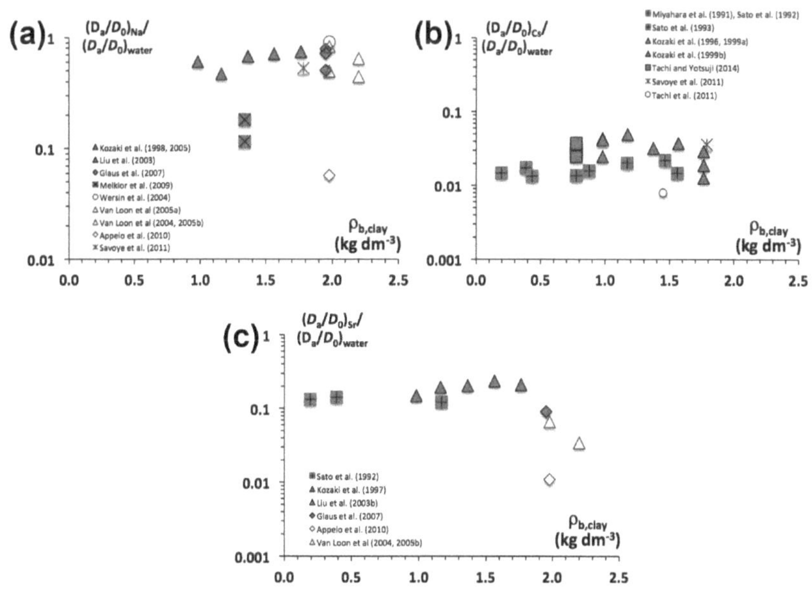

FIGURE 6.10 Experimental data on $(D_a/D_0)_{anion}/(D_a/D_0)_{water}$ as a function of $\rho_{b,clay}$ for (a) Na^+, (b) Cs^+, and (c) Sr^{2+} in bentonite (filled red symbols (dark gray in print versions)) and in clay-rocks (open blue symbols (light gray in print versions)).

of bentonite equilibrated with a 0.1 mol L^{-1} NaCl solution. The dashed lines in Figure 6.7 were calculated using K_d values chosen to bracket reported α values in bentonite. Results yield $K_d = 3–11$ $dm^3\,kg^{-1}$ for Na^+, 50–1000 $dm^3\,kg^{-1}$ for Cs^+, and 70–220 $dm^3\,kg^{-1}$ for Sr^{2+}. The large range of K_d values associated with Cs^+ diffusion data may reflect the existence of different adsorption sites on clay mineral surfaces with very different affinities for Cs^+ (Figure 6.2). This may complicate the interpretation of Cs^+ diffusion data by causing significant adsorption competition effects (Jakob et al., 2009) and a strong impact of cesium concentration on Cs^+ diffusion properties (Gimmi and Kosakowski, 2011). The α values of cations in clay-rocks are roughly one order of magnitude lower than in bentonite, with significant scatter. This difference is consistent with the lower specific surface area of the clay fraction in clay-rocks than in bentonite (because of the collapsed interlayer spaces of illite) and, also, with the higher ionic strengths used for diffusion measurements in clay-rocks (0.1–0.24 mol L^{-1}, in the case of clay-rocks data reported for Cs^+ in Figure 6.7).

Experimental data on the salinity dependence of the α values of cations (Figure 6.8) show significant scatter and no clear dependence on $\rho_{b,clay}$. Average reported values of $d(\log \alpha)/d(\log I)$ for cations in bentonite at 0.1 mol L^{-1} NaCl are -0.8 ± 0.3 for Na^+, -1.0 ± 0.2 for Cs^+, and -1.2 ± 0.2 for Sr^{2+}. The first two values are consistent with that expected based on ion exchange theory for a homovalent exchange reaction ($d(\log \alpha)/d(\log I) = -1$) (Glaus et al., 2007). The average reported value for Sr^{2+} is significantly less negative than the value of -2 expected for a heterovalent Sr^{2+}–Na^+ exchange reaction, perhaps indicating that Sr^{2+} adsorption is modulated by the formation of ion pairs such as $SrHCO_3^+$ or $SrSO_4$ (Cole et al., 2000; Appelo et al., 2010).

Experimental data on D_a/D_0 for cations in bentonite in the direction parallel to compaction show a strong dependence on $\rho_{b,clay}$ (Figure 6.9) as also reported in the case of water and anions. Data on the ratio $(D_a/D_0)_{cation}/(D_a/D_0)_{water}$ show that the presence of clay minerals has a greater impact on the D_a values of cations than on those of water (Figure 6.10). The ratio $(D_a/D_0)_{cation}/(D_a/D_0)_{water}$ is essentially identical in bentonite and clay-rocks and is not significantly influenced by $\rho_{b,clay}$ within the precision of experimental data. For comparison, models that rely on EDL theories with the assumption that Stern layer species (surface complexes) are immobile predict that $(D_a/D_0)_{cation}/(D_a/D_0)_{water}$ increases with $\rho_{b,clay}$ (Shainberg and Kemper, 1966; Kim et al., 1993), whereas models that account for the slower diffusivity of water molecules located in direct contact with clay mineral surfaces predict a slight increase in $(D_a/D_0)_{cation}/(D_a/D_0)_{water}$ with $\rho_{b,clay}$ (Bourg et al., 2007; Bourg and Sposito, 2010).

As $\rho_{b,clay}$ and I approach zero, experimental data on $(D_a/D_0)_{cation}/(D_a/D_0)_{water}$ should yield an estimate of the ratio of cation diffusion coefficients on the clay mineral surfaces and in bulk liquid water (Bourg et al., 2007, 2008; Bourg and Sposito, 2010). The average values of $(D_a/D_0)_{cation}/(D_a/D_0)_{water}$ reported in Figure 6.10 at $\rho_{b,clay} < 1.2$ kg dm^{-3} and $I = 0$ mol L^{-1} equal 0.54 ± 0.13 for

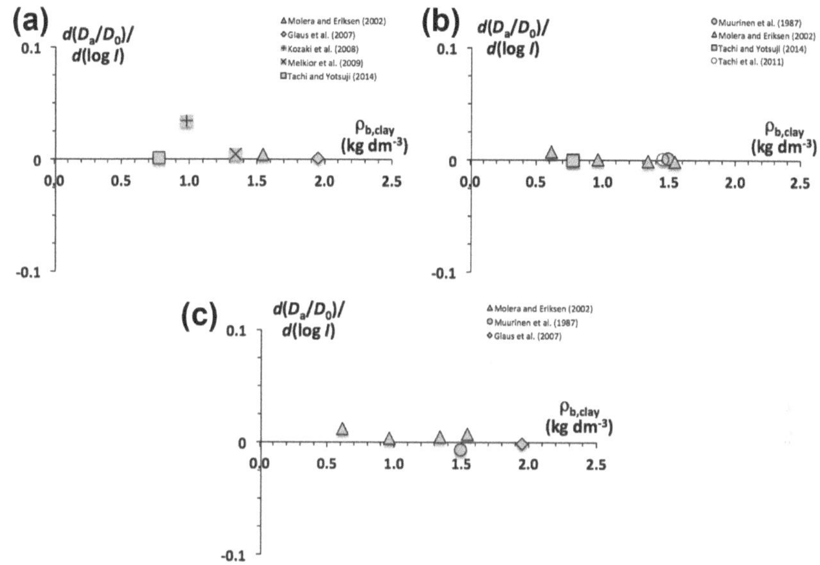

FIGURE 6.11 Experimental data on $d(D_a/D_0)/d(\log I)$ for (a) Na^+, (b) Cs^+, and (c) Sr^{2+}, plotted as a function of $\rho_{b,clay}$ in bentonite (filled red symbols (gray in print versions)) and clay-rocks (open blue symbols (light gray in print versions)).

Na^+, 0.03 ± 0.01 for Cs^+, and 0.13 ± 0.01 for Sr^{2+}. These values are roughly consistent with data on the electrical conductivity of dilute smectite dispersions, according to which adsorbed Na^+, Cs^+, and Ca^{2+} ions are 0.55, 0.15, and 0.15 times as mobile as the same ions in bulk water, respectively (Cremers, 1968). These values also are consistent with MD simulations of Na^+/Ca^{2+}-smectite external basal surfaces (where Na^+ outer-sphere surface complexes, the main adsorbed Na^+ species, diffuse 0.48 ± 0.08 times as fast as in bulk liquid water (Bourg and Sposito, 2011a)) and of Na^+-smectite interlayer nanopores (where Na^+, Cs^+, and Sr^{2+} diffuse 0.24 ± 0.14, 0.06 ± 0.03, and 0.13 ± 0.07 times as fast as in bulk liquid water, on average, in the two- and three-layer hydrates (Bourg and Sposito, 2010)). These results clearly demonstrate that adsorbed Na^+ ions retain a significant mobility (van Schaik et al., 1966; Gimmi and Kosakowski, 2011), in contradiction with the assumptions of several modeling studies (Jo et al., 2006; Leroy et al., 2006). The much lower mobility of adsorbed Cs^+ versus Na^+ may be related to significant differences in adsorption behavior, as Na^+ and Cs^+ tend to form outer- and inner-sphere surface complexes, respectively, on clay mineral basal surfaces (Marry et al., 2008a; Bourg and Sposito, 2011b).

Salinity has essentially no impact on the D_a/D_0 values of cations as shown in Figure 6.11. (Salinity impacts D_e, however, because of its effect on α.) This behavior contrasts with that of anions (Figure 6.6(c)). This difference between the salinity dependence of D_a for cations and anions is consistent with

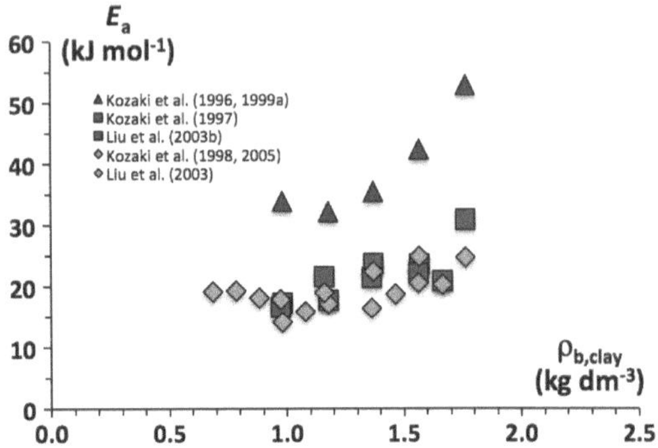

FIGURE 6.12 Experimental data on the activation energy of diffusion E_a as a function of $\rho_{b,clay}$ for Na^+ (yellow diamonds (light gray in print versions)), Cs^+ (green triangles (dark gray in print versions)), and Sr^{2+} (blue squares (gray in print versions)) in bentonite.

Gouy–Chapman theory, according to which salinity has a much stronger impact on the characteristic length scale associated with anion exclusion than on that associated with cation adsorption, a phenomenon known as counterion condensation (Sposito, 2004). Experimental data on the activation energy of diffusion of cations (Figure 6.12) indicate that the *T*-dependence of D_a in bentonite is significantly greater than in bulk liquid water at $\rho_{b,clay} > 1.3$ kg dm^{-3}, as also observed for anions in Figure 6.6(d). MD simulations suggest that this increase in E_a results from cation diffusion occurring predominantly in interlayer nanopores (two- and three-layer hydrates) at $\rho_{b,clay} > 1.3$ kg dm^{-3} and predominantly on external basal surfaces of clay mineral particles at lower degrees of compaction (Holmboe and Bourg, 2014).

6.5 FUTURE RESEARCH OPPORTUNITIES

The microstructure of bentonite and clay-rocks remains poorly known at length scales ranging from individual nanopores to pore networks. Consequently, the relationships between microstructure and the macroscopic parameters for diffusion processes remain poorly constrained. This is one of the reasons for the existence of many different conceptual models of diffusion in natural and engineered clay barriers. Advanced imaging techniques should focus on elucidating the microstructure of clay-rich media at in situ conditions as a function of dry bulk density, clay mineral mass fraction, clay mineralogy, and ionic strength. Little agreement exists regarding fundamental phenomena that occur within individual pores, such as the structure of the EDL and the mobility of water and ions near clay mineral surfaces. Clearly, additional information is needed to help constrain existing models of the pore scale basis of D_a and D_e in clay barriers.

The experimental results reviewed in this chapter clearly contradict several key features of existing macroscopic scale diffusion models. For example, surface water and adsorbed cations retain a significant mobility (they are not immobile) and anions do not diffuse only in bulk-liquid-like water.

Therefore, the pore scale structure of the EDL and the dynamics of water and ions in the EDL should be more accurately characterized using techniques such as QENS, X-ray reflectivity, and MD simulations. Future studies should focus particularly on determining the mobility of water and ions as a function of distance from clay mineral surfaces and the relative importance of short- (surface complexes) and long-range (diffuse ion swarm) interactions in screening the negative surface charge density of clay minerals. These data are essential to developing advanced models of the pore-scale basis of D_a that can describe, for instance, the interesting phenomena (possibly related to the disappearance of large pores or of the connected network of large pores) that occur during anion diffusion at $\rho_{b,clay}$ values in the range of 1.0–1.3 kg dm^{-3}, where the D_a/D_0 values of anions are sensitive to salinity and where $(D_a/D_0)_{anion}/(D_a/D_0)_{water}$ values are sensitive to $\rho_{b,clay}$. In addition, additional efforts should be focused on developing models that can describe diffusion results in both bentonite and clay-rocks and that are based on the same underlying concepts. As shown in this chapter, strong similarities (for example, on the anisotropy of G) and significant differences (related, in part, to differences in clay mineralogy) exist between compacted bentonite and clay-rocks. Finally, additional experimental and theoretical efforts should be focused on demonstrating that diffusion parameters measured in the laboratory are relevant in situ in geologic formation.

REFERENCES

Aertsens, M., De Cannière, P., Moors, H., 2003. Modelling of silica diffusion experiments with ^{32}Si in Boom Clay. J. Contam. Hydrol. 61, 117–129.

Alonso, U., Missana, T., García-Guriérrez, M., Patelli, A., Siitari-Kauppi, M., Rigato, V., 2009. Diffusion coefficient measurements in consolidated clay by RBS micro-scale profiling. Appl. Clay Sci. 43, 477–484.

Altmann, S., Tournassat, C., Goutelard, F., Parneix, J.-C., Gimmi, T., Maes, N., 2012. Diffusion-driven transport in clayrock formations. Appl. Geochem. 27, 463–478.

Appelo, C.A.J., Wersin, P., 2007. Multicomponent diffusion modeling in clay systems with application to the diffusion of tritium, iodide, and sodium in Opalinus Clay. Environ. Sci. Technol. 41, 5002–5007.

Appelo, C.A.J., Van Loon, L.R., Wersin, P., 2010. Multicomponent diffusion of a suite of tracers (HTO, Cl, Br, I, Na, Sr, Cs) in a single sample of Opalinus Clay. Geochim. Cosmochim. Acta 74, 1201–1219.

Barbour, S.L., Hendry, M.J., Wassenaar, L.I., 2012. In situ experiment to determine advective-diffusive controls on solute transport in a clay-rich aquitard. J. Contam. Hydrol. 131, 79–88.

Berend, I., Cases, J.-M., François, M., Uriot, J.-P., Michot, L., Masion, A., Thomas, F., 1995. Mechanism of adsorption and desorption of water vapor by homoionic montmorillonites: 2. The Li^+, Na^+, K^+, Rb^+ and Cs^+-exchanged forms. Clays Clay Miner. 43, 324–336.

Birgersson, M., Karnland, O., 2009. Ion equilibrium between montmorillonite interlayer space and an external solution—consequences for diffusional transport. Geochim. Cosmochim. Acta 73, 1908–1923.

Bock, H., Dehandschutter, B., Martin, C.D., Mazurek, M., de Haller, A., Skoczylas, F., Davy, C., 2010. Self-Sealing Fractures in Argillaceous Formations in the Context of Geological Disposal of Radioactive Waste. NEA Report No. 6184, Nuclear Energy Agency, Organisation for Economic Co-operation and Development.

Boudreau, B.P., 1996. The diffusive tortuosity of fine-grained unlithified sediments. Geochim. Cosmochim. Acta 60, 3139–3142.

Boudreau, B.P., Meysman, F.J.R., 2006. Predicted tortuosity of muds. Geology 34, 693–696.

Bourg, I.C., 2004. Tracer Diffusion of Water and Inorganic Ions in Compacted Saturated Sodium Bentonite. (Ph.D. thesis). University of California, Berkeley, pp. 384.

Bourg, I.C., Sposito, G., Bourg, A.C.M., 2006. Tracer diffusion in compacted, water-saturated bentonite. Clays Clay Miner. 54, 364–376.

Bourg, I.C., Sposito, G., Bourg, A.C.M., 2007. Modeling cation diffusion in compacted water-saturated sodium bentonite at low ionic strength. Environ. Sci. Technol. 41, 8118–8122.

Bourg, I.C., Sposito, G., Bourg, A.C.M., 2008. Modeling the diffusion of Na^+ in compacted water-saturated Na-bentonite as a function of pore water ionic strength. Appl. Geochem. 23, 3635–3641.

Bourg, I.C., Sposito, G., 2010. Connecting the molecular scale to the continuum scale for diffusion processes in smectite-rich porous media. Environ. Sci. Technol. 44, 2085–2091.

Bourg, I.C., Sposito, G., 2011a. Molecular dynamics simulations of the electrical double layer on smectite surfaces contacting concentrated mixed electrolyte (NaCl-$CaCl_2$) solutions. J. Colloid Interface Sci. 360, 701–715.

Bourg, I.C., Sposito, G., 2011b. Ion exchange phenomena. In: Huang, P.M., Li, Y., Sumner, M.E. (Eds.), Handbook of Soil Science, Properties and Processes, second ed. CRC Press (Chapter 16).

Calvet, R., 1973. Hydratation de la montmorillonite et diffusion des cations compensateurs. Ann. Agron. 24, 77–217.

Cavé, L., Al, T., Xiang, Y., Vilks, P., 2009. A technique for estimating one-dimensional diffusion coefficients in low-permeability sedimentary rock using X-ray radiography: comparison with through-diffusion measurements. J. Contam. Hydrol. 103, 1–12.

Cebula, D.J., Thomas, R.K., White, J.W., 1981. Diffusion of water in Li-montmorillonite studied by quasielastic neutron scattering. Clays Clay Miner 29, 241–248.

Chang, F.-R.C., Skipper, N.T., Sposito, G., 1995. Computer simulation of interlayer molecular structure in sodium montmorillonite hydrates. Langmuir 11, 2734–2741.

Chapman, S.W., Parker, B.L., 2005. Plume persistence due to aquitard back diffusion following dense nonaqueous phase liquid source removal or isolation. Water Resour. Res. 41, W12411.

Chen, Z., Montavon, G., Ribet, S., Guo, Z., Robinet, J., David, K., Tournassat, C., Grambow, B., Landesman, C., 2014. Key factors to understand in-situ behavior of Cs in Callovo-Oxfordian clay-rock (France). Chem. Geol. 387, 47–58.

Choi, J.-W., Oscarson, D.W., 1996. Diffusive transport through compacted Na- and Ca-bentonite. J. Contam. Hydrol. 22, 189–202.

Churakov, S.V., Gimmi, T., 2011. Up-scaling of molecular diffusion coefficients in clays: a two-step approach. J. Phys. Chem. C 115, 6703–6714.

Cole, T., Bidoglio, G., Soupioni, M., O'Gorman, M., Gibson, N., 2000. Diffusion mechanisms of multiple strontium species in clay. Geochim. Cosmochim. Acta 64, 385–396.

Comans, R.N.J., Haller, M., De Preter, P., 1991. Sorption of cesium on illite: non-equilibrium behaviour and reversibility. Geochim. Cosmochim. Acta 55, 433–440.

Cormenzana, J.L., García-Gutiérrez, M., Missana, T., Junghanns, Á., 2003. Simultaneous estimation of effective and apparent diffusion coefficients in compacted bentonite. J. Contam. Hydrol. 61, 63–72.

Cremers, A., 1968. Ionic Movement in a Colloidal Environment. (Ph.D. thesis). University of Louvain, Leuven, Belgium.

Descostes, M., Blin, V., Bazer-Bachi, F., Meier, P., Grenut, B., Radwan, J., Schlegel, M.L., Buschaert, S., Coelho, D., Tevissen, E., 2008. Diffusion of anionic species in Callovo-Oxfordian argillites and Oxfordian limestones (Meuse/Haute-Marne, France). Appl. Geochem. 23, 655–677.

De Soto, I.S., Ruiz, A.I., Ayora, C., García, R., Regadio, M., Cuevas, J., 2012. Diffusion of landfill leachate through compacted natural clays containing small amounts of carbonates and sulfates. Appl. Geochem. 27, 1202–1213.

Devineau, K., Bihannic, I., Michot, L., Villiéras, F., Masrouri, F., Cuisinier, O., Fragneto, G., Michau, N., 2006. In situ neutron diffraction analysis of the influence of geometric confinement on crystalline swelling of montmorillonite. Appl. Clay Sci. 31, 76–84.

Dykhuizen, R.C., Casey, W.H., 1989. An analysis of solute diffusion in rocks. Geochim. Cosmochim. Acta 53, 2797–2808.

Eriksen, T.E., Jansson, M., 1996. Diffusion of I^-, Cs^+, and Sr^{2+} in Compacted Bentonite – Anion Exclusion and Surface Diffusion. Swedish Nuclear Fuel and Waste Management Co., Stockholm (Sweden).

Eriksen, T.E., Jansson, M., Molera, M., 1999. Sorption effects on cation diffusion in compacted bentonite. Eng. Geol. 54, 231–236.

Ferrage, E., Lanson, B., Sakharov, B.A., Drits, V.A., 2005. Investigation of smectite hydration properties by modeling experimental X-ray diffraction patterns: Part I. Montmorillonite hydration properties. Amer. Mineral. 90, 1358–1374.

Ferrage, E., Sakharov, B.A., Michot, L.J., Delville, A., Bauer, A., Lanson, B., Grangeon, S., Frapper, G., Jiménez-Ruiz, M., Cuello, G.J., 2011. Hydration properties and interlayer organization of water and ions in synthetic Na-smectite with tetrahedral layer charge. Part 2. Toward a precise coupling between molecular simulations and diffraction data. J. Phys. Chem. C 115, 1867–1881.

Gajo, A., Loret, B., 2007. The mechanics of active clays circulated by salts, acids and bases. J. Mech. Phys. Solids 55, 1762–1801.

García-Gutiérrez, M., Cormenzana, J.L., Missana, T., Mingarro, M., 2004. Diffusion coefficients and accessible porosity for HTO and ^{36}Cl in compacted FEBEX bentonite. Appl. Clay Sci. 26, 65–73.

Gherardi, F., Xu, T., Pruess, K., 2007. Numerical modeling of self-limiting and self-enhancing caprock alteration induced by CO_2 storage in a depleted gas reservoir. Chem. Geol. 244, 103–129.

Gimmi, T., Kosakowski, G., 2011. How mobile are sorbed cations in clays and clay rocks? Environ. Sci. Technol. 45, 1443–1449.

Glaus, M.A., Baeyens, B., Bradbury, M.H., Jakob, A., Van Loon, L.R., Yaroshchuk, A., 2007. Diffusion of ^{22}Na and ^{85}Sr in montmorillonite: evidence of interlayer diffusion being the dominant pathway at high compaction. Environ. Sci. Technol. 41, 478–485.

Glaus, M.A., Frick, S., Rossé, R., Van Loon, L.R., 2010. Comparative study of tracer diffusion of HTO, $^{22}Na^+$ and $^{36}Cl^-$ in compacted kaolinite, illite and montmorillonite. Geochim. Cosmochim. Acta 74, 1999–2010.

Glaus, M.A., Frick, S., Rossé, R., Van Loon, L.R., 2011. Consistent interpretation of the results of through-, out-diffusion and tracer profile analysis for trace anion diffusion in compacted montmorillonite. J. Contam. Hydrol. 123, 1–10.

González Sánchez, F., Van Loon, L.R., Gimmi, T., Jakob, A., Glaus, M.A., Diamond, L.W., 2008. Self-diffusion of water and its dependence on temperature and ionic strength in highly compacted montmorillonite, illite and kaolinite. Appl. Geochem. 23, 3840–3851.

González Sánchez, F., Gimmi, T., Jurányi, F., Van Loon, L., Diamond, L.W., 2009. Linking the diffusion of water in compacted clays at two different time scales: tracer through-diffusion and quasielastic neutron scattering. Environ. Sci. Technol. 43, 3487–3493.

Greathouse, J.A., Refson, K., Sposito, G., 2000. Molecular dynamics simulation of water mobility in magnesium-smectite hydrates. J. Am. Chem. Soc. 122, 11459–11464.

Hendry, M.J., Barbour, S.L., Boldt-Leppin, B.E.J., Reifferscheid, L.J., Wassenaar, L.I., 2009. A comparison of laboratory and field based determinations of molecular diffusion coefficients in a low-permeability geologic medium. Environ. Sci. Technol. 43, 6730–6736.

Hicher, P.Y., Wahyudi, H., Tessier, D., 2000. Microstructural analysis of inherent and induced anisotropy in clay. Mech. Cohes.-Frict. Mater. 5, 341–371.

Higashihara, T., Kinoshita, K., Sato, S., Kozaki, T., 2004. Electromigration of sodium ions and electro-osmotic flow in water-saturated, compacted Na-montmorillonite. Appl. Clay Sci. 26, 91–98.

Higashihara, T., Shibuya, H., Sato, S., Kozaki, T., 2005. Activation energy for diffusion of helium in water-saturated, compacted Na-montmorillonite. Eng. Geol. 81, 365–370.

Holmboe, M., Wold, S., Jonsson, M., 2010. Colloid diffusion in compacted bentonite: micro-structural constraints. Clays Clay Miner. 58, 532–541.

Holmboe, M., Wold, S., Jonsson, M., 2012. Porosity investigation of compacted bentonite using XRD profile modeling. J. Contam. Hydrol 128, 19–32.

Holmboe, M., Bourg, I.C., 2014. Molecular dynamics simulations of water and sodium diffusion in smectite interlayer nanopores as a function of pore size and temperature. J. Phys. Chem. C 118, 1001–1013.

Jakob, A., Pfingsten, W., Van Loon, L., 2009. Effects of sorption competition on caesium diffusion through compacted argillaceous rock. Geochim. Cosmochim. Acta 73, 2441–2456.

Jansson, M., Eriksen, T.E., 2004. In situ anion diffusion experiments using radiotracers. J. Contam. Hydrol. 68, 183–192.

Jenny, H., Overstreet, R., 1939. Surface migration of ions and contact exchange. J. Phys. Chem. 43, 1185–1196.

Jo, H.Y., Benson, C.H., Edil, T.B., 2006. Rate-limited cation exchange in thin bentonitic barrier layers. Can. Geotech. J. 43, 370–391.

Johnson, R.L., Cherry, J.A., Pankow, J.F., 1989. Diffusive contaminant transport in natural clay: a field example and implications for clay-lined waste disposal sites. Environ. Sci. Technol. 23, 340–349.

Joseph, C., Van Loon, L.R., Jakob, A., Steudtner, R., Schmeide, K., Sachs, S., Bernhard, G., 2013. Diffusion of U(VI) in Opalinus Clay: influence of temperature and humic acid. Geochim. Cosmochim. Acta 109, 74–89.

Jougnot, D., Revil, A., Leroy, P., 2009. Diffusion of ionic tracers in the Callovo-Oxfordian clay-rock using the Donnan equilibrium model and the formation factor. Geochim. Cosmochim. Acta 73, 2712–2726.

Kato, H., Muroi, M., Yamada, N., Ishida, H., Sato, H., 1995. Estimation of effective diffusivity in compacted bentonite. Mater. Res. Soc. Symp. Proc. 353, 277–284.

Keller, L.M., Schuetz, P., Erni, R., Rosell, M.D., Lucas, F., Gasser, P., Holzer, L., 2013. Characterization of multi-scale microstructural features in Opalinus Clay. Micropor. Mesopor. Mater. 170, 83–94.

Kemper, W.D., Maasland, D.E.L., Porter, L.K., 1964. Mobility of water adjacent to mineral surfaces. Soil Sci. Soc. Am. Proc. 28, 164–167.

Ketzer, J.M., Carpentier, B., Le Gallo, Y., Le Thiez, P., 2005. Geological sequestration of CO_2 in mature hydrocarbon fields. Basin and reservoir numerical modeling of the Forties Field, North Sea. Oil Gas Sci. Technol. 60, 259–273.

Kim, H.-T., Suk, T.-W., Park, S.-H., Lee, C.-S., 1993. Diffusivities for ions through compacted Na-bentonite with varying dry bulk density. Waste Manage. 13, 303–308.

Kosakowski, G., Churakov, S.V., Thoenen, T., 2008. Diffusion of Na and Cs in montmorillonite. Clays Clay Miner. 56, 190–206.

Kozaki, T., Sato, H., Fujishima, A., Sato, S., Ohashi, H., 1996. Activation energy for diffusion of cesium in compacted sodium montmorillonite. J. Nucl. Sci. Technol. 33, 522–524.

Kozaki, T., Sato, H., Fujishima, A., Saito, N., Sato, S., Ohashi, H., 1997. Effect of dry density on activation energy for diffusion of strontium in compacted sodium montmorillonite. Mater. Res. Soc. Symp. Proc. 465, 893–900.

Kozaki, T., Fujishima, A., Sato, S., Ohashi, H., 1998. Self-diffusion of sodium ions in compacted montmorillonite. Nucl. Technol. 121, 63–69.

Kozaki, T., Sato, H., Sato, S., Ohashi, H., 1999a. Diffusion mechanism of cesium ions in compacted montmorillonite. Eng. Geol. 54, 223–230.

Kozaki, T., Sato, Y., Nakajima, M., Kato, H., Sato, S., Ohashi, H., 1999b. Effect of particle size on the diffusion behavior of some radionuclides in compacted bentonite. J. Nucl. Mater. 270, 265–272.

Kozaki, T., Inada, K., Sato, S., Ohashi, H., 2001. Diffusion mechanism of chloride ions in sodium montmorillonite. J. Contam. Hydrol. 47, 159–170.

Kozaki, T., Fujishima, A., Saito, N., Sato, S., Ohashi, H., 2005. Effects of dry density and exchangeable cations on the diffusion process of sodium ions in compacted montmorillonite. Eng. Geol. 81, 246–254.

Kozaki, T., Liu, J., Sato, S., 2008. Diffusion mechanism of sodium ions in compacted montmorillonite under different NaCl concentration. Phys. Chem. Earth 33, 957–961.

Kozaki, T., Sawaguchi, T., Fujishima, A., Sato, S., 2010. Effect of exchangeable cations on apparent diffusion of Ca^{2+} ions in Na- and Ca-montmorillonite mixtures. Phys. Chem. Earth 35, 254–258.

Laird, D.A., Shang, C., 1997. Relationship between cation exchange selectivity and crystalline swelling in expanding 2:1 phyllosilicates. Clays Clay Miner. 45, 681–689.

Lange, K., Rowe, R.K., Jamieson, H., 2009. Diffusion of metals in geosynthetic clay liners. Geosynthetics Int. 16, 11–27.

Lavastre, V., Jendrzejewski, N., Agrinier, P., Javoy, M., Evrard, M., 2005. Chlorine transfer out of a very low permeability clay sequence (Paris Basin, France): ^{35}Cl and ^{37}Cl evidence. Geochim. Cosmochim. Acta 69, 4949–4961.

Leroy, P., Revil, A., Coelho, D., 2006. Diffusion of ionic species in bentonite. J. Colloid Interface Sci. 296, 248–255.

Li, Y.-H., Gregory, S., 1974. Diffusion on ions in sea water and in deep-sea sediments. Geochim. Cosmochim. Acta 38, 703–714.

Liu, J., Yamada, H., Kozaki, T., Sato, S., Ohashi, H., 2003a. Effect of silica sand on activation energy for diffusion of sodium ions in montmorillonite and silica sand mixture. J. Contam. Hydrol. 61, 85–93.

Liu, J., Kozaki, T., Horiuchi, Y., Sato, S., 2003b. Microstructure of montmorillonite/silica sand mixture and its effects on the diffusion of strontium ions. Appl. Clay Sci. 23, 89–95.

Loomer, D.B., Scott, L., Al, T.A., Mayer, K.U., Bea, S., 2013. Diffusion-reaction studies in low permeability shale using X-ray radiography with cesium. Appl. Geochem. 39, 49–58.

Madsen, F.T., Kahr, G., 1993. Diffusion of ions in compacted bentonite. In: Proceedings of the 1993 International Conference on Nuclear Waste Management and Environmental Remediation. Volume 1: Low and intermediate level radioactive waste management. The American Society of Mechanical Engineers, Praha, 239–246.

Maes, N., Moors, H., Dierckx, A., De Cannière, P., Put, M., 1999. The assessment of electromigration as a new technique to study diffusion of radionuclides in clayey soils. J. Contam. Hydrol. 36, 231–247.

Malikova, N., Cadéne, A., Marry, V., Dubois, E., Turq, P., Zanotti, J.-M., Longeville, S., 2005. Diffusion of water in clays – microscopic simulation and neutron scattering. Chem. Phys. 317, 226–235.

Manjanna, J., Kozaki, T., Sato, S., 2009. Fe(III)-montmorillonite: basic properties and diffusion of tracers relevant to alteration of bentonite in deep geological disposal. Appl. Clay Sci. 43, 208–217.

Marry, V., Turq, P., 2003. Microscopic simulations of interlayer structure and dynamics in bihydrated heteroionic montmorillonites. J. Phys. Chem. B 107, 1832–1839.

Marry, V., Rotenberg, B., Turq, P., 2008a. Structure and dynamics of water at a clay surface from molecular dynamics simulation. Phys. Chem. Chem. Phys. 10, 4802–4813.

Marry, V., Malikova, N., Cadène, A., Dubois, E., Durand-Vidal, S., Turq, P., Breu, J., Longeville, S., Zanotti, J.-M., 2008b. Water diffusion in a synthetic hectorite by neutron scattering—beyond the isotropic translational model. J. Phys.: Condens. Matter 20, 104205.

Marry, V., Dubois, E., Malikova, N., Durand-Vidal, S., Longeville, S., Breu, J., 2011. Water dynamics in hectorite clays: influence of temperature studied by coupling neutron spin echo and molecular dynamics. Environ. Sci. Technol. 45, 2850–2855.

Martens, E., Maes, N., Weetjens, E., Van Gompel, M., Ravestyn, L.V., 2010. Modelling of a large-scale in-situ migration experiment with 14C-labelled natural organic matter in boom clay. Radiochimica Acta Int. J. Chem. Aspects Nucl. Sci. Technol. 98, 695–701.

Mazurek, M., Alt-Epping, P., Bath, A., Gimmi, T., Waber, H.N., Buschaert, S., De Cannière, P., De Craen, M., Gautschi, A., Savoye, S., Vinsot, A., Wemaere, I., Wouters, L., 2011. Natural tracer profiles across argillaceous formations. Appl. Geochem. 26, 1035–1064.

Melkior, T., Yahiaoui, S., Thoby, D., Motellier, S., Barthes, V., 2007. Diffusion coefficients of alkaline cations in Bure mudrock. Phys. Chem. Earth 32, 453–462.

Melkior, T., Gaucher, E.C., Brouard, C., Yahiaoui, S., Thoby, D., Clinard, C., Ferrage, E., Guyonnet, D., Tournassat, C., Coelho, D., 2009. Na^+ and HTO diffusion in compacted bentonite: effect of surface chemistry and related texture. J. Hydrol. 370, 9–20.

Michot, L.J., Delville, A., Humbert, B., Plazanet, M., Levitz, P., 2007. Diffusion of water in a synthetic clay with tetrahedral charges by combined neutron time-of-flight measurements and molecular dynamics simulations. J. Phys. Chem. C 111, 9818–9831.

Mills, R., 1973. Self-diffusion in normal and heavy water in the range 1–45°. J. Phys. Chem. 77, 685–688.

Molera, M., Eriksen, T., 2002. Diffusion of $^{22}Na^+$, $^{85}Sr^{2+}$, $^{134}Cs^+$ and $^{57}Co^{2+}$ in bentonite clay compacted to different densities: experiments and modeling. Radiochim. Acta 90, 753–760.

Molera, M., Eriksen, T., Jansson, M., 2003. Anion diffusion pathways in bentonite clay compacted to different dry densities. Appl. Clay Sci. 23, 69–76.

Montavon, G., Alhajji, E., Grambow, B., 2006. Study of the interaction of Ni^{2+} and Cs^+ on MX-80 bentonite; effect of compaction using the "capillary method". Environ. Sci. Technol. 40, 4672–4679.

Montavon, G., Guo, Z., Tournassat, C., Grambow, B., Le Botlan, D., 2009. Porosities accessible to HTO and iodide on water-saturated compacted clay materials and relation with the forms of water: a low field proton NMR study. Geochim. Cosmochim. Acta 73, 7290–7302.

Moridis, G.J., 1999. Semianalytical solutions for parameter estimation in diffusion cell experiments. Water Resour. Res. 35, 1729–1740.

Motellier, S., Devol-Brown, I., Savoye, S., Thoby, D., Alberto, J.-C., 2007. Evaluation of tritiated water diffusion through the Toarcian clayey formation of the Tournemire experimental site (France). J. Contam. Hydrol. 94, 99–108.

Muurinen, A., Penttilä-Hiltunen, P., 1987. Diffusion mechanisms of strontium and cesium in compacted sodium bentonite. Mater. Res. Soc. Symp. Proc. 84, 803–810.

Muurinen, A., Penttilä-Hiltunen, P., Uusheimo, K., 1989. Diffusion of chloride and uranium in compacted sodium bentonite. Mater. Res. Soc. Symp. Proc. 127, 743–748.

Muurinen, A., Olin, M., Uusheimo, K., 1990. Diffusion of sodium and copper in compacted sodium bentonite at room temperature. Mater. Res. Soc. Symp. Proc. 176, 641–647.

Nakashima, Y., Mitsumori, F., Nakashima, S., Takahashi, M., 1999. Measurement of self-diffusion coefficients of water in smectite by stimulated echo ^1H nuclear magnetic resonance imaging. Appl. Clay Sci. 14, 59–68.

Nakashima, Y., 2001. Pulsed field gradient proton NMR study of the self-diffusion of H_2O in montmorillonite gel: effects of temperature and water fraction. Amer. Mineral. 86, 132–138.

Nakashima, Y., 2002. Diffusion of H_2O and I^- in expandable mica and montmorillonite gels: contribution of bound H_2O. Clays Clay Miner. 50, 1–10.

Nakashima, Y., 2003a. Diffusion of H_2O in smectite gels: obstruction effects of bound H_2O layers. Clays Clay Miner. 51, 9–22.

Nakashima, Y., 2003b. Diffusivity measurement of heavy ions in Wyoming montmorillonite gels by X-ray computed tomography. J. Contam. Hydrol. 61, 147–156.

Nakashima, Y., 2004. Nuclear magnetic resonance properties of water-rich gels of Kunigel-V1 bentonite. J. Nucl. Sci. Technol. 41, 981–992.

Nakashima, Y., Mitsumori, F., 2005. H_2O self-diffusion restricted by clay platelets with immobilized bound H_2O layers: PGSE NMR study of water-rich saponite gels. Appl. Clay Sci. 28, 209–221.

Nakazawa, T., Takano, M., Nobuhara, A., Torikai, Y., Sato, S., Ohashi, H., 1999. Activation energies of diffusion of tritium and electrical conduction in water-saturated compacted sodium montmorillonite. Proceedings of the 7th International Conference on Radioactive Waste Management and Environmental Remediation (ICEM 99), The American Society of Mechanical Engineers.

Neuzil, C.E., 1994. How permeable are clays and shales? Water Resour. Res. 30, 145–150.

Nye, P.H., 1979. Diffusion of ions and uncharged solutes in soils and soil clays. Adv. Agron. 31, 225–272.

Oscarson, D.W., Dixon, D.A., Hume, H.B., 1996. Mass transport through defected bentonite plugs. Appl. Clay Sci. 11, 127–142.

Palut, J.-M., Montarnal, P., Gautschi, A., Tevissen, E., Mouche, E., 2003. Characterisation of HTO diffusion properties by an in-situ tracer experiment in Opalinus Clay at Mont Terri. J. Contam. Hydrol. 61, 203–218.

Parker, B.L., Cherry, J.A., Chapman, S.W., 2004. Field study of TCE diffusion profiles below DNAPL to assess aquitard integrity. J. Contaminant Hydrology 74 (1), 197–230.

Patriarche, D., Michelot, J.-L., Ledoux, E., Savoye, S., 2004a. Diffusion as the main process for mass transport in very low water content argillites. 1. Chloride as a natural tracer for mass transport-Diffusion coefficient and concentration measurements in interstitial water. Water Resour. Res. 40, W01516.

Patriarche, D., Ledoux, E., Michelot, J.-L., Simon-Coinçon, R., Savoye, S., 2004b. Diffusion as the main process for mass transport in very low water content argillites. 2. Fluid flow and mass transport modeling. Water Resour. Res. 40, W01517.

Poinssot, C., Baeyens, B., Bradbury, M.H., 1999. Experimental and modelling studies of caesium sorption on illite. Geochim. Cosmochim. Acta 63, 3217–3227.

Revil, A., Leroy, P., Titov, K., 2005. Characterization of transport properties of argillaceous sediments. Application to the Callovo-Oxfordian argillite. J. Geophys. Res. 110, B06202.

Robinet, J.-C., Sardini, P., Coelho, D., Parneix, J.-C., Prêt, D., Sammartino, S., Boller, E., Altmann, S., 2012. Effects of mineral distribution at mesoscopic scale on solute diffusion in a clay-rich rock: example of the Callovo-Oxfordian mudstone (Bure, France). Water Resour. Res. 48, W05554.

Robinson, R.A., Stokes, R.H., 1959. Electrolyte Solutions. second ed. Academic Press, pp. 559.

Sato, H., 2008. Thermodynamic model on swelling of bentonite buffer and backfill materials. Phys. Chem. Earth 33, S538–S543.

Sato, H., Ashida, T., Kohara, Y., Yui, M., Sasaki, N., 1992. Effect of dry density on diffusion of some radionuclides in compacted sodium bentonite. J. Nucl. Sci. Technol. 29, 873–882.

Sato, H., Ashida, T., Kohara, Y., Yui, M., 1993. Study on retardation mechanism of H-3, Tc-99, Cs-137, Np-237 and Am-241 in compacted sodium bentonite. Mater. Res. Soc. Symp. Proc. 294, 403–408.

Sato, H., Suzuki, S., 2003. Fundamental study on the effect of an orientation of clay particles on diffusion pathway in compacted bentonite. Appl. Clay Sci. 23, 51–60.

Sato, H., Miyamoto, S., 2004. Diffusion behaviour of selenite and hydroselenide in compacted bentonite. Appl. Clay Sci. 26, 47–55.

Savoye, S., Page, J., Puente, C., Imbert, C., Coelho, D., 2010. New experimental approach for studying diffusion through an intact and unsaturated medium: a case study with Callovo-Oxfordian argillite. Environ. Sci. Technol. 44, 3698–3704.

Savoye, S., Goutelard, F., Beaucaire, C., Charles, Y., Fayette, A., Herbette, M., Larabi, Y., Coelho, D., 2011. Effect of temperature on the containment properties of argillaceous rocks: the case study of Callovo-Oxfordian claystones. J. Contam. Hydrol 125, 102–112.

Savoye, S., Lacour, J.-L., Fayette, A., Beaucaire, C., 2013. Mobility of zinc in the Callovo-Oxfordian claystone. Proc. Earth Planet. Sci. 7, 774–777.

Seida, Y., Terashima, M., Tachi, Y., Iijima, K., Nakazawa, T., Yamada, M., Yui, M., 2010. Sorption and diffusion of Eu in sedimentary rock in the presence of humic substance. Radiochimica Acta Int. J. Chem. Aspects Nucl. Sci. Technol. 98, 703–709.

Shackelford, C.D., 1991. Laboratory diffusion testing for waste disposal – a review. J. Contam. Hydrol. 7, 177–217.

Shackelford, C.D., Moore, S.M., 2013. Fickian diffusion of radionuclides for engineered containment barriers: diffusion coefficients, porosities, and complicating issues. Eng. Geol. 152, 133–147.

Shainberg, I., Kemper, W.D., 1966. Hydration status of adsorbed cations. Soil Sci. Soc. Am. Proc. 30, 707–713.

Sposito, G., 2004. The Surface Chemistry of Natural Particles. Oxford University Press, pp. 242.

Steefel, C.I., Carroll, S., Zhao, P., Roberts, S., 2003. Cesium migration in Hanford sediment: a multisite cation exchange model based on laboratory transport experiments. J. Contam. Hydrol. 67, 219–246.

Suzuki, S., Sato, H., Tachi, Y., 2003. A technical problem in the through-diffusion experiments for compacted bentonite. J. Nucl. Sci. Technol. 40, 698–701.

Suzuki, S., Sato, H., Ishidera, T., Fuji, N., 2004. Study on anisotropy of effective diffusion coefficient and activation energy for deuterated water in compacted sodium bentonite. J. Contam. Hydrol. 68, 23–37.

Swift, A.M., Anovitz, L.M., Sheets, J.M., Cole, D.R., Welch, S.A., Rother, G., 2014. Relationship between mineralogy and porosity in seals relevant to geologic CO_2 sequestration. Environ. Geosci. 21, 39–57.

Tachi, Y., Seida, Y., Doi, R., Xia, X., Yui, M., 2009. Sorption and diffusion of Cs in Horonobe-URL's sedimentary rock: Comparison and model prediction of retardation parameters from sorption and diffusion experiments. Mater. Res. Soc. Symp. Proc. 1124, 573–579.

Tachi, Y., Nakazawa, T., Ochs, M., Yotsuji, K., Suyama, T., Seida, Y., Yamada, N., Yui, M., 2010. Diffusion and sorption of neptunium (V) in compacted montmorillonite: effects of carbonate and salinity. Radiochimica Acta 98, 711–718.

Tachi, Y., Yotsuji, K., Seida, Y., Yui, M., 2011. Diffusion and sorption of Cs^+, I^- and HTO in samples of the argillaceous Wakkanai formation from the Horonobe URL, Japan: clay-based modeling approach. Geochim. Cosmochim. Acta 75, 6742–6759.

Tachi, Y., Yotsuji, K., 2014. Diffusion and sorption of Cs^+, Na^+, I^- and HTO in compacted sodium montmorillonite as a function of porewater salinity: integrated sorption and diffusion model. Geochim. Cosmochim. Acta 132, 75–93.

Tachi, Y., Yotsuji, K., Suyama, T., Ochs, M., 2014. Integrated sorption and diffusion model for bentonite. Part 2: porewater chemistry, sorption and diffusion modeling in compacted systems. J. Nucl. Sci. Technol 51, 1191–1204.

Tanaka, S., Noda, N., Sato, S., Kozaki, T., Sato, H., Hatanaka, K., 2011. Electrokinetic study of migration of anions, cations, and water in water-saturated compacted sodium montmorillonite. J. Nucl. Sci. Technol. 48, 454–462.

Tomioka, S., Kozaki, T., Takamatsu, H., Noda, N., Nisiyama, S., Kozai, N., Suzuki, S., Sato, S., 2010. Analysis of microstructural images of dry and water-saturated compacted bentonite samples observed with X-ray micro CT. Appl. Clay Sci. 47, 65–71.

Tournassat, C., Chapron, Y., Leroy, P., Bizi, M., Boulahya, F., 2009. Comparison of molecular dynamics simulations with triple layer and modified Gouy-Chapman models in a 0.1 M NaCl - montmorillonite system. J. Colloid Interface Sci. 339, 533–541.

Tournassat, C., Appelo, C.A.J., 2011. Modelling approaches for anion-exclusion in compacted Na-bentonite. Geochim. Cosmochim. Acta 75, 3698–3710.

Tournassat, C., Grangeon, S., Leroy, P., Giffaut, E., 2013. Modeling specific pH dependent sorption of divalent metals on montmorillonite surfaces. A review of pitfalls, recent achievements and current challenges. Am. J. Sci. 313, 395–451.

Tsai, S.-C., Ouyang, S., Hsu, C.-N., 2001. Sorption and diffusion behavior of Cs and Sr on Jih-Hsing bentonite. Appl. Radiat. Isotop. 54, 209–215.

Van Loon, L.R., Soler, J.M., Jakob, A., Bradbury, M.H., 2003. Effect of confining pressure on the diffusion of HTO, $36Cl^-$ and $125I^-$ in a layered argillaceous rock (Opalinus Clay): diffusion perpendicular to the fabric. Appl. Geochem. 18, 1653–1662.

Van Loon, L.R., Soler, J.M., Müller, W., Bradbury, M.H., 2004. Anisotropic diffusion in layered argillaceous rocks: a case study with Opalinus Clay. Environ. Sci. Technol. 38, 5721–5728.

Van Loon, L.R., Eikenberg, J., 2005. A high-resolution abrasive method for determining diffusion profiles of sorbing radionuclides in dense argillaceous rocks. Appl. Rad. Isot. 63, 11–21.

Van Loon, L.R., Jakob, A., 2005. Evidence for a second transport porosity for the diffusion of tritiated water (HTO) in a sedimentary rock (Opalinus Clay – OPA): application of through- and out-diffusion techniques. Transp. Porous Media 61, 193–214.

Van Loon, L.R., Müller, W., Iijima, K., 2005a. Activation energies of the self-diffusion of HTO, $^{22}Na^+$ and $^{36}Cl^-$ in a highly compacted argillaceous rock (Opalinus Clay). Appl. Geochem. 20, 961–972.

Van Loon, L.R., Baeyens, B., Bradbury, M.H., 2005b. Diffusion and retention of sodium and strontium in Opalinus Clay: comparison of sorption data from diffusion and batch sorption measurements, and geochemical calculations. Appl. Geochem. 20, 2351–2363.

Van Loon, L.R., Glaus, M.A., Müller, W., 2007. Anion exclusion effects in compacted bentonites: towards a better understanding of anion diffusion. Appl. Geochem. 22, 2536–2552.

Van Loon, L.R., Glaus, M.A., 2008. Mechanical compaction of smectite clays increases ion exchange selectivity for cesium. Environ. Sci. Technol. 42, 1600–1604.

Van Loon, L.R., Baeyens, B., Bradbury, M.H., 2009. The sorption behaviour of caesium on Opalinus Clay: a comparison between intact and crushed material. Appl. Geochem. 24, 999–1004.

van Schaik, J.C., Kemper, W.D., 1966. Chloride diffusion in clay–water systems. Soil Sci. Soc. Am. Proc. 30, 22–25.

van Schaik, J.C., Kemper, W.D., Olsen, S.R., 1966. Contribution of adsorbed cations to diffusion in clay–water systems. Soil Sci. Soc. Am. Proc. 30, 17–22.

Wang, X., Liu, X., 2004. Effect of pH and concentration on the diffusion of radiostrontium in compacted bentonite—a capillary experimental study. Appl. Radiat. Isot. 61, 1413–1418.

Wang, X., Tan, X., Chan, C., Chen, L., 2005. The concentration and pH dependent diffusion of ^{137}Cs in compacted bentonite by using capillary method. J. Nucl. Mater. 345, 184–191.

Wang, Q., Tang, A.M., Cui, Y.-J., Barnichon, J.-D., Ye, W.-M., 2013. Investigation of the hydro-mechanical behavior of compacted bentonite/sand mixture based on the BExM model. Comput. Geotech. 54, 46–52.

Wersin, P., Van Loon, L.R., Soler, J.M., Yllera, A., Eikenberg, J., Gimmi, T., Hernán, P., Boisson, J.-Y., 2004. Long-term diffusion experiment at Mont Terri: first results from field and laboratory data. Appl. Clay Sci. 26, 123–135.

Wersin, P., Soler, J.M., Van Loon, L., Eikenberg, J., Baeyens, B., Grolimund, D., Gimmi, T., Dewonck, S., 2008. Diffusion of HTO, Br^-, I^-, Cs^+, $^{85}Sr^+$ and $^{60}Co^{2+}$ in a clay formation: results and modelling from an in situ experiment in Opalinus Clay. Appl. Geochem. 23, 678–691.

Wilson, J., Cuadros, J., Cressey, G., 2004. An in situ time-resolved XRD-PSD investigation into Na-montmorillonite interlayer and particle rearrangement during dehydration. Clays Clay Miner. 52, 180–191.

Wittebroodt, C., Savoye, S., Frasca, B., Gouze, P., Michelot, J.-L., 2012. Diffusion of HTO, $^{36}Cl^-$ and $^{125}I^-$ in Upper Toarcian argillite samples from Tournemire: effects of initial iodide concentration and ionic strength. Appl. Geochem. 27, 1432–1441.

Yaroshchuk, A.E., Glaus, M.A., Van Loon, L.R., 2009. Determination of diffusion and sorption parameters of thin confined clay layers by direct fitting of through-diffusion flux. J. Colloid Interface Sci. 337, 508–512.

Chapter 7

Gas Transfer Through Clay Barriers

A. Amann-Hildenbrand,[a] B.M. Krooss,[a] J. Harrington,[b] R. Cuss,[b] C. Davy,[c] F. Skoczylas,[c] E. Jacops[a,d,e] and N. Maes[d]

[a] *Energy and Mineral Resources Group, Institute of Geology and Geochemistry of Petroleum and Coal, Aachen, Germany;* [b] *British Geological Survey, Nottingham, UK;* [c] *Ecole Centrale de Lille/LML UMR CNRS 8107, Cité Scientifique, Villeneuve d'Ascq Cedex, France;* [d] *SCK●CEN, Belgian Nuclear Research Centre, Expert Group, Waste & Disposal, Mol, Belgium;* [e] *KU Leuven, Department of Earth & Environmental Sciences, Heverlee, Belgium*

Chapter Outline

Developments in Clay Science, Vol. 6C. http://dx.doi.org/10.1016/B978-0-08-100027-4.00007-3

7.1 INTRODUCTION

Gas transport through clay-rocks can occur by different processes that can be basically subdivided into pressure-driven *flow* of a *bulk gas* phase and *transport of dissolved gas* either by molecular diffusion or advective water flow (Figure 7.1, Marschall et al., 2005). The relative importance of these transport mechanisms depends on the boundary conditions and the scale of the system. Pressure-driven volume flow ("Darcy flow") of gas is the most efficient transport mechanism. It requires, however, pressure gradients sufficiently large to overcome capillary forces in the typically water-saturated rocks (purely gas-saturated argillaceous rocks are not considered in the present context). These pressure gradients may form as a consequence of the gravity field (buoyancy, compaction) or by gas generation processes (thermogenic, microbial, radiolytic). Dissolved gas may be transported by water flow along a hydraulic gradient. This process is not affected by capillary forces but constrained by the solubility

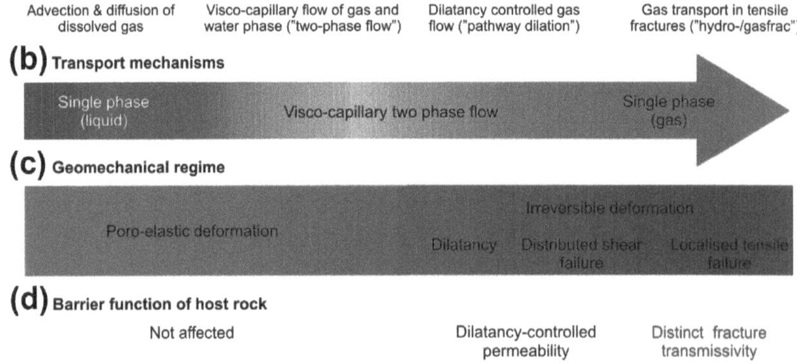

FIGURE 7.1 Classification and analysis of gas transport processes in Opalinus clay: (a) phenomenological description based on the microstructural model concept; (b) basic transport mechanisms; (c) geomechanical regime; and (d) effect of gas transport on the barrier function of the host rock. *Reprinted with permission from IFP, Figure 2 in Marschall et al. (2005).*

of the gas. It has much lower transport efficiency than bulk gas phase flow. Molecular diffusion of dissolved gas, finally, is occurring essentially without constraints, ubiquitously and perpetually. Effective diffusion distances are, however, proportional to the square root of time, which limits the relevance of this transport process to the range of tens to hundreds of meters on a geological timescale (millions of years).

Process understanding and the quantification of the controlling parameters, like diffusion coefficients, capillary gas breakthrough pressures, and effective gas permeability coefficients, is of great importance for upscaling purposes in different research disciplines and applications. During the past decades, gas migration through fully water-saturated geological clay-rich barriers has been investigated extensively (Thomas et al., 1968; Pusch and Forsberg, 1983; Horseman et al., 1999; Galle, 2000; Hildenbrand et al., 2002; Marschall et al., 2005; Davy et al., 2009; Harrington et al., 2009, 2012a, 2014). All of these studies aimed at the analysis of experimental data determined for different materials (rocks of different lithotype, composition, compaction state) and pressure/temperature conditions. The clay-rocks investigated in these studies, ranged from unconsolidated to indurated clays and shales, all characterized by small pores (2–100 nm) and very low hydraulic conductivity ($K < 10^{-12}$ m s^{-1}) or permeability coefficients ($k < 10^{-19}$ m^2).

Studies concerning radioactive waste disposal include investigations of both the natural host rock formation and synthetic/engineered backfill material at a depth of a few hundred meters (IAEA, 2003, 2009). Within a geological disposal facility, hydrogen is generated by anaerobic corrosion of metals and through radiolysis of water (Rodwell et al., 1999; Yu and Weetjens, 2009). Additionally, methane and carbon dioxide are generated by microbial degradation of organic wastes (Rodwell et al., 1999; Ortiz et al., 2002; Johnson, 2006; Yu and Weetjens, 2009). The focus of carbon capture and storage (CCS) studies is on the analysis of the long-term sealing efficiency of lithologies above depleted reservoirs or saline aquifers, typically at larger depths (hundreds to thousands of meters). During the last decade, several studies were published on the sealing integrity of clay-rocks to carbon dioxide (Hildenbrand et al., 2004; Li et al., 2005; Hangx et al., 2009; Harrington et al., 2009; Skurtveit et al., 2012; Amann-Hildenbrand et al., 2013). In the context of petroleum system analysis, a significant volume of research has been undertaken regarding gas/oil expulsion mechanisms from source rocks during burial (Tissot and Pelet, 1971; Appold and Nunn, 2002), secondary migration (Luo et al., 2008) and the capillary sealing capacity of caprocks overlying natural gas accumulations (Berg, 1975; Schowalter, 1979; Krooss, 1992; Schlömer and Krooss, 2004; Li et al., 2005; Berne et al., 2010). Recently, more attention has been paid to investigations of the transport efficiency of shales in the context of shale oil/gas production (Bustin et al., 2008; Amann-Hildenbrand et al., 2012; Eseme et al., 2012; Ghanizadeh et al., 2013, 2014). Analysis of the migration mechanisms within partly unlithified strata becomes important when exploring the origin of overpressure zones, subseafloor gas domes and gas seepages (Hovland and Judd, 1988; Boudreau, 2012).

The conduction of experiments and data evaluation/interpretation requires a profound process understanding and a high level of experience. The acquisition and preparation of adequate samples for laboratory experiments usually constitutes a major challenge and may have serious impact on the representativeness of the experimental results. Information on the success/failure rate of the sample preparation procedure should therefore be provided. Sample specimens "surviving" this procedure are subjected to various experimental protocols to derive information on their gas transport properties.

The present overview first presents the theoretical background of gas diffusion and advective flow, each followed by a literature review (Sections 7.2 and 7.3). Different experimental methods are described in Sections 7.4.1 and 7.4.2. Details are provided on selected experiments performed at the Belgian Nuclear Research Center (SCK-CEN, Belgium), Ecole Centrale de Lille (France), British Geological Survey (UK), and at RWTH-Aachen University (Germany) (Section 7.4.3). Experimental data are discussed with respect to different petrophysical parameters outlined above: (1) gas diffusion, (2) evolution of gas breakthrough, (3) dilation-controlled flow, and (4) effective gas permeability after breakthrough. These experiments were conducted under different pressure and temperature conditions, depending on sample type, burial depth, and research focus (e.g., radioactive waste disposal, natural gas exploration, or carbon dioxide storage). The interpretation of the experimental results can be difficult and sometimes a clear discrimination between different mechanisms (and the controlling parameters) is not possible. This holds, for instance, for gas breakthrough experiments, where the observed transport can be interpreted as intermittent, continuous, capillary- or dilation-controlled flow. Also, low gas flow rates through samples on the length-scale of centimeters can be equally explained by effective two-phase flow or diffusion of dissolved gas.

7.2 DIFFUSIVE TRANSPORT OF GAS IN SOLUTION

7.2.1 Theoretical Background

7.2.1.1 General Description of Diffusion

Diffusion is a mass transport due to the random thermal motion of particles of molecular size (atoms, molecules, ions). This process can be represented by a molecular random walk, known as the Brownian motion. Diffusion causes a molecular flux along a concentration gradient and increases the spreading of a concentration front or peak with time. The random motion of particles results in increasing entropy (Grathwohl, 1998).

Fick's first law of diffusion relates the diffusive flux F (mol m^{-2} s^{-1}) to the concentration gradient $\partial C/\partial x$ (Crank, 1975):

$$F = -D\frac{\partial C}{\partial x} \tag{7.1}$$

D denotes the diffusion coefficient (m^2 s^{-1}), which is a measure for mobility of the particles, C is the concentration (mol m^{-3}), and x is the position (m). The mass conservation equation

$$\frac{\partial C}{\partial t} = D\frac{\partial^2 C}{\partial^2 x} \tag{7.2}$$

is commonly termed Fick's second law of diffusion, where t is time (s).

7.2.1.2 Gas Diffusion in Water

Gases dissolve in liquids to different extent. Henry's Law (Mackay and Shiu, 1981) states that at a constant temperature, the amount of gas that dissolves in a volume of liquid is directly proportional to the partial pressure p_i (Pa) of the gas i in equilibrium with the liquid:

$$p_i = k_H C_{i\_liquid} \tag{7.3}$$

Here, C_{i_liquid} (mol m^{-3}) is the concentration of the dissolved gas in the liquid and k_H (Pa m^3 mol^{-1}) is the Henry constant. Various other formulations of Henry's law exist and attention should be paid to the choice of the correct units and parameters. Henry's law is a limiting law valid for low partial pressures (<0.5–1 MPa) and dissolved gas concentrations below 3 mol%. At very high pressures, the relationship between the gas partial pressure and the concentration of the dissolved gas is no longer linear (Prausnitz et al., 1999).

7.2.1.3 Gas Diffusion in Porous Media

Under the influence of a concentration gradient a net diffusive flux of the dissolved gas will occur. The diffusion coefficient of dissolved gas in bulk water or bulk aqueous solution is denoted as D_0. Experimental values for D_0 are available in the literature (c.f. Boudreau, 1996).

Diffusion of gas in water-saturated porous media is described in the literature in different ways (Krooss, 1992; Horseman et al., 1996; Berne et al., 2010). The reader is therefore advised to check the individual definitions before using published diffusion coefficients. Terms and definitions used in this chapter are outlined below.

The diffusive flux of a gas (i) in a water-saturated porous medium with porosity ϕ can be described in terms of the effective diffusion coefficient (D_{eff}).

$$F = -D_{eff}\frac{\partial C_{i\_bulk\_rock}}{\partial x} \tag{7.4}$$

Here the porous medium is treated as a macroscopically homogeneous material and $C_{i\_bulk\_rock}$ is the "bulk rock" volume concentration in a representative element of volume of this material. If the gas is only present in the dissolved state in the pore water (no adsorption on mineral surfaces or organic matter) the bulk volume concentration is

$$C_{i\_bulk\_rock} = \phi \cdot C_{i\_liquid} \tag{7.5}$$

Fick's first law is then rewritten as

$$F = -D_{eff} \frac{\partial C_{i\_bulk\_rock}}{\partial x} = -\phi * D_p \frac{\partial C_{i\_liquid}}{\partial x} \tag{7.6}$$

Here D_p is the "pore diffusion coefficient" that contains the effects of the pore structure on the molecular diffusion process. These effects comprise the narrowing (constriction) of transport pathways (constrictivity, δ) due to pore throats and the increased length of the effective transport path (l_{tp}) relative to the Euclidian distance (l) between two positions in the porous medium (tortuosity $\tau = l_{tp}/l$). The "pore diffusion coefficient" is commonly defined as

$$D_p = D_0 \delta / \tau^2 \tag{7.7}$$

It is thus related to the effective and the bulk fluid diffusion coefficients according to

$$D_{eff} = D_0 \phi \delta / \tau^2 = D_p \phi \tag{7.8}$$

It is evident from Eqn (7.6) that for the calculation of diffusive transport D_{eff} has to be used in combination with the *bulk rock concentration gradient* whereas D_p and D_0 have to be used in combination with the liquid phase concentration gradient.

Apart from constrictivity and tortuosity the diffusion of molecules through porous media (rocks) can be retarded by physicochemical interactions with mineral or organic matrix components. These effects involving reversible (ad-/desorption, precipitation/dissolution) and irreversible (chemisorption, reaction) effects can be taken into account by introducing the "apparent diffusion coefficient" ($D_{app} = D_p/R$) where R is the retardation factor (Bruggeman et al., 2009; Jacops et al., 2013).

7.2.2 Literature Review

Diffusive transport of gas through water-saturated rocks on the length-scales of geological systems (>10 m) has only a very low transport capacity as compared with convective (pressure-driven) transport (Krooss, 1992). However diffusion may become the dominant transport mechanism on the 10–1000-m scale if there is no pressure-driven volume flow over extended periods of geological time. Relevant distances for diffusive transport may be estimated using the relationship

$$\frac{D_{eff} \cdot t}{l^2} = 0.1 \tag{7.9}$$

Thus, for an effective diffusion coefficient of 10^{-10} m^2 s^{-1} and a diffusion time of 1 Ma the diffusion distance is in the range of 200 m (177.6 m).

During the past decades research on diffusion in geological systems was mainly focused on three topics

- The role of diffusion in primary migration of petroleum, i.e., transport of gas/oil within source rocks and expulsion from these source rocks (Leythaeuser et al., 1982; Krooss, 1988; Thomas and Clouse, 1990a,b,c; Stainforth and Reinders, 1990)
- Losses from natural gas reservoirs through clay overburden/caprocks (Antonov et al., 1958; Stklyanin et al., 1968; Stklyanin and Litvinova, 1971; Nesterow and Uschatinskij, 1972; Price et al., 1981; Leythaeuser et al., 1982; Krooss and Leythaeuser, 1988; Krooss et al., 1992a,b; Schlömer and Krooss, 1997)
- Long-term safety of natural and geoengineered barriers, i.e., (radioactive) waste disposal (Horseman et al., 1996)

Factors influencing the effective diffusion coefficients, such as temperature, tortuosity/constrictivity, and retardation effects of the geologic medium have been widely discussed in the literature. Table 7.1 lists published diffusion coefficients for gases in water-saturated rocks of different lithotypes and at different temperatures.

Diffusion coefficients in pure water and brines (D_0) can be considered as baseline values for diffusion in water-saturated sediments and sedimentary rocks (Horseman et al., 1996). While D_0 values range from approximately 10^{-9} m^2 s^{-1} at 0°C to 10^{-8} m^2 s^{-1} at 160°C (Wilke and Chang, 1955; Ferrell and Himmelblau, 1967), effective diffusion coefficients in sedimentary rocks are lowered by one to three orders of magnitudes (Krooss, 1992). This decrease in diffusivity is attributed to the tortuosity and constrictivity of the transport pore system and/or the interaction of the diffusing gas with mineral surfaces and the organic matter of carbonaceous rocks. Schlömer and Krooss (2004) showed that an increase of effective stress from 10 to 40 MPa resulted in a decrease of the effective diffusion coefficients of sedimentary rocks by 30–40%. Thomas and Clouse (1990a,b) attributed enhanced diffusive fluxes of hydrocarbon gases through cores with organic coatings to the higher solubility of hydrocarbons in the organic coating, i.e., increased bulk concentration gradient. Effective diffusion coefficients of coated and uncoated samples were similar, indicating that the enhanced diffusive flux was not due to an increase in molecular mobility of the diffusing species but to a higher bulk volume concentration resulting in a higher transport capacity. Several studies have documented that effective diffusion coefficients of hydrocarbon gases in sedimentary rocks decrease with increasing total organic carbon (TOC) contents (Schlömer and Krooss, 1997, 2004; Zhang and Krooss, 2001). This is attributed to retardation effects by adsorption of gas on/in dispersed organic material, which is more pronounced for methane than for nitrogen. As adsorption capacity decreases with increasing temperature, the

TABLE 7.1 Summary of Diffusion Coefficients for a Variety of Rock Types and Gases. The Diffusion Coefficients Refer to Those Used in the Publications COx, Callovian-Oxfordian

Lithotype/Formation	Reference	Gas	Type	Measured D ($m^2 s^{-1}$)	ϕ (%)	T (°C)
Pure water	Wilke and Chang (1955), Ferrell and Himmelblau (1967)	–	D_0	10^{-9} to 10^{-8}		0–160 10–55 (He, H_2)
Boom clay	Jacops et al. (2013)	CH_4, He, H_2, N_2	D_{app}	$12.2–10^{-10}$	38	21
Opalinus clay	Gómez-Hernández (2000)		D_{app}	7×10^{-10}	30	
Opalinus clay	Rübel et al. (2002)		D_{app}	$3.5\ (\pm 1.3) \times 10^{-11}$		Room
COx	Bigler et al. (2005)		D_p	2.41×10^{-10}	15.4	20
COx	Rebour et al. (1997)		D_{app}	$5 \pm 1 \times 10^{-11}$	23	50
Toarcian/Domerian shale	Bensenouci et al. (2011)		D_p	$2.4–12.7 \times 10^{-11}$	3.5–11	
COx	Didier (2012)	H_2	D_p	$\sim 10^{-12}$ (91% water-saturated)		23
Clay-rocks (North German Basin)	Schlömer and Krooss (1997)	CH_4	D_{eff}	$1.0–24.4 \times 10^{-10}$	0.4–1.9	150
Fanglomerate (North German Basin)				4.46×10^{-10}	6.6	
Jurassic claystones and siltstones Smørbukk field, Haltenbanken, Norway				$1.4 \times 10^{-11}–1.14 \times 10^{-10}$	0.7–6.7	
Boom clay	Jacops et al. (2013)		D_{app}	$2.42–10^{-10}$	38	21
Clay-rocks and tight sandstones	Schlömer and Krooss (2004)	N_2	D_{eff}	$0.18–18.2 \times 10^{-11}$	0.7–10.5	90
				$0.39–21.6 \times 10^{-11}$		
Muderong shale	Busch et al. (2008)	CO_2		$3.0–4.8 \times 10^{-11}$ (increasing)	20	45–50
Emscher Marl	Wollenweber et al. (2010)			$1.2–7.8 \times 10^{-11}$ (increasing)	15	21–28

retardation effect is reduced and effective diffusion coefficients increase. Additionally, it was found that diffusion results in an isotopic fractionation of the gases, due to differences in mobility of the diffusing isotopic species (D_{eff}). Solubility fractionation was found to be negligible (Zhang and Krooss, 2001; Schlömer and Krooss, 2004). The diffused methane was significantly depleted in the heavier carbon isotope (^{13}C), especially during the initial nonstationary state. The effect was larger for shale samples with higher TOC contents.

Diffusion experiments with CO_2 on sedimentary rocks provided indications that the pore morphology of the rock samples changed in the course of repetitive tests. For a shale and marlstone, Busch et al. (2008) and Wollenweber et al. (2010) observed an increase of effective diffusion coefficient by a factor of 1.6 and 6.5, respectively. This was attributed to irreversible alteration of the pore system by mineral reactions or sorptive CO_2 uptake resulting in a retardation effect during the first diffusion test. In the follow-up tests, the reactivity of the rock samples was obviously reduced as indicated by reduced bulk CO_2 volume equilibrium concentrations and higher effective diffusion coefficients (Busch et al., 2008).

In the context of radioactive waste disposal in clay host rocks, the relative generation and transport rates of molecular hydrogen (H_2) are important parameters for the estimation of pressure build-up. The availability of diffusion coefficients for H_2 in water-saturated clays/clay-rocks required for these calculations is still very limited. Due to the experimental difficulties associated with the determination of H_2 diffusion coefficients (Volckaert et al., 1995; Ortiz et al., 2002; Didier, 2012; Jacops et al., 2013), the diffusion coefficients of He are often used as a first approximation. Recently, significant progress has been made in the determination of reliable diffusion coefficients for H_2 in clay-rocks and shales (Jacops et al., 2013).

7.3 ADVECTIVE FLOW

7.3.1 Theoretical Background

7.3.1.1 Single-Phase Viscous Flow (Darcy Flow)

Apart from the diffusional process, viscous pressure-driven transport of the gas specimen must be considered. In a fully water-saturated porous medium, dissolved gas molecules will be carried along with the water flood. Transport of the water phase is described by Darcy's law for incompressible media, relating the flow rate (Q, $m^3\ s^{-1}$) to the absolute or intrinsic permeability coefficient (k, m^2), fluid viscosity (η, Pa s), and the pressure gradient ($\Delta p/\Delta x$, Pa):

$$Q = -\frac{k}{\eta} \cdot A \cdot \frac{\Delta p}{\Delta x} \tag{7.10}$$

The permeability coefficient k represents a characteristic parameter of the rock specimen.

In hydrology, the hydraulic head (Δh) is commonly used to express potential gradients. It denotes the height of a static water column resulting in the pressure Δp according to

$$\Delta h = \frac{\Delta p}{\rho_{water} \cdot g} \tag{7.11}$$

In this context, flow properties are described by the hydraulic conductivity (K, m/s), which relates hydraulic head gradient to the volume flow rate:

$$Q = -K \cdot A \cdot \frac{\Delta h}{\Delta x} \tag{7.12}$$

Assuming a water density of 1000 kg/m^3 and a viscosity of 10^{-3} Pa s, the relationship between hydraulic conductivity, K, and intrinsic permeability, k, is

$$K = k \cdot 9.81 \cdot 10^6 \sim k \cdot 10^7$$

The pressure-driven flow of the gas phase is described by Darcy's law for compressible media, which accounts for the expansion of the gas upon transfer from the upstream to the downstream pressure level. The conventional volume-based form of Darcy's law is first converted to a mass-based form using the ideal gas law, where the amount of mass (dn) moving through the sample is calculated from either the volume (dV) or pressure (dp) change with time.

$$\frac{dn}{dt} = \frac{p_2}{R \cdot T}\frac{dV_2}{dt} = \frac{V_2}{R \cdot T}\frac{dp_2}{dt} = -\frac{k_{gas} \cdot A}{\eta \cdot R \cdot T}\frac{\left(p_2^2 - p_1^2\right)}{2 \cdot dx} \tag{7.13}$$

Here p_1 and p_2 are the pressures on the up- and downstream side of the specimen, k_{gas} (m^2) is the apparent gas permeability at a given mean pressure, $(p_1 + p_2)/2$.

To obtain the absolute/intrinsic permeability coefficient of the rock, one usually has to account for the slip-flow effect or Knudsen diffusion, which is related to the pressure- and temperature-dependent mean free path of the gas molecules. The apparent gas permeability coefficients increase linearly with the reciprocal mean pressure, as the proportion of gas molecules colliding with the pore walls increases. The "Klinkenberg plot" of the apparent permeability coefficients versus reciprocal mean pressure yields the intercept k_{inf_gas} for "infinite" mean gas pressure. This value is considered to represent the intrinsic permeability, k, of the porous medium in the absence of electroosmotic counter-pressures (Klinkenberg, 1941; Weber and Stanjek, 2012).

7.3.1.2 Two-Phase Fluid System

The flow of gas in a partly water-saturated rock is controlled by two processes:

1. Viscocapillary two-phase flow: this is represented by gas flow within the original porosity by the displacement of pore water (drainage) by the gas, resulting in partial desaturation of the rock matrix. As gas pressure increases, water is displaced from successively smaller pores. Increasing desaturation of the rock matrix results in a corresponding increase in gas conductivity and the likelihood of capillary failure increases.
2. Dilatant flow: upon mechanical failure of the pore network, gas flow occurs along pressure-induced pathways along previous or reactivated fractures on the micro and macroscale (tensile fracturing) or within the matrix by "dilatancy" (opening/closure of pores).

The movement of gas in these instances is a dynamic process. When the gas pressure drops below a critical value (often referred to as the capillary threshold pressure), advective gas flow ceases. Similarly, changes in effective stress or resaturation of the medium with water, can result in a reduction, or even a cessation, of gas flow (Hildenbrand et al., 2003; Davy et al., 2012, 2013; Liu et al., 2012).

7.3.1.2.1 Capillary Pressure-Controlled Desaturation, Drainage/ Imbibition

The effective saturation and distribution of each fluid phase within a rigid porous medium is generally associated with capillary effects, i.e., the actual pore-size distribution, fluid pressures, and wetting characteristics (interfacial tension and contact angles). A special form of the Young–Laplace equation for cylindrical capillaries, also known as Washburn equation (Washburn, 1921; Dullien, 1979), describes the relationship between the (excess) capillary pressure (p_c) and the (equivalent) capillary radius (r) invaded by the nonwetting phase at this particular pressure:

$$p_c = p_{nw} - p_w = 2\gamma \cos\theta / r \tag{7.14}$$

Here γ is the liquid/gas interfacial tension (Nm^{-1}), θ the wetting angle, and r is the radius (m) of a cylindrical pore.

This equation can be used to describe the purely capillary-controlled invasion of a nonwetting (gas) phase into a fully water-saturated medium (Figure 7.2). As the gas pressure increases up to a critical gas/water pressure difference ($p_{nw} - p_w$), the capillary entry pressure (p_{c_entry}), the gas phase starts to displace water from the pore system. Upon further gas pressure increase, the system will be successively drained as the gas phase displaces the wetting phase from smaller pores. When finally a continuous nonwetting phase (backbone) has developed through the pore network, viscous gas flow is initiated (threshold or breakthrough pressure, $p_{c\_breakthrough}$). Higher capillary

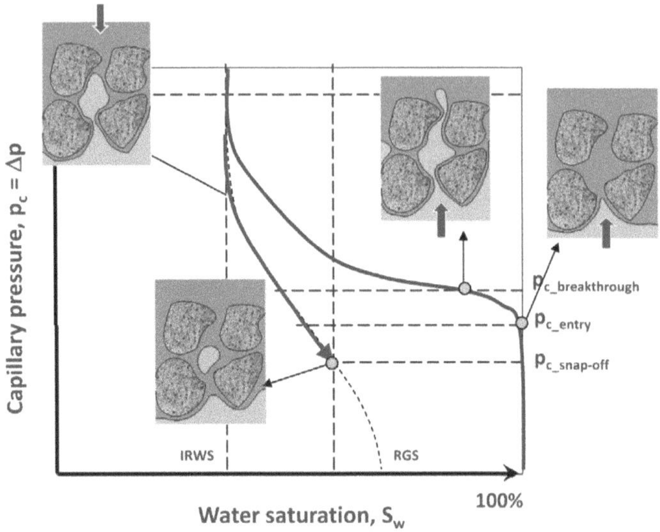

FIGURE 7.2 Conceptual sketch of capillary processes in clay-rocks. The graph shows capillary pressure (p_c) as a function of water saturation, S_w. The blue line marks the drainage path, where the nonwetting fluid displaces the wetting phase (increasing capillary pressures). The red line describes the imbibition path, where the wetting phase spontaneously reimbibes (decreasing capillary pressures). Starting at 100% water saturation the nonwetting phase (gas) starts to displace water from the largest capillary when the entry pressure (p_{c_entry}) has been overcome. Upon further increase in capillary pressure the gas will break through at the outlet of the sample ($p_{c\_breakthrough}$). Desaturation increases with increasing capillary pressure until irreducible water saturation (IRWS) is reached. On the imbibition path water imbibes into the pore system until the last interconnecting capillary path is blocked. This pressure is termed here snap-off pressure ($p_{c\_snap\text{-}off}$). *Reprinted with permission from Elsevier, Figure 1 in Busch and Amann-Hildenbrand (2013).*

pressures lead to further desaturation and correspondingly higher gas flow rates until maximum desaturation (irreducible water saturation, IRWS) is reached. A decrease in capillary pressure leads to the reimbibition of water, successively reducing the interconnected flow pathways of the nonwetting phase until flow stops (final snap-off pressure, $p_{c\_snap\text{-}off}$, or shut-in pressure). Trapped amounts of the nonwetting phase within the pore network lead to a saturation hysteresis between the first drainage and imbibition process as well as after successive drainage/imbibition cycles (repetitive gas pressure buildup, gas pressure release, and decrease).

Upon desaturation, the porous medium contains two immiscible fluids, a wetting phase and a non-wetting phase, each at a different saturation. The extent to which the fluids are in contact with the pore walls is determined by the interfacial tension, γ, between the fluid phases (solid, liquid, gas) and the contact angle, Θ, (measured in the denser fluid phase). Generally, wetting fluids occupy the smallest pores and are in direct contact with the pore walls. Nonwetting fluids tend to minimize their contact area with the pore walls.

Usually, the mineral phases are considered water wet, with water being the wetting fluid (Schowalter, 1979; Pentland, 2010). However, rocks may be of intermediate wettability, depending on their composition (e.g. mineral and organic phases), surface structure and chemistry, fluid types, and geologic evolution (Honarpour et al., 1987; Borysenko et al., 2009; Bikkina, 2011).

Two different types of immiscible fluid flow concepts are known: (1) both fluid phases progress simultaneously through all capillary paths (funicular flow), where the wetting fluid remains in contact with the solid pore walls, and (2) the separate channel flow concept, where each fluid progresses through its own interconnected paths (Scheidegger, 1974; Dullien, 1979; Qingjie et al., 2002). In practice, two-phase flow through a pore network is a combination of the above two processes.

The hysteresis phenomenon is related to two different processes: water retention within the pore network by capillary forces and adsorption (Daïan, 2010; M'Jahad, 2012). The first desaturation cycle (drainage) can be described by (1) the initial viscous water displacement, ending at the IRWS, and finally (2) by desorption down to (theoretically) full drying. Upon resaturation (imbibition), natural porous materials do not regain the same water saturation. This phenomenon is independent of the type of resaturation, i.e., change of capillary pressure ($p_{nw} - p_w$) or relative humidity (relative water vapor pressure, $p_{partial}/p_{saturation}$) and is a consequence of the complex arrangement of small pore throats and larger pore bodies (Diamond, 2000; Espinosa and Franke, 2006; Ishida et al., 2007). During resaturation, nonwetting phase bubbles are trapped within the pore space due to the existence of small pore throats, the so-called bottle neck effect (snap-off) (Kleppe et al., 1997; Diamond, 2000; Luo et al., 2008). Therefore, water saturation at a given pressure is higher during first desaturation than along any subsequent drainage/imbibition process.

7.3.1.2.2 Effective/Relative Gas Permeability

In a two-phase fluid system, both saturation and permeability to each fluid phase (i), will change as function of the capillary pressure. By definition, the conductivity of each phase depends on the prevailing fluid saturation. One does no longer consider permeability to be a characteristic rock property. Instead, each fluid phase is characterized by an effective permeability (k_{eff_i}) as function of its saturation (S_i). Normalized to the intrinsic permeability, k, the conductivity to each phase is expressed by its relative permeability (k_{ri}). In general, multiphase fluid flow is modeled by combining the continuity equation for each phase with the generalized form of Darcy's law and constitutive equations for relative permeabilities and capillary pressures between phases (Parker, 1989). The continuity equation for each phase may be written as

$$\frac{\partial}{\partial t}(\phi \rho_i S_i) + \nabla \cdot (\rho_i \boldsymbol{q}_i) = \gamma_i \tag{7.15}$$

where ϕ is the medium porosity, ρ_i is the phase density for phase i, S_i is the saturation of phase i, $\boldsymbol{q}_i$ is the Darcy velocity vector for phase i, and γ_i is a source-sink term for mass transfer between the phases and any external sources. Darcy's law for phase i may be written as

$$\boldsymbol{q}_i = -\frac{k_{ri}\boldsymbol{k}}{\eta_i}\nabla(p_i + \rho_i g z) \tag{7.16}$$

where k_{ri} is the relative permeability of phase i, $\boldsymbol{k}$ is the intrinsic permeability tensor, η_i is the dynamic viscosity of phase i, p_i is the pressure in phase i, and g is the acceleration due to gravity. To complete the description of the system, it is necessary to define the relative permeabilities and capillary pressures as functions of the phase saturations (e.g., Honarpour et al., 1987). These may be defined either with tabulated data or explicit functional forms such as van Genuchten (1980) and may also incorporate hysteretic effects (Kool and Parker, 1987).

7.3.1.2.3 Dilatancy

A robust and comprehensive mathematical framework describing the dilatant response of clay-rocks to the passage of gas is as yet unavailable. A number of approaches have been proposed using expressions describing the movement of gas through individual or multiple capillaries, or, discrete fractures as reviewed in Rodwell et al. (1999). Since this time some progress has been made in the representation of dilatancy. Work by Romero et al. (2012) on laboratory scale measurements lead to the development of a void ratio-dependent permeability expression (Senger et al., 2014). An alternative approach proposed by Gerard et al. (2014) used an embedded preexisting fracture in a continuum finite element model. A hydromechanical coupling between the fracture aperture, permeability, and the retention properties along the fracture was included. While the solution to this model is somewhat arbitrary, dependent on the initial parameterization of the fracture(s), a better fit to data was obtained compared with that using standard viscocapillary approaches. Other methods such as those based on Griffith theory have been proposed (Rodwell et al., 1999), in which a crack will propagate when the decrease in strain energy just balances the increase in surface energy. However, while all of these approaches have some merit, in the continued absence of appropriate phenomenological-based models, outputs from these approaches should be treated with caution, in particular, when used in a predictive manner. While Darcy flow is a macroscopic phenomenon that can be understood from the fluid pressures alone, flow by dilatant pathway creation is essentially a microscopic process that couples fluid pressures to the stress–strain characteristics of the material at the pore scale. Solving such a coupled problem and scaling it up to describe the macroscale effects remains a very difficult mathematical problem.

7.3.2 Literature Review

7.3.2.1 Pure Capillary-Controlled Flow

In the context of the exploration and production of natural gas/oil fields, capillary-controlled flow through the sedimentary column is generally considered to be the most relevant transport process. For high-permeable reservoirs and petroleum carrier systems, oil and gas migration within water-saturated rocks has been extensively analyzed in the past, theoretically and experimentally (Berg, 1975; Schowalter, 1979; Dembicki and Anderson, 1989; Hirsch and Thompson, 1995; Kleppe et al., 1997). Gas transport is usually described by capillary pressure and relative permeability curves as function of water saturation (Honarpour et al., 1987). Recently, research shifted more to the analysis of tight gas sandstones (Cui et al., 2009; Cluff and Byrnes, 2010) and gas/oil shales (Amann-Hildenbrand et al., 2012; Ghanizadeh et al., 2013). As these rocks contain much smaller pores, capillary resistance to gas flow is much larger. Capillary pressures required for desaturation are much higher and the percentage of immobile water increases (Byrnes, 1996).

A comprehensive literature review on capillary gas breakthrough pressures ($p_{c\_breakthrough}$) is given in (Busch and Amann-Hildenbrand, 2013). Values compiled from N_2 and He gas breakthrough experiments cover $p_{c\_breakthrough}$-values ranging from 5×10^{-3} MPa for sandstones (k $\sim$ 10^{-13} m^2) to 30 MPa for clay-rocks (10^{-21} m^2) (Figure 7.3). The capillary snap-off pressure detected after spontaneous imbibition is lower than the breakthrough pressure measured on the drainage path. Both critical pressures increase with decreasing permeability.

After breakthrough, the conductivity to the gas phase is strongly influenced by the prevailing water content. Relative gas permeability decreases dramatically with increasing water saturation, as does the relative water permeability with decreasing water contents. As this effect becomes more pronounced for less conductive rocks, the gas–water relative permeability crossover occurs at successively lower relative permeabilities. This effect creates water saturation regions with virtually no-flow, i.e., where relative gas and water permeabilities are less than 2% (Byrnes, 1996; Shanley et al., 2004; Cluff and Byrnes, 2010). In clay-rich rocks, gas flow rates detected after gas breakthrough can be extremely low, corresponding to relative gas permeability coefficients down to 0.001% (Amann-Hildenbrand et al., 2013).

One effect discussed in literature is the importance of the slip-flow or Klinkenberg effect, which causes gas flow velocity and permeability to increase with decreasing mean gas pressures and decreasing pore width. In tight lithologies, which are characterized by small pores down to the nanometer range, slip-flow is therefore highly relevant—for dry rocks but also with increasing water saturations. Increasing slip factors are associated with those systems, where the effective pore width is reduced by water films along the pore walls (funicular flow) (Qingjie et al., 2002; Li and Horne, 2004). Decreasing slip factors are assumed to result from channel flow types, where

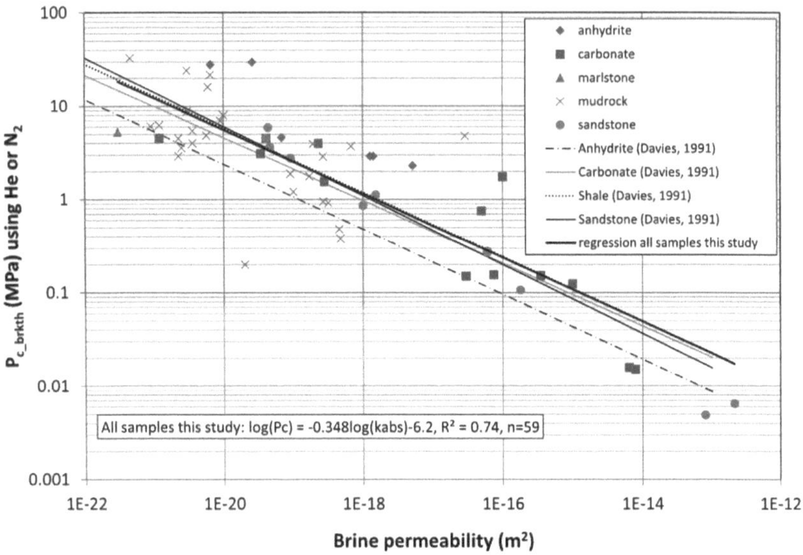

FIGURE 7.3 Drainage capillary breakthrough pressure (p_{c_brkth}) versus brine permeability on sample plugs using N_2 and Helium. *From Purcell, 1949; Thomas et al., 1968; Ibrahim et al., 1970; Schowalter, 1979; Harrington and Horseman, 1999; Hildenbrand et al., 2002, 2004; Li et al., 2005; Marschall et al., 2005; Al-Bazali et al., 2005, 2009; Ito et al., 2010; Carles et al., 2010; Amann-Hildenbrand et al., 2013. The trend lines by Davies (1991) are based on previously published data sets too, where unfortunately details of the underlying data are not given. Reprinted with permission from Elsevier, Figure 10 in Busch and Amann-Hildenbrand (2013).*

gas and water flow through separate paths. With increasing water saturations, the smallest "slip-prone" pores are blocked with water and gas flow occurs in successively larger pores (Estes and Fulton, 1956; Abbas et al., 1999; Qingjie et al., 2002). From an experimental point of view, small saturation changes are very difficult to determine. Due to spontaneous water imbibition upon gas pressure release, water/gas saturation should be determined during the experiment, i.e., under external stress conditions and prevailing fluid pressures. Additionally, the expected quantities tend to be below the detection limit of any petrophysical method. Based on imbibition gas breakthrough experiments (residual pressure technique), Hildenbrand et al. (2002) calculated effective transport porosities to the gas phase after gas breakthrough according to a simple capillary bundle model (assuming cylindrical pores). Corresponding maximum cumulative gas-filled effective transport porosity was below 0.01%.

7.3.2.1.1 Evidence for Dilatancy-Controlled Gas Flow

In initially saturated clay-rocks with extremely narrow interparticle spaces, the capillary threshold pressure required to initiate gas flow can simply be too large for the gas to be able to penetrate and desaturate the clay (Harrington and

Horseman, 1999). Indeed, Hedberg (1974), in a paper examining the role of gas in overpressure within source rocks, quotes a translation from Tissot and Pelet (1971) in which they state "The displacement of an oil or gas phase from the center of a finely grained argillaceous matrix goes against the laws of capillarity and is in principle impossible. The barrier can, however, be broken in one way. The pressure within the fluids formed in the pores of the source-rock increases constantly as products of the evolution of kerogen are formed. If this pressure comes to exceed the mechanical resistance of the rock, microfissures will be produced which are many orders of size greater than the natural (pore) channel of the rock, and will permit the escape of an oil or gas phase, until the pressure has fallen below the threshold which allows the fissures to be filled and a new cycle commences." This approach to the conceptualization of immiscible gas flow succinctly sums up modern thinking on the topic and is of potential relevance to all storage and extraction industries dealing with low-permeability clay-based systems. Mandl and Harkness (1987) support this hypothesis and suggest that hydrocarbon migration can only occur through thick, continuous water-wet rocks of low permeability through the process of fracturing, forming what they refer to as "dykelets."

Several studies performed on clay materials revealed that bulk gas flow is associated with dilatant flow. In an early study, Horseman and Harrington (1994), attributed migration of gas through Boom clay to the formation of pressure-induced dilatant pathways. Gas flow was linked to an approach first proposed by Lambe (1960), which recognized the importance of "water" within the system and the ability of thin water films to conduct stress when strongly associated with the clay substrate. Horseman and Harrington (1994) also reported intermittent or "burst" type flow during gas migration experiments (Horseman and Harrington, 1998). This combined with minimal desaturation of test samples indicated that the processes first identified for source rocks (Tissot and Pelet, 1971) also apply to less-indurated materials such as Boom clay (Volckaert et al., 1995; Horseman et al., 1996; Ortiz et al., 1996; Sen et al., 1996). In addition, tests on compacted bentonite exhibited a critical gas breakthrough pressure of a similar order of magnitude to that of the swelling pressure (Pusch and Forsberg, 1983; Pusch et al., 1985). Detailed studies by Horseman et al. (1997) and Horseman and Harrington (1997) reported a series of experiments and theoretic analyses in which gas flow was accompanied by dilation of the clay fabric. Initial gas breakthrough pressures were found to be equal to or slightly in excess of total stress and again, no measurable desaturation of the material was observed, even after prolonged injection times. Horseman and Harrington (1997) suggested that gas did not migrate through the original pore space of the water-saturated bentonite and that gas permeability was a dependent variable, rather than a material property, linked to the number, width, and aperture distributions of dilatant features. This hypothesis was further developed by Harrington and Horseman (1999) summarizing a series of observations on natural and engineered materials.

Studies on subsea hydrocarbon seepages (unlithified strata) suggest capillary displacement pressures are often so large, that gas pressure required to initiate flow can approach or even exceed the total stress (Clayton and Hay, 1992; Judd and Sim, 1998). Donohew et al. (2000) examined gas migration behavior through unconfined clay pastes of varying moisture content and mineralogy under ambient conditions. In all experiments, gas flow was accompanied by the creation of dilatant, preferential pathways, the morphology of which was related to the plasticity and density of the clay. Hovland et al. (2005) examined pockmark formation in the Nyegga area of the North Sea and concluded that the observed features were formed by sudden "catastrophic" fluid flow. Post failure, microseepages were also noted suggesting only partial self-sealing of the initial pockmark pathway occurred. The development of dilatant pathways is a well-recognized phenomenon in hydrate studies and is commonly referred to as "grain-displacing hydrate formation" (Holland et al., 2008). These authors presented a series of X-ray images showing filamental hydrate structures traversing sediment cores recovered from the Godavari Basin, India. Similar to the processes described by Donohue et al. (2000), gas flow and subsequent hydrate formation in these weak sediments occurred through the creation of new porosity and was not associated with flow within the original fabric of the sediment.

Further work on laboratory samples of compact bentonite (Harrington and Horseman, 2003) clearly showed that advective gas flow was associated with the development of multiple dilatant pathways. These features were shown to vary temporally and spatially within the clay with gas pressure, total stress and pore water pressure integrally linked once the gas entry pressure had been reached. For fully water-saturated bentonite/sand plugs, gas migration was at gas pressures equal or slightly greater than the bentonite/sand swelling pressure (Liu et al., 2012).

Field-scale observations at an underground research facility in Sweden confirmed these results, showing clear hydromechanical coupling during gas flow (Harrington et al., 2007; Cuss et al., 2010, 2011; Graham et al., 2012).

In recent years, significant effort has been placed on examining gas and water flow in more lithified clay-rocks (Rodwell et al., 1999; Harrington et al., 2003, 2009, 2012a; Marschall et al., 2005; Angeli et al., 2009; Skurtveit et al., 2010, 2012; Cuss and Harrington, 2011; Romero et al., 2012; Gerad et al., 2014; Cuss et al., 2014). Marschall et al. (2005) presented borehole data from a gas injection test within the Opalinus clay (OPA), which exhibited a hydromechanical response as recorded by extensiometers during gas injection testing. Later work by Romero et al. (2012) on laboratory scale OPA samples confirmed dilation occurred during gas flow, leading to the development of a void ratio-dependent permeability expression (Senger et al., 2014). Caprock-integrity studies from a number of North Sea hydrocarbon fields (Angeli et al., 2009; Harrington et al., 2009; Skurtveit et al., 2010, 2012) indicate dilatancy may be a common mechanism associated with caprock failure. Laboratory experiments injecting carbon dioxide through samples of Draupne shale show a clear increase in volume during gas flow (Angeli et al., 2009; Skurtveit et al., 2010, 2012). Simultaneous measurements of P- and S-wave acoustic velocity show a change in behavior as CO_2 moves through the sample, and is related to mechanical changes in the sample. Laboratory tests on Nordland shale samples from the Sleipner CO_2 inject site were described and numerically modeled by Harrington et al. (2009). Their analysis suggested anisotropy to gas flow was greater than that of water and that gas movement primarily occurred through the development of pressure-induced pathways.

Gerard et al. (2014) described gas flow through Callovian-Oxfordian (COx) clay-rock using an embedded fracture model. While the solution to this model is somewhat arbitrary, dependent on the initial parameterization of the fracture(s), a better fit to laboratory data was achieved compared with that obtained using a standard viscocapillary approach.

7.4 EXPERIMENTS

From the above literature review it is evident, that gas transport in initially water-saturated clay-rich rocks is very restricted. High differential (fluid, effective) pressures or concentration gradients are usually required to obtain effective flow rates that can be detected and quantified on the laboratory scale. Relatively small samples (millimeter to centimeter range) are mounted in experimental setups capable of mimicking in situ pressure and temperature conditions. Data evaluation aims at obtaining valid key-parameters, like diffusion coefficients, critical capillary entry pressures and effective gas permeability coefficients, that can be used for upscaling to the field scale. Well-known and controlled initial and boundary conditions are a prerequisite for proper evaluation and interpretation of the data.

7.4.1 Diffusion Experiments

Diffusivities can be measured by either taking advantage of the original ("naturally occurring") gas content of rock samples or by applying gas from an external source.

In the first approach, samples or rock volumes are degassed in vacuum containers or directly within the borehole. The derivation of diffusion coefficients by this procedure relies on distinct symmetry assumptions (Gómez-Hernández, 2000; Bigler et al., 2005). Diffusion coefficients can also be estimated from large-scale concentration profiles of natural tracers, such as He and Ar (Rubel et al., 2002; Bensenouci et al., 2011) or light hydrocarbons (Leythaeuser et al., 1980) outgassing from geological formations. Both methods have two major disadvantages: the technique is only applicable for gases that are naturally present in the clay (basically limited to He, Ar, and hydrocarbon gases) and the experimental results, obtained from the outgassing approach have to be interpreted with caution. Thus, the initial and boundary conditions are usually not known and have to be estimated. Cores start to degas immediately after or even during sampling, and they also may take up gas from the atmosphere. The uncertainties in the initial and boundary conditions directly control the uncertainty in the diffusion coefficients obtained from these procedures.

Laboratory experiments applying external gas concentration gradients are either conducted as in- or through-diffusion techniques under well-constrained boundary conditions (pressure, temperature, water saturation, gas phases, and concentration potentials). Different types of through-diffusion experiments have been described in literature (Krooss and Schaefer, 1987; Rebour et al., 1997; Jacops et al., 2013). Cylindrical samples of known dimensions are usually placed into diffusion cells with gas source/sampling reservoirs on both sides. The chemical potential gradient across the sample is established by charging one compartment with the gas of interest. The gas molecules dissolve in the pore water and start to diffuse through the pore system. The gas concentration at the outlet face (sampling reservoir) or the incremental amounts of gas diffused through the sample are determined periodically and solved for effective diffusion coefficients (D_{eff}) and other likely retardation processes. The pressure gradient across the sample must be maintained at (or close to) zero throughout the experiment to ensure that no convective transport (pressure-driven fluid flow) interferes with the diffusion process.

"In-diffusion" experiments are performed such that one or both faces of the water-saturated cylindrical sample plug are exposed to the gas phase at a starting time ($t = 0$) and the concentration in the adjacent reservoirs is monitored over time (Volckaert et al., 1995).

Depending on the definitions and evaluation procedures, different diffusion coefficients are obtained (e.g., effective, pore, apparent; Horseman et al., 1996). Transient diffusion curves (cumulative diffused amount of substance vs time) recorded up to near-steady state diffusive flow (Antonov, 1970; Krooss and Schaefer, 1987), for example, contain information on retardation effects (adsorption/desorption equilibria, "chromatographic effects"). If these are not explicitly taken into account this can result in an underestimation of the "real" steady state diffusion coefficients.

Solution functions of the differential equations for through-diffusion (e.g., diffusion through a plane parallel sheet) and in-diffusion (e.g., diffusion into a semi-infinite medium with zero initial concentration) cases under various boundary conditions are provided by Crank (1975).

7.4.2 Two-Phase Flow Experiments

Reviews of different laboratory methods examining gas entry and breakthrough pressure measurements are found in Egermann et al. (2006) and Boulin et al. (2011, 2013). The step-by-step, racking, dynamic, and residual methods (imbibition) were compared. While the racking and dynamic methods provide values for the capillary entry pressure (in homogeneous media equal to breakthrough pressure), the step-by-step method also yields information on capillary gas breakthrough pressures. The residual method yields a value for the capillary snap-off pressure on the imbibition path.

In many studies, the step-by-step pressure increase method is used, although longer experimental durations are often involved (e.g., Thomas et al., 1968; Schowalter, 1979; Horseman et al., 1997; Harrington and Horseman, 1999; Li et al., 2005; Davy et al., 2012; Cuss et al., 2014). It allows for the detection of the capillary entry and breakthrough pressure by increasing gas pressure gradually or in distinct pressure steps. This method allows different successive fluid flow phases: water expulsion (due to gas pressure), intermittent (discontinuous), and continuous gas flows, not provided by the other methods. Several experiments were successfully performed on clay formations like Boom clay (Mol, HADES) or COx clay-rock from Bure (Harrington and Horseman, 1999; Marschall et al., 2005; Zhang and Rothfuchs, 2008; Harrington et al., 2012a; Cuss et al., 2014). The experiments were performed either in an open system, with analysis of the effluent (gas, water), or in a closed system, with interpretation of the pressure decay.

The dynamic method is based on measuring the change in downstream liquid flow rate under given upstream gas pressure. After gas entry, the water outflow rate decreases as the driving pressure difference is reduced by capillary forces. The method is fast (only a few days) but requires a sufficient sample length and carefully chosen fluid pressure conditions. The experiments fail, e.g., in cases where breakthrough occurs nearly instantaneous (too high pressure differences, short sample plugs) or when differences in flow rate before and after gas entry are insufficiently small (M'Jahad, 2012). The racking method is similar. Here the water pressure is reduced on the outflow side and the pressure decay is recorded and interpreted (Boulin et al., 2011).

In contrast, the residual approach relies on imbibition, i.e., water phase resaturation of the pore network as gas pressure decreases, to close conductive pathways (Hildenbrand et al., 2002). The experiments are performed in a closed and volume-calibrated system and are interpreted through analysis of the pressure transients (Amann-Hildenbrand et al., 2013). The experiments are somewhat faster than those of the step-by-step approach (hours to weeks). The residual pressure difference corresponds to the final capillary "snap-off" pressure on the imbibition path at which the system is again dynamically sealed. It is not representative for the capillary breakthrough/threshold pressure on the drainage path (e.g., Boulin et al., 2011).

7.4.3 Examples

The experiments described below were conducted on sample sets of three different origins: Oligocene (Rupelian) Boom clay from the Belgian HADES underground laboratory (SCK-CEN), Opalinus clay from the Mont Terri Rock Laboratory (Switzerland), and COx clay-rock from the Meuse/Haute-Marne Underground Research Laboratory site (ANDRA, France). While the Boom clay is a plastic clay with no evidence of mineral transformations, the Opalinus clay and COx are consolidated clay-rocks. Their associated mineralogical, petrophysical, and rock mechanical properties are summarized in Table 7.2. Data variability reflects differences in mineralogical composition (i.e., clay-rich vs carbonate-rich areas) and anisotropy. Additionally, rock mechanical properties are strongly controlled by moisture content and the experimental procedure (e.g., drained vs undrained conditions, rate of loading) (Chiarelli et al., 2003; Zhang and Rothfuchs, 2004).

7.4.3.1 Example I—Through-Diffusion

Jacops et al. (2013) used the through-diffusion technique (Shackelford, 1991; Van Loon et al., 2003) to determine simultaneously the diffusion coefficients of two counter-diffusing gases in water-saturated clay. Gas compositions and concentrations were measured by gas chromatography. Several diffusion experiments with a suite of gases (He, Ne, Ar, Xe, CH_4, C_2H_6) were performed on two samples of the lower Oligocene Boom clay formation from the Mol-1 well (drilled close to the community of Mol in the northeast of Belgium). The sample plugs were oriented perpendicular and parallel to the bedding plane, respectively. The measuring temperature was $21 \pm 2°C$ and the gas pressure in both compartments 1 MPa.

The effective diffusion coefficients (D_{eff}) obtained in this study are listed in Table 7.3. For both samples the effective diffusion coefficient decreases with increasing molar mass and size of the gas molecule. This decrease partly reflects the decrease of the diffusion coefficient in the pore fluid (D_0) with increasing molecular mass. In addition, it is influenced by the pore geometry ("pore structure") and pore-size distribution of the transport pathways. The anisotropy of the clay samples is evident from the significant differences between effective diffusion coefficients measured perpendicular and parallel to bedding. Currently under investigation are (1) the influence of the microstructural properties resulting in limited accessibility of portions of the pore space (size-exclusion effects) for individual gases and (2) adsorption effects. Samples were characterized by mercury injection porosimetry, low-pressure N_2 adsorption (Brunauer–Emmett–Teller method), computed tomography, micro computed tomography. Furthermore mineralogical composition, bulk density, grain-size distribution, and water content were determined.

In the context of radioactive waste disposal in clay formations, reliable values for diffusion coefficients of dissolved hydrogen (H_2) are essential for a correct evaluation of the balance between generated and dissipated gas. These are required mainly to estimate the pressure buildup in storage formations over time. Diffusion experiments with hydrogen can be strongly affected by the conversion of hydrogen into methane by methanogenic microorganisms

TABLE 7.2 Mineralogical, Petrophysical, and Rock Mechanical Charactersitics of Boom Clay, Opalinus Clay, and the Callovian-Oxfordian Argillite

Lithology (location)		Boom Clay, (HADES, Mol-Dessel Nuclear Site)	Opalinus Clay (Mont Terri and Weinland)	Callovian-Oxfordian Clay-Rock (Meuse/Haute-Marne, Bure site)
Deposition age	[Ma]	30	180	150–160
Depth below surface	[M]	190–290	1000 (65–120Ma) 650 (at present)	420–600
Reference number		1, 5, 7, 10	1, 2, 6, 7, 10	1, 3, 4, 7, 8, 10
Mineralogy				
Clay mineral fraction	[%]	23–60	40–80	25–60
Illite	[%]	10–50	15–30	17
Illite/smectite mixed layer	[%]	5–50	5–20	15–36
Kaolinite	[%]	5–20	15–37	2
Chlorite	[%]	0–10	3–10	2
Chlorite/smectite ML	[%]	0.5		
Quartz	[%]	15–60	10–32	18–31
K-feldspar	[%]	1–11	0.6	<2
Albite	[%]	1–10	0.3	
Carbonates	[%]	1–5	4–50	23–44

Continued

TABLE 7.2 Mineralogical, Petrophysical, and Rock Mechanical Charactersitics of Boom Clay, Opalinus Clay, and the Callovian-Oxfordian Argillite—cont'd

Lithology (location)		Boom Clay, (HADES, Mol-Dessel Nuclear Site)	Opalinus Clay (Mont Terri and Weinland)	Callovian-Oxfordian Clay-Rock (Meuse/Haute-Marne, Bure site)
Pyrite	[%]	1–5	1.1	<2
Iron oxides	[%]		0.5	<2
Natural organic matter	[%]	1–5	0.6–1.5	<1
Petrophysical Characteristics				
Bulk dry density	[kg m^{-3}]	2026	2280–2320	2250–2310
Grain/skeletal density	[kg m^{-3}]	2650	2700–2770	2700
Water content	[%]	19–27	5–10	4–8
Porosity WC	[%]	37	12–21	11–19
Porosity (Hg or He-pycnometry)	[%]	36–40	10–24.7	10–18
Specific surface (N_2-BET)	[m^2 g^{-1}]	44	24–43	
Rock Mechanical Properties				
Plastic limit	[%]	23–29	Not applicable for stiff materials	Not applicable for stiff materials
Liquid limit	[%]	55–80		
Plasticity index	[%]	32–51		

Friction angle, drained (undrained)	[°]	18–25 (2–4)	23–25	
Cohesion, drained (undrained)	[MPa]	0.3 (0.5–1.3)	1–5.5	
Young's modulus, E normal to bedding*	[MPa]	0.3[10], 300[7], 2[1]	2011–3500	1760–2630[11], 6000[7], 2000–8000[11], 4000–1,0000[12]
Poisson ratio	–	0.125	0.24	0.3 0.1–0.25[12]
Drained bulk modulus	[MPa]	200–2000		1490–2260
Shear modulus	[MPa]		800–1600	
Swelling pressure, normal (parallel) to bedding	[MPa]	0.92	1.2 (0.5)	
Uniaxial compressive strength normal (parallel) to bedding	[MPa]	2	23.1–28.1 (4–17)	20–50

BET, Brunauer–Emmett–Teller method.
** References only listed individually for strongly deviating values.*
Data were compiled from various papers: 1. Boisson (2005), 2. Bossart (2005), 3. Gaucher et al. (2004), 4. Harrington et al. (2012a), 5. Maes et al. (2008), 6. Marschall et al. (2005), 7. Volckaert et al., (2004), 8. Wenk et al. (2008), 9. Wileveau and Bernier (2008), 10. Li et al. (2007), 11. Zhang and Rothfuchs (2004), 12. Chiarelli et al. (2003).

TABLE 7.3 Origin of the Boom Clay Samples and Effective Diffusion Coefficients Measured for Different Gases on These Samples

	Sample 84b	Sample 127b
Depth (m BDT*)	233.02–233.12	275.62–272.72
Orientation with respect to bedding plane	⊥	//
D_{eff} He (10^{-10} m^2 s^{-1})	4.37	7.18
D_{eff} Ne (10^{-10} m^2 s^{-1})	1.89	2.34
D_{eff} Ar (10^{-10} m^2 s^{-1})	0.75	1.53
D_{eff} Xe (10^{-10} m^2 s^{-1})	0.60	0.72
D_{eff} CH_4 (10^{-10} m^2 s^{-1})	0.83	1.42
D_{eff} C_2H_6 (10^{-10} m^2 s^{-1})	0.46	0.61

** Reference depth is expressed as "m below drilling table." The drilling table was located at 29.73 m Tweede Algemene Waterpassing—second general leveling.*

(Volckaert et al., 1995; Ortiz et al., 2002). At the Belgian Nuclear Research Center (SCK-CEN), a sterilization protocol which combines (1) heat treatment and gamma irradiation of the entire setup, (2) sterilization of the gas phase over a filter, and (3) the use of a microbial inhibitor was successfully applied. First experiments conducted with this protocol yielded effective diffusion coefficients for H_2 ranging from 5.4 to 7.5×10^{-10} m^2 s^{-1} for Boom clay samples oriented parallel to the bedding plane. Didier (2012) studied H_2 diffusion in saturated COx clay-rock and found comparable values (from 10^{-11} and 10^{-12} m^2 s^{-1} decreasing with applied P).

7.4.3.2 *Example II—Self-Sealing Versus Healing, Breakthrough Characteristics*

Several experiments were performed on an artificially macrocracked and "resealed" COx sample in order assess its "self-healing" ability. Within about two weeks for a 10–30-mm thick sample, COx clay-rock is able to reseal and recover a water permeability similar to intact material (10^{-21} m^2), i.e., similar to that of plugs without macroscopically visual fractures. The self-healing capacity, where the fractured sample loses all memory regarding the macrocrack (Bock et al., 2010), should be reflected by unchanged gas breakthrough characteristics. However, in comparison to the intrinsic (water) permeability, gas breakthrough occurred at comparatively lower pressures. The breakthrough pressure was observed to be well below the total applied stress (Davy et al., 2009, 2013), which is related to a mixture of capillary and poromechanical response (Davy et al., 2013).

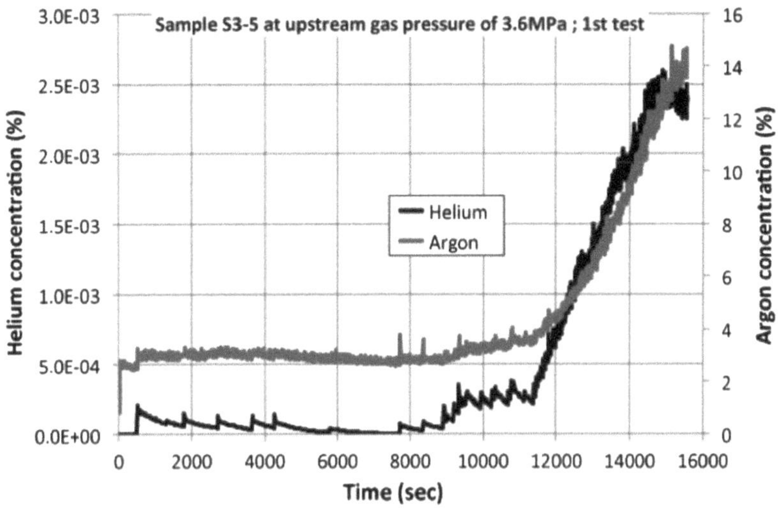

FIGURE 7.4 Second He, Ar gas breakthrough experiment on a fractured, water resaturated Callovian-Oxfordian (COx) sample ("resealed," plug length = 9.5 mm). Gas peaks frequency increases in the course of the experiment, ultimately ending in the continuous passage of helium and argon (at 12,000 s). The relatively higher Ar-concentration is caused by Ar-background concentrations (detection at the outflow side at atmospheric pressure).

The nature of gas breakthrough and its evolution with experimental time is shown in Figure 7.4 for a sample of 9.5-mm thickness. The experiment was performed at confining pressures of 12 MPa and clearly indicates that breakthrough does not occur suddenly but rather time-dependent. The initial phase is strongly nonstationary and occurs in a discontinuous manner. A first gas peak was detected at an upstream pressure (p_1) of 2.96 MPa. At a somewhat higher gas pressure (3.6 MPa) several He-peaks were sporadically recorded, with the peak frequency increasing with time ("discontinuous" flow). A similar breakthrough behavior was shown in a repetitive experiment (after water resaturation). However, He-peaks occurred more readily and at a higher frequency, which is likely caused by the "reactivation" of previously used pathways. The final breakthrough occurred after more than 3 h, resulting in significant He- and Ar-concentrations (gas pressure remained unchanged, Figure 7.4).

The "discontinuous" or "burst type" flow type can either be related to pure capillary snap-off-controlled gas passage, which requires pores of sufficiently varying size (Rossen, 2000) or the intermittent opening and closure of microcracks or tensile widening of the pore system. It is likely a mixture of several processes occurring simultaneously.

7.4.3.3 Example III—Dilation-Controlled Gas Flow

Data from a series of long-term laboratory experiments examining gas flow in COx clay-rock clearly showed flow was accompanied by directly measured

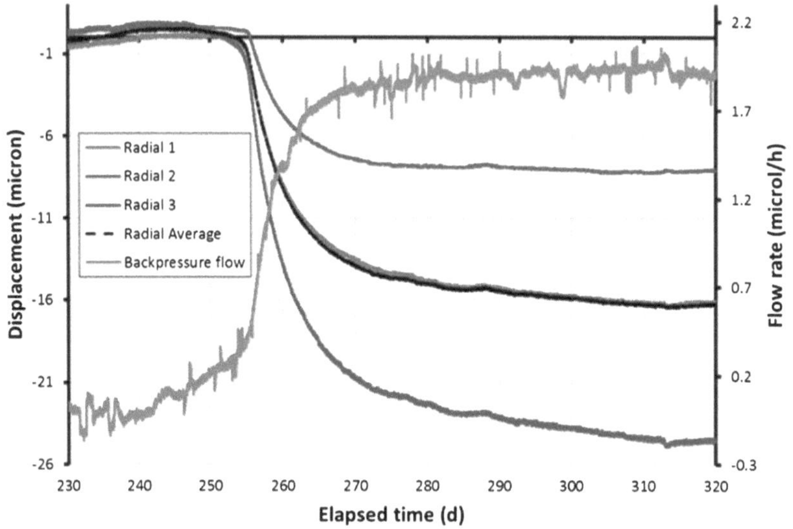

FIGURE 7.5 Strain data from triaxial test undertaken on a sample of Callovian-Oxfordian clay showing coupling between sample volume and gas outflow. Negative strain (radial 1–3 and average) indicates dilation of the sample.

dilation of the original fabric at gas pressures below that of the minimum principal stress (Cuss and Harrington, 2011, 2012; Harrington et al., 2012a, 2014; Cuss et al., 2014). Figure 7.5 is from a triaxial test performed on COx clay-rock under in situ conditions with radial 1–3 strains measured directly at mid-plane of the sample at a 120° separation. The data clearly show that gas flow is accompanied by a small, but well-defined volume increase of the sample which cannot be explained by poroelastic compressibility. As outflow evolves toward steady state, dilation increases indicating gas permeability is a dependent variable, integrally linked to the number, location, and aperture distributions of conductive pathways (Harrington and Horseman, 1999).

In this example, dilatancy is manifested as a change in volumetric strain. Cuss et al. (2014) note that the observed increase in radial strain is nonuniform, suggesting localized flow occurs within the sample. Posttest measurements of sample weight and volume indicate no discernible desaturation, similar to observations by other researchers investigating clay materials (Pusch and Forsberg, 1983; Pusch et al., 1985; Horseman and Harrington, 1994, 1997; Harrington and Horseman, 1999, 2003; Horseman et al., 2004). Posttest visual observations of COx samples submerged in glycerol and gently heated to promote degassing suggest that gas flow occurs through a localized network of pathways, bypassing much of the clay matrix (Harrington et al., 2014, Figure 7.6). Within COx samples, Harrington et al. (2012a, 2014) and Cuss et al. (2014) suggest that the coupling of gas flow and mechanical variables results in the development of significant time-dependent effects, impacting

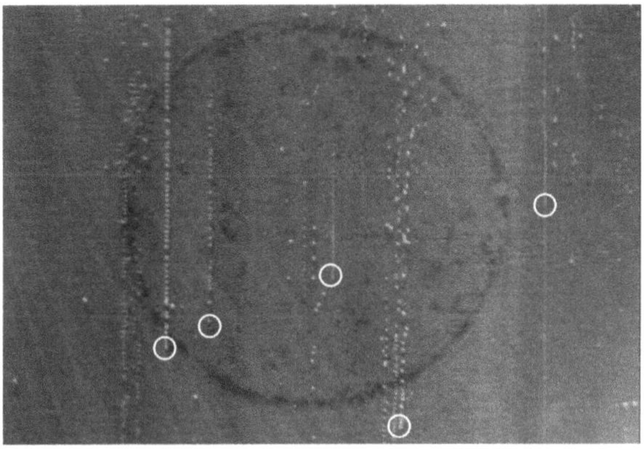

FIGURE 7.6 Posttest degassing of a Callovian-Oxfordian sample (downstream end) submerged and gently heated in glycerol, indicating localized gas flow through the clay-rock. The dark staining is a residual imprint of the backpressure filter. Bubble streams are seen to emerge from multiple locations, five of which are highlighted by the white circles. *Figure 8 in Harrington et al., (2012a).*

many aspects of COx samples behavior, from gas breakthrough time to the control of deformation processes.

Similarly, images showing the traces of previous "dilatant" features within a sample of Boom clay were presented by Harrington et al. (2012b) in which a mixture of gold and titanium oxide nanopowders were injected (within a carrier gas) through the material. Posttest scanning electron microscopy imaging showed layers of clay material draped around aggregates of nanopowder as well as a line of aggregates left along the trace of a sealed fracture. The dimension and shape of the aggregates along with the morphology of the fabric conclusively demonstrated that the Boom clay must have dilated to accommodate the gas flow.

7.4.3.4 Example IV—Gas Transport Velocities

In the context of numerous experiments aiming at the analysis of gas transport before and after breakthrough, it was shown that detected flow rates can often be explained by two different processes, pressure-driven volume flow and/or diffusion (Hildenbrand et al., 2002; Amann-Hildenbrand et al., 2012, 2013). During the last decade, different rock types were tested, including strongly consolidated clay-rocks (e.g. Boom clay, Opalinus clay), tight sandstones, and limestones. As these studies were aimed at estimating the natural (methane) or artificial (carbon dioxide) storage efficiency, all tests were performed under relatively high confining pressures and temperatures (up to 40 MPa, 55°C) representative for depths below 1500 m. With the fluid (gas/water) pressures much below these pressures, hydrofracturing of the sample plugs during testing was considered unlikely and capillary-controlled flow was assumed to occur within the clay-rock matrix and/or preexisting microfractures (Hildenbrand et al., 2002).

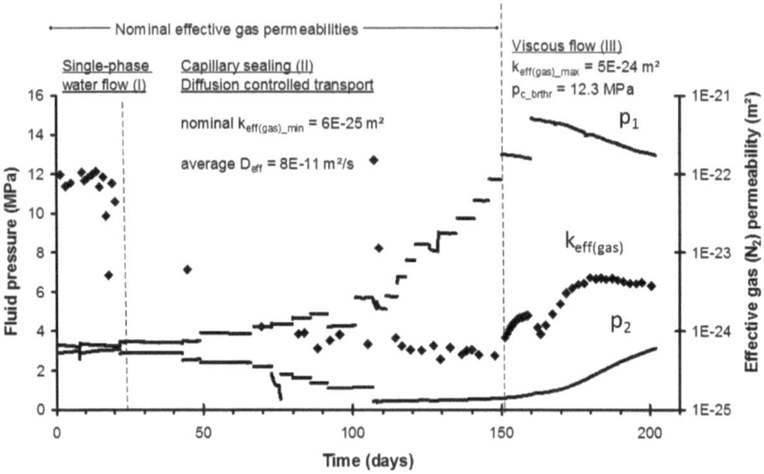

FIGURE 7.7 Gas (N_2) breakthrough experiment performed on an Opalinus clay at a confining pressure of 30 MPa and 45°C. The maximum fluid (gas) pressure on the high pressure, p_1, side is 14 MPa. The differential pressure was step-wise increased until the pressure in the downstream pressure compartment clearly increased, which was interpreted as the onset of viscous gas flow (III). After that, the flow rates increased up to an effective gas permeabilities of only 5×10^{-24} m^2 (diamonds). Mass transport prior to breakthrough is attributed to pure (I) water displacement and (II) diffusion—here the term "nominal effective gas permeability" is used.

Figure 7.7 shows the pressure recordings on an Opalinus clay sample (plug length 11 mm, 10% porosity) with specification of the prevailing transport process: (1) The initial stage of the experiment is characterized by the displacement of remaining water. Here, pressure equilibration is controlled by pure (single-phase) water displacement, indicating an intrinsic permeability of $k = 3 \times 10^{-21}$ m^2. (2) The next phase is controlled by capillary forces. The gas phase is in contact with the sample surface at differential pressures insufficient to cause breakthrough, and no pressure equilibration occurs across the sample ($p_{1/2} =$ constant). This is denoted as the capillary sealing regime, where diffusion is the only transport process occurring. The average effective diffusion coefficient is 8×10^{-11} m^2 s^{-1}, the corresponding nominal effective gas permeability is 6×10^{-25} m^2. (3) Gas breakthrough occurs only at significantly higher gas pressure differences ($\Delta p = 12.3$ MPa), manifesting itself by increasing effective gas permeability coefficients. The decline after peak permeability (5×10^{-24} m^2) is characteristic for the imbibition process, where transport is gradually restricted by the backflow of water. This decline proceeds down to flow rates associated with diffusion-controlled flow (final capillary snap-off).

During experiments on samples with very low conductivity, the authors observed very long pressure equilibration times. This holds for the equilibration to the effective stress conditions as well as for the applied fluid pressure gradient. For shales and clay-rocks considerable time (up to several days or even months) may be required to reach stable (gas) flow rates (Thomas et al., 1968; Horseman and Harrington, 1997; Harrington and Horseman, 1999; Amann-Hildenbrand et al., 2013; Cuss et al., 2014).

7.5 FINAL REMARKS AND CONCLUSIONS

In 1999, Rodwell et al. stated, "there are few problems in geosciences more complex than the quantitative prediction of gas migration fluxes through an argillaceous rock formation." While this statement remains true, considerable progress has been made in our understanding of the fundamental physics governing gas transport through water-saturated porous materials, which is relevant to a broad range of applications, including conventional and unconventional hydrocarbons, underground gas storage, CCS, radioactive waste disposal, landfill, and other engineered liner systems. Under these auspices, numerous laboratory studies have been undertaken, providing process descriptions of gas diffusion, viscous flow, and the poromechanical response of the material.

Diffusion of the dissolved gas phase in the pore water is generally associated with low-transport efficiencies (especially over long distances) but is considered a ubiquitous process. Depending on the porous network (tortuosity, constrictions etc.) effective diffusion coefficients (D_{eff}) are one to three orders of magnitude lower than those within the pure water phase ($D_{0(N_2,CH_4)}$ $\sim 10^{-9}-10^{-8}$ m^2 s^{-1}; 0–160°C). Pressure-driven transport of dissolved gas is dependent upon aqueous solubility and the intrinsic permeability of the shale for the aqueous phase. In compact clay-rocks and engineered barriers, these values are generally in the nano- to sub nano-Darcy range (1 nano-Darcy $= 10^{-21}$ m^2).

The pressure at which advective gas transport is initiated through a water-saturated rock is controlled by capillary-scale forces. In weak clays, clay-rocks and shales, this can occur through the creation of pressure-induced dilatant pathways whose permeability is then dependent on the aperture, width, and distribution functions of these features. Under these circumstances, standard viscocapillary models of two-phase flow (based on simple concepts of phase displacement) provide a poor representation of the underlying physics governing multiphase flow. In compact clay-rocks and engineered barriers, gas permeability may be as low as 10^{-24} m^2. At such low values, it is no longer possible to differentiate experimentally between diffusive and advective flux.

To this day, clay-rock formations (in their diversity) remain a significant challenge to the numerical modeler. While significant advances have been made, additional phenomenological relationships describing the coupling between gas flow and stress, burial history, mineralogy, petrology, geochemistry, and thermal loading are required to improve model representation. The fine-grained nature of clay-rocks often results in a nanoscale control of many of these processes and parameters. In addition, the full description of heterogeneity and anisotropy and the upscaling of processes to reservoir studies remains a vibrant and challenging area of research.

SYMBOLS AND ABBREVIATIONS

p	Pressure (Pa)
p_1, p_2	Experimental up- and downstream pressure (Pa)
V	Volume (m^3)
$p_{mean} = (p_1 + p_2)/2$	Fluid mean pressure (Pa)
Δp	Differential pressure (Pa)
Δh	Hydraulic head (m)
p_c	Capillary pressure ($p_c = p_{nw} - p_w = \Delta p$) (Pa)
p_w, p_{nw}	Pressure of the wetting and nonwetting fluid (Pa)
p_{c_entry}	Entry pressure, drainage path, gas phase enters pore space (Pa)
$p_{c\_breakthrough}$	Breakthrough pressure, drainage path, gas outflow is detected on (Pa)
$p_{c\_snap\text{-}off}$	Snap-off pressure (after water imbibition) (Pa)
η	Dynamic viscosity (Pa s)
A	Cross-sectional area of the sample plug (m^2)
dx	Sample length (m)
t	Time (s)
Q	Fluid flux ($m^3\ s^{-1}$)
q_i	Darcy velocity of phase i ($m\ s^{-1}$)
k	Intrinsic (water) permeability coefficient (m^2)
k_{gas}	Apparent gas permeability (varies with p_{mean}) (m^2)
k_{gas_inf}	Intrinsic, slip-flow corrected, gas permeability ($k \sim k_{gas\_inf}$) (m^2)
$k_{eff(gas)}$	Apparent effective gas permeability (two-phase flow) (m^2)
k_H	Henry constant ($Pa\ m^3\ mol^{-1}$)
k_{ri}	Relative permeability to phase i
ϕ	Porosity (fraction or %)
ρ	Density ($kg\ m^{-3}$)
S_i	Saturation of phase i
γ	Interfacial tension ($N\ m^{-1}$)
γ_i	Source term ($kg\ m^{-3}\ s^{-1}$)
θ	Contact angle (measured in the denser fluid phase) (°)
r	Pore radius (m)
C	Concentration ($mol\ m^{-3}$)
C_{fluid}	Concentration in water phase ($mol\ m^{-3}$)
C_{bulk}	Bulk concentration ($C_{fluid}*\phi$) ($mol\ m^{-3}$)
$C_{i\ liquid}$	Concentration of gas i in the liquid phase ($mol\ m^{-3}$)
C_{i_liquid}	Volume concentration of gas component i in the liquid phase ($mol\ m^{-3}$)
$C_{i\_bulk\_rock}$	Bulk volume concentration of gas component i ($mol\ m^{-3}$)
D_{eff}	Effective diffusion coefficient (related to bulk concentration gradient) ($m^2\ s^{-1}$)
D_p	Pore diffusion coefficient within porous medium ($m^2\ s^{-1}$)
D_{app}	Term used when considering a retardation factor R, $D_{app} = D_p/R$ ($m^2\ s^{-1}$)
D_0	Diffusion coefficient in pure water phase ($m^2\ s^{-1}$)
F	Diffusive flux ($mol\ m^{-2}\ s^{-1}$)
δ	Constrictivity
τ	Tortuosity
R	Retardation factor, retardation by physicochemical interactions

REFERENCES

Abbas, A., Carcasses, M., Olliver, J.-P., 1999. Gas permeability of concrete in relation to its degree of saturation. Mater. Struct. 32, 3–8.

Al-Bazali, T.M., Zhang, J., Sharma, M.M., 2005. Measurement of the sealing capacity of shale caprocks. In: SPE Annual Technology Conference and Exhibition. October, 9–12, Dallas, Texas.

Al-Bazali, T.M., Zhang, J., Chenevert, M.E., Sharma, M.M., 2009. An Experimental Investigation on the Impact of Capillary Pressure, Diffusion Osmosis, and Chemical Osmosis on the Stability and Reservoir Hydrocarbon Capacity of Shales, Offshore Europe. Society of Petroleum Engineers, Aberdeen, UK.

Amann-Hildenbrand, A., Ghanizadeh, A., Krooss, B.M., 2012. Transport properties of unconventional gas systems. Mar. Pet. Geol. 31, 90–99.

Amann-Hildenbrand, A., Bertier, P., Busch, A., Krooss, B.M., 2013. Experimental investigation of the sealing capacity of generic clay-rich caprocks. Int. J. Greenh. Gas Con. 123, 20–33.

Angeli, M., Solday, M., Skurtveit, E., Aker, E., 2009. Experimental percolation of supercritical CO_2 through a caprock. Energy Procedia 1, 3351–3358.

Antonov, P.L., Gladyscheva, G.A., P., K.W., 1958. Die Diffusion von Kohlenwasserstoffen durch Steinsalz (Diffusion of hydrocarbons through salt rocks). Z. Angew. Geol. 8, 387.

Antonov, P.L., 1970. Results of the investigation of diffusion permeability of sedimentary rocks for hydrocarbon gases (in Russian). Trudy VNIIYaGG 8, 51–79.

Appold, M.S., Nunn, J.A., 2002. Numerical models of petroleum migration via buoyancy-driven porosity waves in viscously deformable sediments. Geofluids 2, 233–247.

Bensenouci, F., Michelot, J., Matray, J., Savoye, S., Lavielle, B., Thomas, B., Dick, P., 2011. A profile of helium-4 concentration in pore-water for assessing the transport phenomena through an argillaceous formation (Tournemire, France). Phys. Chem. Earth, Parts A/B/C 36, 1521–1530.

Berg, R.R., 1975. Capillary pressure in stratigraphic traps. AAPG Bull. 59, 939–956.

Berne, P., Bachaud, P., Fleury, M., 2010. Diffusion properties of carbonated caprocks from the Paris Basin. Oil Gas Sci. Technol. Rev. IFP 65, 473–484.

Bigler, T., Ilhy, B., Lehmann, B., Waber, H., 2005. Helium Production and Transport in the Low-Permeability Callovo-Oxfordian Shale at the Site Meuse/Haute Marne, France. NAGRA, Wettingen.

Bikkina, P.K., 2011. Contact angle measurements of CO_2-water-quartz/calcite systems in the perspective of carbon sequestration. Int. J. Greenh. Gas Con. 5, 1259–1271.

Bock, H., Dehandschutter, B., Martin, C.D., Mazurek, M., Haller, A.D., Skoczylas, F., Davy, C.A., 2010. Self-sealing of Fractures in Argillaceous Formations in the Context of Geological Disposal of Radioactive Waste – Review and Analysis. OECD Nuclear Energy Agency report n.6184. ISBN:978-92-64-99095-1.

Boisson, J.-Y., 2005. Clay Club Catalogue of Characteristics of Argillaceous Rocks. Nuclear Energy Agency, NEA No. 4436. ISBN:92-64-01067-X. https://www.oecd-nea.org/rwm/reports/2005/nea4436-argillaceous-catalogue.pdf, p. 72.

Borysenko, A., Clennell, B., Sedev, R., Burgar, I., Ralston, J., Raven, M., Dewhurst, D., Liu, K., 2009. Experimental investigations of the wettability of clays and shales. J. Geophys. Res. 114, B07202.

Bossart, P., 2005. Characteristics of the Opalinus Clay at Mont Terri. http://www.mont-terri.ch, 26.

Boudreau, P., 1996. Diagenic Models and Their Interpretation—Modelling Transport and Reactions in Aquatic Sediments. Springer, Berlin.

Boudreau, B.P., 2012. The physics of bubbles in surficial, soft, cohesive sediments. Mar. Pet. Geol. 38, 1–18.

Boulin, P.F., Bretonnier, P., Vassil, V., Samouillet, A., Fleury, M., Lombard, J.M., 2011. Entry pressure measurements using three unconventional experimental methods. In: Society of Core Analysts, Austin, Texas, USA, p. 12.

Boulin, P.F., Bretonnier, P., Vassil, V., Samouillet, A., Fleury, M., Lombard, J.M., 2013. Sealing efficiency of caprocks: experimental investigation of entry pressure measurement methods. Mar. Pet. Geol. 48, 20–30.

Bruggeman, C., Maes, N., Aertsens, M., De Canniere, P., 2009. Tritiated Water Retention and Migration Behaviour in Boom Clay – SFC1 Level 5 Report: First Full Draft. NIRAS-ONDRAF, Brussels.

Busch, A., Amann-Hildenbrand, A., 2013. Predicting capillarity of mudrocks. Mar. Pet. Geol. 45, 208–223.

Busch, A., Alles, S., Gensterblum, Y., Prinz, D., Dewhurst, D.N., Raven, M.D., Stanjek, H., Krooss, B.M., 2008. Carbon dioxide storage potential of shales. Int. J. Greenh. Gas Con. 2, 297–308.

Bustin, A.M.M., Bustin, R.M., Cui, X., 2008. Importance of fabric on the production of gas shales. In: Society of Petroleum Engineers (SPE), Unconventional Gas Conference, Keystone, Colorado, USA. SPE Paper 114167.

Byrnes, A.P., 1996. Reservoir characteristics of low-permeability sandstones in the Rocky Mountains. Mount. Geol. 34, 39–51.

Carles, P., Bachaud, P., Lasseur, E., Berne, P., Bretonnier, P., 2010. Confining properties of carbonated dogger caprocks (Parisian Basin) for CO_2 storage purpose. Oil Gas Sci. Technol. - Revue d'IFP 65 (3), 461–472.

Chiarelli, A.S., Shao, J.F., Hoteit, N., 2003. Modeling of elastoplastic damage behavior of a claystone. Int. J. Plasticity 19, 23–45.

Clayton, C.J., Hay, S.J., 1992. Gas migration mechanisms from accumulation to surface. Bull. Geol. Soc. Denmark 41, 1223.

Cluff, R.M., Byrnes, A.P., 2010. Relative permeability in tight Gas Sandstone reservoirs – the "Permeability Jail" model. In: SPWLA 51st Annual Logging Symposium, Perth, Australia, p. 16.

Crank, J., 1975. The Mathematics of Diffusion. Clarendon Press, Oxford.

Cui, X., Bustin, A.M.M., Bustin, R.M., 2009. Measurements of gas permeability and diffusivity of tight reservoir rocks: different approaches and their applications. Geofluids 9, 208–223.

Cuss, R.J., Harrington, J.F., Noy, D.J., 2010. Large Scale Gas Injection Test (Lasgit) Performed at the Aspo Hard Rock Laboratory. Summary Report 2008. Svensk Karnbranslehantering AB (SKB) Technical Report TR-10–38. SKB, Stockholm, Sweden, p. 109.

Cuss, R.J., Harrington, J.F., 2011. Update on Dilatancy Associated with Onset of Gas Flow in Callovo-Oxfordian Claystone. Progress Report on Test SPP_COx-2. British Geological Survey commissioned report CR/11/110, p. 34.

Cuss, R.J., Harrington, J.F., Noy, D.J., Wikman, A., Sellin, P., 2011. Large scale gas injection test (Lasgit): results from two gas injection tests. Phys. Chem. Earth 36, 1729–1742.

Cuss, R.J., Harrington, J.F., 2012. Final Report of FORGE WP4.1.1: The Stress Path Permeameter Experiment Conducted on Callovo-Oxfordian Claystone. British Geological Survey Commissioned Report, CR/12/140, p. 116.

Cuss, R.J., Harrington, J., Giot, R., Auvray, C., 2014. Experimental Observations of Mechanical Dilation at the Onset of Gas Flow in Callovo-Oxfordian Claystone. Clays in Natural and Engineered Barriers for Radioactive Waste Confinement. 400. Geological Society Special Publications, London, United Kingdom (Geological Society of London).

Daïan, J.F., 2010. Equilibre et transferts en milieux poreux, première partie: Etats d"équilibre. p. 188.

Davies, P.B., 1991. Evaluation of the Role of Threshold Pressure in Controlling Flow of Waste-Generated Gas into Bedded Salt at the Waste Isolation Pilot Plant (WIPP). Sandia National Laboratories, Albuquerque, New Mexico.

Davy, C.A., Skoczylas, F., Lebon, P., Dubois, T., 2009. Gas migration properties through an argillite/bentonite interface. Appl. Clay Sci. 42, 639–648.

Davy, C.A., M'Jahad, F., Skoczylas, S., Talandier, J., Ghayaza, M., 2012. Evidence of discontinuous and continuous gas migration through undisturbed and self-sealed COx claystone. In: 5th International Meeting on Clays in Natural and Engineered Barriers for Radioactive Waste Management, Montpellier, 22–25th October, 2012.

Davy, C.A., M'Jahad, F., Skoczylas, S., Talandier, J., 2013. Gas migration through COx claystone and implications for self-healing. In: Proceedings of the 5th Biot's Conference on Poromechanics (10 Pages), Vienna (Austria), 10–12th July 2013.

Dembicki, J.R.H., Anderson, M.J., 1989. Secondary migration of oil: experiments supporting efficient movement of separate, buoyant oil phase along limited conduits. Am. Assoc. Pet. Geol. Bull. 73, 1018–1021.

Diamond, S., 2000. Mercury porosimetry: an inappropriate method for the measurement of pore size distributions in cement-based materials. Cem. Concr. Res. 30, 1517–1525.

Didier, M., 2012. Étude du transfert réactif de l'hydrogène au sein de l'argilite. l'Institut des Sciences de la Terre. Université de Grenoble, Grenoble, p. 275.

Donohew, A.T., Horseman, S.T., Harrington, J.F., 2000. Chapter 18: gas entry into unconfined clay pastes between the liquid and plastic limits. In: Environmental Mineralogy. Mineralogical Society Series No. 9, London, pp. 369–394.

Dullien, F.A.L., 1979. Porous Media: Fluid Transport and Pore Structure, second ed. Academic Press, New York.

Egermann, P., Lombard, J.-M., Bretonnier, P., 2006. A fast and accurate method to measure threshold capillary pressure of caprocks under representative conditions. In: International Symposium of the Society of Core Analysts, Trondheim, Norway, pp. 1–14.

Eseme, E., Krooss, B.M., Littke, R., 2012. Evolution of petrophysical properties of oil shales during high-temperature compaction tests: implications for petroleum ex-pulsion. Mar. Pet. Geol. 31 (1), 110–124.

Espinosa, R.M., Franke, L., 2006. Influence of the age and drying process on pore structure and sorption isotherms of hardened cement paste. Cem. Concr. Res. 36, 1969–1984.

Estes, R.K., Fulton, P.F., 1956. Gas slippage and permeability measurements. Trans. Am. Inst. Min. Metal. Eng. 207, 338–342.

Ferrell, R.T., Himmelblau, D.M., 1967. Diffusion coefficients of nitrogen and oxygen in water. J. Chem. Eng. Data 12, 111–115.

Galle, C., 2000. Gas breakthrough pressure in compacted Fo-Ca clay and interfacial gas overpressure in waste disposal context. Appl. Clay Sci. 17, 85–97.

Gaucher, E., Robelin, C., Matray, J.-M., Négrel, G., Gros, Y., Heitz, J.-F., Vinsot, A., Rebours, H., Cassagnabère, A., Bouche, A., 2004. ANDRA underground research laboratory: interpretation of the mineralogical and geochemical data acquired in the Callovo-Oxfordian formation by investigative drilling. Phys. Chem. Earth 29, 55–77.

Gerard, P., Harrington, J., Charlier, R., Collin, F., 2014. Modelling of localised gas preferential pathways in claystone. Int. J. Rock Mech Min. Sci. 67, 104–114.

Ghanizadeh, A., Amann-Hildenbrand, A., Gasparik, M., Gensterblum, Y., Krooss, B.M., 2013. Lithological controls on matrix permeability of organic-rich shales: an experimental study. Energy Procedia 40, 127–136.

Ghanizadeh, A., Gasparik, M., Amann-Hildenbrand, A., Gensterblum, Y., Krooss, B.M., 2014. Experimental study of fluid transport processes in the matrix system of the European organic-rich shales: I. Scandinavian Alum Shale. Mar. Pet. Geol. 51, 79–99.

Gómez-Hernández, J.J., 2000. Technical Note 2000-40: FM-C Experiment: Part A) Effective Diffusivity and Accessible Porosity Derived from In-situ He-4 Tests Part B) Prediction of HE-3 Concentration in a Cross-Hole Experiment.

Graham, C.C., Harrington, J.F., Cuss, R.J., Sellin, P., 2012. Gas migration experiments in bentonite: implications for numerical modelling. Mineral. Mag. 76 (8), 3279–3292.

Grathwohl, P., 1998. Diffusion in Natural Porous Media: Contaminant Transport, Sorption/desorption and Dissolution Kinetics. Kluwer Academic Publishers.

Hangx, S.J.T., Spiers, C.J., Peach, C.J., 2009. The mechanical behaviour of anhydrite and the effect of deformation on permeability development – implications for caprock integrity during geological storage of CO_2. Energy Procedia 5358–5363.

Harrington, J.F., Horseman, S.T., 1999. Gas transport properties of clays and mudrocks. In: Aplin, A.C., Fleet, A.J., Macquaker, J.H.S. (Eds.), Muds and Mudstones: Physical and Fluid Flow Properties, vol. 158. Geol. Soc. London, Spec. Pub, pp. 107–124.

Harrington, J.F., Horseman, S.T., 2003. Gas Migration in KBS-3 Buffer Bentonite: Sensitivity of Test Parameters to Experimental Boundary Conditions. Report TR-03–02. Svensk Kärbränslehantering AB (SKB), Stockholm, Sweden.

Harrington, J.F., Horseman, S.T., Noy, D.J., 2003. Measurements of Water and Gas Flow in Opalinus Clay Using a Novel Guard-Ring Methodology. Commissioned Report CR/03/32 (Commercial – In Confidence). British Geological Survey, Environmental Protection Programme.

Harrington, J.F., Birchall, D.J., Noy, D.J., Cuss, R.J., 2007. Large Scale Gas Injection Test (Lasgit) Performed at the Aspö Hard Rock Laboratory: Summary Report 2007. British Geological Survey Commissioned Report, CR/07/211.

Harrington, J.F., Noy, D.J., Horseman, S.T., Birchall, J.D., Chadwick, R.A., 2009. Laboratory study of gas and water flow in the Nordland Shale, Sleipner, North Sea. In: Grobe, M., Pashin, J.C., Dodge, R.L. (Eds.), Carbon dioxide sequestration in geological media—State of the science: AAPG Studies in Geology, vol. 59, pp. 521–543.

Harrington, J.F., de la Vaissière, R., Noy, D.J., Cuss, R.J., Talandier, J., 2012a. Gas flow in Callovo-Oxfordian clay (COx): results from laboratory and field-scale measurements. Mineral. Mag. 76, 3303–3318.

Harrington, J.F., Milodowski, A.E., Graham, C.C., Rushton, J.C., Cuss, R.J., 2012b. Evidence for gas-induced pathways in clay using a nanoparticle injection technique. Mineral. Mag. 76 (8), 465–474.

Harrington, J.F., Cuss, R.C., Noy, D.J., Talandier, J., 2014. Processes governing advective gas flow in the Callovo-Oxfordian claystone (COx). In: Fourth EAGE Shale Workshop. http://dx.doi.org/10.3997/2214-4609.20140026.

Hedberg, H.D., 1974. Relationship of methane generation to undercompacted shales, diapirs and mud volcanoes. Bull., Am. Assoc. Pet. Geol. 58, 661–673.

Hildenbrand, A., Schlömer, S., Krooss, B.M., 2002. Gas breakthrough experiments on fine-grained sedimentary rocks. Geofluids 2, 3–23.

Hildenbrand, A., Krooss, B.M., Schlömer, S., Littke, R., 2003. Dynamic gas leakage through fine-grained seal lithologies. In: EAGE Conference 8–11 September 2003: Fault and Top Seals. What Do We Know and Where Do We Go? Montpellier, France, pp. 0–15, 11–10.

Hildenbrand, A., Schlömer, S., Krooss, B.M., Littke, R., 2004. Gas breakthrough experiments on pelitic rocks: comparative study with N_2, CO_2 and CH_4. Geofluids 4, 61–80.

Hirsch, L.M., Thompson, A.H., 1995. Minimum saturations and buoyancy in secondary migration. AAPG Bull. 79.

Holland, M., Schultheiss, P., Roberts, J., Druce, M., 2008. Observed gas hydrate morphologies in marine sediments. In: Proceedings of the 6th International Conference on Gas Hydrates (ICGH 2008), Vancouver, British Columbia, Canada, July 6–10, 2008.

Honarpour, M., Koederitz, L., Harvey, H.A., 1987. Relative Permeability of Petroleum Reservoirs, second ed. CRC Press, Inc., Florida.

Horseman, S.T., Harrington, J.F., 1994. Migration of Repository Gases in an Overconsolidated Clay. British Geological Survey. Technical report WE/94/7.

Horseman, S.T., Higgo, J.J.W., Alexander, J., Harrington, J.F., 1996. Water, Gas and Solute Movement in Argillaceous Media. Rept. No. CC-96/1 to OECD/NEA Working Group on Measurement and Physical Understanding of Groundwater Flow through Argillaceous Media. Nuclear Energy Agency, OECD, Paris. ISBN:92-64-1488-1.

Horseman, S.T., Harrington, J.F., 1997. Study of Gas Migration in Mx80 Buffer Bentonite. British Geological Survey. Technical Report WE/97/7.

Horseman, S.T., Harrington, J., Sellin, P., 1997. Gas migration in MX80 buffer bentonite. Mat. Res. Soc. Symp. Proc. 465, 1003–1010.

Horseman, S.T., Harrington, J.F., 1998. Evidence for threshold, pathways and intermittent flow in argillaceous rocks. In: Proceedings of a Joint NEA/EC Workshop on Fluid Flow through Faults and Fractures in Argillaceous Formations, Berne, Switzerland, 1996. NEA/OECD, Paris, pp. 85–103.

Horseman, S.T., Harrington, J.F., Sellin, P., 1999. Gas migration in clay barriers. Eng. Geol. 54, 139–149.

Horseman, S.T., Harrington, J.F., Sellin, P., 2004. Water and gas flow in Mx80 bentonite buffer clay. Mat. Res. Soc. 807, 715–720.

Hovland, M., Judd, A.G., 1988. Seabed Pockmarks and Seepages: Impact on Geology, Biology and the Marine Environment. Graham and Trotman, London.

Hovland, M., Svensen, H., Forsber, C.F., Johansen, H., Fichler, C., Fosså, J.H., Jonsson, R., Rueslåtten, H., 2005. Complex pockmarks with carbonate-ridges off mid-Norway: products of sediment degassing. Mar. Geol. 218, 1991–2206. http://dx.doi.org/10.1016/j.margeo.2005.04.005.

IAEA, 2003. The Long-Term Storage of Radioactive Waste: Safety and Sustainability, p. 24.

IAEA, 2009. Borehole Disposal Facilities for Radioactive Waste. International Atomic Energy Agency, p. 118.

Ibrahim, M.A., Tek, M.R., Katz, D.L., 1970. Threshold pressure in gas storage. In: Pipeline Research Committee. American Gas Association at the University of Michigan, Michigan.

Ishida, T., Maekawa, K., Kishi, T., 2007. Enhanced modeling of moisture equilibrium and transport in cementitious materials under arbitrary temperature and relative humidity history. Cem. Concr. Res. 37, 565–578.

Ito, D., Akaku, K., Okabe, T., Takahashi, T., Tsujia, T., 2010. Measurement of threshold capillary pressure for seal rocks using the step-by-step approach and the residual pressure approach. In: GHGT10. Amsterdam, the Netherlands.

Jacops, E., Volckaert, G., Maes, N., Weetjens, E., Govaerts, J., 2013. Determination of gas diffusion coefficients in saturated porous media: He and CH_4 diffusion in Boom clay. Appl. Clay Sci. 83–84, 217–223.

Johnson, L., 2006. Gas Production and Transport in the Near Field of SF and HLW Repositories in Clay and Crystalline Rocks: Processes, Uncertainties and Performance Assessment Aspects. D-No 5.1.6 of the NF-PRO project. EU Contract No F16W-CT-2003-02389, Luxembourg.

Judd, A.G., Sim, R.H., 1998. Shallow gas migration mechanisms in deep water sediments. In: Proc. S.U.T. Conference on Offshore Site Investigation and Foundation Behaviour, pp. 163–173.

Kleppe, J., Delaplace, P., Lenormand, R., Hamon, G., Chaput, E., 1997. Representation of capillary pressure hysteresis in reservoir simulation. In: Proceedings – SPE Annual Technical Conference and Exhibition Sigma, pp. 597–604.

Klinkenberg, L.J., 1941. The permeability of porous media to liquids and gases. Dril. Prod. Prac. 200–213.

Kool, J.B., Parker, J.C., 1987. Development and evaluation of closed-form expressions for hysteretic soil hydraulic properties. Water Resour. Res. 23, 105–114.

Krooss, B.M., Schaefer, R.G., 1987. Experimental measurements of the diffusion parameters of light hydrocarbons in water-saturated sedimentary rocks. A new experimental procedure. Org. GeoChem. 11, 193–199.

Krooss, B.M., 1988. Experimental investigation of the molecular migration of C1-C6 hydrocarbons: kinetics of hydrocarbon release from source rocks. Org. GeoChem. 13, 513–523.

Krooss, B.M., Leythaeuser, D., 1988. Experimental measurements of the diffusion parameters of light hydrocarbons in water-saturated sedimentary rocks: II. Results and geochemical significance. Org. GeoChem. 12, 91–108.

Krooss, B.M., 1992. Diffusive loss of hydrocarbons through cap rock. Erdöl & Kohle – Erdgas – Petrochemie/Hydrocarbon Technol. 45, 387–396.

Krooss, B.M., Leythaeuser, D., Schaefer, R.G., 1992a. The quantification of diffusive hydrocarbon losses through cap rocks of natural gas reservoirs—A reevaluation. AAPG Bull. 76, 403–406.

Krooss, B.M., Leythaeuser, D., Schaefer, R.G., 1992b. The quantification of diffusive hydrocarbon losses through cap rocks of natural gas reservoirs—a reevaluation: Reply. AAPG Bull. 76, 1842–1846.

Lambe, T.W., 1960. A mechanistic picture of shear strength in clay. In: Proc ASCE Res. Conf. Shear Strength of Cohesive Soils, pp. 550–580.

Leythaeuser, D., Schaefer, R.G., Yükler, A., 1980. Diffusion of light hydrocarbons through near-surface rocks. Nature 284, 522–525.

Leythaeuser, D., Schaefer, R.G., Yukler, A., 1982. Role of diffusion in primary migration of hydrocarbons. AAPG Bull. 66, 408–429.

Li, K., Horne, R.N., 2004. Experimental study of gas slippage in two-phase flow. In: SPE Western Regional Meeting. Society of Petroleum Engineers, Bakersfield, California, pp. 409–415.

Li, S., Dong, M., Li, Z., Huang, S., Qing, H., Nickel, E., 2005. Gas breakthrough pressure for hydrocarbon reservoir seal rocks: implications for the security of long-term CO_2 storage in the Weyburn field. Geofluids 5, 326–334.

Li, X., Bernier, F., Vietor, T., Lebon, P., 2007. European Commission: TIMODAS Deliverables; FI6W-CT-036449. European Commission, Brussels, Belgium, 106 p.

Liu, J., Davy, C.A., Skocylas, F., Talandier, J., 2012. Gas migration through compacted bentonite/sand under water and gas pressure. In: 5th International Meeting, Montpellier, 22–25 October 2012.

Luo, X.R., Yan, J.Z., Zhou, B., Hou, P., Wang, W., Vasseur, G., 2008. Quantitative estimates of oil losses during migration, part II: measurement of the residual oil saturation in migration pathways. J. Pet. Geol. 31, 179–190.

Mackay, D., Shiu, W.Y., 1981. A critical review of Henry's law constants for chemicals of environmental interest. J. Phys. Chem. Ref. Data 10, 1175–1199.

Maes, N., Salah, S., Jacques, D., Aertsens, M., Van Gompel, M., Cannière, D., Velitchkova, N., 2008. Retention of Cs in Boom clay: comparison of data from and diffusion experiments on intact clay cores. Phys. Chem. Earth 33.

Mandl, G., Harkness, R.M., 1987. Hydrocarbon migration by hydraulic fracturing. In: Deformation of Sediments and Sedimentary Rocks, vol 29. Geol. Soc. Special Pub, pp. 39–53.

Marschall, P., Horseman, S., Gimmi, T., 2005. Characterisation of gas transport properties of the Opalinus clay, a potential host rock formation for radioactive waste disposal. Oil Gas Sci. Technol. – IFP Int. Workshop Rev. IFP 60 (1), 121–139.

M'Jahad, S., 2012. Etude de l'impact de la fissuration sur les propriétés de rétention d'eau et de transport de gaz des bétons, de l'argilite et des interfaces argilite/béton. Application au stockage géologique des déchets radioactifs (PhD thesis (in French)), Ecole Centrale de Lille and Université des Sciences et Technologie de Lille, France.

Nesterow, I.I., Uschatinskij, I.N., 1972. Abschirmeigenschaften der tonigen Gesteine über Erdöl- und Erdgaslagern in den mesozoischen Ablagerungen der Westsibirischen Tiefebene (Sealing properties of claystones above petroleum and natural gas reservoirs in the Mesozoic sediments of the West Siber). Z. Angew. Geol. 18, 548–555.

Ortiz, L., Volcharet, G., De Canniere, P., Aertsens, M., Horseman, S.T., Harrington, J.F., Impey, M., Einchcomb, S., 1996. MEGAS – modelling and experiments on gas migration in repository host-rocks. In: Proc. EU PEGASUS Meeting at Rapolina-terme, Italy, 14–15 June 1995. Nuclear Science and Technology Series, pp. 127–147. EUR 16746 EN, Luxembourg.

Ortiz, L., Volckaert, G., Mallants, D., 2002. Gas generation and migration in Boom clay, a potential host rock formation for nuclear waste storage. Eng. Geol. 64, 287–296.

Parker, J.C., 1989. Multiphase flow and transport in porous media. Rev. GeoPhys. 27, 311–328.

Pentland, C.H., 2010. Measurements of Non-wetting Phase Trapping in Porous Media Department of Earth Science and Engineering. Royal School of Mines Imperial College London, London, p. 154.

Prausnitz, J., Lichtenthaler, R., de Azevedo, E., 1999. Molecular Thermodynamics of Fluid Phase Equilibria, third ed. Prentice Hall, New Jersey.

Price, L.C., Blount, C.W., Mac, G.,D., Wenger, L.M., 1981. Methane solubility in brines with application to the geopressured resource. In: Proceedings of the 5th US Gulf Coast Geopressured Geothermal Energy Conference, Baton Rouge, pp. 205–214.

Purcell, W.R., 1949. Capillary pressures – their measurement using mercury and the calculation of permeability therefrom. Pet. Trans. AIME 186, 39–48.

Pusch, R., Forsberg, T., 1983. Gas Migration through Bentonite Clay. SKB Technical Report 83–71. Svensk Karnbranslehantering AB, Stockholm, Sweden.

Pusch, R., Ranhagen, L., Nilsson, K., 1985. Gas Migration through Mx-80 Bentonite. Nagra Technical Report NTB 85–36. Nagra, Wettingen, Switzerland.

Qingjie, L., Baohua, L., Xianbing, L., Shouguo, Y., 2002. The effect of water saturation on gas slip factor by pore scale network modeling. In: SCA 2002 Symposium, Monterey, California.

Rebour, V., Billiotte, J., Deveughele, M., Jambon, A., le Guen, C., 1997. Molecular diffusion in water-saturated rocks: a new experimental method. J. Contam. Hydrol. 28, 71–93.

Rodwell, W.R., Harris, A.W., Horseman, S.T., Lalieux, P., Müller, W., Ortiz Amaya, L., Pruess, K., 1999. Gas Migration and Two-Phase Flow through Engineered and Geological Barriers for a Deep Repository for Radioactive Waste. European Commission, Report EUR19122EN.

Romero, E., Senger, R., Marschall, P., 2012. Air injection laboratory experiments on Opalinus clay. In: Experimental Techniques. 3rd EAGE Shale Workshop, Barcelona, Spain, 23–25 January 2012, Results and Analyses.

Rossen, W.R., 2000. Snap-off in constricted tubes and porous media. Colloids Surf. A 166, 101–107.

Rubel, A., Sonntag, C., Lippmann, J., Pearson, F., Gautschi, A., 2002. Solute transport in formations of very low permeability: profiles of stable isotope and dissolved noble gas contents of pore water in the Opalinus Clay, Mont Terri, Switzerland. Geochim. Cosmochim. Acta 66, 1311–1321.

Scheidegger, A.E., 1974. The Physics of Flow through Porous Media. University of Toronto Press, Toronto.

Schlömer, S., Krooss, B.M., 1997. Experimental characterisation of the hydrocarbon sealing efficiency of cap rocks. Mar. Pet. Geol. 14, 565–580.

Schlömer, S., Krooss, B.M., 2004. Molecular transport of methane, ethane and nitrogen and the influence of diffusion on the chemical and isotopic composition of natural gas accumulations. Geofluids 4, 81–108.

Schowalter, T.T., 1979. Mechanics of secondary hydrocarbon migration in entrapment. AAPG Bull. 63, 723–760.

Sen, M.A., Horseman, S.T., Harrington, J.F., 1996. Further Studies on the Movement of Gas and Water in an Overconsolidated Clay: Contribution to MEGAS Phase 2 Report (PEGASUS Project). British Geological Survey, Technical Report WE/96/8 To: Commission of European Communities (CEC), Brussels.

Senger, R., Romero, E., Ferrari, A., Marschall, P., 2014. Characterization of Gas Flow through Low-Permeability Claystone: Laboratory Experiments and Two-Phase Flow Analyses, vol. 400. Geol. Soc. Special Pub. http://dx.doi.org/10.1144/SP400.15.

Shackelford, C., 1991. Laboratory diffusion testing for waste disposal – a review. J. Contam. Hydrol. 7, 177–217.

Shanley, K.W., Cluff, R.M., Robinson, J.W., 2004. Factors controlling prolific gas production from low-permeability sandstone reservoirs: implications for resource assessment, prospect development, and risk analysis. AAPG Bull. 88, 1083–1121.

Skurtveit, E., Aker, E., Soldal, M., Angeli, M., Hallberg, E., 2010. Influence of micro fractures and fluid pressure on sealing efficiency of caprock: a laboratory study on shale. In: Conference on Greenhouse Gas Technologies (GHGT), vol. 10. RAI, Amsterdam, The Netherlands.

Skurtveit, E., Aker, E., Soldal, M., Angeli, M., Wang, Z., 2012. Experimental investigation of CO_2 breakthrough and flow mechanisms in shale. Pet. GeoSci. 18, 3–15. http://dx.doi.org/10.1144/1354-079311-016.

Stainforth, J.G., Reinders, J.E.A., 1990. Primary migration of hydrocarbons by diffusion through organic matter networks, and its effect on oil and gas generation. Organic GeoChem. 16, 361–374.

Stklyanin, Y.I., Breeva, T.F., Vasileeva, L.B., Domnina, N.A., Kotsonis, A.N., Kondratov, L.C., Antoshina, I.N., Zavyalova, L.M., Krishchenko, V.I., Shlepova, A.I., 1968. Conditions, Parameters and Pathways of Ascending Natural Gas Migration from Oil and Gas Source Beds and Several Theoretical Migrational Models. Trudy VNIIYaGG. Nedra, Moscow. pp. 58–131.

Stklyanin, Y.I., Litvinova, V.N., 1971. On the parameters of methane diffusion through a water-saturated core. Geologija Nefti i Gaza 15, 19–22.

Thomas, L.K., Katz, D.L., Tek, M.R., 1968. Threshold pressure phenomena in porous media. Soc. Pet. Eng. J. 243, 174–184.

Thomas, M.M., Clouse, J., 1990a. Primary migration by diffusion through kerogen: I. Model experiments with organic-coated rocks. Geochim. Cosmochim. Acta 54, 2775–2779.

Thomas, M.M., Clouse, J.A., 1990b. Primary migration by diffusion through kerogen: II. Hydrocarbon diffusivities in kerogen. Geochim. Cosmochim. Acta 54, 2781–2792.

Thomas, M.M., Clouse, J.A., 1990c. Primary migration by diffusion through kerogen: III. Calculation of geologic fluxes. Geochim. Cosmochim. Acta 54, 2793–2797.

Tissot, B., Pelet, R., 1971. Nouvelles données sur les mécanismes de genèse et de migration du pétrole: simulation mathématique et application à la prospection. In: Proceedings of the 8th World Petroleum Congress, Moscow, pp. 35–46.

van Genuchten, M.Th., 1980. A closed form equation for predicting the hydraulic conductivity of unsaturated soils. Soil Sci. Soc. Am. J. 44, 892–898.

Van Loon, L., Soler, J., Bradbury, M., 2003. Diffusion of HTO, Cl-36(-) and I-125(-) in Opalinus clay samples from Mont Terri – effect of confining pressure. J. Contam. Hydrol. 73–83.

Volckaert, G., Ortiz, L., De Canniere, P., Put, M., Horseman, S.T., Harrington, J.F., Fioravante, V., Impey, M., 1995. MEGAS – Modelling and Experiments on Gas Migration in Repository Rocks. Final Report Phase 1, Contract No. F12-CT91-0076. Nuclear Science and Technology Series EUR 16235 EN, Luxembourg.

Volckaert, G., Bernier, F., Sillen, X., Van Geed, M., Mayor, J.-C., Göbel, I., Blümling, P., Fried, B., Su, K., 2004. Similarities and differences in the behaviour of plastic and indurated clays. In: EURADWASTE '04. In: http://cordis.europa.eu/fp6-euratom/ev_euradwaste04_proceedings.htm. Luxembourg, 29.03.2004–31.03.2004, 11.

Washburn, 1921. The dynamics of capillary flow. Phys. Rev. 17.

Weber, C., Stanjek, H., 2012. Development of diffuse double layers in column-wicking experiments: implications for pH-dependent contact angles on quartz. J. Colloid Interf. Sci. 387, 270–274.

Wenk, H.-R., Voltolini, M., Mazurek, M., Van Loon, L.R., Vinsot, A., 2008. Preferred orientations and anisotropy in shales: Callovo-Oxfordian shale (France) and Opalinus caly (Switzerland). Clays Clay Min. 56, 285–306.

Wileveau, Y., Bernier, F., 2008. Similarities in the hydromechanical response of Callovo-Oxfordian clay and Boom clay during gallery excavation. Phys. Chem. Earth Part A: Solid Earth Geodesy 33, 343–349.

Wilke, C.R., Chang, P., 1955. Correlation of diffusion coefficients in dilute solutions. Am. Inst. Chem. Eng. J. 1, 264–270.

Wollenweber, J., Alles, S., Busch, A., Krooss, B.M., Stanjek, H., Littke, R., 2010. Experimental investigation of the CO_2 sealing efficiency of caprocks. Int. J. Greenh. Gas Con. 4, 231–241.

Yu, L., Weetjens, E., 2009. Summary of Gas Generation and Migration, Current State-of-the-Art. Mol, Belgium.

Zhang, T., Krooss, B.M., 2001. Experimental investigation on the carbon isotope fractionation of methane during gas migration by diffusion through sedimentary rocks at elevated temperature and pressure. Geochim. Cosmochim. Acta 65, 2723–2742.

Zhang, C., Rothfuchs, T., 2004. Experimental study of the hydro-mechanical behaviour of the Callovo-Oxfordian argillite. Appl. Clay Sci. 26, 325–336.

Zhang, C.L., Rothfuchs, T., 2008. Damage and sealing of clay rocks detected by measurements of gas permeability. Phys. Chem. Earth 33, S363–S373.

Chapter 8

Semipermeable Membrane Properties and Chemomechanical Coupling in Clay Barriers

Julio Gonçalvès,[a] Pierre M. Adler,[b] Philippe Cosenza,[c] Aliaksei Pazdniakou[b] and Ghislain de Marsily[b]
[a] *University Aix-Marseille, UMR-7330 CEREGE, Aix en Provence, France;*
[b] *Sorbonne Universités, UPMC Univ. Paris 06, UMR-7619 METIS, Paris cedex, France;*
[c] *University of Poitiers, CNRS, UMR 7285 IC2MP-HydrASA, ENSIP, Poitiers, France*

Chapter Outline

Developments in Clay Science, Vol. 6C. http://dx.doi.org/10.1016/B978-0-08-100027-4.00008-5

TABLE OF NOTATION

Symbols	Definition	Units
A	Hamaker constant	J
A_S	Specific surface area	$m^2\ kg^{-1}$
α	Chemomechanical coupling parameter	–
α_C	Bulk compressibility of the porous medium	Pa^{-1}
$\boldsymbol{\alpha_p}$	Electroosmotic coupling tensor	$m^2\ s^{-1}\ V^{-1}$
b	Half-pore size	m
b_i	Ion i mobility	$m^2\ s^{-1}\ V^{-1}$
B	Skempton coefficient	–
β	Chemo-hydraulic coupling parameter	–
$\boldsymbol{\beta_l}$	Electroosmotic coupling tensor	$m^2\ s^{-1}\ V^{-1}$
β_l	Compressibility of water	Pa^{-1}
β_S	Compressibility of solid grains	Pa^{-1}
C	Drained compressibility (or inverse of the drained bulk modulus)	Pa^{-1}
C_{chem}	Chemomechanical interaction parameter	Pa^{-1}
c_i	Concentration of ion i at the pore scale	ions m^{-3}
c^+	Concentration of the cation in the porosity	ion m^{-3}
c^-	Concentration of the anion in the porosity	ion m^{-3}
c_f	Concentration of a salt in the equilibrium solution	mol L^{-1} or ion m^{-3}
$<c_s>$	Mean concentration of a salt in the pore fluid	ion m^{-3}
$<c_w>$	Mean concentration of the water in the pore fluid	molecules m^{-3}
χ_T	Coefficient of thermal expansion of water	K^{-1}
D_{eff}	Effective diffusion coefficient	$m^2\ s^{-1}$
D_i	Diffusion coefficient of specie i at pore scale	$m^2\ s^{-1}$
D_p	Diffusion coefficient of the porous medium	$m^2\ s^{-1}$

Symbols	Definition	Units
e	Elementary electric charge	V
E	Electric field	$V\ m^{-1}$
E_Y	Young modulus	Pa
ϵ_{el}	Permittivity of the fluid	$F\ m^{-1}$
ε	Volumetric strain	–
$\boldsymbol{\varepsilon}$	Strain tensor	–
ε_c	Osmotic efficiency or reflection coefficient	–
ε_ψ	Electroosmotic coefficient	$Pa\ V^{-1}$
ε_T	Thermoosmotic coefficient	$Pa\ K^{-1}$
f	Helmholtz free energy per unit porous medium volume	$J\ m^{-3}$
F_F	Formation factor	–
$\mathbf{F}$	Body force vector	N
ϕ	Ion flux	$mol\ m^{-2}\ s^{-1}$
φ	Macropores porosity	–
$\mathbf{g}$	Acceleration due to gravity vector	$m^2\ s^{-1}$
G	Shear modulus	Pa
γ	Purely chemical constant	Pa^{-1}
γ_B	Biot's coefficient	–
γ_w	Specific weight of water	$N\ m^{-3}$
γ_1	Empirical coefficient	Pa
γ_2	Empirical coefficient	M
h	Hydraulic head	M
H	Specific enthalpy of the pore fluid	$J\ m^{-3}$
I	Electrical current density	$A\ m^{-2}$
$\mathbf{j}_i$	Flux of specie i at pore scale	ions or mole $m^2\ s^{-1}$
$J_V = q$	Specific discharge	$m\ s^{-1}$
J_s^D	Solute velocity relative to the solvent	$m\ s^{-1}$
$J_\pm$	Velocity of the cation in excess as compared with the anion	$m\ s^{-1}$
J_Q	Heat flux	$J\ m^{-2}\ s^{-1}$
$\mathbf{k}$	Intrinsic permeability tensor	m^2
$\mathbf{K}$	Tenseur de conductivit hydraulique	$m\ s^{-1}$
k_B	Boltzmann constant	$J\ K^{-1}$
k_T	Thermoosmotic permeability	$m^2\ s^{-1}\ K^{-1}$
m_s	Mass increment of solute per unit reference volume of porous medium	ions m^{-3}
m_w	Mass increment of water per unit reference volume of porous medium	molecules m^{-3}
μ_s	Chemical potential of solute (typically NaCl salt)	$J\ m^{-3}$
μ_f	Chemical potential of the equilibrium solution	$J\ m^{-3}$
μ_w	Chemical potential of water	$J\ m^{-3}$
ν	Drained Poisson's ratio	–
$\mathbf{p}$	Generalized "pressure" tensor	Pa
p_f	Fluid hydrostatic pressure	Pa
p_N	Normal component to clay surface of the generalized "pressure" tensor	Pa

Symbols	Definition	Units
p_T	Tangential component to clay surface of the generalized "pressure" tensor	Pa
π	Osmotic pressure	Pa
π_D	Disjoining pressure	Pa
π_D^e	Electrostatic component of the disjoining pressure	Pa
π_D^{vdw}	Attractive van der Walls component of the disjoining pressure	Pa
π_D^s	Solid surface hydration component of the disjoining pressure	Pa
$\mathbf{\Pi}$	Macroscopic physicochemical stress tensor	Pa
ψ	Electrical potential	V
ψ_0	Electrical potential at the clay surface	V
ψ_β	Electrical potential at the β plane	V
ψ_d	Electrical potential at the *d* plane	V
ψ_b	Electrical potential at the half-pore size distance from the clay surface	V
$\mathbf{q}$	Specific discharge vector	$m\ s^{-1}$
Q_0	Surface charge density at clay surface	$C\ m^{-2}$
Q_β	Surface charge density in the Stern layer	$C\ m^{-2}$
Q_S	Surface charge density in the diffuse layer	$C\ m^{-2}$
Q_V	Macroscopic volumetric excess charge	$C\ m^{-3}$
θ	Relative ionic content	–
R	Perfect gas constant	$J\ mol^{-1}\ K^{-1}$
ρ	Charge density at pore scale	$C\ m^{-3}$
ρ_f	Density of the fluid	$kg\ m^{-3}$
ρ_S	Density of the solid	$kg\ m^{-3}$
S_s	Specific storage coefficient	Pa^{-1}
S_S^{chem}	Chemomechanical specific storage coefficient	M^{-1}
S_{sp}	Specific entropy	$J\ m^{-3}$
S_σ	Unconstrained specific storage	Pa^{-1}
σ	Mean stress	Pa
$\boldsymbol{\sigma}$	Total stress tensor	Pa
$\boldsymbol{\sigma}'$	Effective stress tensor	Pa
$\boldsymbol{\sigma_e}$	Conductivity tensor	$S\ m^{-1}$
T	Temperature	K
$\mathbf{u}$	Displacement vector	M
v_w	Molar volume of water	$m^3\ mole^{-1}$
v_s	Molar volume of salt	$m^3\ mole^{-1}$
$\mathbf{v}$	Velocity vector at pore scale	$m\ s^{-1}$
ϖ	Specific discharge of salt with respect to the solvent	$m\ s^{-1}$
ω	Porosity	–
ω_D	Ratio between an apparent	–
ω_{eff}	Effective porosity chemical diffusivity and the hydraulic diffusivity	–
z_i	Valence of ion *i*	–
ζ_f	Fluid content	–
$\mathbf{1}$	Unit tensor	–

8.1 INTRODUCTION

Quantitative models simulating the hydrodynamic functioning in large sedimentary systems often use a simplifying assumption of vertical leakage across clay formations. The pressure (hydraulic head) values of the aquifers on both sides of the clay-rocks and a calibrated vertical hydraulic conductivity in these clay-rocks (or aquitard) allow vertical leakage to be identified. The aquitard is then considered as a "black box," whose "internal" pressure field or hydraulic head is not explicitly solved. Therefore, in a pseudo-3D approach, the 2D groundwater flow is calculated within aquifers while the vertical leakage across semipermeable formations is estimated using 1D Darcy's law, generally in steady-state (see Wei et al., 1990; Castro et al., 1998). The models take into account two fundamental clay characteristics. The first one is hydraulic conductivity anisotropy, i.e., a hydraulic conductivity along the strata, which is generally higher than the perpendicular one. The second one is the scale effect often producing regional hydraulic conductivities that may be three orders of magnitude higher than the values measured on samples. This effect was convincingly established when the first large-scale numerical simulations were made of sedimentary basin behaviors (see Pierre Shale's clays (Bredehoeft et al., 1983; Neuzil, 1986)). This scale effect was attributed to the regional heterogeneity of hydraulic conductivity in shales, more specifically to the existence of fractures or even fault networks (Neuzil, 1986).

With the exception of the work by Bredehoeft and Papadopoulos (1980), Bredehoeft et al. (1983), or Neuzil (1982, 1986, 1993), the hydrodynamics of clays had, until these last two decades, attracted only limited attention from the hydrogeologists. The currently renewed theoretical and experimental interest in clays has largely been caused by increasing research efforts, started in the 1990s, on radioactive waste storage in clay formations. For the first time, pore-pressure measurements were conducted inside such formations and showed unexplained pressure anomalies, mostly overpressures (excess hydraulic heads), i.e., pressures (hydraulic heads) in excess as compared to the hydrostatic ones or to those measured in the surrounding aquifers. This was unexpected since, until then, a simple linear hydraulic head profile in steady-state was assumed to be based on the hypothesis of a purely Darcian vertical leakage flow across clay formations. In this context, a major characteristic of clay materials concerns the applicability of Darcy's law in hydrogeological models. As discussed below, for clay media, Darcy's law is incomplete. Indeed, coupled transport processes occur, i.e., other driving forces than the so-called "conjugate" ones (e.g., the pressure gradient for the water flow) are involved in flow and transport. All the classical driving forces in transport processes (chemical potential, electrical-potential, or temperature gradients) can cause a fluid motion, solute or heat transport.

To identify all the driving forces, all the flows (solutes, fluid, electrical current, and possibly heat) have to be solved, hence the notion of coupling. The microscopic origin of the coupled flows is briefly illustrated. To understand and predict the coupling phenomena, one has to return to the pore scale to examine the physical processes specific to clayey media. Various publications in the literature show that a focus on the pore scale makes it possible to establish the coupling coefficients that multiply the nonconjugate driving forces. The major consequence is the reevaluation of the formulation classically used in the calculations of transport in porous media. This reevaluation is based on theoretical developments concerning coupled flows as well as much experimental work (Young and Low, 1965; Kemper and Rollins, 1966; Dirksen, 1969; Olsen, 1972; Bresler, 1973; Mitchell et al., 1973; Srivastava and Avasthi, 1975; Bolt, 1979; Neuzil, 2000; Cey et al., 2001; Keijzer et al., 2001; Kooi et al., 2003; Malusis et al., 2003; Dufreche et al., 2005; Rosanne et al., 2006; Garavito et al., 2007). Here, the focus is primarily on fluid flow, including transient fluid-flow calculations, hydro-chemo-mechanical couplings. The processes of mass and heat transport are discussed only briefly. More discussion on mass transport (especially diffusion) can be found in Chapter 6 in this volume.

8.2 TRANSPORT PROCESSES IN CLAY-ROCK FORMATIONS

8.2.1 Reevaluation of Darcy's Law for Clay-Rocks

Calculations of water flow in porous media use the classical Darcy formulation:

$$q = -\frac{\mathbf{k}}{\eta_f}\left(\nabla p_f + \rho_f g \nabla z\right), \tag{8.1}$$

where p_f is the fluid pressure (Pa), ∇z is the vector (0,0,1) if the z-axis is vertical ascending, $\mathbf{k}$ is the intrinsic permeability tensor (m^2), η_f and ρ_f are the dynamic viscosity (Pa s) and the fluid density ($kg\,m^{-3}$), and g ($m\,s^{-2}$) is the acceleration due to gravity. The validity of Darcy's law for clays was widely questioned in the 1960s and the existence of a possible threshold effect, although controversial, was proposed (Mitchell, 1993). This filtration pressure (or head) gradient threshold was discussed and attributed to experimental errors by Mitchell (1993). More importantly, the invalidity of Darcy's law, or more precisely, the fact that this law only incompletely describes water filtration through clay materials, was clearly established experimentally. However, this discrepancy with regard to Darcy's law can be explained more convincingly by the existence of other driving forces than just the pressure gradient. In the field of soil sciences or petroleum engineering, for example, numerous experiments at the sample scale showed the existence of chemical osmosis, i.e., a flow of water associated with a salt-concentration gradient

(Kemper, 1960; McKelvey and Milne, 1960; Kemper and Evans, 1963; Young and Low, 1965; Kemper and Rollins, 1966; Olsen, 1969; Bresler, 1973; Mitchell et al., 1973) in clay materials. Among these experimental studies, only the work by Young and Low (1965) considered a natural material, the others dealt with cleaned and/or remolded materials. In these experiments, a fluid flow from the low- toward the high-salinity zones identifies an osmotic permeability (Fritz and Marine, 1983; Fritz, 1986). The study by Neuzil (2000) clearly showed the existence of in situ chemical osmotic processes in clay formations. This important result underlines the weak representativeness of Darcy's law in its classical formulation for such media, and has motivated experimental studies (at sample scale or in situ) to characterize the osmotic behavior of clays, thus improving the knowledge of their properties for confining radioactive wastes (Cey et al., 2001; Keijzer et al., 2001; Kooi et al., 2003; Malusis et al., 2003; Noy et al., 2004; Garavito et al., 2007). Although the main focus of the scientific community was on chemical osmosis, the other coupled processes (see Table 8.1) such as electroosmosis (Olsen, 1969, 1972; Greenberg et al., 1973; Yeung and Mitchell, 1993; Rosanne et al., 2004, 2006), i.e., a fluid flow under the influence of an electrical-potential gradient, or thermoosmosis, i.e., a fluid flow driven by a temperature gradient (Dirksen, 1969; Srivastava and Avasthi, 1975) were identified experimentally in clayey materials. Thermoosmosis is however less well-known because of a weaker theoretical understanding than is the case for the other processes. This theoretical gap prevents us, for instance, from explaining the dispersion of the few available values of the thermoosmotic permeability (Soler, 2001). As shown by Soler (2001), the characterization of these coupled processes is crucial because it can affect our understanding of the transport processes and of the pressure fields, which are now being accurately measured in some clay formations.

TABLE 8.1 Terminology of Coupled Fluxes (de Marsily, 1986)

	Gradients			
Flux	∇h or ∇p_f	$\nabla \psi$	∇c_f	∇T
Fluid	Darcy	Electroosmosis	Chemical osmosis	Thermal osmosis
Electrical	Rouss (streaming current)	Ohm	Diffusion current	Seebeck effect
Solutes	Ultrafiltration	Electrophoresis	Fick	Soret effect
Heat	Heat filtration	Peltier effect	Dufour effect	Fourier

The limited validity of Darcy's law in clay media thus results from its incomplete formulation that neglects a priori additional driving forces for fluid flow. The identification and introduction of such driving forces can be done, for example, within the framework of the theory of irreversible thermodynamic processes (de Groot and Mazur, 1962; Katchalsky and Curran, 1967).

8.2.2 Experimental Evidence for a Limited Validity of Darcy's Law in Clay-Rocks

8.2.2.1 Origin of Non-Darcian Fluxes

Understanding the origin of the osmotic flows requires a brief return to the microscopic or even nanometric scale, i.e., the pore scale. These osmotic flows can be explained by considering the electrochemical interactions between the surface of clay mineral particles and the electrolyte at near neutral pH. Basal surfaces of clay minerals are negatively charged. This negative charge is balanced by cations, which are attracted to the surface and are termed compensating cations. They are distributed between a compact layer, the Stern layer, and a more scattered one, the diffuse layer. In the diffuse layer, the concentration of the cations decreases almost exponentially toward the middle of the pores. The equilibrium solution, a very important notion in interface sciences, is the solution at a certain distance of the charged solid surface where the electrical field becomes negligible. The equilibrium solution is also the solution in local equilibrium with the fluid contained in the pores, i.e., the solution contained in a macropore or contained in a measurement chamber isolated in a borehole or of an aquifer in direct contact with the clay formation. In the following, p_f (Pa) and c_f ($\mathrm{mol\,L^{-1}}$) represent the fluid pressure and the salinity of this solution, respectively.

Such interactions are reproduced by electrical models, which simulate the electrical-potential distribution and the related ionic distributions by solving the Poisson–Boltzmann equation (see Gonçalvès et al., 2007).

Flows corresponding to chemical osmosis have been recognized historically and described for biological membranes (see discussion by Gonçalvès et al., 2004). A membrane is a material that restricts ionic transport, but not the movement of neutral species such as water. A perfect membrane completely prevents ionic transport. The negative-surface charge of clay minerals causes anionic exclusion, and thus a limitation of the ions transport between two reservoirs separated by this type of material. Indeed, to ensure the electroneutrality of the reservoirs (zero electrical charge), under a salinity gradient, cations and anions must be transported together from one reservoir toward the other, and thus through the clay. Consequently, the transport restriction applying to the anions also applies to the cations limiting the total ionic transport. Therefore, according to the above definitions, the anionic exclusion caused by the surface charge of the clay particles confers a membrane

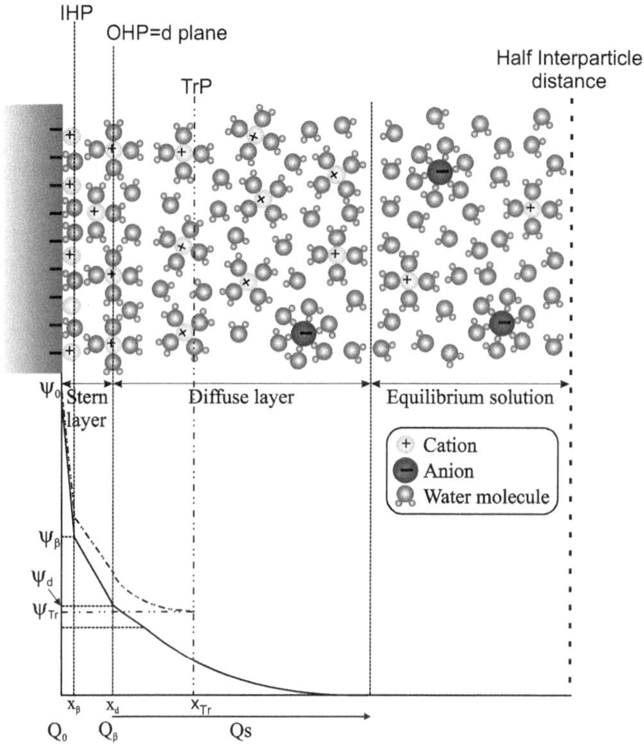

FIGURE 8.1 Schematic representation of the electrochemical interactions at the surface of one clay mineral particle. This figure also shows the evolution of the electric potential as a function of the distance to this surface. The terms IHP and OHP indicate "Inner Helmholtz Plane" and "Outer Helmholtz Plane." Q_0, Q_β, and Q_S ($C\,m^{-2}$) indicate charge densities at the surface of the particle, in the Stern layer, and in the diffuse layer, respectively. ψ_0, ψ_β, and ψ_d (V) indicate the electrical potential at the surface and at the planes β (IHP) and d (OHP), respectively. TrP is the truncation plane situated at a distance x_{Tr} from the particle surface.

behavior on the clay-rocks. However, in such media, ionic transport is only partially restricted. Consequently, clays are often described as imperfect or semipermeable membranes.

The electroosmotic process is also associated with the presence of the cations at the clay mineral particle surface. When a fluid flows due to a pressure gradient, the cations are driven along the flowpath, thus establishing an electrical-potential gradient, the streaming potential. In turn, this electrical-potential gradient creates a backward opposite force for the cations. Therefore, a fluid flow associated with a pressure gradient causes the development of an electrical-potential gradient. According to the laws of irreversible thermodynamics, the reciprocal process, i.e., a fluid flow caused by an electrical-potential gradient and its associated current density, must exist. This fluid flow corresponds to the electroosmotic process.

The only theoretical background for the thermoosmotic flow was presented by Derjaguin et al. (1987). These authors used some concepts of the theory of irreversible thermodynamic processes and considered the conjugate process, i.e., the mechanocaloric effect to justify the existence of thermoosmosis. The interaction between a charged surface and an electrolyte in contact with this surface causes an important change of the water structure in terms of the Hydrogen-bond (H Bond) network. This modification of the structure of the H Bonds produces an alteration in the specific enthalpy of the solution in the porosity of a clay-rock. If H ($J\,m^{-3}$) denotes the specific enthalpy of the equilibrium solution, the specific enthalpy of the water film in the pores is $H+\Delta H$, where ΔH represent "the excess" of specific enthalpy due to surface interactions. An isothermal fluid flow through charged porous media associated with a pressure gradient would thus produce an "excess" of heat flow creating a temperature gradient. This effect is termed the mechanocaloric effect. Nonequilibrium thermodynamics predicts that the conjugate process must exist, i.e., a fluid flow caused by a temperature gradient.

8.2.2.2 Coupled Flux Equations

The thermodynamics of irreversible processes provides the theoretical background for the existence of coupled flows (de Groot and Mazur, 1962; Katchalsky and Curran, 1967; Bolt, 1979). In this framework, the authors express the entropy dissipation function for one unit volume of the system:

$$T\frac{dS_{sp}}{dt}=\sum_{i=1}^{n}J_iX_i, \tag{8.2}$$

where T is the temperature (K), S_{sp} is the specific entropy, J_i and $X_{i=1,n}$ are the flows and the driving thermodynamics forces (such as the pressure gradient ∇p_f). However, in fact Eqn (8.2) relates to the isothermal case (Bolt, 1979). The force X_i is then called the conjugate driving force associated with the flow J_i (e.g., the pressure gradient of fluid flow). The entropy production corresponds to the dissipated energy together with the movement of species (water, solutes). According to the second law of thermodynamics, this production of entropy has to be positive. A straightforward and practical relation between flows and forces respecting the requirement that $\frac{dS_{sp}}{dt}>0$ writes

$$J_i=\sum_{j=1}^{n}L_{ij}X_j, \tag{8.3}$$

where L_{ij} are the so-called phenomenological coefficients. These relations are in line with the empirical laws, for example, Darcy's or Fick's laws, which stipulate that the flows of water and solutes are proportional to their conjugate driving forces, i.e., ∇p_f and the chemical potential of the solution $\nabla \mu_f$, respectively. But, accounting for the complete set of driving forces predicted by thermodynamics leads to more general relations. For instance, Bolt (1979) introduced three driving forces in isothermal conditions, the pressure gradient ∇p_f, the osmotic pressure gradient (related to the chemical potential gradient of the solution $\nabla \mu_f$) $\nabla \pi$, and the electrical-potential gradient $\nabla \psi$. The latter includes the effect of the double electrical layer and the macroscopic disturbance associated with the movement of the cations along the flowpath (streaming potential). The corresponding flows are the specific discharge of fluid (Darcy's law) J_V (noted q hereafter, m s^{-1}), the solute velocity relative to the solvent (m s^{-1}), $J_s^D = \phi / c_f - J_V$ where ϕ is the flow of ions (mole m^{-2} s^{-1}), c_f (mole L^{-1}) is the salinity (or the concentration of the equilibrium solution), and finally the velocity (m s^{-1}) of the cation in excess as compared with the anion $J_\pm$. If the temperature-gradient effect is taken into account, the flow of heat J_Q (J m^{-2} s^{-1}) should be added to the system described by Bolt (1979), and the coupled flow system writes

$$\begin{pmatrix} J_V \\ J_s^D \\ J_\pm \\ J_Q \end{pmatrix} = -\begin{vmatrix} L_{11} & L_{12} & L_{13} & L_{14} \\ L_{21} & L_{22} & L_{23} & L_{24} \\ L_{31} & L_{32} & L_{33} & L_{34} \\ L_{41} & L_{42} & L_{43} & L_{44} \end{vmatrix} \cdot \begin{pmatrix} \nabla p_f \\ \nabla \pi \\ \nabla \psi \\ \dfrac{\nabla T}{T} \end{pmatrix} \tag{8.4}$$

Note that System (8.4) can be written differently to obtain more practical coupled-fluxes equations corresponding to measurements (see Revil and Leroy, 2004; Revil et al., 2011), e.g., ionic flow ϕ or the electrical current density $I = e \times J_\pm$, where e is the elementary charge. In fact, the authors (Katchalsky and Curran, 1967; Bolt, 1979) usually assume isothermal flow in porous media thus simplifying System 4, which is then reduced to three equations. Although elegant, the thermodynamic analysis briefly outlined here does not provide predictive expressions for the coupling coefficients. At this stage, experiments are the only way to obtain the desired coupling coefficients values. Such experiments which consist in establishing exclusive pressure, electrical, or chemical (solute concentration) potential gradients across a sample to measure the corresponding osmotic flow are discussed in Section 8.2.4.

8.2.3 Standard Hydrogeological Approach: Nondiagonal Terms are Assumed Negligible

8.2.3.1 Pressure Diffusion Equation

If all $L_{ij,j\neq i}$ coefficients can be neglected, the classical laws for the transport processes are recovered. For fluid flow, Darcy's law in its classical form Eqn (8.1) applies to porous media and no coupling is considered. Usually, in sedimentary environments, the principal flow directions are X and Y within the stratum and a direction Z perpendicular to the stratum, often close to the vertical axis. The intrinsic permeability tensor expressed in these main directions (assumed to be its principal ones) then becomes:

$$\mathbf{k} = \begin{vmatrix} k_X & 0 & 0 \\ 0 & k_Y & 0 \\ 0 & 0 & k_Z \end{vmatrix}$$

Finally, if the fluid density can be considered constant, the hydraulic head,

$$h = \frac{p_f}{\rho_f g} + z, \tag{8.5}$$

can be used in Eqn (8.1) with the z-axis oriented vertically upward yielding

$$q = -\mathbf{K}\nabla h \tag{8.6}$$

with

$$\mathbf{K} = \frac{\rho_f g \mathbf{k}}{\eta_f}, \tag{8.7}$$

the tensor of hydraulic conductivity (m s^{-1}). Equation (8.1) or Eqn (8.6) is combined with the mass balance equation,

$$\nabla \cdot (\rho_f q) + \frac{\partial(\rho_f \omega)}{\partial t} = 0, \tag{8.8}$$

where ω is the total porosity. Generally, in clays, the fluid is confined and a pressure variation produces a mechanical deformation of the solid skeleton as well as a slight variation of the fluid density. In order to express the second term on the left-hand side of Eqn (8.8) as a function of the pressure, a hydromechanical coupling and equations of state for the liquid and the solid must be introduced. Such a development is described within the framework of the poroelasticity by de Marsily (1986), which gives

$$\nabla \cdot (\rho_f q) = -\frac{S_S}{g} \frac{\partial p_f}{\partial t}, \tag{8.9}$$

where S_S (m^{-1}) is the specific storage coefficient, which takes into account the deformation of the porous medium and the compressibility of the water and of the solid grains:

$$S_S = \rho_f g \omega \left(\beta_l - \beta_s + \frac{\alpha_c}{\omega} \right), \tag{8.10}$$

where β_l and β_s (Pa^{-1}) are the water and solid grain compressibilities, respectively, and α_c (Pa^{-1}) is the bulk compressibility of the porous medium. If temperature variations in the state equation are taken into account,

$$\frac{d\rho_f}{\rho_f} = \beta_l dp_f - \chi_T dT, \tag{8.11}$$

where χ_T is the coefficient of thermal expansion of water (K^{-1}), a second term on the right-hand side of Eqn (8.9) must be added, which leads to

$$\nabla \cdot (\rho_f q) = -\frac{S_S}{g}\frac{\partial p_f}{\partial t} + \rho_f \omega \chi_T \frac{\partial T}{\partial t} \tag{8.12}$$

The second term on the right-hand side of Eqn (8.12) is very often neglected by assuming a steady-state for the heat transport in sedimentary basins. However, the thermal expansion effect has been considered in basin simulations as a possible cause of overpressures. This process was called the aquathermal effect by Luo and Vasseur (1992). In fact, the density of the fluid depends on both the temperature and the salinity. If this influence is great, and if a transient state for heat and solute transport can be suspected, then fluid flow, heat and mass transport are coupled via the dependence of the fluid density on temperature and salinity and the three transport equations should be solved simultaneously. The equations usually used for mass and heat transport in porous media are not described here but can be found in de Marsily (1986). In fact, in the hydrogeological community, the term "couplings" generally refers exclusively to the hydro-chemo-mechanical effects caused by the dependence of the fluid density on the temperature and salinity. If density variations can be neglected, Darcy's law combined with the conservation equation (without source terms) leads to the diffusivity equation, which writes

$$\nabla \cdot (\mathbf{K}\nabla h) = S_S \frac{\partial h}{\partial t}. \tag{8.13}$$

This equation is the fundamental one that is solved on a hydrogeologic domain subject to its boundary conditions. To calculate exactly the flow of water in clay-rich layers, the parameters involved in these two fundamental Eqn (8.9) or Eqn (8.13) must be identified and their possible spatial heterogeneity taken into account. In the following section, the identification of the hydrodynamic properties of clay-rocks is briefly considered, i.e., the hydraulic conductivity and the specific storage coefficient (which is required for transient calculations).

8.2.3.2 Identification of Clay-Rock Hydrodynamic Properties

The hydrodynamic parameters (hydraulic conductivity, specific storage coefficient) can be determined by various methods, which were analyzed by Neuzil (1986) for clay materials. At the sample scale, the hydraulic conductivity is a "by-product" of consolidation experiments with oedometric cells. A mechanical load is applied to a sample causing its consolidation. According to the theory developed by Terzaghi (see de Marsily, 1986), this consolidation is associated with fluid expulsion, which depends on the hydraulic conductivity. The significance of this hydraulic conductivity is however uncertain. If the sample is taken perpendicularly to the stratification, and drained in the vertical direction, the hydraulic conductivity is measured in the direction perpendicular to the main alignment of clay mineral particles (vertical hydraulic conductivity). This alignment is due to the rearrangement of the clay mineral particles perpendicularly to the maximal lithostatic stress during the natural compaction of the formation. This phenomenon is enhanced for clay particles characterized by a flattened shape. Conversely, if the same sample under the same vertical load is drained horizontally, the horizontal hydraulic conductivity is determined. Another method consists in reproducing Darcy's historical experiment at constant pressure difference and thus to use a permeameter. However, large pressure differences and small sample sizes must be used to obtain a flow of water through these low-permeability media. It is often preferable to use a variable hydraulic head apparatus (Hsieh et al., 1981; Neuzil et al., 1981). In such a device, a variation of pressure or hydraulic head is applied on one side of the saturated sample and the pressure evolution is monitored, which makes it possible to identify the hydraulic conductivity and the specific storage coefficient by using the diffusivity or pressure diffusion Eqn (8.9) or Eqn (8.13) in transient state.

At the formation scale, hydraulic pulse tests in closed measurement chambers isolated by packers in boreholes constitute the most reliable experiments to determine the hydrodynamic properties in low-permeability formations. These experiments use an instantaneous pressure variation of the volume of fluid contained in a measurement chamber caused by injection or subtraction of a small volume of fluid. The pressure response in the chamber is then recorded and hydraulic parameters are inferred by inverse methods. In clay media, the low filtration rate can be detected in closed chambers because of the low compressibility of water (or in fact of the chamber). Analytical solutions have been proposed to interpret such hydraulic pulse tests, for example, by Bredehoeft and Papadopoulos (1980) based on the work of Cooper et al. (1967) or more recently the model by Johns (1998) or Gonçalvès (2008). A modification of the solution from Bredehoeft and Papadopoulos (1980) was proposed by Neuzil (1982) to take into account the compressibility of the chamber, which is different from that of the water. Numerical or analytical solutions are used to infer the hydraulic conductivity and the specific storage coefficient, both at sample scale or in situ. But, at small values of the specific storage coefficient, the reliability of the parameters obtained by hydraulic pulses is weak (Bredehoeft and Papadopoulos, 1980; Neuzil, 1986). An abundant literature characterizes the distribution of the hydraulic conductivity of clay-rocks measured at sample scale or in situ. The first synthesis was proposed by Neuzil (1994). Since then, additional data have been collected in the framework of research programs on the feasibility of nuclear-waste storage in geological formations and particularly in clay-rocks. In Europe, several underground research laboratories (URL) have been developed during the last two decades, supplying a considerable quantity of data on clay-rocks (Opalinus clay, in Switzerland; Boom clay, in Belgium; the

Callovian–Oxfordian formation (COx) and the Toarcian formation at Tournemire, in France). These additional data are shown in Figure 8.2.

The dispersion of the data can be attributed to the mineralogy, the fraction of nonclay minerals (quartz, calcite), the chemistry of the fluids in the porosity, and the petrofabric of the materials (Mitchell, 1993; Neuzil, 1994). Two important characteristics of the hydraulic conductivity of clays are often mentioned: the scale effect (hydraulic conductivity increasing with scale) and the anisotropy. Surprisingly, Neuzil (1994) noted that the scale effect could be weaker than expected. This author compared sample-scale hydraulic conductivities to local geological formation scale values (by in situ pulse tests) and to the hydraulic conductivities deduced from calibrated hydrogeological models at regional or intermediate scales. Differences of two to three orders of magnitude are often found between the regional scale (100 km) and the sample scale, smaller differences are observed for the intermediate scales (100 m–10 km). This scale effect is likely to depend on the presence and connectivity of a fracture network in the formation. Consequently, the measurements at the sample scale might be representative of the formation for blocks of few kilometers in size. Recent data illustrating this scale effect

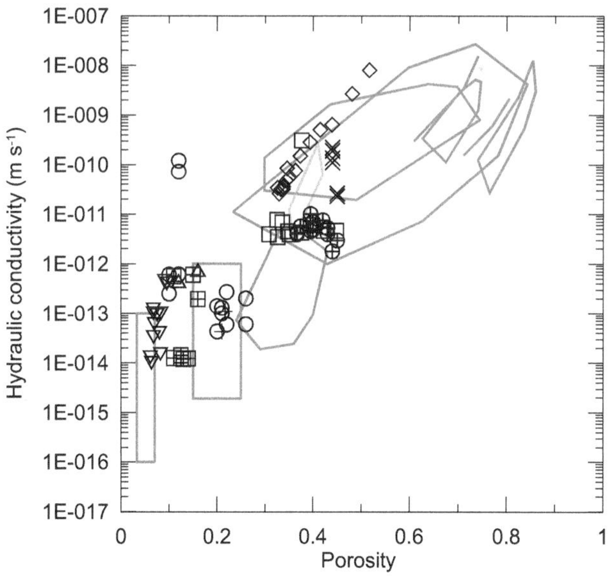

FIGURE 8.2 Hydraulic conductivity **K** (m s^{-1}) and porosity data for different clay-rocks. Callovian–Oxfordian, Bure (⊞) (Delay et al., 2006; Homand et al., 2006), Toarcian, Tournemire (∇) (Boisson et al., 2001), Boom clay (□) (Aertsens et al., 2004), Maquoketa Formation (○) (Hart et al., 2006), Opalinus clay (△) (Marschall et al., 2005; Fernandez-Garcia et al., 2007), ODP Leg (+) (Clennel et al., 1999), ocean floor clays (Yang and Aplin, 2007), London clay and Gault clay (X) (Ratman et al., 2005), and Snakebite formation (⊕) (Shaw and Hendry, 1998). The gray lines correspond to the data used by Neuzil (1994).

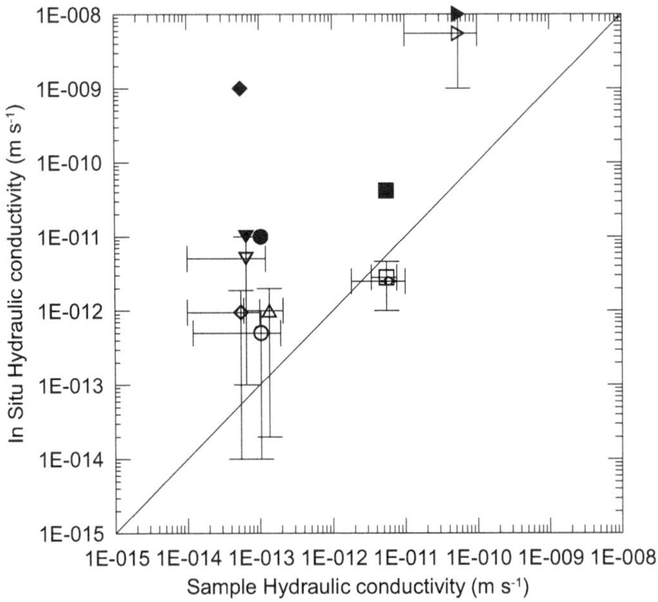

FIGURE 8.3 Scale effect for clay-rocks: Callovian–Oxfordian (○) (Delay et al., 2006; Distinguin and Lavanchy, 2007), Opalinus clay (△) (Croise et al., 2004), Toarcian, Tournemire (▽) (Boisson et al., 2001), Boom clay (□) (Aertsens et al., 2004; Garavito et al., 2007), Pierre Shale (◇) (Neuzil, 1994; Garavito et al., 2007), Lacustrine formation (Mexico) (▷) (Rivera et al., 1991; Vargas and Ortega-Guerrero, 2004), Snakebite formation (○) (Shaw and Hendry, 1998). The same plain symbols are used for the hydraulic conductivity inferred from hydrogeological regional models. A mean value and error bars indicating the dispersion of the data are shown.

(Figure 8.3) shows similar values for hydraulic conductivities measured on samples (0.1 m) and in situ (1–10 m) with the notable exception of two formations, i.e., the Toarcian shales at Tournemire (France) and the lacustrine aquitard of Mexico (Mexico City).

For these two formations, a difference of several orders of magnitude exists for the same two scales. In addition, in situ values almost correspond to values determined at the regional scale by calibration of hydrogeological models (Rivera et al., 1991; Jost et al., 2004). Interestingly, a network of fractures was clearly identified for these two formations (Mexico, Tournemire) thus providing a reasonable explanation for the difference between the values obtained on samples and those obtained in situ. The second major characteristic of these materials is their permeability anisotropy. In clay-rocks, the clay mineral layers can be distributed as scattered layers or more generally in the form of piles of layers called particles. Particles may be grouped into aggregates of particles. An important characteristic of the particles is their flattened shape described by the aspect ratio, i.e., the ratio between the "particle diameter" and its thickness. During formation burial and compaction, particles rearrange themselves perpendicularly to the lithostatic load i.e., in the direction of the stratification. The process increases the tortuosity of the flowlines perpendicular to the stratification, thus decreasing the vertical hydraulic conductivity. This rearrangement is invoked to explain the anisotropy of the hydraulic conductivity. Recent experiments with pure clay minerals and a raw material showed that the anisotropy ratio $a_z = K_X/K_Z$ does not exceed a value of 10 (Clennel et al., 1999). For pure clay minerals, Clennel et al. (1999) measured a maximum anisotropy of 8 for smectites, which are characterized by a high aspect ratio and a minimum anisotropy value of 2 for kaolinites (low aspect ratio). Anisotropy ratios determined on clay-rocks fall mainly between 1 and 10, highly consistent with the measurements on pure materials (Figure 8.4).

This suggests that the anisotropy ratio of several orders of magnitude at the scale of geological formations (thickness of 100 m) can only be explained by mineralogical variations (for example, the smectite content) together with the presence of some weakly permeable beds within a shale. Indeed, the harmonic mean classically used to obtain an equivalent vertical hydraulic conductivity for the whole thickness is controlled by the lowest value (de Marsily, 1986).

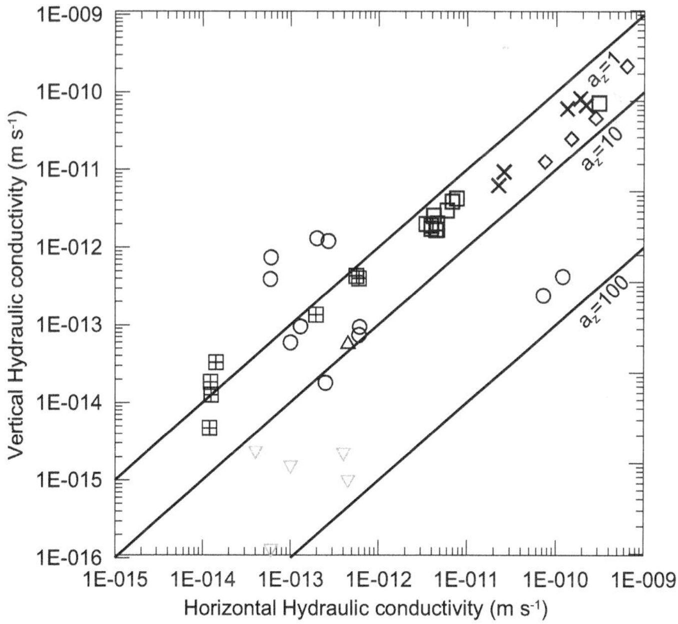

FIGURE 8.4 Hydraulic conductivity anisotropy for shales: Callovian–Oxfordian (⊞) (Delay et al., 2006; Homand et al., 2006), Toarcian, Tournemire (▽) (Boisson et al., 2001), Boom clay (□) (Aertsens et al., 2004), Opalinus clay (△) (Marschall et al., 2005), Maquoketa formation (○) (Hart et al., 2006), ODP Leg (◇) (Clennel et al., 1999), London clay and Gault clay (X) (Ratman et al., 2005).

Transient flow problems in porous media require that the second fundamental variable in hydrogeology, i.e., the specific storage coefficient, be determined. A theoretical expression for this coefficient found in the general framework of the elasticity is discussed by de Marsily (1986). This development is based on Terzaghi's consolidation theory, which relates the volume change of the porous medium to the effective stress:

$$\boldsymbol{\sigma}' = \boldsymbol{\sigma} - \gamma_B p_f \mathbf{1}, \tag{8.14}$$

where $\boldsymbol{\sigma}'$ and $\boldsymbol{\sigma}$ are the tensors of effective and total stress (Pa), respectively, and γ_B is the non-dimensional Biot coefficient (see Cosenza et al., 2002). In the development described by de Marsily (1986), this coefficient is considered constant and equal to 1, whereas for clay-rock formations its value may be much lower than 1 according to Cosenza et al. (2002). The decrease of Biot's coefficient with the total porosity for various materials (Giraud and Rousset, 1996; Cosenza et al., 2002) is shown in Figure 8.5.

Empirical relations between Biot's coefficient and the total porosity were proposed such that of Krief et al. (1990) represented in Figure 8.5. Within the framework of small deformations, i.e., of low porosity variations during the flow, a constant value for Biot's coefficient is considered although it is related

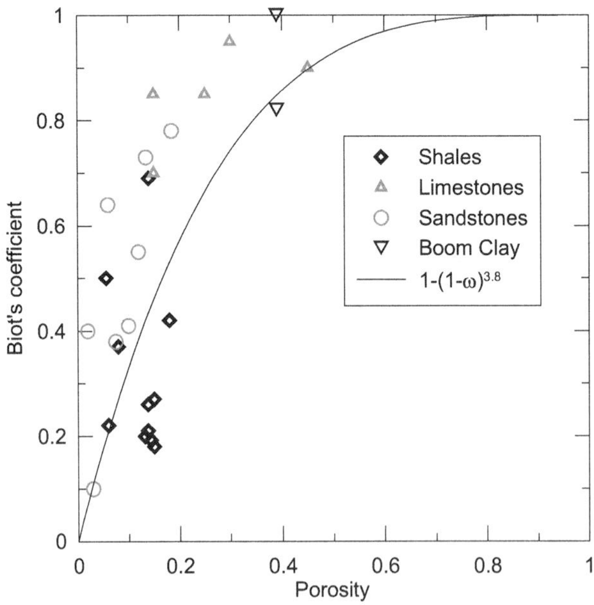

FIGURE 8.5 Biot's coefficient as a function of the total porosity for clay-rocks, carbonates, and sandstones after Cosenza et al. (2002), and the Boom clay (Giraud and Rousset, 1996). The model depicted in the figure is by Krief et al. (1990).

to the porosity. The introduction of this coefficient into the development proposed by de Marsily (1986) leads to

$$S_S = \rho_f g \omega \left(\beta_l - \beta_S + \frac{\gamma_B}{\omega} \frac{3(1-2\nu)}{E_Y} \right), \tag{8.15}$$

where coefficient α_c in Eqn (8.10) is replaced by its theoretical expression ($\alpha_c = 3(1-2\nu)/E_Y$ (de Marsily, 1986)), ν and E_Y are the nondimensional Poisson coefficient and the Young modulus (Pa), respectively. Similarly to Biot's coefficient, the Young modulus depends on the total porosity. This coefficient decreases when the porosity increases (see Munro, 2004). According to Munro (2004), the Young modulus can be described by the following relation:

$$E_Y = E_Y^0 (1-\omega)^n, \tag{8.16}$$

where E_Y^0 (Pa) and n (dimensionless) are two parameters to be optimized. By using some measurements for shales available in the literature (Giraud and Rousset, 1996; Cosenza et al., 2002; Benhamidaa et al., 2005), values of 10 GPa for E_Y^0 and 8 for n can be identified. These values are used in Eqn (8.16), which is introduced into Eqn (8.15). As the Poisson coefficient is weakly dependent on porosity, an average value of 0.3 is used. With these expressions, it is possible to reproduce quite satisfactorily the available data for the specific storage coefficients as shown in Figure 8.6.

Note that, only the hydromechanical couplings were considered here. The well-known swelling properties of clay minerals (particularly for smectites), which require a hydro-chemo-mechanic development have, so far, not been taken into account. The coupling involving the concept of disjoining or swelling pressure will be discussed in Section 8.4.

8.2.4 When Off-Diagonal Terms Are Not Negligible: Formalism and Experimental Evidence

In clay media, the assumption of negligible non-diagonal coefficients for the fluid flow can be highly questionable unless it is based on coupling coefficients estimates. Especially in the 1970s, numerous historical experiments pointing to, among others, the existence of chemical osmosis in clays were reported (see data collected by Bresler (1973)). These data consisted mostly of remolded and thus nonnatural materials. But, the in situ study in the Pierre Shale by Neuzil (2000) definitely confirmed the existence of such couplings in natural geological media. However, when these couplings are taken into account, two simplifying assumptions are made for natural environments such as clay-rocks (Bolt, 1979): isothermal fluxes and zero macroscopic current density $I = J_{\pm} \times e$ (see Eqn (8.4)) (Bolt, 1979; Revil et al., 2005). The first assumption makes it possible to eliminate one flux equation and one driving force (∇T). The second assumption enables simplifying the system of coupled

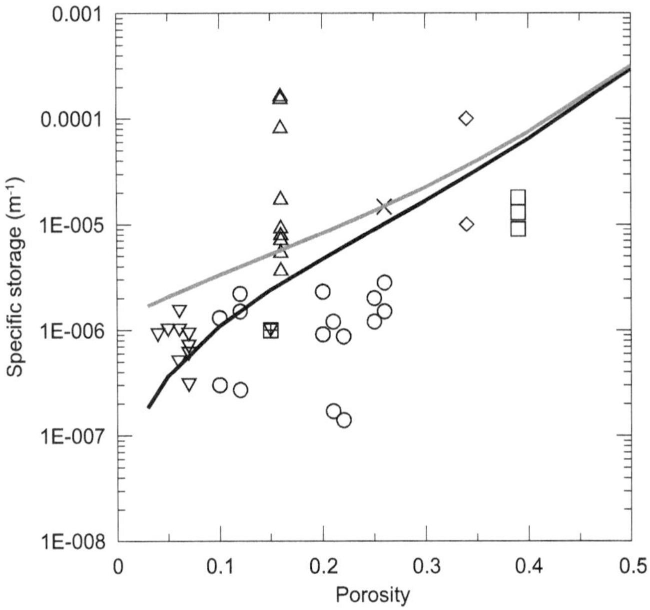

FIGURE 8.6 Specific storage coefficient as a function of the total porosity for different clay-rocks (same symbols as in Figure 8.3). The black curve represents the values calculated from Eqn (8.15) by using a Biot coefficient γ_B depending on the porosity and the gray curve represents the same model but with $\gamma_B = 1$. In these two calculations, a Poisson coefficient of 0.3 and a Young modulus depending on porosity (see text) are considered.

flow equations. Electrokinetic effects (electrical potential) are apparently absent from most coupled flow applications. But, in fact they are intrinsically part of the hydraulic and chemoosmotic parameters, which are thus apparent and not absolute. This was shown experimentally by Heister et al. (2005) through chemical osmosis experiments in short-circuited ($\nabla\psi = 0$) and non-short-circuited conditions. These authors obtained slightly different values for the chemical osmotic coefficient. Consequently, only the experiments in short-circuited conditions (the reservoirs on each side of the sample are connected by a conductor, $\nabla\psi = 0$) provide the absolute hydraulic and osmotic conductivities. But, chemical osmotic studies, for instance, are generally conducted in nonshort-circuiting conditions, suggesting that most of the available data are apparent values. This electrokinetic effect was already discussed by Olsen (1960) as early as the 1960s and later by Bolt (1979). On the basis of measurements and theoretical calculations, these authors claimed that electrokinetic effects are fairly small for the chemical osmotic efficiency and the hydraulic conductivity and show very similar absolute and apparent values. More recently, Revil and Pessel (2002) and Revil et al. (2005) proposed some theoretical expressions for the apparent hydraulic conductivity with the assumption of a null macroscopic current density and discussed the fact that most measured values are apparent ones. Basically, with the previously introduced general formulation (Eqn (8.4)), a relation between the electrical-potential gradient and the other gradients can be obtained with the second hypothesis:

$$\nabla\psi = \frac{L_{31}}{L_{33}}\nabla p_f + \frac{L_{32}}{L_{33}}\nabla\pi, \tag{8.17}$$

which can be introduced into the remaining transport equations (fluid, solutes, see Eqn (8.4)). Under the above-mentioned simplifying assumptions, after some algebraic processing not detailed here (see Katchalsky and Curran, 1967; Revil and Leroy, 2004; Revil et al., 2005, 2011), and by introducing a more familiar formulation, the system of coupled equations is conveniently reduced to

$$q = -\frac{\mathbf{k}}{\eta_f}\left(\nabla p_f + \rho_f g \nabla z\right) + \frac{\mathbf{k}}{\eta_f}\varepsilon_c \nabla\pi \tag{8.18}$$

$$\phi = (1-\varepsilon_c) q c_f - D_{eff}\nabla c_f \tag{8.19}$$

where ε_c is the osmotic efficiency coefficient (dimensionless), D_{eff} is the effective diffusion coefficient (see discussion below), ϕ is the flow of solutes in mole m^{-2} s^{-1} and with

$$\nabla\pi = \nu_f RT \nabla c_f \tag{8.20}$$

which arises from the van't Hoff relation and where ν_f is the number of separate ions (2 for NaCl), R is the gas constant (8.32 m^3 Pa K^{-1} mol^{-1}). For transient-state calculations, Eqns (8.18) and (8.19) are combined with mass conservation equations. The generalized use of the isothermal and zero (negligible) macroscopic electrical density current assumptions together with the study by Neuzil (2000) explain why over the last decade most experimental and numerical work in natural clay-rocks had considered the topic of chemical osmosis.

The effective diffusion coefficient D_{eff} (m^2 s^{-1}) in Eqn (8.19) corresponds to $\omega_{eff} D_p$ with D_p, the diffusion coefficient of the porous media including the effect of the tortuosity (de Marsily, 1986) and ω_{eff}, the effective or accessible porosity. The importance of effective porosity and its evaluation are analyzed in Chapter 6 in this volume. In Eqn (8.19), the kinematic dispersion classically introduced in hydrogeology is not considered. But it is generally admitted that mass transport is dominated by molecular diffusion in low-permeability formations such as shales (Cey et al., 2001; Patriarche et al., 2004). Here, essentially the osmotic coefficients are discussed, but some measurements and theoretical expressions for the effective diffusion coefficient D_{eff} appearing in Eqn (8.19) have been presented in the literature (see Revil and Leroy, 2004). The effective diffusion coefficient depends on the electrochemical interactions between the porous-media solution and the clay mineral surfaces (see Revil et al., 2005).

The osmotic efficiency ε_c appearing in Eqns (8.18) and (8.19) characterizes the membrane behavior of clay. It is defined as the ratio L_{21}/L_{11} (Bolt, 1979). As stated in Section 8.2.2.1, the membrane properties of clay material stems from the negative charge of the clay mineral particle surfaces at natural pH, which causes some anionic exclusion. In the case of a perfect membrane, $\varepsilon_c = 1$ and the effective diffusion coefficient is negligible or null producing almost zero ionic transport (see Eqn (8.19)). More generally, the osmotic efficiency coefficient takes a value between 0 and 1, explaining the semipermeable behavior of natural media. In the framework of the two above-mentioned simplifying assumptions, apparent hydraulic conductivity and chemical osmotic efficiency, and effective diffusion coefficient are the only parameters to identify. Focusing here on the chemical osmotic efficiency, this coefficient as well as the other hydrochemical ones can be determined experimentally. Different experimental devices to estimate the osmotic efficiency are proposed in the literature. The basic protocol is to impose a concentration difference across a core sample and to measure an osmotic flow. These experiments are generally performed on remolded samples of clay-rocks (Kemper and Rollins, 1966; Keijzer and Loch, 1999; Malusis and Shackelford, 2002). Few studies concern undisturbed samples of natural clay-rocks (Young and Low, 1965; Cey et al., 2001; Horseman et al., 2007). Furthermore, almost all the published experiments were performed on thin disk-shaped samples (often no more than 1 cm thick (Kemper and Rollins, 1966; Keijzer et al., 2001; Horseman et al., 2007)). Generally, the confinement of the samples is obtained by using a piston (anisotropic confinement), to prevent it from swelling. The studies by Keijzer and Loch (1999), Keijzer et al. (2001), and Horseman et al. (2007) performed on disk-shaped samples were carried out under isotropic confinement. Generally, steady-state horizontal flow is interpreted. A concentration difference is maintained constant between two reservoirs one on each side of a sample. A pressure rise is obtained in the high-concentration reservoir corresponding to a chemical osmotic fluid flow from the low- to the high-concentration zone. When the pressure stabilizes, it is possible to write $\varepsilon_c = \Delta p_f/\Delta\pi$, where Δp_f is the pressure difference measured between the two reservoirs at equilibrium; $\Delta\pi$ is the imposed osmotic pressure difference, which can be approximated with van't Hoff formula. Alternatively, transient experiments make possible the identification of the full set of parameters (k, ε_c, D_{eff}). During such experiments, the concentration difference is not maintained constant. After the pressure rise, the concentration difference decreases due to ionic diffusion, and a return fluid flow from the high- to the low-concentration reservoir creates a long-lasting pressure decrease (controlled by the diffusion). The interpretation is more complex and requires the solution of Eqns (8.18) and (8.19), combined with mass conservation equations (see Rousseau-Gueutin et al., 2009). Recently, an alternative experimental device limiting the fluid leaks was developed to improve the accuracy of osmotic efficiency measurements (Rousseau-Gueutin et al., 2009). A cylindrical sample several centimeters in diameter and height was placed under isotropic confinement. This allows an accurate estimate of the osmotic coefficient for undisturbed samples, under experimental conditions reproducing exactly, on a small sample, those of an in situ chemical pulse test (see below). During such tests, the solution (formation pore fluid) contained in a small isolated chamber located in the center of the sample is replaced instantaneously by a solution at a different concentration. Then the pressure response caused by an incoming or a diverging osmotic fluid flow toward or from the chamber is measured and interpreted. This apparatus was used to measure the osmotic efficiency of the COx formation located at the French URL at Bure. Recently, another alternative experimental apparatus based on the chemomechanical processes was developed (Sarout and Detournay, 2011). A clay sample equilibrated with a salt solution is placed in a triaxial cell, then the fluid in the cell is changed for a higher concentration one. The sample deformation associated with chemical–mechanical coupling is monitored and the osmotic coefficient is obtained by inversion.

In addition to sample-scale experiments, in situ tests were performed following the same protocol as the pioneering experiment reported by Neuzil (2000). The fluid in contact with a clay-rock in an isolated chamber limited by packers in a borehole is replaced by a solution with a different concentration from that of the geological formation. The pressure evolution in the chamber is recorded in a similar way as in hydraulic pulse tests. This method was discussed by Gonçalvès (2008) who proposes an analytical solution for the interpretation of such chemical pulse tests. The replacement of the chamber fluid by a fluid of higher concentration causes a convergent fluid flow toward the chamber and thus a pressure increase. The subsequent pressure disturbance dissipation is controlled by the ionic diffusion from the chamber. This pressure signal depends in fact on all the parameters in Eqns (8.18) and (8.19), but also on the volume and the compressibility of the chamber. By inversion of this signal, the hydraulic and chemical coupling coefficients can be identified. The chemical osmotic coefficients of the COx clay-rock were determined in this way (Rousseau-Gueutin et al., 2010). In situ results are very close to the values obtained by on samples with the same experimental geometry underlining the microscopic control of chemical osmosis. The values obtained for the COx as well as a collection of available values for other natural clay formations are shown in Figure 8.7 as a function of the product $b\sqrt{c_f}$ similarly to Bresler's abacus (Bresler, 1973).

Theoretical expressions of the osmotic efficiency show a strong dependence of this coefficient on the half-pore size b and on the salinity c_f (see Section 8.3.1 and discussion by Neuzil and Provost (2009)). The effective pore size, i.e., the average value representative of macroscopic volume of porous medium, was estimated by using the simple mass balance equation given by Neuzil (2000) valid in a plane-parallel conceptual model for the porous media and which is written as

$$\omega = (1 - \omega)A_S\rho_S b, \tag{8.21}$$

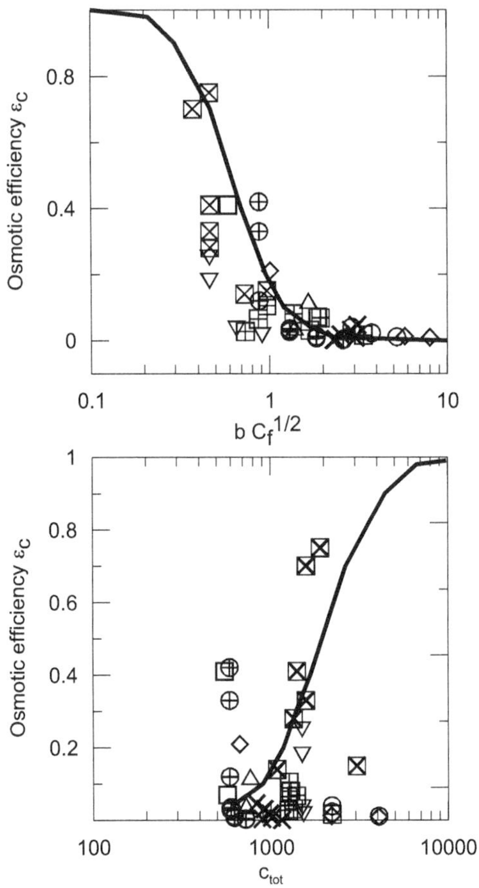

FIGURE 8.7 Osmotic efficiency data as a function of $b\sqrt{(c_f)}$ where b is in nm and c_f in mol L^{-1} and Bresler's curve (top) and c_{tot} with the transposed Bresler's curve (bottom) measured for different clay-rocks. Callovian–Oxfordian, Bure (⊞) (Rousseau-Gueutin et al., 2009, 2010), Toarcian, Tournemire (▽) (Tremosa, 2010), Boom clay (□) (Garavito et al., 2007), "C1" and China Arco shales (○) (Al-Bazali, 2005), Opalinus clay (△) (Noy et al., 2004; Horseman et al., 2007), Pierre Shale (◇) (Al-Bazali, 2005; Garavito et al., 2006), Wakkanai formation (×) (Takeda et al., 2013), Bearpaw formation (⊕) (Cey et al., 2001), and Bulldog, North sea, Pierre II, Muderong shales (⊠) (Bunger et al., 2014). The maximum concentration was used since osmotic efficiency decreases with salinity, the lower value controls the measured osmotic flux. Half-pore sizes were calculated using reference SSA and effective porosity (see references in Gonçalvès et al., 2012) except for C1 and Arco shales where the experimental pore size was used directly. Porosities of 0.37, 0.12, 0.29, and 0.2 for Bulldog, North sea, Pierre II, Muderong shales, respectively, were assumed based on the references and the SSA listed in Bunger et al. (2014).

where A_S is the specific surface area (SSA, $m^2\,kg^{-1}$), ρ_S is the density of the solid ($kg\,m^{-3}$). The significance of this pore size should be carefully analyzed. Indeed, only pores efficient in conducting fluid flow, i.e., those in contact with the external surfaces of clay aggregates, should be considered for hydrodynamic calculations. This excludes the interlayer space, i.e., the pores between clay mineral layers (interlayer distance as short as 1 nm or less) where fluid is almost immobile (Mitchell, 1993). Consequently, only petrophysical properties related to external surfaces should be used. From this standpoint, a porosity measurement by mercury porosimetry, which does not explore pores with less than 2–3 nm, satisfies the above-mentioned constraint. The use of "external" SSA (measured e.g., using nitrogen adsorption) and porosity values in Eqn (8.21) provides an estimate of the required representative pore size $2b$, which is in contact with the external surfaces. A pore size $2b$ of 8 nm was considered for the Bearpaw formation. For this formation, the same value as that of Pierre Shale's clay-rock was considered because the Bearpaw is stratigraphically equivalent to the latter and shows the same hydraulic conductivity. A relative agreement is shown between Bresler's curve and the data obtained from in situ or sample tests for clays (Figure 8.7). However, this curve seems to overestimate the osmotic efficiency for natural materials, but the uncertainty on the pore size could partly explain the observed discrepancy.

In order to facilitate the comparisons with the other osmotic effects and especially thermoosmosis (see below), the data are also shown as a function of the total ionic content c_{tot} in mole m^{-3} (anions and cations in the Stern and diffuse layers). c_{tot} is obtained using a simple Donnan equilibrium model (Mitchell, 1993) $c_{tot} = (Q_V^2/4Na^2e^2 + c_f^2)^{1/2}$, where Na is Avogadro's number and e the elementary charge, Q_V ($C\,m^{-3}$) is the macroscopic volumetric excess charge due to the surface charge and is obtained from petrophysical parameters since $Q_V = (1 - \omega)\rho_s CEC \times 96.3/\omega = CEC \times 96.3/A_s b$, where CEC is the cation exchange capacity. Note that more than 90% of the ionic content is located in the Stern layer (Revil and Leroy, 2004). The variable c_{tot} contains all the control parameters for the coupling coefficients: geometry (ω), surface charge (CEC), and salinity (c_f) and was previously used to describe thermoosmotic data (Gonçalvès et al., 2012). To translate Bresler's curve into the (ε_c, c_{tot}) space, $b\sqrt{c_f}$ values were converted into c_{tot} using a clear correlation between these two variables obtained by Monte Carlo simulations based on plausible values for ω, CEC, c_f, and a relation between A_S and CEC (Neuzil and Provost, 2009).

Whether or not the general assumption $I = 0$ is used (conservation equation for I if not), the explicit consideration of the electroosmosis requires the introduction of four additional processes and their relevant parameters: Rouss, Ohm, diffusion current, electroosmosis. Only the latter is discussed here (see Paszkuta et al., 2006; Rosanne et al., 2006 for the remaining coefficients). The introduction for electroosmosis of a similar formalism as that for chemical osmosis results in

$$q = -\frac{\mathbf{k}}{\eta_f}\left(\nabla p_f + \rho_f g \nabla z\right) + \frac{\mathbf{k}}{\eta_f}\varepsilon_c \nabla \pi - \frac{\mathbf{k}}{\eta_f}\varepsilon_\psi \nabla \psi \tag{8.22}$$

Some experimental ε_ψ data were collected based on measurements of the fluid flow, caused by an electric potential gradient in isothermal and isochemical conditions. These data concern a natural indurated clay-rock (COx, France) and remolded materials (COx, montmorillonite, and mica). In order to describe the behavior at low electrical surface charge, some data for kaolinite were also added. As for chemical osmosis, the data are presented in Figure 8.8 as a function of c_{tot} to facilitate flow comparisons.

Most of the researchers' attention has been focused on chemical and electroosmosis but, what can be said about thermoosmosis? A formulation similar to that of chemical osmosis writes

$$q = -\frac{k}{\eta_f}\left(\nabla p_f + \rho_f g \nabla z\right) + \frac{k}{\eta_f}\varepsilon_c \nabla \pi - \frac{k}{\eta_f}\varepsilon_T \nabla T \tag{8.23}$$

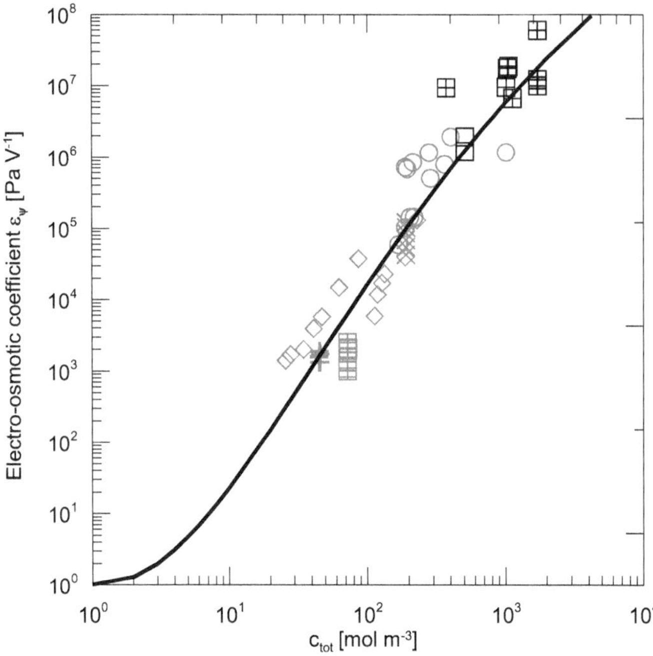

FIGURE 8.8 Electroosmotic coefficient (ε_ψ) as a function of c_{tot}. Callovian–Oxfordian (⊞) (Revil et al., 2005), Boom clay (□) (Rosales et al., 2014). In gray symbols, remolded materials: Callovian–Oxfordian (⊞), mica (+), and montmorillonite (X) (Rosanne et al., 2006); kaolinite (◇), and illite (○) (Olsen, 1960).

Although for clay materials, a case of flow directed from the cold zone toward a warmer zone was described by Dirksen (1969), most experiments show a flow toward the lower temperature zones ($\varepsilon_T > 0$). In a recent study Gonçalvès et al. (2012), the available data for $k_T = \varepsilon_T k$ were collected, and an interesting relation between ε_T (Pa K^{-1}) and the half-pore size b was identified. This is shown in Figure 8.9 reproduced from Gonçalvès et al. (2012) (see this reference for more details). The pore size is calculated with Eqn (8.21). Due to the scarcity of data, some values for pure clays were considered. When the porosity of the materials was not indicated, relations between hydraulic conductivity and porosity of the pure materials were used. For the COx clay-rock, the estimate was based on a remolded material with a porosity of 0.3 while the

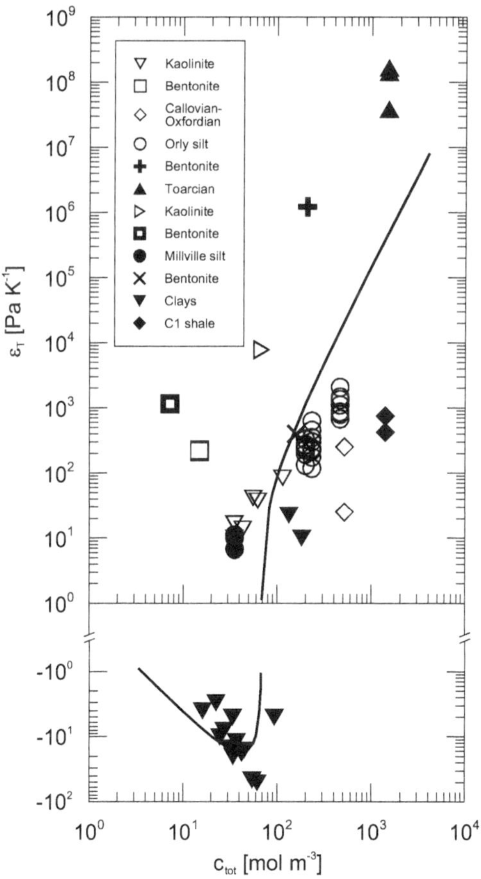

FIGURE 8.9 Relation between measured $k_T/k = \varepsilon_T$ and c_{tot} (see text). *(Reproduced from Gonçalvès et al. (2012) (see this reference, Table 1, for material properties).)* The dark line is the best fit on the data.

natural material porosity is 0.15. The only known in situ measurements on a natural material were conducted in the Toarcian clay-rock at Tournemire (Tremosa et al., 2010).

The variable ε_T increases strongly when the pore size drops to a few nm. This observation is consistent with the interpretation of the thermoosmotic process, which is described by Derjaguin et al. (1987), i.e., an effect of the change of the structure of the water in the interfacial zones. Some layers of water whose structure is the most strongly modified are near the clay surfaces. The smaller the pore size, the larger is the proportion of these altered layers of water. Another variable which could influence the value of the thermoosmotic permeability is the temperature. Only the study by Dirksen (1969) includes a semiquantitative discussion of this influence. This author made a sensitivity analysis on the average temperature during the experimental determination of k_T for a sodium bentonite. The moderate values of ε_T for this material are almost constant when the temperature increases, suggesting that this quantity varies little with the temperature. The present simple analysis, based on a few available data, suggests that the pore size could be a fundamental descriptive variable for the thermoosmotic conductivity. In addition, the diffusive heat transport (Fourier's law of thermal conductivity) may be weakly affected by electrochemical effects since the solid phase accounts for most of the bulk thermal conductivity of the porous media (de Marsily, 1986; Tremosa et al., 2010).

8.2.5 Importance of Coupled Terms in Transport Equations

The relative importance of osmotic flows in natural conditions should be analyzed first. Consider the following plausible values for the pressure difference $\Delta p_f = 10^5$ Pa, the salinity difference $\Delta c_f = 1\ \mathrm{g\,L^{-1}}$ of NaCl, the electrical-potential difference $\nabla\psi = 1$ mV, and the temperature difference $\Delta T = 1$ K across a clay stratigraphic layer. These values, the absolute values of the fluxes $q_p = -\frac{k}{\eta}\nabla p_f$, $q_\pi = \frac{k}{\eta}\varepsilon_c\nabla\pi$, $q_\psi = -\frac{k}{\eta}\varepsilon_\psi\nabla\psi$, and $q_T = -\frac{k}{\eta}\varepsilon_T\nabla T$, and the curves approximately fitting the coupling coefficients data (ε_c, ε_ψ, ε_T) in Figures 8.7–8.9, are used to calculate the ratios q_π/q_p, q_ψ/q_p, and q_T/q_p as functions of c_{tot}. The results are shown in Figure 8.10.

Despite the uncertainty inherent in the data dispersion, the calculations suggest that in isothermal and natural conditions, chemical osmosis could be the primary osmotic flow (twice the electroosmotic flow) while under non-isothermal conditions, thermoosmosis might play a prominent role. The first conclusion supports the use of the simple coupled flow system (Eqns (8.18) and (8.19)) in isothermal conditions. The second conclusion was recently discussed in terms of pressure perturbation within a clay-rock formation and thus as a potential cause of abnormal fluid pressures (Gonçalvès et al., 2012).

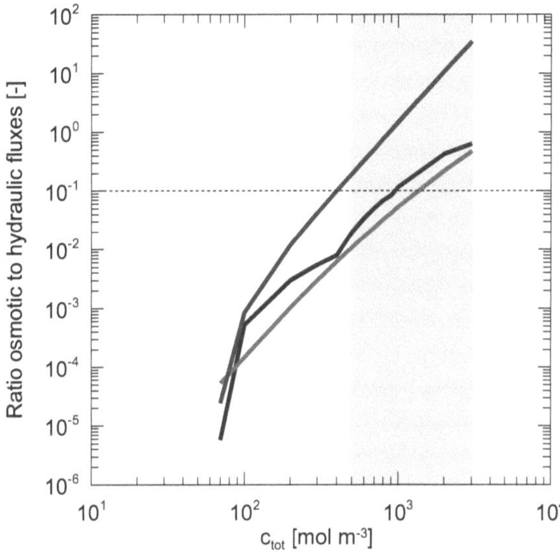

FIGURE 8.10 Calculated ratios of the unit osmotic flows to the hydraulic (purely Darcian flow): red (gray in print versions) line for thermoosmosis, blue (dark gray in print versions) line for chemical osmosis, and green (light gray in print versions) line for electroosmosis. In gray, the area corresponding to the most natural conditions for c_{tot}.

The possible importance of coupled flow phenomena is crucial in the interpretation of the pressure or hydraulic head profiles, which are now currently measured in clay formations, especially in URL. Indeed, the interpretation of pressure anomalies in clay formations is still being debated (Neuzil, 2013). If the transport properties of these materials are clearly identified, the profiles can be properly and accurately interpreted. Whether or not, the couplings (nondiagonal terms) are taken into account leads to divergent interpretations of the pressure or hydraulic head profiles, which can be measured. This is illustrated in Figure 8.11. An idealized case is considered here, i.e., that of a clay formation 100 m thick with a linear salinity profile (0–5 g L^{-1} of NaCl from the top at $z = 100$ m toward the bottom of the formation at $z = 0$ m in Figure 8.11, a constant hydraulic head at the limits and a constant hydraulic conductivity $K = 10^{-13}$ m s^{-1}. Here, consider a formulation of Darcy's law with the hydraulic head, which includes a chemical osmosis term (see Gonçalvès et al. (2004)):

$$q = \underbrace{-K\frac{\partial h}{\partial z}}_{q_h} + \underbrace{\frac{\varepsilon_c K \nu RT}{\rho_f g}\frac{\partial c_f}{\partial z}}_{q_\pi} \tag{8.24}$$

In the hydrogeological interpretations presented below whose aim is to determine the leakage flow direction, the hydraulics term q_h is systematically

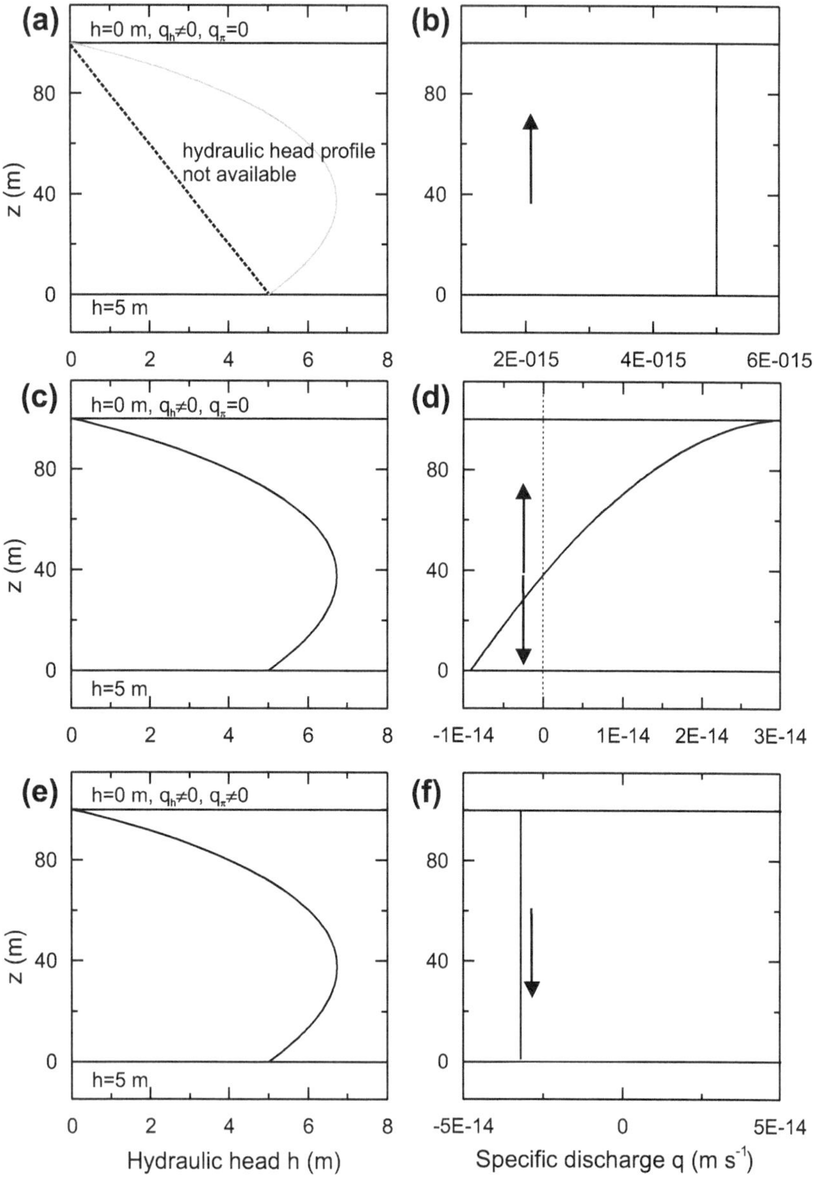

FIGURE 8.11 Various interpretations of a hydraulic head profile. The left column indicates the available information for the hydrogeological interpretation, i.e., the knowledge of the field pressure in the clay or not (figure on the top) and consideration or not of the osmotic term ($q_\pi \neq 0$ or $q_\pi = 0$). The right column indicates the resulting velocity and the leakage flow.

taken into account, which is not the case for the chemical osmosis term q_π. The nonlinearity of Eqn (8.24) should be noticed because the osmotic efficiency ε_c depends on the concentration c_f. The left column of Figure 8.11 shows the available information within a layer of a clay formation and the type of interpretation considered (absence or presence of a chemical osmotic coupling). The right column represents the velocity fields and the direction of the interpreted leakage. Three scenarios are represented:

- The hydraulic head profile is not available, only the hydraulic heads on both boundaries of the formation are accessible (Figure 8.11(a)). Very often in such a case, the hydrogeologist determines Darcy's flow by using only the term q_h. Therefore, an identified leakage flow is represented by the arrow Figure 8.11(b). This interpretation corresponds to a steady-state represented by the linear hydraulic head profile, indicated by the dotted line in Figure 8.11(a).
- The hydraulic head profile is available and osmotic flows are neglected (Figure 8.11(c)). In this case, the interpretation indicates divergent flows toward the boundaries of the formation (Figure 8.11(d)). It is important to note that the resulting divergent flows correspond to a transient regime.
- The same hydraulic head profile can be interpreted with Eqn (8.24). This option is detailed below.

In fact the hydraulic head profile presented in the left column of Figure 8.11 is obtained by a generalized Darcy's law (Eqn (8.24)), the dependence of the osmotic efficiency on the salinity given by Bresler's curve (Figure 8.7) and a pore size $2b = 4$ nm. This profile corresponds in fact to a steady-state (q constant) and the parabolic shape of the hydraulic head profile is associated with the nonlinearity of ε_c. Consequently, this curve could also have been interpreted as the result of an osmotic process at steady-state causing a downward vertical leakage (toward the high-salinity zone). This simple example illustrates the importance, at least locally, of the identification of the coupling coefficients for the interpretation of the leakage directions, for example, in research programs on nuclear-waste storage. This example was also discussed by Gonçalvès et al. (2012) including the thermoosmotic effect for the COx formation.

Based on a set of experimental data on chemical osmosis collected by Neuzil and Provost (2009) for different clay-rocks, large osmotically induced overpressures of up to tens of MPa were predicted by these authors. However, according to them, such large overpressures have not been found in sedimentary basins. Two reasons explaining this disparity between predictions and observations were proposed and analyzed by Tremosa et al. (2012). Accordingly, two natural causes for lower than expected osmotic pressure were investigated: the effect of the complex composition of natural water, including both monovalent and divalent cations, on the osmotic efficiency and the resulting abnormal pressures and the presence of steady-state rather than transient-state salinity distributions. The latter was exemplified above showing weak and persistent excess-pressure (or excess-head) profiles. Divalent cations such as Ca^{2+} are naturally present in the clay-rock pore water. To account for their presence, an electrical triple-layer model accounting for multiionic solutions was developed and used to calculate the osmotic efficiency with different proportions of monovalent and divalent cations (Tremosa et al., 2012). The calculated decrease of the osmotic efficiency when Ca^{2+} is introduced in a Na^{+} clay mineral system produces a noticeable decrease in the ability of the clay-rocks to generate overpressures even at low divalent content (Tremosa et al., 2012).

Another influence should also be discussed here: the importance of osmotic flows for mass transport. This is an important concern especially in the context of a safety analysis of a nuclear-waste repository in a clay-rock formation. The influence of convection in mass transport can be identified by calculating the Peclet number ($q \times L/D_{eff}$, L the characteristic length of the system). When considering only chemical osmosis, Peclet numbers are generally low (see Rousseau-Gueutin et al. (2010), Tremosa et al. (2012)), pointing to a mass transport still dominated by molecular diffusion. Only in the case of strong thermoosmotic coefficients can high values of the Peclet number be obtained as discussed by Tremosa et al. (2012). However, this conclusion should be carefully considered due to the large dispersion and uncertainty on thermoosmotic coefficients. But, in the context of a nuclear-waste storage site with great heat dissipation, the existence of this effect has to be ascertained.

8.3 PREDICTIVE MODELS FOR HYDRODYNAMICAL COUPLING TERMS USING CONTINUOUS AND/OR GRANULAR MEDIA PHYSICS

8.3.1 Chemical Osmotic Models

Extending further the thermodynamic analysis outlined in Section 8.2.2.2, some authors, such as Katchalsky and Curran (1967) or Bolt (1979), write molecular-scale mechanical equilibrium equations using average species concentrations in the charged porous medium. Assuming a mechanical equilibrium between friction and driving forces and applying Onsager's reciprocity principle ($L_{ij} = L_{ji}$) provides analytical expressions for the coupling coefficients, and theoretical expressions for the "osmotic type" forces acting at distance, to be used in Navier–Stokes calculations. For instance, the driving force for chemical osmotic flow writes

$$F_\pi = -\left(1 - \frac{\langle c^- \rangle}{c_f}\right) \nabla \pi, \tag{8.25}$$

where the term $\left(1 - \frac{\langle c^- \rangle}{c_f}\right)$ is called the exclusion factor, and the symbol <> indicates the average value. Note that the local c^- varies perpendicularly to a charged particle, i.e., to the flow direction driven by the osmotic pressure gradient. According to Bolt (1979), by considering average values of the concentrations in the porosity, these first expressions could be crude approximations. Alternatively, such coupling parameters expressions can be identified by solving the Navier–Stokes equation using the previously identified external forces (e.g., Eqn (8.25) with the local c^- and no more $<c^->$). Therefore, for a simple salt (e.g., NaCl), the chemical osmotic coupling term (using a parallel-plane geometrical conceptual model for the porous medium) is given by

$$L_{12} = \frac{-k}{\eta} \frac{\int_0^b (1 - c^-/c_f)(bx - x^2/2)dx}{\int_0^b (bx - x^2/2)dx}, \tag{8.26}$$

and,

$$\varepsilon_c = \frac{\int_0^b (1 - c^-/c_f)(bx - x^2/2)dx}{\int_0^b (bx - x^2/2)dx}, \tag{8.27}$$

where b (corresponding to $x_{Tr} - x_d$ in Figure 8.1) is the average effective half-pore size (mobile water), x is the distance from the beginning of the diffuse layer, c^+ and c^- are the cation and anion concentrations in the diffuse layer, which must be calculated by an electrical model, and c_f is the equilibrium solution concentration. Note here that the integration interval in Eqns (8.26) and (8.27) refers to the mobile water and thus represents the effective porosity where flow of water and solutes occur. For practical use at a macroscopic scale, an average value of b on the representative elementary volume (REV) has to be used in the required electrical model (de Marsily, 1986). In case of weak variations of ionic concentrations, average values of these concentrations in the interparticle space can be used according to Bolt (1979). Equation (8.27) was used by Bresler (1973) to calculate ε_c values using a truncated diffuse double-layer model (truncated DDL). These calculated values together with osmotic efficiency data were exploited to obtain adjusted curves as a function of the product $b\sqrt{c_f}$ discussed above. The so-called Bresler curve for sodium clay minerals in Figure 8.7 is the most popular abacus in the literature. Such theoretical estimates of the osmotic efficiency coefficient using Eqn (8.27) or similar expressions (Revil et al., 2005) were performed by Gonçalvès et al. (2007), Rousseau-Gueutin (2008), Rousseau-Gueutin et al. (2008), and Tremosa et al. (2012) with an electrical model different from that used by Bresler (Truncated Triple-Layer Model).

The above-mentioned historical studies (Katchalsky and Curran, 1967; Bolt, 1979) provide the theoretical basis for the second option, alternative to measurements, to establish the coupled flow parameters. This approach makes use of the equations that are valid at the pore scale, i.e., the Stokes equations for the fluid flow and the Nernst–Planck equations for the movements of ions. These equations, written at the microscopic scale, are then homogenized to obtain theoretical expressions for the coupling coefficients at the REV scale for hydrogeological applications. These analytical expressions involve average electrochemical properties of charged porous media. Such an approach was followed by Coelho et al. (1996), Gupta et al. (2008, 2006), Moyne and Murad (2002) or Revil and Leroy (2004), Revil et al. (2005) where more detailed descriptions are available. An example of this approach is given in the next section for the case of electroosmosis.

8.3.2 Electroosmosis

If an electrolyte flows due to a pressure gradient ∇p_f (or an electrical-potential gradient $\nabla\psi = \mathbf{E}$) through a porous medium whose walls are electrically charged, it induces a net electric current density (or a net seepage velocity). When the electric field and pressure field gradients are sufficiently small, the fluxes **I** and **q** are linearly related to these gradients by the following relations

$$\mathbf{I} = \boldsymbol{\sigma}_e \cdot \mathbf{E} - \boldsymbol{\alpha}_p \cdot \nabla P, \tag{8.28a}$$

$$\mathbf{q} = \boldsymbol{\beta}_e \cdot \mathbf{E} - \frac{\mathbf{k}}{\mu_f} \cdot (\nabla P + \rho_f \mathbf{g}), \tag{8.28b}$$

where $\boldsymbol{\sigma}_e$ and $\mathbf{k}$ are the conductivity and intrinsic permeability tensors, respectively; μ_f and ρ_f are the fluid dynamic viscosity and density, and $\mathbf{g}$ is the acceleration due to the gravity vector. $\boldsymbol{\alpha}_p$ and $\boldsymbol{\beta}_e$ are the electroosmotic coupling tensors (with $\boldsymbol{\alpha}_p = \boldsymbol{\beta}_e^t$ by virtue of Onsager's irreversible thermodynamic theorem (de Groot and Mazur, 1962)).

$\boldsymbol{\sigma}_e$, $\mathbf{K}$, $\boldsymbol{\alpha}_p$, and $\boldsymbol{\beta}_e$ are tensors when the medium is anisotropic. For isotropic porous media, these four tensors become spherical and can be expressed as

$$\mathbf{K} = K\mathbf{1}, \quad \boldsymbol{\sigma}_e = \sigma_e \mathbf{1}, \quad \boldsymbol{\alpha}_\mathbf{p} = \alpha_p \mathbf{1}, \quad \boldsymbol{\beta}_e = \beta_e \mathbf{1}, \tag{8.29}$$

where $\mathbf{1}$ is the unit tensor. The four scalars K, σ_e, α_p, and β_e correspond to permeability, conductivity, and electroosmotic coupling coefficients, respectively. These scalars are used most of the time below, since only isotropic media are treated here. In this case, $\alpha_p = \beta_e = \varepsilon\psi\frac{k}{\mu_f}$ reducing the problem to the knowledge of three parameters: σ_e, k, and α_p.

At the pore scale, the electric potential, the ion concentrations, the solvent velocity and its pressure are governed by the Poisson–Boltzmann equation, the convection–diffusion equation for each ion, and the Stokes equation supplemented by introducing an electrical force. The macroscopic tensors $\boldsymbol{\sigma}_e$, $\mathbf{k}$, and $\boldsymbol{\alpha}_p$ can be derived by solving the microscopic equations in a unit cell whose content is characteristic of the pore structure of the medium.

8.3.2.1 Electrokinetic Equations

This section presents a brief introduction to all the necessary equations describing the electrokinetic phenomena. Consider an N-component Newtonian electrolyte of density ρ_f, flowing with velocity $\mathbf{v}(\mathbf{r},t)$ in the interstices of a porous material. Let $\psi(\mathbf{r},t)$ be the electric potential prevailing within the solute; the flux $\mathbf{j}_i$ of each ith ion species, composing the solute, is given by the following constitutive equation (Levich, 1962)

$$\mathbf{j}_i = -D_i \nabla c_i - e z_i b_i c_i \nabla \psi + c_i \mathbf{v}, \quad i = 1, 2, \ldots, N \tag{8.30}$$

where c_i is the ion concentration in number of molecules per unit volume, z_i the ion algebraic valence (> or < 0), and e the absolute charge of the electron (1.602×10^{-19} C). D_i and b_i are the ion's diffusivity and electric mobility, related by the Stokes–Einstein equation

$$D_i = b_i k_B T \tag{8.31}$$

where k_B is Boltzmann's constant and T the absolute temperature. The ionic flux $\mathbf{j}_i$ and the concentration c_i obey the continuity equation

$$\frac{\partial c_i}{\partial t} + \nabla \cdot \mathbf{j}_i = 0. \tag{8.32}$$

When the characteristic Reynolds number is much lower than unity, the velocity field is governed by Stokes equations

$$\mu_f \nabla^2 \mathbf{v} = \nabla p_f + \mathbf{f}, \quad \nabla \cdot \mathbf{v} = 0, \tag{8.33}$$

where μ_f is the constant dynamic viscosity of the electrolyte, p_f is the fluid pressure, and $\mathbf{f}$ is the electric volumetric force density

$$\mathbf{f} = \rho \nabla \psi + \rho_f \mathbf{g} \tag{8.34}$$

where the electric charge density ρ (C m^{-3}) can be expressed as

$$\rho = e \sum_{i=1}^{N} c_i z_i. \tag{8.35}$$

Finally, the electric potential ψ is given by Poisson's equation

$$\nabla^2 \psi = -\frac{\rho}{\epsilon_{el}} = -\frac{e}{\epsilon_{el}} \sum_{i=1}^{N} c_i z_i \tag{8.36}$$

where ϵ_{el} is the fluid permittivity. Equations (8.32)–(8.36) must be solved subject to the following boundary conditions on the liquid–solid interface S_i

$$\boldsymbol{n} \cdot \mathbf{j}_i = 0 \tag{8.37a}$$

$$\mathbf{v} = 0 \tag{8.37b}$$

$$\psi = \zeta \text{ or } \boldsymbol{n} \cdot \nabla \psi = -\frac{\sigma_s}{\epsilon_{el}} \tag{8.37c}$$

where $\boldsymbol{n}$ is the outer normal to S_i, σ_s (C m^{-2}) the surface charge density, and ζ the zeta potential (V). The two boundary conditions, i.e., the Dirichlet or Neuman conditions are described extensively in the literature.

The general system of Eqns (8.32)–(8.35) is coupled and nonlinear. This feature generally complicates its solution, when an external electric or pressure gradient is applied to the domain. A technique, generally accepted in most studies, is to work close to equilibrium, i.e., when no external gradient is imposed. First, the system at equilibrium is analyzed. Second, small perturbations are added to the concentrations, the potential, the pressure, and the velocity. For convenience, the system is made dimensionless (see Gupta et al., 2008). In the following, dimensionless variables denoted by$'$ are obtained by dividing the physical quantities by: $(\epsilon_{el}/\mu_f)(k_bT/e)^2\kappa$ for velocity, κ^{-2} for permeability, k_BT/e for electrical potential, $\epsilon_{el}k_BT/\mu_f e$ for α_p where $\kappa = \sqrt{\frac{2c_f e^2}{\epsilon_{el}k_BT}}$ is the Debye length for a monovalent symmetric salt considered here.

8.3.2.2 Resolution of the Electrokinetic Equations

Except for simple cases such as the plane channel configuration, this system, even when linearized, cannot be solved analytically.

The porous medium is discretized into elementary cubes of size a. The unit cell in which the system (32, 33, 36) is solved, is made of $N_{cx} \times N_{cy} \times N_{cz}$ elementary cubes, which are either solid or liquid.

First, recall from Coelho et al. (1996) a few details on the discretization and resolution of the system (32, 33, 36). The flow field is solved by using an artificial compressibility with a staggered marker-and-cell mesh. For a given pressure field, the flow velocity is calculated and then introduced into the continuity equation. Furthermore, the pressure field is modified to minimize the local velocity divergence. These steps are repeated until the required accuracy is reached. Equation (8.36) is nonlinear and it can be solved by using a nonlinear conjugate gradient method. Equation (8.32) is solved by a classical conjugate gradient method; the flow field is determined by an elementary Jacobi method.

The general calculation scheme includes, as a first step, the calculations of the equilibrium potential $\psi^{0'}$ since it does not depend on the other quantities.

Then, a macroscopic pressure gradient is assumed to be applied to the medium. Since Eqns (8.32) and (8.33) are coupled, an iteration scheme is used. Equation (8.32) is first solved for $\mathbf{v}' = 0$. Then, Eqn (8.33) is solved for $\mathbf{v}'$ with the obtained value of the concentration. Equation (8.32) is solved again with the value of $\mathbf{v}'$. This loop is continued until the concentration and velocity fields converge.

The same scheme is used when a macroscopic electric gradient is applied to the medium.

Once these systems are solved, the macroscopic quantities can be calculated and the proportionality tensors in Eqn (8.28) can be determined. Consider a macroscopic pressure gradient ∇p_f applied to the porous medium, which induces an ionic flow with a local velocity **v**. The current density **I** can be expressed as

$$\mathbf{I} = -\boldsymbol{\alpha}_p \cdot \nabla p_f = \frac{1}{\tau_f} \int\limits_{\tau_f} \rho \mathbf{v} dv, \tag{8.38}$$

where ρ is already defined by Eqn (8.35) and τ_f is the pore volume. The coupling electroosmotic tensor $\boldsymbol{\alpha}_p$ is derived from this relation.

8.3.2.3 The Length Scale Λ

A characteristic length scale Λ, which can be determined for any porous medium, was introduced by Johnson et al. (1986). It is defined by

$$\Lambda = 2 \frac{\int_{\tau_f} \mathbf{E}^2(\mathbf{r}) dv}{\int_{S_i} \mathbf{E}^2(\mathbf{r}) ds}, \tag{8.39}$$

where $\mathbf{E}(\mathbf{r})$ is the local electric field in a porous medium filled by an electrolyte of constant conductivity and submitted to a macroscopic potential difference. τ_f is the pore volume and S_i the corresponding solid interface. According to Johnson et al. (1986), $2/\Lambda$ is an effective surface-to-pore volume ratio eliminating contributions from the isolated regions of the pore space that do not contribute to transport.

This length scale Λ has proved to be very useful in many ways. As suggested by Johnson et al. (1986), Pengra et al. (1995) showed that the permeability K can be determined via Λ

$$K = \frac{\Lambda^2}{8F_F} \tag{8.40}$$

where F_F is the formation factor. This was recently confirmed by systematic calculations of Valfouskaya et al. (2005) who computed K, F_F, and Λ for a variety of porous media. Moreover, the authors showed numerically that Λ is indeed approximately equal to twice the inverse of the surface-to-volume ratio

$$\Lambda \approx 2\frac{\tau_f}{S_i}. \tag{8.41}$$

Within the limits of small potentials, the numerical simulations by Coelho et al. (1996) showed that the electroosmotic coefficient α_p was better fitted with a slightly different length scale Λ' defined by

$$\Lambda' = \sqrt{12.5KF_F} \tag{8.42}$$

It is important to notice that this property was verified by the experiments by where K, F_F, ζ, and α_p were measured independently. This property is very important for practical applications since it is relatively difficult to measure α_p.

8.3.2.4 Estimates of the Electroosmotic Coupling Tensors

The length scale Λ' can be used as a basis to estimate α_p. The average ionic concentration $<c>$ across the channel (mobile fluid, excluding the Stern layer) can be determined and approximated as shown by Gupta et al. (2008). Then, a macroscopic double-layer thickness $\overline{\kappa}^{-1}$ based on $<c>$ is introduced

$$\overline{\kappa}^{-1} = \sqrt{\frac{\epsilon_{el} kT}{e^2 <c>}} \tag{8.43}$$

With a good approximation, $\overline{\kappa}$ can be estimated by the formula

$$\overline{\kappa}\Lambda'\sqrt{3/8} \approx \left[\kappa\Lambda'\sqrt{3/8}\,E\tan\left(\kappa\Lambda'\sqrt{3/8}E/2\right)\right]^{1/2} \tag{8.44}$$

Gupta et al. (2008) showing that this relation is in good agreement with the corresponding numerical evaluations.

A simple generalization of the plane channel formula yields

$$\kappa\Lambda'\sqrt{3/2}\alpha'_p \approx -\frac{E^2}{2}\int_0^{\kappa\Lambda'\sqrt{3/8}} y'\left(\kappa\Lambda'\sqrt{3/2}-y'\right) \times \left[1+\tan^2\left(\frac{E}{2}\left(\kappa\Lambda'\sqrt{3/8}-y'\right)\right)\right]\mathrm{d}y' \tag{8.45}$$

Numerical and experimental results follow this formula quite well as shown in Figure 8.12. More details are provided in Gupta et al. (2008).

8.3.3 Thermoosmosis: From Molecular Concepts to a Macroscopic Model

The physical principles of thermoosmosis, at the molecular and pore scales, were established by Derjaguin and Sidorenkov (1941) and Churaev (2000) as the so-called mechanocaloric effect. When a quasi-isothermal pressure gradient is prescribed through a porous membrane, the fluid flow causes a temperature-gradient buildup. This heat flux is attributed to a specific enthalpy of the fluid film in the pores, which differs by a quantity ΔH J m^{-3} at the pore scale from that of the bulk fluid (Derjaguin and Sidorenkov, 1941; Churaev, 2000). This difference is associated with an alteration of the H-bond network of the water molecules and thus of the intermolecular bond energy due to fluid–solid interactions. If the H Bond is weakened by these interactions, i.e., its concentration decreases, heat absorption occurs when the fluid enters into the pore space (Churaev, 2000). Conversely, heat is released at the outlet generating a temperature gradient. The counterpart of the mechanocaloric effect outlined above, i.e., thermoosmosis, derives from Onsagers reciprocity

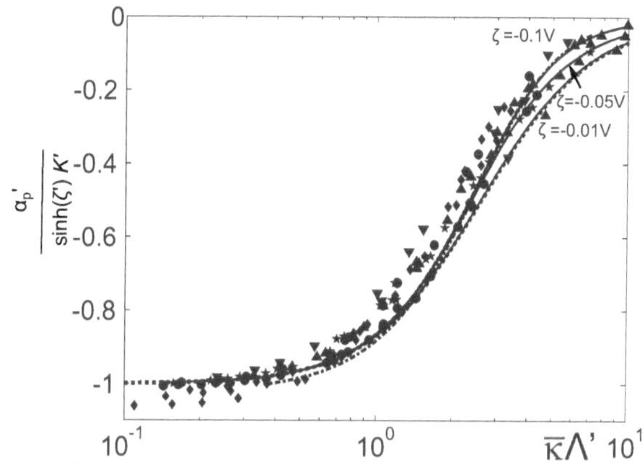

FIGURE 8.12 $\frac{\alpha_p'}{\sinh(\zeta')K'}$ versus $\bar{\kappa}\Lambda'$ for various porous structures and $\zeta = -0.01$, -0.05, and -0.1 V. Data are for semianalytical solution (solid line), nonlinear approximation Eqn (8.45) for a plane channel only for $\zeta = -0.1$ and -0.05 V (dashed lines), linear approximation for small values of ζ (dotted line), numerical results for simple cubic packing of spheres (▲), simple cubic packing of ellipsoids (▼), monodisperse spheres (♦), reconstructed porous media (●), and vugular porous media (★). The dotted line of the linear approximation is very close to the solid line corresponding to the semianalytical solution for $\zeta = -0.01$ V.

principle. This thermodynamic principle requires that the reciprocal process exists, i.e., a fluid flow caused by a temperature gradient.

In a recent study (Gonçalvès et al., 2012), a new formalism for estimating the thermoosmotic permeability was proposed. It is based on the above-mentioned theoretical analysis at the molecular and pore scale, then upscaled. For this purpose, H-bond concentrations in the bulk and in the pore space solutions were estimated as functions of the ionic content of the solutions and of the average H-bond numbers per water molecule in the bulk solution, in the hydration shells of the ions and in the highly structured water of the Stern layer. In the pore space, a simple Donnan equilibrium model is used to identify the ionic concentrations. The macroscopic excess enthalpy (ΔH) due to surface-charged solid-pore fluid is then expressed by introducing the energy to break 1 mol of H bond. The thermoosmotic permeability

$$k_T = \frac{\Delta H}{T} k \text{ or } \varepsilon_T = \frac{\Delta H}{T} \tag{8.46}$$

is estimated only from surface-charge density, temperature, pore size, and salinity. With highly plausible values for the model parameters, a fairly good agreement (order of magnitude and sign) with the data is found. This is a noteworthy improvement since, so far, an empirical range of $10^{-13} - 10^{-17}$ Pa m^2 K^{-1} for k_T has been used in natural media (Soler, 2001). The molecular control for coupled flow processes is here treated directly at the macroscopic scale.

8.4 COUPLED HYDRO-CHEMO-MECHANICAL BEHAVIOR IN CLAY-ROCKS

Due to the large characteristic time associated with transport processes in clay-rocks, hydraulic and chemical boundary conditions are likely to be time varying. Consequently, transient-state transport calculations involving a combination of mass balance and transport equations are often required. The mass balance equation involving a purely hydromechanical coupling and introducing a storage coefficient was discussed in Section 8.2.3.2. But, time variations of chemical boundary conditions (solute concentrations), which can cause a swelling or shrinking of a clay formation, was not discussed so far. The introduction of electrochemical reactivity of clays requires further extending the previous hydromechanical coupling to a hydro-chemo-mechanical coupling. This introduces a "chemical" storage coefficient, which is discussed in this section.

8.4.1 Surface Forces and Disjunction Pressure

Langmuir (1938) and Derjaguin et al. (1987) showed that the pressure in water films between two charged particles differs from the hydrostatic pressure, i.e., the water pressure at the same depth in the reservoir containing these particles. They attributed this difference to electrostatic and molecular interactions between charged surfaces and the electrolyte present in the interplate space. According to Derjaguin et al. (1987), this "departure from the hydrostatic law" only occurs if the interfacial zones, here the diffuse layers, overlap. If no overlap exists (i.e., large pore size or high-concentration solutions), the concept of pressure is equivalent to that in the bulk or equilibrium solution. Thus, this difference in pressure, named disjoining pressure in colloidal sciences, is due to the action of surfaces forces induced by the overlap. These surfaces forces are of two types: repulsive and attractive. Repulsive and attractive forces per unit surface area define a positive and negative pressure, respectively. A negative pressure is a traction in a mechanical sense. The balance between these two kinds of forces per unit surface area gives the sign of the disjoining pressure. Furthermore, noting that the surface forces at work in overlap conditions are mostly perpendicular to the surface (in the direction of the electric field), it is easy to argue that the "pressure" in thin films is no longer a scalar but becomes a tensor. This result was theoretically established by Derjaguin et al. (1987) by writing a generalized Maxwell stress tensor, i.e., the stress tensor for a body under the influence of an electromagnetic field. Thus, this writing leads to the consideration of a generalized "pressure" in which the hydrostatic pressure and electrostatic and molecular forces per unit surface area are encapsulated. This generalized "pressure" whose anisotropy is imposed by the surface forces is no longer a fluid pressure in the strict sense but has the significance of a local stress. For modeling purposes, Derjaguin et al. (1987) refined their initial definition and developed the mathematical expression of the disjoining pressure. First, they interpreted the disjoining pressure as the difference between the pressure normal to the solid (clay mineral) surfaces in the film, i.e., the direction of surface forces, and the pressure in the equilibrium solution. This interpretation allowed the formulation, in a second stage, of the local volumetric force balance within a film of water between two charged surfaces embedded in a bulk (equilibrium) solution:

$$grad(p) + \rho\, grad(\psi) + grad\left(\frac{1}{2}\frac{d\epsilon_{el}}{d\rho_f}E^2\right) = 0 \tag{8.47}$$

with $p(x)$ (Pa), the hydrostatic pressure (the x direction is perpendicular to the charged surfaces), ρ the charge density (C m^{-3}) in the film (Van Olphen, 1963), ψ (V) is the electrical potential, ρ_f the density of the fluid (Kg m^{-3}), ϵ_{el} (F m^{-1}) is the permittivity of the fluid, and E the electric field (V m^{-1}). The second term on the left-hand side of Eqn (8.47) is the electric force acting on the volume charge and the third term is the electrostriction term (force due to the orientation of water molecules subjected to an external electric field). An algebraic treatment, which is detailed in Gonçalvès et al. (2010) is used to obtain the following relationships for the components p_N and p_T of a generalized "pressure" tensor **p** normal and tangential to the surface, respectively:

$$p_N = p_f + \pi_D^e \tag{8.48}$$

$$p_T = p_f + \frac{1}{\epsilon_{el}}\frac{d\epsilon_{el}}{d\rho_f}\rho_f\,\pi_D^e + \left(\frac{1}{\epsilon_{el}}\frac{d\epsilon_{el}}{d\rho_f} + 1\right)\int_0^{\psi}\rho d\psi \tag{8.49}$$

where ψ is the electrical potential at a distance z from the surface (virtually $\psi = 0$ at an infinite distance) and with

$$\pi_D^e = -\int_0^{\psi_b}\rho d\psi \tag{8.50}$$

where π_D^e (Pa) is the electrostatic component of the disjoining pressure, ψ_b (V) is the electrical potential at the mid-plane, and p_f is the hydrostatic pressure in the equilibrium solution. This component is independent of x and depends only on the distance between clay mineral layers. In the case of adjacent surfaces of equivalent charge and for a monovalent and symmetric salt (e.g., NaCl), the component π_D^e is given by

$$\pi_D^e = -2k_BTc_f\left(\cosh\left(\frac{e\psi_b}{k_BT}\right) - 1\right) = k_BT\left(c^+ + c^- - 2c_f\right)_{x=b} \tag{8.51}$$

where k_B is Boltzmann's constant (1.38×10^{-23} J K^{-1}), T is temperature (K), c^+ and c^- (ions m^{-3}) are the concentrations of cations and anions in the porosity, respectively, c_f is the concentration of these ions in the equilibrium solution, b is the half-distance between clay mineral layers. In addition to the electrostatic component described above, other components associated with complex physicochemical interactions must be considered. Derjaguin et al. (1987) introduced an attractive van der Waals term corresponding to a weak attraction between the molecules of the adjacent surfaces. This component can be written as follows:

$$\pi_D^{vdw} = -\frac{A}{6\pi(2b)^3} \tag{8.52}$$

where A is Hamaker's constant dependent on the mineralogical nature of the surface and $2b$ is the distance between the two surfaces. For mica and water films, a value of A of 2×10^{-20} J has been reported (Derjaguin et al., 1987).

These two repulsive and attractive components defined by Eqns (8.50) and (8.52) are used in the so-called DLVO (Derjaguin–Landau–Verview–Overbek, see Van Olphen, 1963; Mitchell, 1993) theory of the stability of colloidal dispersions in which the electrostatic and molecular components are assumed to be linearly additive:

$$\pi_D = \pi_D^e + \pi_D^{vdw} \tag{8.53}$$

However, the DLVO theory cannot explain some experimental results as claimed by McBride (1997) and also previously recognized by Derjaguin and Churaev (1974). This is why an additional term besides the electrostatic and molecular components has been proposed (Derjaguin and Churaev, 1974; Derjaguin et al., 1987). This term often called the structural component or the adsorption component is attributable to the solid surface hydration with the presence of several monolayers of water molecules with a modified structure of the H-Bond network as compared to the equilibrium solution. This has long been reported in the literature through the observation of the alteration of water properties such as density or viscosity in the first nanometer from the surface (see Martin, 1960; Mitchell, 1993).

The experimental data demonstrate that the local forces associated with this solid surface hydration depend exponentially on the interlayer thickness $2b$ (Marcelja and Radic, 1976; Churaev and Derjaguin, 1985):

$$\pi_D^s = \gamma_1 \exp\left(-\frac{2b}{\gamma_2}\right) \tag{8.54}$$

where γ_1 and γ_2 are coefficients that have to be experimentally determined. For quartz and mica, values in the range 7×10^5 to 10^7 Pa for γ_1 and 0.8–1 nm for γ_2 are reported (Derjaguin et al., 1987). The term π_D^s can also be added to the components π_D^e and π_D^{vdw} in Eqn (8.53) to yield the following general relationships from Eqns (8.48) and (8.49):

$$p_N = p_f + \pi_D^e + \pi_D^{vdw} + \pi_D^s \tag{8.55}$$

$$p_T = p_f + \frac{1}{\epsilon_{el}}\frac{d\epsilon_{el}}{d\rho_f}\rho_f\,\pi_D^e + \left(\frac{1}{\epsilon_{el}}\frac{d\epsilon_{el}}{d\rho_f} + 1\right)\int_0^{\psi}\rho d\psi + \pi_D^{vdw} + \pi_D^s \tag{8.56}$$

Thus, these relationships strongly suggest that the local pore-pressure in clay materials is not a simple hydraulic parameter, e.g., the hydrostatic pressure classically used by hydrogeologists. It is likely a tensorial quantity, which stems from electrostatic interactions, van der Waals forces, and complex physicochemical structural forces; this disjoining pressure must be taken into account in the hydrodynamic formalism used to quantitatively describe the movement of water in clays and clay-rocks at the macroscopic scale.

8.4.2 Implication of the Disjunction Pressure Term for the Storativity: The Electrochemical Specific Storage Coefficient

To our knowledge, the effects of the disjoining pressure on the macroscopic behavior of clay barriers have been investigated according to two classes of theoretical approaches. In the first class, the disjoining pressure, a microscopic concept by nature, is upscaled by using theoretical techniques, which allow propagating microscopic governing equations to a larger scale. These techniques include mixture theory, volume averaging, and homogenization. All the results obtained from these upscaling techniques applied to clay geomaterials confirm that the conventional effective stress tensor, $\boldsymbol{\sigma}'$, should be rewritten as follows (Hueckel, 1992; Murad et al., 1995; Murad and Cushman, 1996; Bennethum et al., 1997; Moyne and Murad, 2002, 2003; Dormieux et al., 2006; Giraud et al., 2007; Ortega et al., 2007; Cariou et al., 2013):

$$\boldsymbol{\sigma}' = \boldsymbol{\sigma} - \gamma_B p_f \mathbf{1} - \boldsymbol{\Pi} \tag{8.57}$$

where $\boldsymbol{\sigma}$ is the total stress tensor (Pa), γ_B is Biot's coefficient (−) (recall $0 \le \gamma_B \le 1$), which can be a tensorial quantity when a clay-rock is considered (Cosenza et al., 2002), $\mathbf{1}$ is the unit tensor, $\boldsymbol{\Pi}$ is the (macroscopic) physicochemical stress tensor (Pa) related to the disjoining pressure. More recently, in the framework of a microporomechanical approach and by assuming a very weak normal stiffness of the clay mineral particles compared to the elasticity of the clay mineral layer solid component, an isotropic swelling (isotropic distribution of the clay mineral particles orientations), and no shear component for $\mathbf{\Pi}$, Dormieux et al. (2006) and Cariou et al. (2013) obtained a simplified expression for the macroscopic $\mathbf{\Pi}$ tensor:

$$\mathbf{\Pi} = g(\varphi)p_N\mathbf{1} \tag{8.58}$$

where p_N is the (microscopic) normal component of the disjoining pressure and the quantity g is a function of the volume fraction of the interparticle pores named φ, which corresponds to the macropores porosity:

$$g(\varphi) = \frac{(2+\varphi)(1-4\varphi)}{2-3\varphi} \tag{8.59}$$

Note that Eqn (8.58) keeps its physical sense of a “swelling pressure” if $\varphi < 1/4$.

These upscaling techniques are also able to produce the governing equations of coupled fluid flow in clay geosystems including an electrochemical storage coefficient induced by the existence of the disjoining pressure. However, it should be noted that the induced electrochemical storage coefficient is more easily introduced into the framework of chemoporoelasticity whose governing equations are obtained from thermodynamics of porous continua (Coussy, 2004).

This thermodynamical approach often called Biot’s theory (Biot, 1941, 1972, 1977) forms the second class of theoretical approaches used to infer the macroscopic state equations of reactive clay barriers. This type of approach, by which the disjoining pressure can be accounted for indirectly, aims to directly build macroscopic state equations for engineering purposes. In comparison with upscaling techniques, it does not explicitly discriminate between the solid and the fluid phases: its main advantage is clearly its relative simplicity. Moreover, although its starting point is not the microscopic scale, i.e., that of the clay particle, it allows the user to include some microscopic data and patterns as explained below.

Linear chemoporoelasticity established from a thermodynamical viewpoint was used for the first time in the context of a reactive clay geomaterial by Sherwood (1993) whose formulation was extended by Dormieux et al. (1995), Heidug and Wong (1996), Loret et al. (2002), Gajo et al. (2002), and Coussy (2004) to account for nonlinear effects (e.g., irreversible deformation, stress-dependent parameters). More recently, Sarout (2003) and Sarout and Detournay (2011) (see also Bunger et al., 2010) modified the previous chemoporoelasticity formalism into a more convenient and operational form whereby the mechanical interpretation of the coupled and linear hydro-chemo-mechanical parameters was clarified from an experimental viewpoint. The notations of these authors will be used in the following.

To obtain a simple expression of an electrochemical storage coefficient, a linear isotropic and isothermal framework is considered: the transformations are isothermal and linear (i.e., infinitesimal and reversible changes in state variables are introduced) and all the physicochemical parameters are isotropic. The thermodynamical approach promoted here leads to the choice of the following state variables for reactive clay barriers saturated by a binary solution: the generalized stress $S = (\boldsymbol{\sigma}, \mu_w, \mu_s)$ and the dual generalized strain $E = (\epsilon, m_w, m_s)$, where $\boldsymbol{\sigma}$ is the total (Cauchy) stress tensor, μ_w and μ_s are the chemical potential of water (solvent) and solute (typically NaCl salt), respectively, ε is the strain tensor, m_w and m_s are the mass increments of water and solute, respectively. Since the changes in state variables are reversible, the dissipation related to the elastic skeleton is zero, i.e., the so-called Clausius–Duhem inequality, which is a direct result of the second law of thermodynamics, degenerates to

$$\boldsymbol{\sigma} d\epsilon + \mu_s dm_s + \mu_w dm_w - df = 0 \tag{8.60}$$

where f is Helmholtz's free energy of the porous system per unit reference volume of porous medium, df is an exact differential. At this stage, as in classical thermostatics, the postulate of local state is introduced. This postulate stipulates that the current state of internal energy of a homogenized system in any evolution is irrespective of the rate of evolution and is characterized by the same state variables as the ones characterizing equilibrium state (Coussy, 2004). In other words, the free energy f only depends on the set of state variables (ϵ, m_w, m_s). Consequently, the following macroscopic state equations are obtained from Eqn (8.60):

$$\boldsymbol{\sigma} = \frac{\partial f}{\partial \epsilon}; \quad \mu_s = \frac{\partial f}{\partial m_s}; \quad \mu_w = \frac{\partial f}{\partial m_w} \tag{8.61}$$

In the framework of a linear theory, the changes in state variables of the first order are considered: the free energy f exhibits a quadratic dependence on the state variables and the second order derivatives of f are constants, which are the physico-parameters of a linear model. However, in order to obtain a clear physical meaning for these physico-parameters and to easily compare these new parameters to those of the classical poroelasticity, it is convenient to formulate the following changes of state variables:

$$\zeta_f = v_w \Delta m_w + v_s \Delta m_s \tag{8.62}$$

$$\theta = \frac{\Delta m_s}{\langle c_f \rangle} - \frac{\Delta m_w}{\langle c_w \rangle} \tag{8.63}$$

where ζ_f is the fluid content (originally introduced by Biot) (−), θ the relative ionic content (−), v_w and v_s are the molar volumes (i.e., L mol^{-1}) of the water and the salt, respectively, m_w and m_s are the mass increments of the water and salt, respectively, expressed as moles per unit reference volume of the porous medium. The quantities $\langle c_w \rangle$ and $\langle c_f \rangle$ are mean concentrations of, respectively, the water and the salt in the pore fluid, where it is assumed in this linear theory that the water/salt concentrations will show small differences from the mean values. The latter may be taken, for example, as the initial concentrations in the pore fluid.

From the macroscopic state Eqn (8.61) accounting for the new state variables ζ_f and θ and from the thermodynamical expressions of chemical potentials μ_w and μ_s as functions of the (pore) water pressure p_f and the (macroscopic) physicochemical stress Π (no shear component is considered), this approach allows to write the following linear state equations (see details in Sherwood, 1993, 1995; Sarout, 2003; Bunger et al., 2010; Sarout and Detournay, 2011):

$$\begin{pmatrix} \epsilon \\ \zeta_f \\ \theta \end{pmatrix} = \mathbf{C} \begin{pmatrix} \sigma \\ p_f \\ \Pi \end{pmatrix} \tag{8.64}$$

with

$$\mathbf{C} = \begin{pmatrix} C & \gamma_B C & -\alpha\gamma_B C \\ \gamma_B C & S_\sigma & -S_\sigma\beta \\ -\alpha\gamma_B C & -S_\sigma\beta & \gamma \end{pmatrix} \tag{8.65}$$

and

$$2G\boldsymbol{\epsilon} = \boldsymbol{\sigma} - \frac{3\nu}{1+\nu}\sigma\mathbf{1} + \frac{1-2\nu}{1+\nu}\gamma_B\left(p_f - \alpha\Pi\right)\mathbf{1} \tag{8.66}$$

where σ and ϵ are the mean stress $\sigma = \frac{tr(\boldsymbol{\sigma})}{3}$ and the volumetric strain $\epsilon = \frac{tr(\epsilon)}{3} = \frac{\Delta V}{V_0}$, respectively, G is the shear modulus (Pa), ν is the drained Poisson ratio (−). Here some comments can be made: Eqns (8.64) and (8.65) represent the chemoelastic volumetric response relationship, where the first one of these state equations can be obtained from a contraction in Eqn (8.66), the tensorial strain−stress relationship. Equation (8.64) is the generalized effective stress relationship 8.57 considering an elastic solid skeleton and an additional chemical−mechanical coupling parameter α. The symmetric matrix $\mathbf{C}$ has six parameters, which can be organized into three groups. The first group consists of the classical poroelastic parameters, i.e., the Biot coefficient γ_B, the drained compressibility (or the inverse of the drained bulk modulus) C (Pa^{-1}), and the unconstrained specific storage S_σ (Pa^{-1}) often expressed as $S_\sigma = \left(\frac{\partial\zeta_f}{\partial p_f}\right)_{\sigma=0} = \gamma_B C/B$, where B is the Skempton coefficient. The second group corresponds to parameters α and β, which are the chemical−mechanical and the chemical−hydraulic coupling parameters, respectively. As shown by Sarout (2003), these parameters belong to the interval [0,1]. Consider the two limiting cases corresponding either to the absence of chemical interaction or to the immobility of the salt ions. In the first case, in the absence of chemical interaction and following the third state Eqn (8.64), the absence of chemical interaction ($\theta = 0$) implies $\alpha = 0$ (in case of a mechanical loading $d\sigma = 0$) or $\beta = 0$ (in case of a hydraulic loading $dp_f = 0$). In the second case referred to as "perfect ion-exclusion membrane model," the immobility of the salt ions trapped in the pores ($\Delta m_s = 0$ in Eqns (8.62) and (8.63) and thus $\theta = -\zeta_f$) under either mechanical or hydraulic loading implies $\alpha = 1$ and $\beta = 1$. The last parameter γ is a purely chemical constant (Pa^{-1}), which relates θ and Π. Assuming that Π is a pure osmotic pressure and a dilute pore solution, the parameter γ can be written as follows:

$$\gamma = \frac{\omega}{<c_f> (1- <c_f> v_s)RT} \tag{8.67}$$

However, to fully solve a chemoelastic problem, the following additional equations must be introduced:

The momentum balance equation:

$$\nabla \cdot \boldsymbol{\sigma} + \mathbf{F} = 0 \tag{8.68}$$

where **F** is the body force vector ($N\,m^{-3}$).

The kinematic compatibility relationship:

$$\epsilon = \frac{1}{2}(\nabla \mathbf{u} + {}^T\nabla \mathbf{u}) \tag{8.69}$$

where **u** is the displacement vector (m).

The transport equations:

$$\begin{pmatrix} \mathbf{q} \\ \varpi \end{pmatrix} = -\frac{k}{\mu_f}\mathbf{R}\begin{pmatrix} \nabla p_f \\ \nabla \Pi \end{pmatrix} \text{ with } \mathbf{R} = \begin{pmatrix} 1 & -\epsilon_c \\ -\epsilon_c & \dfrac{\omega_D}{SN} \end{pmatrix} \tag{8.70}$$

where **q** is the specific discharge of the solution with respect to the solid ($m\,s^{-1}$), ϖ is the specific discharge of salt with respect to the solvent ($m\,s^{-1}$), ϵ_c is the reflection or osmotic efficiency coefficient in the classical Darcy's law to account for the coupling between hydraulic and chemical thermodynamical driving forces (−) ($0 \le \epsilon_c \le 1$), S is a storage coefficient (−), ω_D gives the ratio between an apparent chemical diffusivity and the hydraulic diffusivity (−) (Sherwood, 1995), $N = RT/v_s$ (Pa) v_s represents the molar volume of the salt ($v_s = 0.03\ L\ mol^{-1}$ for NaCl), T temperature (K), and **R** the perfect gas constant (N m). Equation (8.70) are the generalized Darcy's law and the generalized Fick's law (the first and second transport equation in Eqn (8.70), respectively) including the coupling between hydraulic and chemical thermodynamical driving forces.

The mass balance equations:

$$\nabla \cdot \mathbf{q} = -\frac{\partial \zeta_f}{\partial t}, \quad \nabla \cdot \varpi = -\frac{\partial \theta}{\partial t} \tag{8.71}$$

Finally, the system of equations made of governing Eqns (8.64) to (8.71) comprises 25 independent equations for 25 unknown quantities. They are usually combined to form a set of equations consisting of one Navier-type equation and two coupled diffusion equations. This system is then solved for a class of problems characterized by an irrotational displacement field **u**, in the form:

$$\mathbf{u} = \nabla f_s \text{ thus } \nabla \times \mathbf{u} = \nabla \times \nabla f_s = 0 \tag{8.72}$$

i.e., **u** derived from a scalar potential f_s. The Navier-type equation is obtained by using the compatibility Eqn (8.69) and substituting the constitutive Eqn (8.66) into the equilibrium Eqn (8.68) (with no body forces) (Coussy, 2004 in the classical poroelastic case):

$$2G\frac{1-\nu}{1-2\nu}\nabla(\nabla\cdot\mathbf{u}) - G\nabla \times (\nabla \times \mathbf{u}) = \gamma_B \nabla p_f - \alpha\gamma_B\nabla\Pi \tag{8.73}$$

From Eqn (8.72), Eqn (8.73) may be integrated to yield

$$\epsilon = \frac{\gamma_B}{2G}\frac{1-\nu}{1-2\nu}(p_f - \alpha\Pi) + f(t) \tag{8.74}$$

where in the specific case of infinite or semi-infinite domain problems (conventional in geological contexts), the function $f(t)$ vanishes (since the variables ϵ, p_f, and Π in Eqn (8.74) must tend to 0 when x tends to ∞) (Detournay and Cheng, 1993) as considered in the following for the sake of simplicity (this will not modify the final conclusion). In order to obtain the mathematical form of a chemomechanical storage coefficient called hereafter S_S^{chem}, focus is made on the pore-pressure diffusion equation. For this purpose, first, σ is expressed in the second state equation in Eqn (8.64) in terms of ϵ, p_f, and Π using the first state equation in Eqn (8.64):

$$\zeta_f = \gamma_B\epsilon + \left(\frac{1}{B} - \gamma_B\right)\gamma_B C p_f - \left(\frac{\beta}{B} - \alpha\gamma_B\right)\gamma_B C\Pi \tag{8.75}$$

Then, the volumetric strain, ϵ, can be eliminated from Eqn (8.74):

$$\zeta_f = Sp_f - C_{chem}S\Pi \tag{8.76}$$

where S can be seen as a storage coefficient (Pa^{-1}) defined in poroelasticity and given by

$$S = \gamma_B\left[\frac{\gamma_B}{2G}\frac{1-\nu}{1-2\nu} + \left(\frac{1}{B} - \gamma_B\right)C\right] \tag{8.77}$$

and C_{chem} is a thermodynamical interaction parameter (−) defined by

$$C_{chem} = \alpha - (\alpha - \beta)\frac{S_\sigma}{S} \tag{8.78}$$

Finally combination of Eqn (8.74) with the generalized Darcy's law in Eqn (8.70) and the solution mass balance in Eqn (8.71) leads to the pore-pressure diffusion equation

$$\frac{k}{\mu_f}\left(\nabla^2 p_f - \varepsilon_c \nabla^2 \Pi\right) = S\frac{\partial p_f}{\partial t} - C_{chem} S\frac{\partial \Pi}{\partial t} \tag{8.79}$$

from which the specific storage coefficient Ss (m^{-1}) defined by hydrogeologists is

$$S_S = \gamma_w S = \gamma_w \gamma_B \left[\frac{\gamma_B}{2G}\frac{1-\nu}{1-2\nu} + \left(\frac{1}{B} - \gamma_B\right) C\right] \tag{8.80}$$

where γ_w is the specific weight of water ($N\,m^{-3}$). From Eqn (8.76) or Eqns (8.79) and (8.80), the chemomechanical specific storage coefficient S_S^{chem} (m^{-1}) can be identified as follows:

$$S_S^{chem} = C_{chem} S_S = \gamma_w C_{chem} \gamma_B \left[\frac{\gamma_B}{2G}\frac{1-\nu}{1-2\nu} + \left(\frac{1}{B} - \gamma_B\right) C\right] \tag{8.81}$$

In the particular case of an ideal ion-exclusion membrane model, recall that $\alpha \approx \beta \approx 1$ and thus from Eqn (8.78), $C_{chem} \approx 1$ and $S_S^{chem} \approx S_S$. Moreover, in this particular case, if the salt concentration c_f is used as a state variable (assuming that Π depends only on c_f for a given material), the formulation of the chemomechanical specific storage coefficient obtained by Gonçalvès et al. (2010), $S_S'^{chem}$, is retrieved:

$$S_S'^{chem} = \frac{S_S}{g}\frac{\partial \Pi}{\partial c_f} \tag{8.82}$$

At this stage, it may be desirable to link these formulations of the macroscopic chemomechanical specific storage coefficient to the microscopic relationships described in Section 8.4.1. For this purpose, a few approaches involving the physicochemical stress Π can be used. The simplest one was proposed by Bolt (1956) who suggested the following identity from Eqn (8.51):

$$\Pi = \pi_D^e = -2k_B T c_f \left(\cosh\left(\frac{e\psi_b}{k_B T}\right) - 1\right) = k_B T\left(c^+ + c^- - 2c_f\right)_{x=b} \tag{8.83}$$

where the mid-plane potential ψ_b and the mid-plane concentrations c^+ and c^- are determined by the DLVO theory. The mean distance $2b$ is assumed to be equal to the mean pore size given by Eqn (8.21). Another approach to downscaling Π is to use simple analytical expressions obtained from upscaling techniques. For instance, from Eqn (8.58):

$$\Pi = g(\varphi) p_N \tag{8.84}$$

Figure 8.13 displays $g(\varphi)$ (see Eqn (8.59)) and confirms that $g(\varphi) \approx 1$ when $\varphi \ll 1$. This condition may be satisfied for some clay-rocks characterized by low macropores porosity. If $g(\varphi) \approx 1$ and only the electrostatic interactions are considered, Eqn (8.84) is simplified as follows:

$$\Pi = p_N = \pi_D^e \tag{8.85}$$

and Eqn (8.83) is retrieved, following Bolt's approach (Bolt, 1956). The applicability of the approach promoted by Bolt (1956) was tested on remolded clay samples consisting of specially prepared, very fine grained clay minerals, mostly Na^+-montmorillonite and Ca^{2+}-montmorillonite in 10^{-3} and 10^{-4} M electrolyte solution (Bolt, 1956; Warkentin et al., 1957). These tests have indicated that the theoretical physicochemical pressure values provided using Eqns (8.83) and (8.21) are close to those measured in laboratory: Bolt's approach can be used to estimate physicochemical pressure in a first order approach. Madsen and Muller-Vonmoos (1985) confirmed the interest in using Bolt's approach by showing that Eqn (8.83) successfully predicts swelling pressure measured in Opalinus clay, a Jurassic clay-rock (Figure 8.14). The physicochemical or swelling pressure values of 19 core samples were obtained by measuring the pressures necessary to constrain volume when samples were immersed in saline solution. A fairly satisfactory agreement was found between theoretical estimates and experimental data regarding the complexity of physicochemical behavior and microstructure of a clay-rock (Figure 8.14). However, strong discrepancies between the theoretical predictions from Bolt's approach and the experimental results have been observed in three cases: heteroionic cation systems, clay samples containing particles larger than a few tenths of a micrometer, and very compacted clay samples with swelling clay

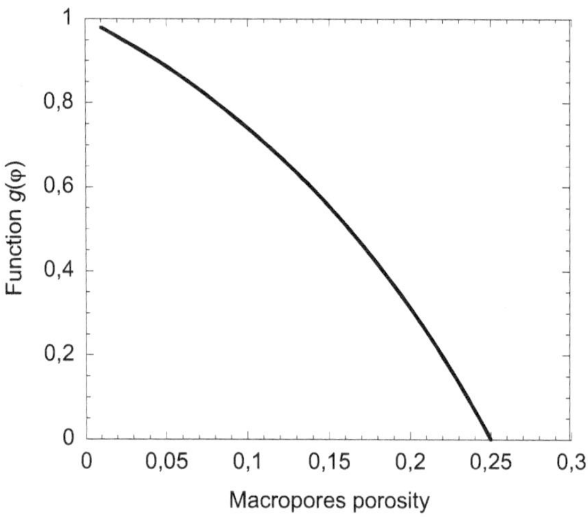

FIGURE 8.13 Function $g(\varphi)$, parameter φ is the macropores porosity.

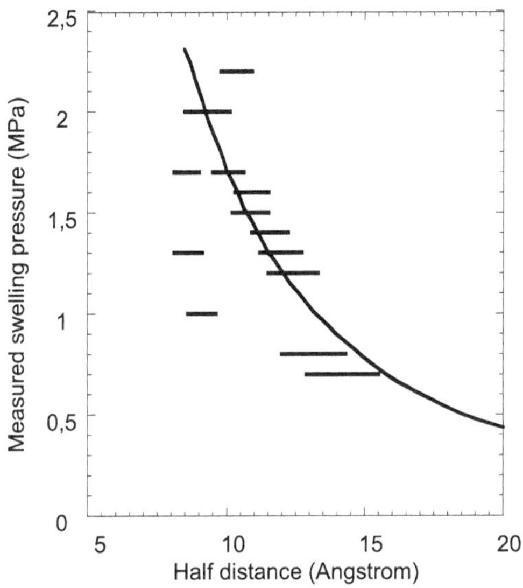

FIGURE 8.14 Physicochemical or swelling pressure as a function of half-distance *(Modified from Madsen and Muller-Vonmoos (1985).)*. The gray line corresponds to the theoretical swelling pressure obtained from Bolt's approach (Bolt, 1956). Half-distance values were calculated from Eqn (8.21) by assuming a saturated rock. The uncertainty bars associated with half-distance values are due to the uncertainties associated with the determination of specific surface area.

minerals (Mitchell, 1993; Schanz and Tripathy, 2009). These disagreements are mainly due to the implicit assumptions associated with Eqns (8.83) and (8.21): the van der Waals and surface hydration components (see Section 8.4.1) are disregarded in Eqn (8.83), the pore volume distribution is assumed to be uniform through a constant mean distance $2b$, and multivalent salts in the interlayer space are not considered.

8.5 CONCLUSION

The analysis of the semipermeable behavior of clay-rocks was mainly exemplified on fluid flow. For clay media, it was clearly established that Darcy's law incompletely describes fluid flow, other driving forces in the context of coupled transport processes are involved: chemical, electrical-potential and thermal gradients. Consequently, the resulting so-called osmotic fluxes must be considered in clay media for hydrodynamic and transport analysis. All available osmotic conductivity data have been collected here, including the most recent values acquired in the course of ongoing research on nuclear-waste confinement in clay-rock formations. Whether or not these data are directly used as abacuses by means of adjusted curves, they can also support some theoretical calculations, some of which have been outlined here.

Predictive models (theoretical expressions or abacuses) for osmotic parameters make it possible to perform some fluid and transport calculations within clay-rock formations. However, due to the nonlinearity and complexity of coupled flow processes, such calculations are limited to the local formation scale ($10-10^3$ m), i.e., the URL scale. Therefore, it would be unrealistic to account for such processes for basin-scale hydrogeological modeling. For such simulations, the classical Darcian formulation should be used while noting that, in fact, equivalent or apparent hydraulic conductivity is used. Returning to the local formation scale, pressure profiles now available across clay-rock formations at various URLs are still poorly understood. Since they are likely to result, at least partly, from coupled transport and hydro-chemo-mechanical processes, an appropriate interpretation of these pressure profiles provides an opportunity to clarify the transport and swelling properties of natural clay-rocks.

On the pore scale, the analysis has so far been mostly based on the classical double-layer theory. It seems interesting to enlarge its framework in order to cope with non-monovalent electrolytes and with electrolytes flowing in channels that are of a nanometric size. Thermoosmosis is the most weakly constrained both experimentally and theoretically compared to chemical and electrical osmosis. From an experimental standpoint, some well-constrained measurements at the sample scale in natural clay-rocks would improve our understanding. Some molecular fluid calculations exploring the thermoosmotic process, also present a promising prospect for extending the exploratory work outlined here. Although challenging, the introduction of a practical hydro-chemo-mechanical coupling would be helpful to interpret the pressure profiles in clay-rock formations. By using suitable upscaling techniques (e.g., mixture theory, volume averaging, homogenization which may combine linear chemoporoelasticity), the microscopic concept of disjoining pressure, p_N, can be generalized into a macroscopic electrochemical specific storage coefficient $S_S^{\prime chem}$. However, as for the classical specific storage coefficient, the writing of $S_S^{\prime chem}$ includes poromechanical parameters, which may be difficult to measure in clay barriers due to their low permeability. Moreover, the mathematical formulation of $S_S^{\prime chem}$ often requires a numerical value of the microscopic disjoining pressure p_N, which must be experimentally validated.

REFERENCES

Aertsens, M., Wemaerea, I., Woutersb, L., 2004. Spatial variability of transport parameters in the Boom clay. Appl. Clay Sci. 26, 37–45.

Al-Bazali, J., 2005. Experimental Study of the Membrane Behavior of Shale during Interaction with Water-Based and Oil-Based Muds (Ph.D. thesis). University of Texas, Austin, USA.

Biot, M.A., 1941. General theory of three-dimensional consolidation. J. Appl. Phys. 12, 155–164.

Biot, M.A., 1972. Theory of finite deformations of porous solids. Indiana Univ. Math. J. 21, 597–620.

Biot, M.A., 1977. Variational Lagrangian-thermodynamics of nonisothermal finite strain mechanics of porous solids and thermomolecular diffusion. Int. J. Solids Struct. 13, 579–597.

Bennethum, L.S., Murad, M.A., Cushman, J.H., 1997. Modified Darcy's law, Terzaghi's effective stress principle and Fick's law for swelling clay soils. Comput. Geotech. 20, 245–266.

Benhamidaa, A., Djeran-Maigre, I., Dumonteta, H., Smaoui, S., 2005. Clay compaction modelling by homogenization theory. Int. J. Rock Mech. Min. Sci. 42, 996–1005.

Boisson, Y., Bertrand, L., Heitz, J.-F., Golvan, Y.M.-L., 2001. In situ and laboratory investigations of fluid flow through an argillaceous formation at different scales of space and time, Tournemire tunnel, southern France. Hydrogeol. J. 12, 108–123.

Bolt, G.H., 1956. Physico-chemical analysis of the compressibility of pure clays. Geotechnique 6, 86–93.

Bolt, G.H., 1979. Soil Chemistry, B. Physico-Chemical Models. Elsevier, Amsterdam.

Bredehoeft, J., Papadopoulos, S., 1980. A method for determining the hydraulic properties of tight formations. Water Resour. Res. 16, 233–238.

Bredehoeft, J.D., Neuzil, C., Milly, P., 1983. Regional flow in the Dakota aquifer: a study of the role of confining layers. U.S. Geol. Surv. Supply Pap. 2237, 1–45.

Bresler, E., 1973. Anion exclusion and coupling effects in nonsteady transport unsaturated soils: I. Theory. Soil Sci. Soc. Am. Proc. 37, 663–669.

Bunger, A.P., Loret, B., Hueckel, T., 2010. The Mandel-Cryer effect in chemoporoelasticity. Int. J. Numer. Analyt. Meth. Geomech. 34, 1479–1511.

Bunger, A., Sarout, J., Kear, J., Delle Piane, C., Detournay, E., Matthew, J., Dewhurst, D.N., 2014. Experimental chemoporoelastic characterization of shale using millimeter-scale specimens. J. Petrol. Sc. Eng. 118, 40–51.

Cariou, S., Dormieux, L., Skoczylas, F., 2013. An original constitutive law for Callovian-Oxfordian argillite, a two-scale double-porosity material. Appl. Clay Sci. 80, 18–30.

Castro, M.C., Goblet, P., Ledoux, E., Violette, S., de Marsily, G., 1998. Noble gases as natural tracers of water circulation in the Paris basin. Part 2. Calibration of a groundwater flow model using noble gas isotope data. Water Resour. Res. 34, 2467–2483.

Cey, B., Barbour, S.L., Hendry, M.J., 2001. Osmotic flow through a Cretaceous clay in southern Saskatchewan, Canada. Can. Geotech. J. 38, 1025–1033.

Churaev, N.V., Derjaguin, B.V., 1985. Inclusion of structural forces in the theory of stability of colloids and films. J. Colloid Interface Sci. 103, 542–553.

Churaev, N., 2000. Liquid and Vapour Flows in Porous Bodies: Surface Phenomena. Taylor & Francis, London.

Clennel, M.B., Dewhurst, D.N., Brown, K.M., Westbrook, G., 1999. Permeability anisotropy of consolidated clays. In: Muds and Mudstones: Physical and Fluid Flow Properties. Geological Society, London. Special Publication 158.

Coelho, D., Shapiro, M., Thovert, J., Adler, P., 1996. Electroosmotic phenomena in porous media. J. Colloid Interface Sci. 181, 169–190.

Cooper, H.H., Bredehoeft, J.D., Papadopulos, I., 1967. Response of a finite-diameter well to an instantaneous charge of water. Water Resour. Res. 3, 263–269.

Cosenza, P., Ghoreychi, M., de Marsily, G., Vasseur, G., Violette, S., 2002. Theoretical prediction of poroelastic properties of argillaceous rocks from in situ specific storage coefficient. Water Resour. Res. 38, 1207.

Coussy, O., 2004. Poromechanics. John Wiley & Sons, Chichester, UK.

Croisé, J., Schlickenrieder, L., Marschall, P., Boisson, J.Y., Vogel, P., Yamamoto, S., 2004. Hydrogeological investigations in a low permeability claystone formation: the Mont Terri rock laboratory. Phys. Chem. Earth 29, 3–15.

Delay, J., Trouiller, A., Lavanchy, J.-M., 2006. Propriétés hydrodynamiques du Callovien-oxfordien dans l'est du bassin de Paris: Comparaison des résultats obtenus selon différentes approches. C. R. Geosci. 338, 892–907.

Derjaguin, B.V., Sidorenkov, G.P., 1941. On thermo-osmosis of liquid in porous glass. Proc. USSR Acad. Sci. 32, 622–626.

Derjaguin, B.V., Churaev, N.V., 1974. Structural component of disjoining pressure. J. Colloid Interface Sci. 49, 249–255.

Derjaguin, B.V., Churaev, N.V., Muller, V.M., 1987. Surface Forces. Consultant Bureau, New York.

Detournay, E., Cheng, A.H.-D., 1993. Fundamentals of poroelasticity. In: Fairhurst, C. (Ed.), Chapter 5 in Comprehensive Rock Engineering: Principles, Practice and Projects, Analysis and Design Method, vol. II. Pergamon Press, London, pp. 113–171.

Dirksen, C., 1969. Thermo-osmosis through compacted saturated clay membranes. Soil Sci. Soc. Am. Proc. 33, 821–826.

Distinguin, M., Lavanchy, J.-M., 2007. Determination of hydraulic properties of the Callovian-Oxfordian argillite at the Bure site: synthesis of the results obtained in deep boreholes using several in situ investigation techniques. Phys. Chem. Earth 32, 379–392.

Dormieux, L., Barboux, P., Coussy, O., Dangla, P., 1995. A macroscopic model of the swelling phenomenon of a saturated clay. Eur. J. Mech. A Solid. 14, 981–1004.

Dormieux, L., Lemarchand, E., Sanahuja, J., 2006. Macroscopic behavior of porous materials with lamellar microstructure. C. R. Mécanique 334, 304–310.

Dufrêche, J.-F., Marry, V., Malíková, N., Turq, P., 2005. Molecular hydrodynamics for electro-osmosis in clays: from Kubo to Smoluchowski. J. Mol. Liq. 118, 145.

Fernandez-Garcia, D., Gomez-Hernandez, J.J., Mayor, J.-C., 2007. Estimating hydraulic conductivity of the Opalinus clay at the regional scale: combined effect of desaturation and edz. Phys. Chem. Earth 32, 639–645.

Fritz, S.J., Marine, I.W., 1983. Experimental support for a predictive osmotic model of clay membranes. Geochim. Cosmochim. Acta 47, 1515–1522.

Fritz, S.J., 1986. Ideality of clay membranes in osmotic processes: a review. Clays Clay Min. 34, 214–223.

Gajo, A., Loret, B., Hueckel, T., 2002. Electro-chemo-mechanical couplings in saturated porous media: elastic-plastic behaviour of heteroionic expansive clays. Int. J. Solids Struct. 39, 4327–4362.

Garavito, A., Kooi, H., Neuzil, C.E., 2006. Numerical modeling of a long-term in situ chemical osmosis experiment in the Pierre Shale, South Dakota. Adv. Water Resour. 29, 481–492.

Garavito, A., De Cannire, I., Kooi, H., 2007. In situ chemical osmosis experiment in the Boom clay at the Mol underground research laboratory. Phys. Chem. Earth 32, 421–433.

Giraud, A., Rousset, G., 1996. Time-dependent behaviour of deep clays. Eng. Geol. 41, 181–195.

Giraud, A., Huynh, Q.V., Hoxha, D., Kondo, D., 2007. Effective poroelastic properties of transversely isotropic rock-like composites with arbitrarily oriented ellipsoidal inclusions. Mech. Mater. 39, 1006–1024.

Gonçalvès, J., Violette, S., Wending, J., 2004. Analytical and numerical solutions for alternative overpressuring processes: application to the Callovian-Oxfordian sedimentary sequence in the Paris basin, France. J. Geophys. Res. 109, B02110. http://dx.doi.org/10.1029/2002JB002278.

Gonçalvès, J., Rousseau-Gueutin, P., Revil, A., 2007. Introducing interacting diffuse layers in TLM calculations. A reappraisal of the influence of the pore size on the swelling pressure and the osmotic efficiency of compacted bentonites. J. Colloid Interface Sci. 316, 92–99.

Gonçalvès, J., 2008. A slug test to assess the osmotic and hydraulic properties of argillaceous formations. Water Resour. Res. http://dx.doi.org/10.1029/2007WR006547.

Gonçalvès, J., Rousseau-Gueutin, P., de Marsily, G., Cosenza, P., Violette, S., 2010. What is the significance of pore pressure in a saturated shale layer? Water Resour. Res. 46, W04514. http://dx.doi.org/10.1029/2009WR008090.

Gonçalvès, J., Tremosa, J., de Marsily, G., 2012. Importance of thermo-osmosis for fluid flow and transport in clay formations hosting a nuclear waste repository. Earth Planet. Sci. Lett. 339–340, 1–10.

Greenberg, J.A., Mitchell, J.K., Witherspoon, P.A., 1973. Coupled salt and water flows in a groundwater Basin. J. Geophys. Res. 78, 6341–6353.

de Groot, S.R., Mazur, P., 1962. Non-equilibrium Thermodynamics. Dover.

Gupta, A., Coelho, D., Adler, P., 2006. Electroosmosis in porous solids for high zeta potentials. J. Colloid Interface Sci. 303, 593–603.

Gupta, A., Coelho, D., Adler, P., 2008. Universal electro-osmosis formulae for porous media. J. Colloid Interface Sci. 319, 549–554.

Hart, D.J., Bradbury, K.R., Feinstein, D.T., 2006. The vertical hydraulic conductivity of an aquitard at two spatial scales. Groundwater 44, 201–211.

Heidug, W.K., Wong, S.W., 1996. Hydration swelling of water? Absorbing rocks: a constitutive model. Int. J. Numer. Analyt. Meth. Geomech. 20, 403–430.

Heister, K., Kleingeld, P., Loch, J., 2005. Quantifying the effect of membrane potential in chemical osmosis across bentonite membranes by virtual short-circuiting. J. Colloid Interface Sci. 286, 294–302.

Homand, F., Shao, J.-F., Giraud, A., Auvray, C., Hoxha, D., 2006. Petrofabric and mechanical properties of mudstones. C. R. Geosci. 228, 882–891.

Horseman, S., Harrington, J.F., Noy, D.J., 2007. Swelling and osmotic flow in a potential host rock. Phys. Chem. Earth 32, 408–420.

Hsieh, P., Tracy, J., Neuzil, C., Bredehoeft, J., Silliman, S., 1981. A transient laboratory method for determining the hydraulic properties of tight rocks—I. Theory. Int. J. Rock Mech. Min. Sci. 18, 245–252.

Hueckel, T.A., 1992. Water-mineral interaction in hygromechanics of clays exposed to environmental loads: a mixture-theory approach. Can. Geotech. J. 29, 1071–1086.

Johns, R., 1998. Pressure solution for sequential hydraulic tests in low-transmissivity fractured and nonfractured media. Water Resour. Res. 34, 889–895.

Johnson, D.L., Koplik, J., Schwartz, L., 1986. New pore-size parameter characterizing transport in porous media. Phys. Rev. Lett. 57, 2564–2567.

Jost, A., Violette, S., Macquar, J.-C., Dromart, G., 2004. Fluid palaeo-circulation generated by the Pyrenean orogeny and its potential role in the formation of lead-zinc deposits in the Cévennes region: a modelling approach. Bull. Soc. Geol. Fr. 175, 663–670.

Katchalsky, A., Curran, P.F., 1967. Nonequilibrium Thermodynamics in Biophysics. Harvard University Press, Cambridge.

Keijzer, T.J.S., Kleingeld, P.J., Loch, J.P.G., 1999. Chemical osmosis in compacted clayey material and the prediction of water transport. Eng. Geol. 53, 151–159.

Keijzer, T.J.S., Loch, J.P.G., 2001. Chemical osmosis in compacted dredging sludge. Soil Sci. Soc. Am. J. 65, 1045–1055.

Kemper, W., 1960. Water and ion movement in thin films as influenced by the electrostatic charge and diffuse layer of cations associated with clay mineral surfaces. Soil Sci. Soc. Am. Proc. 10–16.

Kemper, W.D., Evans, N.A., 1963. Movement of water as effected by free energy and pressure gradients III. Restriction of solutes by membranes. Soil Sci. Soc. Am. Proc. 27, 485–490.

Kemper, W.D., Rollins, J.B., 1966. Osmotic efficiency coefficients across compacted clays. Soil Sci. Soc. Am. Proc. 30, 529–534.

Kooi, H., Garavito, A.M., Bader, S., 2003. Numerical modelling of chemical osmosis and ultrafiltration across clay formations. J. Geochem. Explor. 78–79, 333–336.

Krief, M., Garta, J., Stellingwerff, J., Ventre, J., 1990. A petrophysical interpretation using the velocities of P and S waves (full-waveform sonic). The Log Anal. 31, 355–369.

Langmuir, I., 1938. The role of attractive and repulsive forces in the formation of tactoids, thixotropic gels, protein crystals and coacervates. J. Chem. Phys. 6, 873–896.

Levich, V.G., 1962. Physico-Chemical Hydrodynamics. Prentice Hall, New Jersey.

Loret, B., Hueckel, T., Gajo, A., 2002. Chemo-mechanical coupling in saturated porous media: elastic-plastic behaviour of homoionic expansive clays. Int. J. Solids Struct. 39, 2773–2806.

Luo, X., Vasseur, G., 1992. Contributions of compaction and aquathermal pressuring to geopressure and the influence of environmental conditions. Am. Assoc. Petrol. Geol. Bull. 76, 1550–1559.

Madsen, F.T., Muller-Vonmoos, P., 1985. Swelling pressure calculated from mineralogical properties of a Jurassic opalinum shale, Switzerland. Clays Clay Min. 33, 501–509.

Malusis, M.A., Shackelford, C.D., 2002. Coupling effects during steady-state solute diffusion through a semi-permeable clay membrane. Environ. Sci. Technol. 36, 1312–1319.

Malusis, M.A., Shackelford, C.D., Olsen, H.W., 2003. Flow and transport through clay membrane barriers. Eng. Geol. 70, 235–248.

Marcelja, S., Radic, N., 1976. Repulsion of interfaces due to bound water. Chem. Phys. Lett. 42, 129–130.

Marschall, P., Horseman, S., Gimmi, T., 2005. Characterisation of gas transport properties of the Opalinus clay, a potential host rock formation for radioactive waste disposal. Oil Gas Sci. Technol. Rev. IFP 60, 121–129.

de Marsily, G., 1986. Quantitative Hydrogeology, Groundwater Hydrology for Engineers. Academic Press.

Martin, S., 1960. Absorded water on clay: a review. Clays Clay Min. 9, 28–70.

McBride, M.B., 1997. A critique of diffuse double layer models applied to colloid and surface chemistry. Clays Clay Min. 45, 598–608.

McKelvey, J., Milne, I., 1960. The flow of salt solutions through compacted clays. Clays Clay Min. 9, 248–269.

Mitchell, J.K., Greenberg, J.A., Witherspoon, P.A., 1973. Chemico-osmotic effects in fine-grained soils. J. Soil Mech. Found. Div. 99, 307–322.

Mitchell, J.K., 1993. Fundamentals of Soil Behavior. John Wiley & Sons, New York.

Moyne, C., Murad, M., 2002. Electro-chemo-mechanical couplings in swelling clays derived from a micro/macro-homogenization procedure. Int. J. Solids Struct. 39, 6159–6190.

Moyne, C., Murad, M.A., 2003. Macroscopic behavior of swelling porous media derived from micromechanical analysis. Transport Porous Med. 50, 127–151.

Munro, R.G., 2004. Analytical representations of elastic moduli data with simultaneous dependence on temperature and porosity. J. Res. Natl. Inst. Stand. Technol. 109, 497–503.

Murad, M.A., Bennethum, L.S., Cushman, J.H., 1995. A multi-scale theory of swelling porous media: I. Application to one-dimensional consolidation. Transport Porous Med. 19, 93–122.

Murad, M.A., Cushman, J.H., 1996. Multiscale flow and deformation in hydrophilic swelling porous media. Int. Eng. Sci. 34, 313–338.

Neuzil, C.E., Cooley, C., Silliman, S.E., Bredehoeft, J.D., Hsieh, P.A., 1981. A transient laboratory method for determining the hydraulic properties of "tight" rocks II. Application. Int. J. Rock Mech. Min. Sci. 18, 253–258.

Neuzil, C.E., 1982. On conducting the modified “slug” test in tight formations. Water Resour. Res. 18, 439–441.

Neuzil, C.E., 1986. Groundwater flow in low permeability environments. Water Resour. Res. 22, 1163–1195.

Neuzil, C.E., 1993. Low fluid pressure within the Pierre Shale: a transient response to erosion. Water Resour. Res. 29, 2007–2020.

Neuzil, C.E., 1994. How permeable are clays and shales. Water Resour. Res. 34, 45–50.

Neuzil, C.E., 2000. Osmotic generation of “anomalous” fluid pressures in geological environments. Nature 403, 182–184.

Neuzil, C.E., Provost, A.M., 2009. Recent experimental data may point to a greater role for osmotic pressures in the subsurface. Water Resour. Res. 45. http://dx.doi.org/10.1029/2007WR006450.

Neuzil, C.E., 2013. Can shale safely host U.S. nuclear waste? Eos 94, 261–268.

Noy, D.J., Horseman, S., Harrington, J.F., Bossart, P., Fisch, H.R., 2004. An Experimental and Modelling Study of Chemico-Osmotic Effects in the Opalinus Clay of Switzerland. Tech. Rep. Federal Office for Water and Geology (FOWG).

Olsen, H.W., 1960. Hydraulic flow through saturated soils. In: Proceedings of the Ninth National Conference on Clays and Clay Minerals. Pergamon Press, London, pp. 131–161.

Olsen, H.W., 1969. Simultaneous fluxes of liquid and charge in saturated kaolinite. Soil Sci. Soc. Am. Proc. 33, 338–344.

Olsen, H.W., 1972. Liquid movement through kaolinite under hydraulic, electric and osmotic gradients. Am. Assoc. Petrol. Geol. Bull. 56, 2022–2028.

Ortega, J.A., Ulm, F.J., Abousleiman, Y., 2007. The effect of the nanogranular nature of shale on their poroelastic behavior. Acta Geotech. 2, 155–182.

Paszkuta, M., Rosanne, M., Adler, P.M., 2006. Transport coefficients of saturated compact clays. C. R. Geosci. 338, 908–916.

Patriarche, D., Ledoux, E., Michelot, J.-L., Simon-Coincon, R., Savoye, S., 2004. Diffusion as the main process for mass transport in very low water content argillites: 2. Fluid flow and mass transport modeling. Water Resour. Res. 40. http://dx.doi.org/10.1029/2003WR002700.

Pengra, D.B., Li, S., Li, S.X., Wong, P.Z., 1995. Dynamics in small confining systems II. Mater. Res. Soc. 366.

Ratman, S., Soga, K., Whittle, R.W., 2005. A field permeability measurement technique using a conventional self-boring pressuremeter. Geotechnique 55, 527–537.

Revil, A., Pessel, M., 2002. Electroosmotic flow and the validity of the Darcy equation in silty shale. Geophys. Res. Lett. 29. http://dx.doi.org/10.1029/2001GL013480.

Revil, A., Leroy, P., 2004. Constitutive equations for ionic transport in porous shales. J. Geophys. Res. 109, B03208. http://dx.doi.org/10.1029/2003JB002755.

Revil, A., Leroy, P., Titov, K., 2005. Characterization of transport properties of argillaceous sediments: application to the Callovian-Oxfordian argillite. J. Geophys. Res. 110, B06202. http://dx.doi.org/10.1029/2004JB003442.

Revil, A., Woodruff, W.F., Lu, N., 2011. Constitutive equations for coupled flows in clay materials. Water Resour. Res. 47, W05548. http://dx.doi.org/10.1029/2010WR010002.

Rivera, A., Ledoux, E., de Marsily, G., 1991. Nonlinear modeling of groundwater flow and total subsidence of the Mexico City aquifer-aquitard system. In: Land Subsidence, 200. IAHS Publication, pp. 45–58.

Rosales, E., Loch, J.P.G., Dias-Ferreira, C., 2014. Electro-osmotic transport of nano zero-valent iron in Boom clay. Electrochim. Acta 127, 27–33.

Rosanne, M., Pastzkuta, M., Thovert, J.-F., Adler, P.M., 2004. Electro-osmotic coupling in compact clays. Geophys. Res. Lett. 31, L18614. http://dx.doi.org/10.1029/2004GL020770.

Rosanne, M., Paszkuta, M., Adler, P., 2006. Electrokinetic phenomena in saturated compact clays. J. Colloid Interface Sci. 297, 353–364.

Rousseau-Gueutin, P., 2008. Les processus couplés dans les argilites du Callovien-Oxfordien sur le site de Bure: implications pour les mouvements de fluide et de solutés (Ph.D. thesis). Université Pierre et Marie Curie.

Rousseau-Gueutin, P., Gonçalvès, J., Violette, S., 2008. Osmotic efficiency in Callovian-Oxfordian argillites: experimental vs. theoretical models. Phys. Chem. Earth 33, S106–S113.

Rousseau-Gueutin, P., de Greef, V., Gonçalvès, J., Violette, S., Chancholle, S., 2009. Experimental device for chemical osmosis measurement on natural clay-rock samples maintained at in situ conditions: implications for formation pressure interpretations. J. Colloid Interface Sci. 337, 106–116.

Rousseau-Gueutin, P., Gonçalvès, J., Cruchaudet, M., Violette, S., de Marsily, G., 2010. Hydraulic and chemical pulse tests in a shut-in chamber imbedded in an argillaceous formation: numerical and experimental approaches. Water Resour. Res. 46, W08516. http://dx.doi.org/10.1029/2008WR007371.

Sarout, J., 2003. Experimental Identification of Chemoporoelastic Parameters for Reactive Shales (Master's thesis). University of Minnesota.

Sarout, J., Detournay, E., 2011. Chemoporoelastic analysis and experimental validation of the pore pressure transmission test for reactive shales. Int. J. Rock Mech. Min. Sci. 48, 759–772.

Schanz, T., Tripathy, S., 2009. Swelling pressure of a divalent-rich bentonite: diffuse double-layer theory revisited. Water Resour. Res. 45, W00C12. http://dx.doi.org/10.1029/2007WR006495.

Shaw, R.J., Hendry, M., 1998. Hydrogeology of a thick clay till and cretaceous clay sequence, Saskatchewan, Canada. Can. Geotech. J. 35, 1041–1052.

Sherwood, J.D., 1993. Biot poroelasticity of a chemically active shale. Proc. Roy. Soc. Lond. Ser. A: Math. Phys. Sci. 440, 365–377.

Sherwood, J.D., 1995. Ionic transport in swelling shale. Adv. Colloid Interface Sci. 61, 51–64.

Soler, J.M., 2001. The effect of coupled transport phenomena in the Opalinus clay and implications for radionuclide transport. J. Contam. Hydrol. 53, 63–84.

Srivastava, R., Avasthi, P., 1975. Non-equilibrium thermodynamics of thermo-osmosis of water though kaolinite. J. Hydrol. 24, 111–120.

Takeda, M., Hiratsuka, T., Manaka, M., Finsterle, S., Ito, K., 2013. Experimental examination of the relationships among chemico-osmotic, hydraulic, and diffusion parameters of Wakkanai mudstones. J. Geophys. Res. Solid Earth 119, 4178–4201.

Tremosa, J., 2010. Influence of Osmotic Processes on the Excess-Hydraulic Head Measured in the Toarcian/Domerian Argillaceous Formation of Tournemire (Ph.D. thesis). Université Pierre et Marie Curie, Paris, France.

Tremosa, J., Gonçalvès, J., Matray, J.-M., Violette, S., 2010. Estimating thermo-osmotic coefficients in clay-rocks: II. In situ experimental approach. J. Colloid Interface Sci. 342, 175–184.

Tremosa, J., Gonçalvès, J., Matray, J.-M., 2012. Natural conditions for more limited osmotic abnormal fluid pressures in sedimentary basins. Water Resour. Res. 48, W04530. http://dx.doi.org/10.1029/2011WR010914.

Valfouskaya, A., Adler, P.M., Thovert, J.-F., Fleury, M., 2005. Nuclear-magnetic-resonance diffusion simulations in porous media. J. Appl. Phys. 97, 083510.

Van Olphen, H., 1963. An Introduction to Clay Colloid Chemistry for Clay Technologists, Geologists and Soil Scientists. Wiley Interscience, New York.

Vargas, C., Ortega-Guerrero, A., 2004. Fracture hydraulic conductivity in the Mexico City clayed aquitard: field piezometer rising-head tests. Hydrogeol. J. 12, 336–344.

Warkentin, B.P., Bolt, G.H., Miller, R.D., 1957. Swelling pressure of montmorillonite. Soil Sci. Soc. Am. J. 21, 495–497.

Wei, H., Ledoux, E., de Marsily, G., 1990. Regional modelling of groundwater flow and salt and environmental tracer transport in deep aquifers in the Paris basin. J. Hydrol. 226, 48–65.

Yang, Y., Aplin, A., 2007. Permeability and petrophysical properties of 30 natural mudstones. J. Geophys. Res. 112. http://dx.doi.org/10.1029/2005JB004243.

Yeung, A.T., Mitchell, J.K., 1993. Coupled fluid, electrical and chemical flows in soil. Gothechnique 43, 121–134.

Young, A., Low, P., 1965. Osmosis in argillaceous rocks. Am. Assoc. Petrol. Geol. Bull. 49, 1004–1008.

Chapter 9

Coupled Thermo-Hydro-Mechanical Behavior of Natural and Engineered Clay Barriers

Jonny Rutqvist
Earth Sciences Department, Lawrence Berkeley National Laboratory, Berkeley, CA, USA

Chapter Outline

9.1 INTRODUCTION

Geological repositories for disposal of high-level nuclear waste generally rely on a multibarrier system to isolate radioactive waste from the biosphere (Ahn and Apted, 2010). The multibarrier system typically consists of the natural barrier system, which includes the repository host rock and its surrounding subsurface environment, and the engineered barrier system (EBS) (Figure 9.1). The bedrock should provide a stable chemical and mechanical environment and impede radionuclides if released. The EBS represents the man-made, engineered materials placed within a repository, including the waste form, waste package (or canister), buffer materials, backfill, and seals. The EBS, including the buffer plays a significant role in the containment and long-term retardation of radionuclide transport if release. The buffer has also the function of providing mechanical stability to the canister as well as mechanical support to the tunnel walls.

Coupled thermo-hydro-mechanical (THM) processes will affect the EBS and the natural barrier system, especially in the early time, during the

Developments in Clay Science, Vol. 6C. http://dx.doi.org/10.1016/B978-0-08-100027-4.00009-7

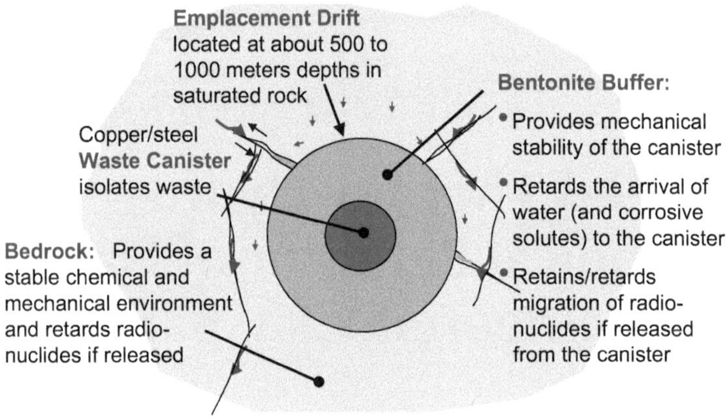

FIGURE 9.1 Different components in a multibarrier geologic nuclear waste repository.

construction, installation, and early postclosure phase when heat release from the decaying waste causes a significant time-varying temperature gradient (Figure 9.2) (Rutqvist et al., 2005). Such thermally-driven coupled THM processes are relatively short-lived from safety assessment perspective, but could potentially give rise to permanent changes such as the formation of a damaged zone around the tunnels that could significantly affect the long-term repository performance (Figure 9.2). The evolution of the excavation damage zone (EDZ), during excavation, installation, and postclosure is critically important for a repository hosted in tight clay, as this could create a transport path for radionuclides if released from a waste package.

A number of performance measures may be defined related to coupled THM processes in a nuclear waste repository, including the evolution in the buffer of (1) temperature and associated maximum temperature, (2) moisture and associated time to full resaturation, and (3) stress and associated possibility of rock failure (Chijimatsu et al., 2005; Rutqvist et al., 2005). The evolution of temperature and saturation are important for the assessment of the performance of the bentonite (Bent) buffer. In general, a maximum temperature may be defined, for example 100 to 150°C, since higher temperatures might induce unwanted chemical changes to the bentonite that in turn could impact the long-term stability of the buffer. The resaturation process would ideally take place uniformly to ensure that the bentonite swells uniformly and thereby prevents high and uneven stressing of the waste package. The stress evolution in both the buffer and rock is important for the structural integrity of the host rock. Any structural changes in the host rock might affect drift stability or rock mass transport properties. Those changes should be quantified to assess their impact on the performance of the repository as a whole.

Over the past thirty years, laboratory and field experiments as well as numerical modeling have resulted in a substantial progress in the understanding

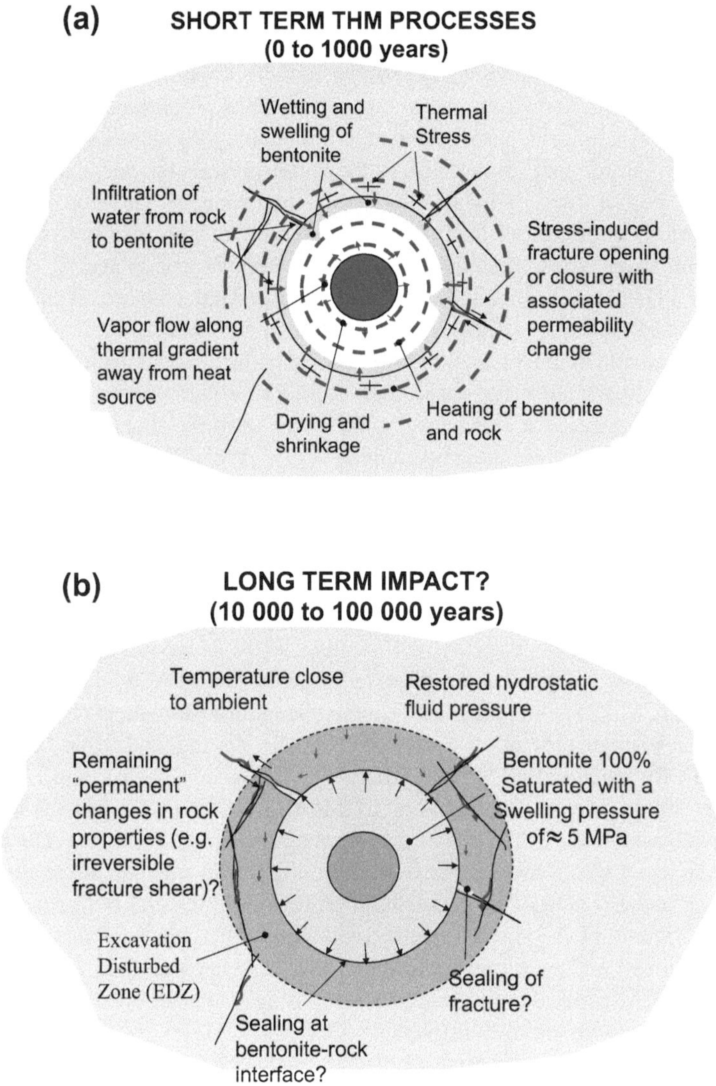

FIGURE 9.2 (a) Coupled thermo-hydro-mechanical (THM) processes driven by heat released from the waste package, and (b) potential long-term impact of early time coupled THM processes.

of coupled THM processes in clay barriers associated with nuclear waste isolation. Clay host rocks are currently being studied in countries where such formations exist and considered an alternative for waste disposal, including France, Switzerland, Germany, Japan, and more recently in Canada and the United States. Coupled THM processes related to waste disposal are best investigated in situ through multiyear heater experiments. Such heater experiments, emulating the heat release from a real waste package, have been conducted at underground research laboratories (URL), including Mont Terri in Switzerland, Meuse/Haute-Marne (Bure) and Tournemire in France, Hades (Mol) in Belgium, and Horonobe in Japan. The THM properties among the host rocks at those site varies widely, from the hard and tight indurated clay-rocks at Mont Terri, Bure, and Tournemire, to the softer plastic clay-rocks at Mol, and the very porous and permeable siliceous mudstone at the Horonobe site. Moreover, THM processes in bentonite-based barriers have also been studied at crystalline rock sites, including heater experiments conducted at Äspö in Sweden (Thomas et al., 2014), Grimsel, Switzerland (Alonso et al., 2005), Kamaishi Mine, Japan (Rutqvist et al., 2001a), and Whiteshell Laboratories, Pinawa, Canada (Gou et al., 2005). A lot can be learned from small-scale laboratory experiments and large-scale field experiments, whereas the long-term performance of a repository system can only be analyzed with a numerical model that simulates the coupled and highly nonlinear behavior of all components (canister, buffer, fractured rock, and water) and their interactions.

National nuclear waste disposal programs as well as international cooperative projects, such as DECOVALEX (DEvelopment of COupled models and their VALidation against EXperiments) have been instrumental in progressing the state-of-the-art in modeling coupled THM processes in geological media. Numerical models developed for modeling THM behavior of clay barriers associated with nuclear waste disposal, include COMPASS (Thomas et al., 1998), ROCMAS (Noorishad and Tsang, 1996; Rutqvist et al., 2001b), CODE-BRIGHT (Olivella et al., 1994), OpenGeoSys (Wang and Kolditz, 2007), THAMES (Ohnishi et al., 1996), ABAQUS (Börgesson, 1996), and FLAC3D (Itasca, 2009). The first five models (COMPASS, ROCMAS, CODE-BRIGHT, OpenGeoSys, and THAMES) are foremost research codes developed in academia and they have been extensively applied in nuclear waste disposal research using constitutive models tailored for both rock and buffer materials. ABAQUS and FLAC3D are commercial codes that have been extensively applied in soil and rock mechanics, and have also been applied for modeling coupled THM processes associated with geologic nuclear waste disposal. In addition, the commercial multiphysics package COMSOL, based on the finite element method, is now being more commonly applied in this field. In this chapter, some applications will be presented in which coupled process modeling were conducted using TOUGH-FLAC (Rutqvist, 2011), a simulator based linking the multiphase flow and heat transport code TOUGH2 (Pruess et al., 2012) with the geomechanical simulator FLAC3D (Itasca, 2009).

This chapter first provides an overview of THM behavior and properties of bentonite-based buffer and backfill materials and clay host rocks based on research finding over the past thirty years. Thereafter, coupled THM modeling simulations are presented that describes short- and long-term coupled THM processes occurring in and around a bentonite-backfilled emplacement tunnel hosted in indurated clay. The coupled modeling enables studying interactions between the buffer and host rock; interactions are strong and persistent for thousands of years. The chapter ends with a discussion on potential links between coupled THM processes and chemistry, and a few concluding remarks.

9.2 THM BEHAVIOR OF BUFFER AND BACKFILL MATERIAL

Bentonite is a main component of the EBS considered in multibarrier nuclear waste disposal concepts worldwide, in various types of host rock formations. Bentonite is a clay with a high content of smectites, having properties very suitable for isolating radioactive waste canisters. It has a very low hydraulic conductivity and very low anion diffusion capacity and low transport capacity of positive charged radionuclides. Thus it can serve as a good buffer around the waste canisters. Mechanically, its high swelling potential and plastic ductile behavior assures self-sealing of openings within the EBS system and of gaps between the bentonite, the rock and the waste package. Also it can provide sufficient bearing capacity to carry the heavy canisters with only insignificant movement. Its thermal conductivity is adequate to conduct heat away from the canister. Further it has good colloid and microbial filtration capability.

When a Bent buffer is installed around waste canisters in a repository, it is initially partially saturated with water. As illustrated in Figure 9.2(a), the heat released from radioactive waste heats the EBS and near-field rock system. Processes involved in the evolution of the Bent buffer include thermally induced distribution of initial pore water in the clay during the early thermal phases. On the outer part of the Bent buffer, water is uptaken from the rock–water interface, with potential swelling of the bentonite first initiated in this region. On the inner part of the Bent buffer, next to the heat-releasing waste package, moisture content decreases (desiccation), with potential shrinkage. These processes are strongly dependent on the coupled THM properties of the buffer material and the properties vary significantly depending on the type and composition of the buffer material.

A number of different bentonite-based buffer and backfill materials exist, most commonly including sodium Bent, but some with calcium Bent, each having different THM characteristics. The Bent buffer may be emplaced as powder and compacted in place (at least in vertical deposition holes) or emplaced using closely fitting prefabricated blocks of highly compacted Bent. Granular (pellets) Bent is another option currently tested for horizontal emplacement at the Mont Terri. Compaction of the buffer to high density assures low permeability, high swelling pressure, and adequate thermal conductivity. The most fundamental and important property of Bent, at least from a THM perspective, is its swelling capacity. An adequate swelling of the buffer in place will assure self-sealing of any voids and will also provide mechanical support to the tunnel walls and thereby contribute to the sealing of the EDZ. In general, sodium Bent is considered to have higher swelling capacity than calcium Bent; however, at high dry-density calcium Bent can have higher swelling pressure (Komine, 2004). Regardless of sodium or calcium Bent, the swelling capacity is strongly dependent on the montmorillonite content. For example, experimental data of swelling pressure for four different types of bentonite as a function of dry-density are reported in Figure 9.3. Sample C and D showing the highest swelling stress are calcium Bent with 80% and 76% montmorillonite, respectively, whereas sample A and B are sodium Bent with 48% and 69% montmorillonite, respectively. There may also be differences regarding the long-term stability of buffer material related to chemical changes affecting the swelling capacity, including impact of potential saline water intrusion or illitization. These are aspects that have to be considered when selecting and designing appropriate buffer and backfill materials for a nuclear waste repository.

The properties of Bent buffer and backfill materials for use in predictive numerical simulations of coupled THM processes are usually determined from laboratory scale experiments, either directly or by back-analysis from numerical modeling of the experiments (e.g., Börgesson et al., 2001). Key THM properties needed for a coupled THM analysis include thermal conductivity, water retention characteristics, permeability, diffusion coefficients, swelling properties, and other mechanical properties such as bulk modulus. One important characteristic of bentonite is that such properties are strongly dependent on both moisture content and density (or compaction). The emplacement of the bentonite is an important engineering task that should

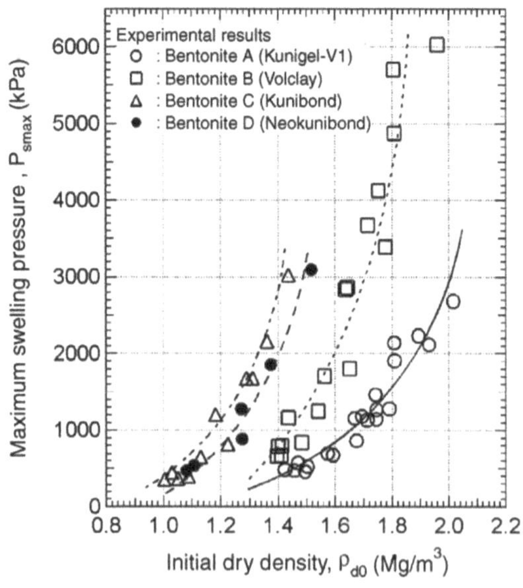

FIGURE 9.3 Relationship between maximum swelling pressure and initial dry density for four different bentonite materials (Komine, 2004).

assure that sufficient and uniform density is achieved. During operation and postclosure, the buffer will be confined within the much stiffer rock and can only expand or contract locally from a designed initial dry density, e.g., 1600 kg/m^3 in some repository concepts.

Thermal conductivity and its variation with water content is one of the key properties for predicting the THM evolution within a Bent buffer. A Bent buffer material can have substantially lower thermal conductivity at dry conditions than at wet conditions and it is very important to have a good estimate of the buffer thermal conductivity to predict the early time temperature evolution and the peak temperature. Thermal conductivity has been measured as a function of water content in compacted samples and is generally well characterized for precompacted Bent blocks and powder Bent that can be compacted in place for vertical deposition holes (e.g., Figure 9.4(a)). For granular Bent, fewer data exist and are not as straightforward to evaluate regarding thermal conductivity (Gaus et al., 2014). Also, Bent pellets when emplaced are

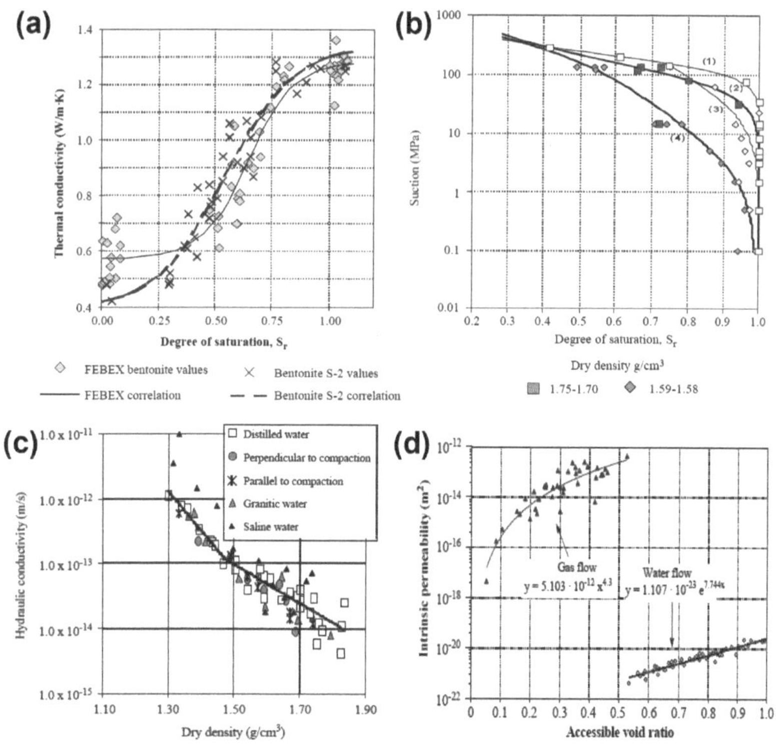

FIGURE 9.4 Laboratory data of FEBEX bentonite for some key THM parameters required for model coupled THM processes (Alonso et al., 2005). (a) Thermal conductivity as function of liquid saturation, (b) water retention curves for different dry densities, (c) hydraulic conductivity as a function of dry density, and (d) intrinsic permeability for gas and liquid flow.

highly compacted and having a saturation of only a few percent. According to experience at the recent HE-E experiment at Mont Terri the pellets quickly took up moisture from ambient air-humidity leading to an initial saturation of about 20% after emplacement of the granular buffer material. Thus, the initial thermal conductivity for such a system would be the thermal conductivity at 20% saturation and at the initial in-place buffer density.

The retention curve and permeability of a Bent buffer are important parameters for the rate of water uptake and the time to full resaturation of the buffer (Rutqvist et al., 2014a). The retention curve and intrinsic permeability can be obtained directly from laboratory experiments conducted at different dry-density (Figure 9.4(b) and (c)), whereas relative permeability can be determined by model calibration against infiltration experiments (e.g., Börgesson et al., 2001). The retention curve depends on dry-density (and porosity) and temperature, indicating other couplings between hydraulic, mechanical, and thermal processes. Such dependencies should be included in the constitutive models and input parameters in a coupled THM analysis of a repository system.

Another process that can significantly impact the evolution of liquid saturation, at least in the early time, is vapor flow along the thermal gradient, or thermal diffusion. In this context, a commonly used approach for modeling thermal diffusion is considered in the classical approach due to Phillip and de Vries (1957). One basic assumption is equivalent to that of Richard's approach for unsaturated flow modeling; the gas pressure p_g is assumed constant and equal to atmospheric pressure throughout the porous medium. As a consequence, vapor transport is assumed to occur only through molecular diffusion driven by a gradient in vapor concentration (density), while advection of vapor with bulk gas flow is neglected. As in the general case, the vapor density, ρ_v, and the relative humidity, r_H, of the moist air in the porous medium are governed by Kelvin's relation (Bear and Bachmat, 1991):

$$\rho_v = \rho_{vS}(T) r_H = \rho_{vS} \exp\left(\frac{p}{\rho_l R_v T}\right) \tag{9.1}$$

where p is pore pressure, T is the temperature, ρ_{vS} is the temperature-dependent saturated vapor density, ρ_l, and R_v is the specific gas constant for water vapor. Assuming thermodynamic equilibrium for pore liquid in contact with its vapor, phase changes occur as evaporation–condensation processes. With the assumption of atmospheric gas pressure (and thus $p = p_l = -p_c$), the liquid water and vapor flux (q_{rl} and q_{rv}) depends respectively on the hydraulic potential and vapor density gradients according to:

$$\mathbf{q}_{rl} = -\rho_l \frac{\mathbf{k} k_{rl}}{\mu_l} (\nabla P - \rho_l g \nabla z) \tag{9.2}$$

and

$$\mathbf{q}_{rv} = -D_v \mathbf{I} \nabla \rho_v \tag{9.3}$$

where **k** is the intrinsic permeability tensor, k_{rl} is relative permeability for liquid flow, D_v is effective molecular diffusion coefficient for vapor, ρ_l and ρ_v are respectively liquid and vapor densities, g is gravity, and z is elevation.

The vapor gradient in Eqn (9.3) is further expanded using Eqn (9.1), leading to the following expression for the vapor flux (Rutqvist et al., 2001b):

$$\mathbf{q}_{rv} = -\rho_l(D_{Pv}\mathbf{I}\nabla P_l + D_{Tv}\mathbf{I}\nabla T) \tag{9.4}$$

where

$$D_{Pv} = \frac{D_v \rho_v}{\rho_l^2 R_v T} \tag{9.5}$$

is the isothermal vapor diffusivity, and

$$D_{Tv} = f_{Tv} D_v \left(\frac{RH}{\rho_l} \frac{\partial \rho_{vS}}{\partial t} - \frac{\rho_v P_l}{\rho_l^2 R_v T^2} \right) \tag{9.6}$$

is the thermal vapor diffusivity.

All the parameters in Eqns (9.5) and (9.6) are known from other constitutive relations except for f_{Tv}, a thermal diffusion enhancement factor (Philip and de Vries, 1957), and the effective molecular diffusion coefficient, D_v, which should be experimentally determined. Thermal diffusivity can be back-calculated from numerical modeling of a thermal gradient test, in which a clay sample is exposed to a thermal gradient and the evolution of moisture content is monitored at several points. Such back-calculation may also be used to back-calculate a tortuously factor. For example, Rutqvist and Tsang (2004), back-calculated a tortuosity factor of 0.8 and a thermal diffusion enhancement factor of 2 for FEBEX Bent. Moreover, the basic diffusion coefficient, D_v, and then also the thermal vapor diffusivity D_{Tv} are temperature depended, with increasing diffusivity at higher temperature. Such temperature dependency on the diffusion coefficients may also be back-calculated from the thermal gradient tests (Börgesson et al., 2001; Rutqvist et al., 2001b).

The assumption of a constant gas pressure would require that gas permeability would be very high; otherwise gas pressure would go up at the hot waste package due to evaporation of liquid water. In fact, multiphase flow modeling results show that to model the typical saturation evolution observed in a Bent buffer surrounding a heat source, the intrinsic gas permeability must be several orders of magnitude higher than the intrinsic permeability for liquid flow. Otherwise, the gas pressure will increase and this in turn will suppress vapor diffusion along the thermal gradient. As a result, drying at the inner parts of the buffer would be prevented. A high intrinsic gas permeability, about six orders higher than the intrinsic permeability for water flow has also been observed in laboratory measurements of clay barriers (Olivella and Gens, 2000). This has also been observed for FEBEX Bent (Figure 9.14(d)). These differences are due to changes in the microstructure of the compacted material when water content changes (Alonso et al., 2005).

Regarding mechanical behavior, the evolution of swelling and the swelling pressure at full saturation of the buffer are perhaps the most important features to predict in a numerical modeling of coupled THM processes in nuclear waste isolation. The simplest approach is to assume a linear elastic model with a moisture swelling strain, $\Delta\varepsilon_{sw}$, being a function of changes in the degree of liquid saturation, S_l (Rutqvist et al., 2001b, 2011)

$$\Delta\sigma'_{sw} = 3K\Delta\varepsilon_{sw} = K\Delta S_l \beta_{sw} \tag{9.7}$$

where $\Delta\sigma'_{sw}$ is the induced swelling stress (an effective stress), K is the bulk modulus, and β_{sw} is a moisture swelling coefficient. The value of the swelling coefficient could be calibrated to produce a desired maximum swelling stress for a given bulk modulus. After full saturation, the pressure-induced deformations and stress would be governed by a conventional effective stress law, where the effective mean stress, σ'_m, is calculated as $\sigma'_m = \sigma_m - p_l$, where σ_m is the (total) mean stress and p_l is the liquid fluid pressure. This is a rational approach for calculating the swelling stress that will prevail after full resaturation of the buffer, but would not accurately represent the real stress-and-strain path to full swelling. In a bentonite, most swelling would occur at high saturation whereas at low saturation (dry conditions), the bentonite becomes very stiff. A better representation of such behavior can be achieved with a nonlinear elastic model, such as a state-surface model. For example, the logarithmic state surface model proposed by Lloret and Alonso (1985) has been applied for bentonite (Rutqvist and Tsang, 2004). In a state-surface model, void ratio (e) is a function of both net mean stress, ($\sigma''_m = \sigma_m - p_g$, where σ_m = total mean stress and p_g is gas pressure) and suction ($s = p_g - p_l$, where p_g and p_l are gas and liquid pressures, respectively).

In Rutqvist and Tsang (2004), the Lloret and Alonso (1985) state-surface model was applied for an analysis of the FEBEX in situ heater experiment at the Grimsel Test Site in Switzerland. The heater experiment included two heaters in a tunnel (2.27 m in diameter) surrounded by a barrier (0.64 m in thickness) made of highly compacted blocks of calcium Bent. The permeability of the surround granitic rock was on the order of $1 \times 10^{-17}\ m^2$, which is much higher than that of the Bent buffer. This means that water was continuously supplied from the higher permeability rock and did not impede the buffer resaturation process.

The state surface model applied for modeling the mechanical behavior of the FEBEX Bent is shown in Figure 9.5(a), and is defined by

$$e = 1.03 - 0.49 \log \sigma''_m - 0.17 \log s + 0.23 \log \sigma''_m \log s \tag{9.8}$$

where the coefficients 1.03, −0.49, −0.17, and 0.23 were determined from oedometer experiments on compacted FEBEX Bent (Rutqvist and Tsang, 2004). For this model the bulk modulus varies with suction according to Figure 9.5(b). It illustrates how the bentonite can change from a very stiff material at dry conditions (high suction) to a very soft one at wet conditions

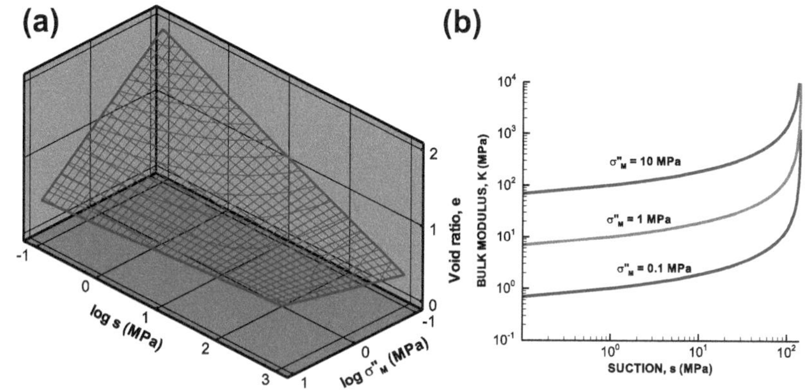

FIGURE 9.5 (a) State-surface model of the FEBEX bentonite in which void ratio is a function of both suction and net mean stress, and (b) variation of bulk modulus with net mean stress and suction.

(low suction). In the field experiment, the Bent buffer was emplaced as compacted blocks to achieve a dry density of 1600 kg/m^3 and a swelling pressure of about 5 MPa at full saturation. Other material input parameters are given in Rutqvist and Tsang (2004), including bentonite properties derived from the laboratory data shown in Figure 9.4.

The model simulation of the FEBEX test was conducted with the ROC-MAS code, which is a finite element code using the Phillip and de Vries (1957) approach for modeling of moisture and heat flow (Rutqvist et al., 2001b). The model and finite element mesh are shown in Figure 9.6, while simulation results are shown in Figure 9.7 with comparison to measured data. The results in Figure 9.7 are quite typical for predicting bentonite THM processes in a heater experiment. Typically the temperature evolution can be quite accurately predicted, whereas the evolution of moisture content may not be as accurately predicted at some locations, and the exact evolution of stress is the most

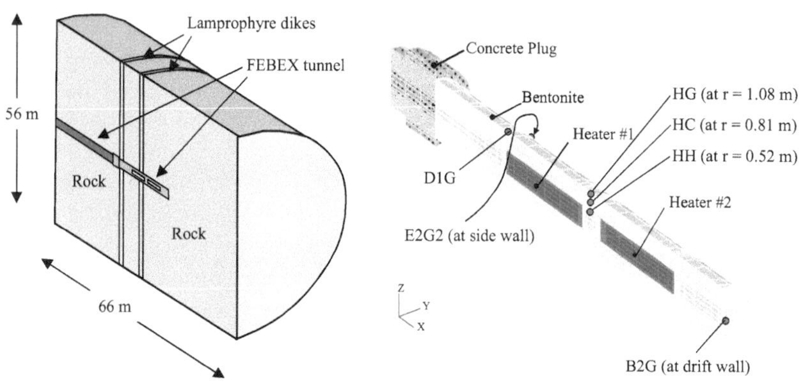

FIGURE 9.6 Finite element model of FEBEX heater experiment (Rutqvist and Tsang, 2004).

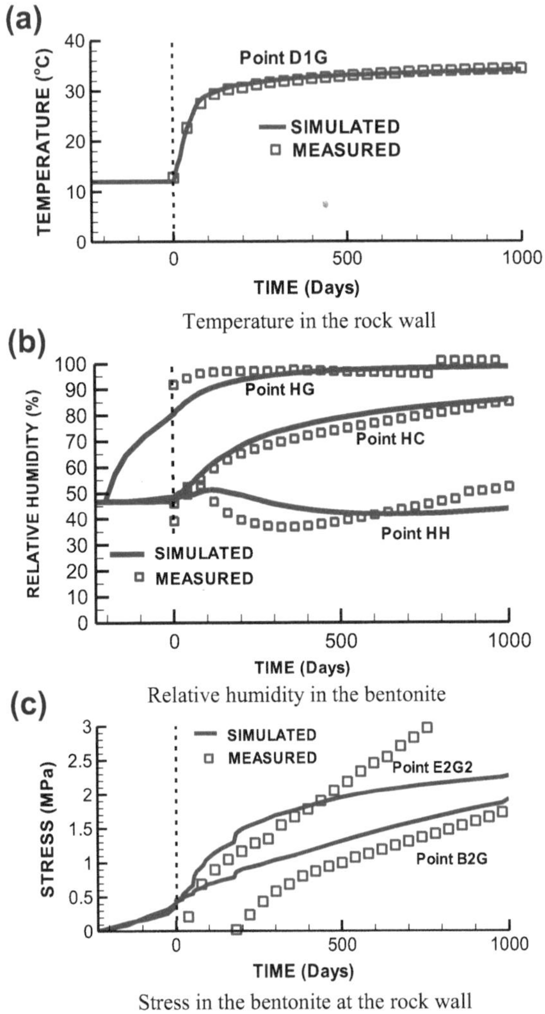

FIGURE 9.7 Comparison of model predictions with measured data during the FEBEX in situ test. (a) Temperature in the rock wall (b) relative humidity in the bentonite (c) stress in the bentonite at the rock wall (Rutqvist and Tsang, 2004). (see point locations in Figure 9.6).

challenging to predict. In this case there were some difference in the stress evolution due to the existence of gaps between the Bent blocks and the ceiling of the tunnel, and there are some other systematic differences. Note though that stress is difficult to measure in the field and this might also be a contributing reason to the disagreement between modeling results and measured data.

More advanced mechanical constitutive models are now being applied for modeling bentonite and backfill material, including the Barcelona Basic

Model (BBM). The BBM was first developed and presented in the early 1990s as an extension of the Modified Cam Clay (MCC) model to unsaturated soil conditions (Alonso et al., 1990). The model can describe many typical features of unsaturated soil mechanical behavior, including wetting-induced swelling or collapse strains, depending on the magnitude of applied stress, as well as the increase in shear strength and apparent preconsolidation stress with suction (Gens et al., 2006). The BBM has been used for modeling of Bent buffer behavior in various national nuclear waste programs in Europe and Japan. For example, the BBM was successfully applied to model coupled THM behavior of unsaturated bentonite associated with the FEBEX in situ heater test at Grimsel (Gens et al., 2009). The BBM has also been applied to other types of bentonite −sand mixtures based on MX-80 Bent (Kristensson and Åkesson, 2008). A thermo-elasto-plastic version of the BBM (Gens, 1995) has also been implemented and applied using TOUGH-FLAC (Rutqvist et al., 2011). The three-dimensional yield surface in p'-q-s space and p'-q-T space is presented in Figure 9.8. Here, p' is net mean stress (i.e., total stress minus gas-phase pressure), q is deviatoric stress (or shear stress), s is suction, and T is temperature (Gens, 1995). Under water-saturated conditions ($s = 0$), the yield surface corresponds to the MCC ellipse (Roscoe and Burgland, 1968), and the size of the elastic domain increases as suction increases. The rate of increase, represented by the loading-collapse (LC) curve, is one of the fundamental characteristics of the BBM (Gens et al., 2006). If applying the BBM model, a large number of input parameters need to be determined, e.g., up to 21 for the full thermo-elasto-plastic version. Among those is the preconsolidation pressure, which should represent the highest mean stress the material ever experienced before. In the case of compacted Bent blocks used in FEBEX, Gens et al. (2009) used a preconsolidation pressure of 12 MPa with the consequence that stress state in the buffer is likely to stay within the elastic

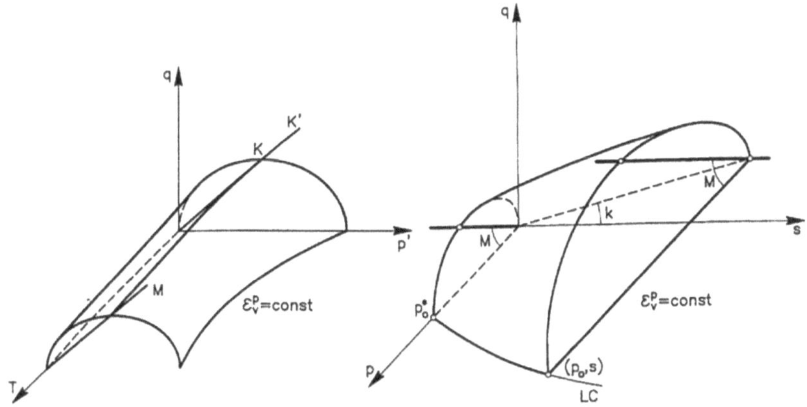

FIGURE 9.8 Three-dimensional representation of the yield surface in the thermo-elasto-plastic Barcelona Basic Model (Gens, 1995).

domain, where the stress and strain is governed by nonlinear elasticity, similar to that of the previously mentioned state-surface model. However, Alonso and Hoffman (2007), in modeling the Mont Terri EB experiment assigned 1 MPa preconsolidation pressure for a granular Bent buffer emplaced in a horizontal tunnel. The experimental results showed locally a clear pore-collapse response (a drop in the swelling stress), but this was not captured in the model despite, using a model that would be suitable for modeling such phenomenon. Overall, the swelling stress evolution was captured reasonably well in the model, with stress increasing to about 1 MPa over the two-year experiment (Alonso and Hoffman, 2007).

A further extension of the BBM is the Barcelona Expansive Model (BExM), which can model some additional features that might be very important for modeling THM processes in a Bent backfilled repository. This includes accumulation of irreversible strain upon suction cycles and effects of microstructural swelling on water permeability. The effect of microstructural swelling on water permeability has been observed to have a significant effect on resaturation of a Bent buffer, for example at the 10-year-long mock-up test performed at CIEMAT, Madrid (Spain), which was a large-scale heating test of a Bent buffer (Sánches et al., 2012). To rigorously model these characteristics of expansive soils, it is necessary to use a dual structure constitutive model that acknowledges the existence of two porosity levels (Gens and Alonso, 1992; Alonso et al., 1999; Sánchez et al., 2005; Gens et al., 2006; Sánchez et al., 2008). Expansive clay materials present a bimodal or trimodal pore size distribution in which the pores having a diameter smaller than 0.1 μm correspond to voids inside clay mineral aggregates and the larger pores correspond to pores between clay mineral aggregates (Figure 9.9). This leads to the differentiation of a microstructure, which is made of the active clay minerals, and a macrostructure, which is formed by the global distribution of clay mineral aggregates and the macropores between them. Fluids can flow solely through the macroporosity (Wang et al., 2013). As the liquid saturation degree of the clay increases (suction decreases), water layers are inserted between the clay

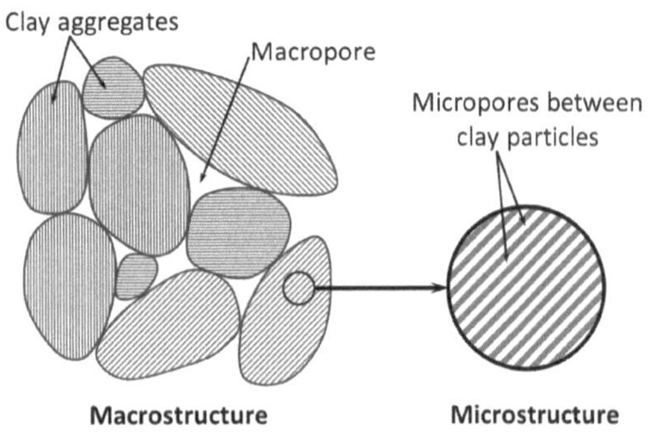

FIGURE 9.9 Schematic representation of the two structural levels considered in the dual structure model. Clay mineral particles are represented by the gray lines.

mineral layers and thus swelling occurs as the volume of clay mineral particles increases. This causes the microporosity to invade the macroporosity as the expansive soil hydrates, which may reduce its permeability by several orders of magnitude (Olivella and Gens, 2000; Alonso and Hoffmann, 2007). A recent modeling with TOUGH-FLAC (Rutqvist et al., 2014b), showed that such micro- and macrostructural interactions, and associated effects on permeability, can be important for the longer-term resaturation process.

9.3 THM BEHAVIOR OF CLAY HOST ROCKS

Clay formations, or clay-rock in their natural state, exhibit very favorable conditions for disposal of radioactive waste, as they generally have a very low and uniform permeability (i.e., 10^{-17} to 10^{-23} m^2), low diffusion coefficients, and good retention capacity for radionuclides (Blümling et al., 2007). The low permeability characteristic is also enhanced by the mechanics of clay formations; brittle fracturing in a clay formation is typically followed by fracture self-sealing once the conditions leading to rock failure have abated. The ability of a clay-rock to provide a low permeability environment over long time periods is affected by conditions of rock stress and stress history, diagenetic alteration, and pore fluid pressure (Nygård et al., 2006). Here clay-rock is defined as a general category encompassing fine-grained sedimentary rock that is either predominately clay-sized particles or clay and silt, including mudstone, claystone, siltstone, and their laminated shale versions.

The tightness, low thermal conductivity, and deformability (compressibility), associated with clay-rock generally results in strong THM couplings, including poroelastic responses and the so-called thermal pressurization, especially when pores are water saturated (Gens et al., 2007; Rutqvist et al., 2014a, Yu et al., 2014). Thermal pressurization occurs when the pore-water is heated in the porous clay in which the solid phase has a much lower thermal expansion coefficient than water. The phenomenon of thermal pressurization is frequently observed during underground heating experiments in tight rocks. An example of simulated and observed pore-pressure responses associated with of the Mont Terri HE-D experiment is shown in Figure 9.10 (Rutqvist et al., 2013). Opalinus clay was heated from two heaters for about one year, in two heating steps, with a threefold power increase in the second step. In Figure 9.10, the comparison of simulated and observed temperature and pressure is made at monitoring point B15, located at about 1−1.5 m from one of the heaters. A generally good agreement with measured temperature was achieved when assigning anisotropic thermal conductivity values of $K_{\|} = 2.15$ W/(m K) along beddings and $K_{\perp} = 1.19$ W/(m K) normal to the beddings, i.e., using an anisotropy factor $K_{\|}/K_{\perp} = 1.8$. Moreover, by assigning a low permeability of 5×10^{-20} m^2, and pore compressibility 1×10^{-9} Pa^{-1}, a reasonably good agreement was also obtained regarding the pressure evolution. In theory, the increase in pore-pressure by thermal

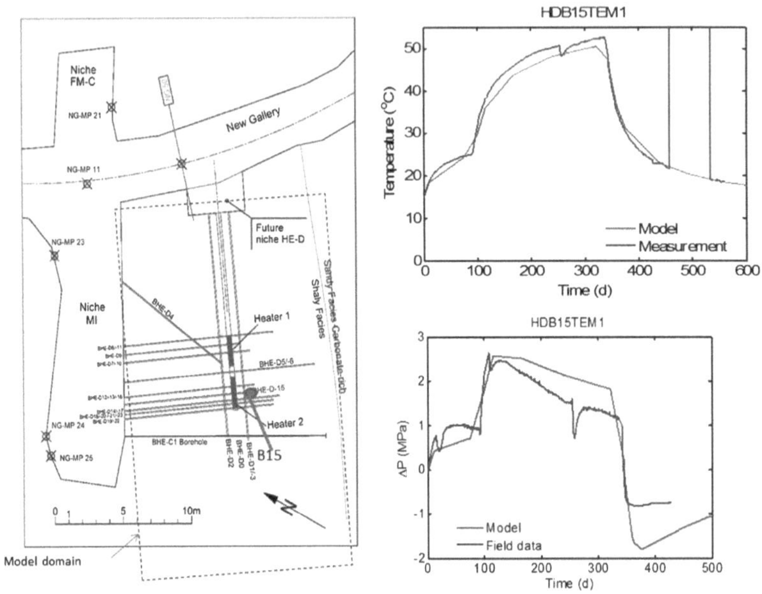

FIGURE 9.10 Simulation results and field monitoring data of temperature and pressure response at the Mont Terri HE-D heater experiment signifying thermal pressurization in Opalinus clay (Rutqvist et al., 2013).

pressurization could be so high that it could induce fracturing or shear slip along existing fractures or along shale beddings. As will be shown in Section 9.4, for a repository hosted in tight clay, the massive heating and associated thermal pressurization may induce elevated fluid pressure that could persist for thousands of years.

The formation and evolution of a disturbed zone around the tunnels and deposition holes is an important issue in nuclear waste isolation in clay-rock. Generally, the disturbance may be divided into excavation disturbed (EdZ) and damaged zones. The EdZ is a zone with hydromechanical and geochemical modifications, without major changes in flow and transport properties, whereas the EDZ is a zone with hydromechanical and geochemical modifications, inducing significant changes in flow and transport properties (Tsang et al., 2005). The formation of the EDZ will be controlled by the initial stress field, the material properties (e.g., material anisotropy), the existence of natural fracture zones or local inhomogeneities of the rock mass, and the geometry of the tunnel. Comprehensive investigations at different sites (e.g., Mont Terri and Hades) have shown that an EDZ occurs in soft or plastic clays as well as in indurated and more brittle clay stones (Blümling et al., 2007). A conceptual model and extent of EDZ in the case of tunneling along the minimum principal stress direction in indurated clay at the Bure site is shown in Figure 9.11. The site investigation shows that shear fractures extending farthest from the tunnel

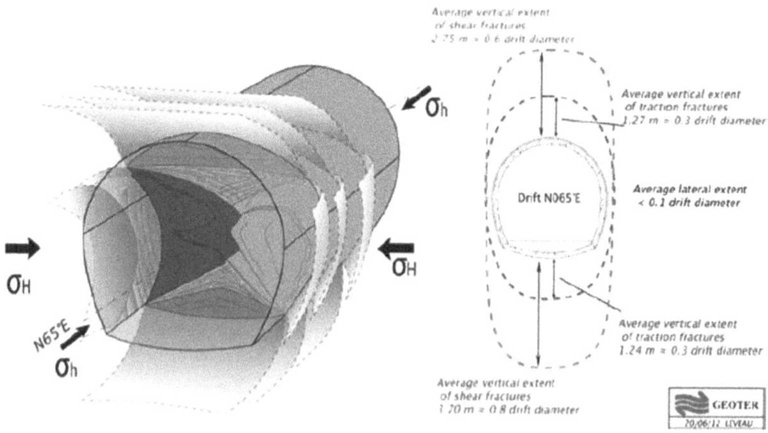

FIGURE 9.11 A conceptual model and extent of excavation damage zone in the case of tunneling along the minimum principal stress direction at the Meuse/Haute-Marne underground research laboratories drifts, Bure (Armand et al., 2014).

are not permeable, whereas close to the walls and the crown of the tunnel, mixed shear and extension fractures unloaded against the free face tunnel wall result in significantly higher permeability, e.g., above $1 \times 10^{-17}\ m^2$.

The long-term behavior of the tunnel near-field can be significantly influenced by adequate support measures and the time-dependent evolution of the EDZ before the emplacement of the waste and the backfilling. The properties of the initial EDZ alter significantly during the transient phase, when the buffer and rock mass are heated by the heat-producing waste and become saturated due to the flow of formation water from the host rock. Experimental results in the laboratory and in situ clearly show that (self-) sealing leads to a significant reduction in the effective hydraulic conductivity of the EDZ with time, thus reducing the potential flow along underground excavations (Blümling et al., 2007). The processes which are responsible for such a reduction in hydraulic conductivity during saturation and water uptake may be related to accelerated creep, disintegration of the rock, swelling, and shear compaction (Blümling et al., 2007). The support of the swelling buffer and backfill on the EDZ can be instrumental in tightening fractures in the EDZ. For example, in situ experiments on fractures at Mont Terri have shown a strong dependency of permeability on effective normal stress, with fractures exhibiting transmissivity values close to that of intact rock if effective normal stress is higher than a few MPa. This means, that a swelling stress of a few MPa within the buffer could reduce the permeability of the EDZ significantly and may assure sealing in the long term.

In the context of the long-term behavior of clay barriers, there is a marked difference between sealing and healing. Sealing indicates solely a reduction in hydraulic properties whereas the mechanical parameters of the damaged rock will still prevail while the healing process would wipe out the memory of the disturbance, resulting in a material that can hardly be distinguished from the original undamaged rock (Tsang et al., 2005). However, numerous laboratory and field investigations on various clay formations have shown no indication of healing (Yu et al., 2014). On the contrary, the reappearance of the fractures or holes in the samples after dismantling confirms that the original mechanical properties cannot be recovered.

The impact of the thermal transient on the evolution of the EDZ was studied in the European Union project TIMODAZ (thermal impact on the damaged zone around a radioactive waste disposal in clay host rocks, 2006–2010). A number of laboratory and small-scale field experiments showed no evidence of temperature-induced opening of fractures, nor of a significant permeability increase of the EDZ (Yu et al., 2014). Instead, thermally induced plasticity, swelling, and creep seem to be beneficial to the sealing of fractures and to the recovery of a very low permeability in the EDZ. However, the sealing of fractures would require fractures being under compression, whereas after excavation, confining stress of the tunnel wall may remain low for long time, especially in the case of slow buffer resaturation. At the same time, thermal pressurization might increase fluid pressure near the tunnel walls. Such transients are studied in the coupled THM numerical model simulations in the next section.

9.4 COUPLED THM EVOLUTION OF ENGINEERED AND NATURAL CLAY BARRIERS IN A NUCLEAR WASTE REPOSITORY

In this section, recent simulation results are presented related to a generic repository case assuming an EBS design with waste emplacement in horizontal tunnels back-filled with Bent as a protective buffer (Figure 9.12). This simulation study was part of the ongoing US Department of Energy, Office of Nuclear Energy's Used Fuel Disposition (UFD) campaign, to investigate current modeling capabilities and to identify issues and knowledge gaps associated with repositories hosted in clay-rock (Rutqvist et al., 2014a). The modeling was conducted with the TOUGH-FLAC simulator, which, as mentioned, is based on linking the multiphase flow and heat transport code TOUGH2 (Pruess et al., 2011) with the geomechanical simulator FLAC3D (Itasca, 2009). TOUGH-FLAC was originally developed and applied for modeling coupled THM processes in unsaturated fractured tuff (Rutqvist et al., 2002). It has more recently been modified for modeling bentonite and clay host rocks, especially within the US DOE's UFD campaign. For the modeling of bentonite, the BBM, and most recently, the BExM have been implemented into TOUGH-FLAC (Rutqvist et al., 2011, 2014b).

In the study presented here, the geomechanical behavior of the bentonite was modeled with the BBM, using properties corresponding to the FEBEX Bent, whereas clay host rock properties were derived from the Opalinus clay at

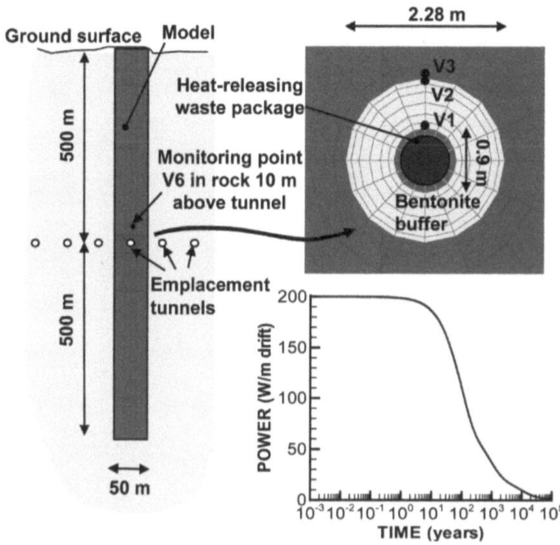

FIGURE 9.12 Model domain, numerical grid, monitoring points, and heat power function for an assumed bentonite back-filled horizontal emplacement drift at 500 m depth in clay host rock (Rutqvist et al., 2014a).

Mont Terri (Gens et al., 2007). The simulations were conducted in a 2D plane strain model as shown in Figure 9.12. Because of the repetitive lateral symmetry, the model extends horizontally to the mid-distance between two emplacement tunnels. In the 2D model, the heat load was scaled and applied as an initial heat load per meter tunnel of 200 W. A complete description of the model, input parameters, and results are given in Rutqvist et al. (2014a).

THM simulation results are shown in Figure 9.13, including evolution of temperature, saturation, fluid pressure, and stress within the buffer. The canister surface temperature peaks at about 91°C after 50 years, whereas the peak temperature at the buffer and rock interface is about 77°C (Figure 9.13(a)). The time to full resaturation of the buffer is 25 years, precedes the thermal peak (Figure 9.13(b)). The fluid pressure increases with temperature and peaks at 8.0 MPa, significantly above the estimated 4.5 MPa hydrostatic pressure (Figure 9.13(c)). This excess fluid pressure above hydrostatic is caused by thermal pressurization, as a result of thermal expansion of the pore fluid that cannot escape in the relatively low-permeability host rock. The excess pressure is initially dissipated through radial and vertical outflow and inflow of water, which lasts for up to 10 000 years until hydrostatic equilibrium pressure is finally attained after about 100 000 years. The magnitude and duration of this excess pressure pulse depends on parameters such as rock permeability, and compressibility of water and rock.

The evolution of total and effective stress in the vertical direction in points V1 and V2 within the buffer, i.e., equivalent to the radial stress at both points is

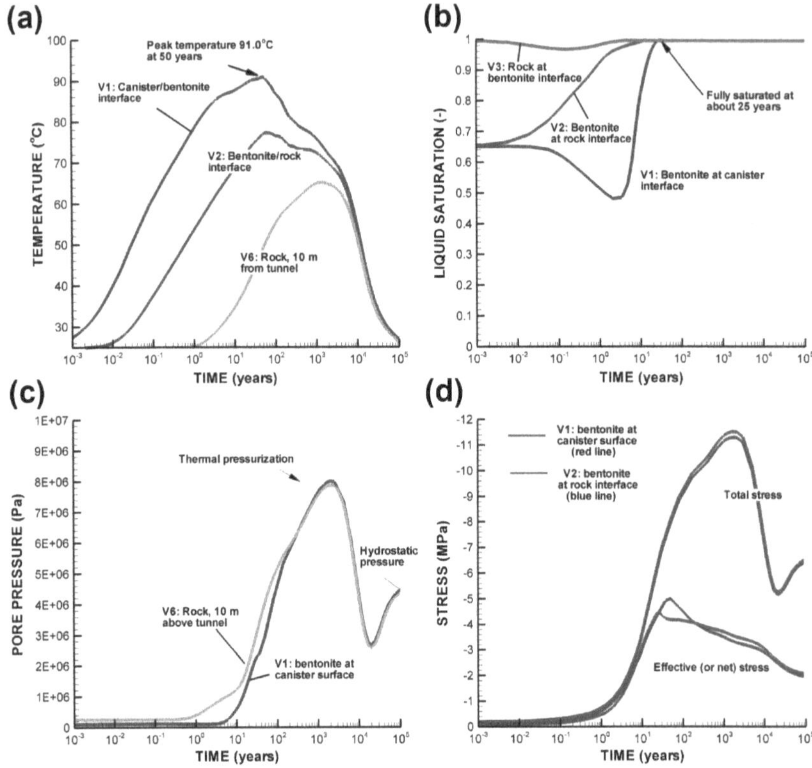

FIGURE 9.13 Simulated evolution of thermo-hydro-mechanical (THM) processes in the buffer: (a) temperature, (b) liquid saturation, (c) fluid pressure, and (d) total and effective (or net) stress in the radial direction. See Figure 9.12 for locations of V1, V2, V3, and V6 (Rutqvist et al., 2014a).

presented in Figure 9.13(b). The evolution of total stress in Figure 9.13(d) results from three components: (1) swelling stress caused by saturation changes, (2) poroelastic stress from fluid pressure changes under saturated conditions, and (3) thermal stress. During the early time, the stress in the buffer increase is mainly a result of swelling stress that occurs as a result of reduced suction during saturation of the buffer and is governed by the BBM. In addition, thermal stress is induced within the buffer along with the temperature increase, and after the buffer is fully saturated, poroelastic stress with increasing liquid fluid pressure in the rock causes a substantial stress increase both in the rock and in the buffer. The fact that the stress evolution in the buffer is strongly affected by the thermal pressurization in the rock, shows that there is a strong interaction between the buffer and the rock coupled THM processes, and thus, the buffer and rock coupled THM processes cannot be analyzed independently.

The model simulation also shows a complex mechanical evolution of the buffer involving a strong variation in the bulk modulus with stress and suction changes (Figure 9.14(a)). The initial bulk modulus is about 2 MPa and peaks at

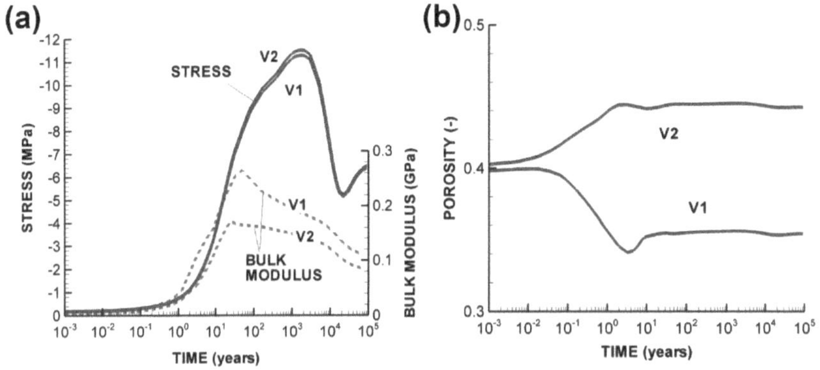

FIGURE 9.14 Simulated evolution of thermo-hydro-mechanical (THM) processes in the buffer when using the Barcelona Basic Model: (a) tangential stress and bulk modulus, and (b) porosity for point V1 and V2 located within the buffer (Rutqvist et al., 2014a).

about 200–300 MPa (0.2–0.3 GPa), i.e., more than a 100-fold increase in stiffness. This complex evolution resulted in a permanent nonuniformity of porosity and buffer density (Figure 9.14(b)). The final porosity of about 0.35 at the inner parts of the buffer (V1) corresponds to a dry density of 1.72 kg/dm^3, whereas the final porosity of about 0.44 at the outer part of the buffer (V2) corresponds to a dry density of 1.48 kg/dm^3. This nonuniformity occurred as a result of thermo-hydro-elastic responses in the buffer governed by the BBM.

The Effective stress path (σ'_1 vs σ'_3) at point V3 at the top of the tunnel, as is the yield surface for the rock matrix is presented in Figure 9.15. Actually, most of the rock failure around the tunnel has already occurred during the excavation, both along bedding planes and in the rock matrix between bedding planes. The stress states move along the yield surface, causing some additional

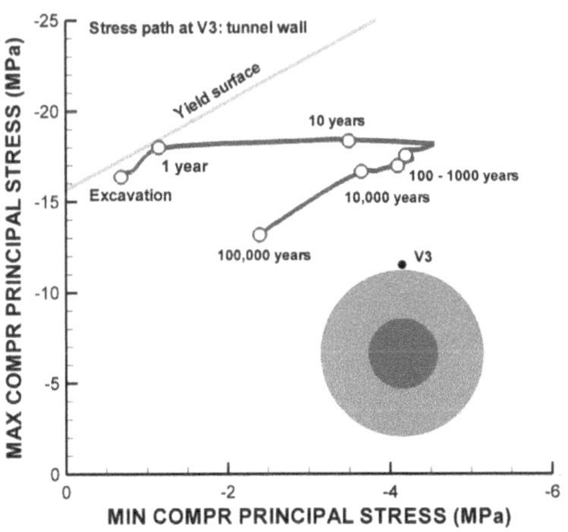

FIGURE 9.15 Effective stress path at point V3 located in the rock just above the tunnel with comparison to the yield surface of a Mohr–Coulomb model (Rutqvist et al., 2014a).

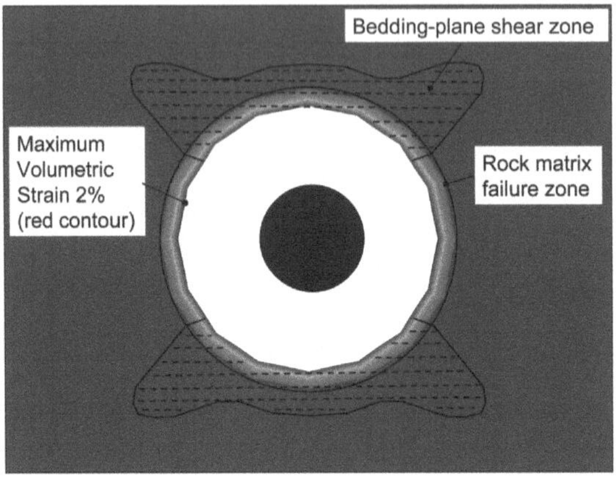

FIGURE 9.16 Volumetric strain contour and extent of failure zones related to horizontal bedding planes and rock matrix (Rutqvist et al., 2014a).

plastic yield until one year has elapsed. After one year, the stress state moves away from the yield surface as a result of increasing minimum principal stress (σ'_3). This minimum principal stress is the radial stress normal to the tunnel wall, which increases as a result of swelling and the associated increase in buffer stress. The stress state then remains within the elastic region over the entire 100 000 year simulation time and no further plastic yield takes place.

The modeling indicated some failure not just on top of the tunnel, but also around the entire tunnel wall, in both the intact rock matrix and along the horizontal bedding planes. The failure zone along bedding planes extends to about 0.4 m into the rock at the top and bottom of the tunnel, whereas there is a 0.15 m thick zone around the entire periphery of the tunnel with both matrix and bedding plane shear failure (Figure 9.16). However, despite an anisotropic failure zone developing as a result of bedding plane orientation, the mechanical response in terms of plastic yield and volumetric strain is uniform around the periphery of the tunnel, with a 2% maximum strain at the tunnel wall. This reflects, first, the fact that the rock can only expand toward the free surface (i.e., the tunnel wall) and second, the fact that the in situ stress was assumed isotropic. Consequently, an EDZ forms as defined in the model simulations as the zone in which more significant plastic strain occurs, in this case 0.15 m into the rock. However, the simulation showed that the buffer swelling had the positive effect of increasing the confining stress on the rock wall and thereby prevents further failure development during the thermal phase.

9.5 LINKS OF THM TO GEOCHEMISTRY

In nuclear waste isolation, THM processes may be linked to geochemistry in a number of ways. One obvious link is that coupled THM processes can impact the transport processes, e.g., the creation and evolution of an EDZ around the tunnel system that could serve as a transport path for radionuclides if released from the waste package. Other links to geochemistry could be time-dependent processes involving subcritical crack growth in the host rock mediated by geochemical changes. The self-sealing and healing processes of fractures, as have been mentioned, might be related to creep and swelling that are inherently chemically mediated mechanical processes. This is related to the long-term stability of the excavations and the associated long-term evolution of the EDZ.

Within the buffer, chemical changes may affect the mechanical properties by illitization (the transformation of smectite to illite), changes in pore-water salinity, and mineral precipitation (e.g., silica precipitation). Illitization and changes in pore-water salinity can lead to loss of swelling capacity (loss of swelling pressure), whereas silica precipitation can lead to cementation, which ultimately could change a buffer from a soft and plastic material to a stiff and brittle material (Zheng et al., 2014; Rutqvist et al., 2014a). Geochemical changes in the form of illitization and cementation by silica precipitation may occur at a substantially higher rate at high temperatures. These are considerations that are the basis for maximum temperature limits for various repository systems applied in national nuclear waste programs. Related to the performance of a repository, a loss of swelling stress could mean a loss of the support load from the buffer and backfill onto the excavations walls, which in turn could mean that fractures do not mechanically close in the EDZ leaving room for a high permeability flow path along the tunnel walls. A cemented buffer by silica precipitation can become stiff and brittle. A brittle material is more prone to tensile fracturing and a stiff cemented buffer will not be able to accommodate plastically external rock movements. This is relevant when considering the possibility of shearing along minor faults and major fractures that may intersect a deposition tunnel.

As a result of such concerns, geomechanical changes affecting the long-term stability and function of the buffer, a maximum temperature limit of 100°C or lower have been adopted in some national nuclear waste disposal programs (e.g., Sweden, France), whereas higher temperature limits are explored by others (e.g., Switzerland). The geomechanical changes that could occur depend on the temperature and the time of exposure to the high temperature. The highest temperature will occur in the inner part of the buffer for some limited time, whereas temperature much below 100°C may occur in the outer part of the buffer and the near field rock. High temperature might be tolerable if occurs on for a relatively short time. Models that couple THM processes with geochemistry could be used to analyze potential long-term effects of such processes. An initial attempt of such modeling is presented in Rutqvist et al. (2014b) and Zheng et al. (2014), involving linking TOUGH-REACT reactive transport simulator with the FLAC3D geomechanical simulator. In this research effort, the link between mechanics and chemistry is provided by the dual-structure model described above.

9.6 CONCLUDING REMARKS

This chapter provided an overview of coupled THM behavior and properties of Bent buffer and backfill materials and clay host rocks based on research findings over the past 30 years. It should be clear from this overview that substantial progress has been made in the understanding of coupled THM processes in clay barriers, though still more work is needed, especially related to the complex mechanical behavior associated with granular Bent. The modeling of the long-term THM behavior showed strong interaction between the bentonite THM evolution and that of the host rock, including thermal pressurization that has a strong impact on the stress evolution within the buffer. Another import THM aspect, not covered in that analysis, is THM processes associated with potential gas generation from the waste packages, an issue that is currently being studied by in situ gas migration experiments at several URLs. In this context the relatively high gas intrinsic permeability (relative to intrinsic permeability for water) is relevant for the release of any build-up of gas pressure. The use of dual-structure models, involving both micro and macro clay-structures, provides more accurate and mechanistically more correct modeling of bentonite, including liquid and gas permeability evolution with saturation, microstructural changes, and bulk strain. Moreover, a dual-structure model provides an avenue for linking mechanical behavior with chemistry. Such chemical–mechanical coupling is relevant for studying the long-term buffer stability and is an area of research deserving further attention, especially when considering higher temperature conditions.

REFERENCES

Ahn, J., Apted, M.J., 2010. Geological Repository Systems for Safe Disposal of Spent Nuclear Fuels and Radioactive Waste, first ed. Elsevier. 792 p.

Alonso, E.E., Gens, A., Josa, A., 1990. A constitutive model for partially saturated soils. Geotechnique 40, 405–430.

Alonso, E.E., Vaunat, J., Gens, A., 1999. Modelling the mechanical behaviour of expansive clays. Eng. Geol. 54, 173–183.

Alonso, E.E., Alcoverro, J., Coste, F., Malinsky, L., Merrien Soukatchoff, V., Kadiri, I., Nowak, T., Shao, H., Nguyen, T.S., Selvadurai, A.P.S., Armand, G., Sobolik, S.R., Itamura, C.M., Stone, C.M., Webb, S.W., Rejeb, A., Tijani, M., Maouche, Z., Kobayashi, A., Kurikami, H., Ito, A., Sugita, Y., Chijimatsu, M., Börgesson, L., Hernelind, J., Rutqvist, J., Tsang, C.F., Jussila, P., 2005. The FEBEX benchmark test. Case definition and comparison of modelling approaches. Int. J. Rock Mech. Min. Sci. 42, 611–638.

Alonso, E.E., Hoffmann, C., 2007. Modelling the field behaviour of a granular expansive barrier. Phys. Chem. Earth 32, 850–865.

Armand, G., Leveau, F., Nussbaum, C., de La Vaissiere, R., Noiret, A., Jaeggi, D., Landrein, P., Righini, C., 2014. Geometry and properties of the excavation-induced fractures at the Meuse/Haute-Marne URL drifts. Rock Mech. Rock Eng. 47, 21–41.

Bear, J., Bachmat, Y., 1991. Introduction to Modeling of Transport Phenomena in Porous Media. Kluwer Academic Publisher, Netherlands, p. 553.

Blümling, P., Bernier, F., Lebon, P., Martin, C.D., 2007. The excavation damaged zone in clay formations time-dependent behavior and influence on performance assessment. Phys. Chem. Earth 32, 588–599.

Börgesson, L., 1996. ABAQUS. In: Stephansson, O., Jing, L., Tsang, C.-F. (Eds.), Coupled Thermo-Hydro-Mechanical Processes of Fractured Media, Developments in Geotechnical Engineering, vol. 79. Elsevier, pp. 565–570.

Börgesson, L., Chijimatsu, M., Nguyen, T.S., Rutqvist, J., Jing, L., 2001. Thermo-hydro-mechanical characterization of a bentonite-based buffer material by laboratory tests and numerical back analyses. Int. J. Rock Mech. Min. Sci. 38, 105–127.

Chijimatsu, M., Nguyen, T.S., Jing, L., De Jonge, J., Kohlmeier, M., Millard, A., Rejeb, A., Rutqvist, J., Souley, M., Sugita, Y., 2005. Numerical study of the THM effects on the near-field safety of a hypothetical nuclear waste repository – BMT1 of the DECOVALEX III project. Part 1: conceptualization and characterization of the problems and summary of results. Int. J. Rock Mech. Min. Sci. 42, 720–730.

Gaus, I., Wieczorek, K., Schuster, K., Garitte, B., Senger, R., Vasconcelos, R., Mayor, J.C., 2014. EBS Behaviour Immediately after Repository Closure in a Clay Host Rock: HE-E Experiment (Mont Terri URL), vol. 400. Geological Society, London, Special Publications, pp. 71–91.

Gens, A., Alonso, E., 1992. A framework for the behaviour of unsaturated expansive clays. Can. Geotech. J. 29, 1013–1032.

Gens, A., 1995. Constitutive laws. In: Gens, A., Jouanna, P., Schrefler, B.A. (Eds.), Modern Issues in Non-saturated Soils. Springer-Verlag, Wien, New York, pp. 129–158.

Gens, A., Sánchez, M., Sheng, D., 2006. On constitutive modelling of unsaturated soils. Acta Geotechnica 1, 137–147.

Gens, A., Garitte, B., Wileveau, Y., 2007. In situ behaviour of a stiff layered clay subject to thermal loading. Observations and interpretation. Geotechnique 57, 207–228, 150.

Gens, A., Sanchez, M., Guimaraes, L., do, N., Alonso, E.E., Lloret, A., Olivella, S., Villar, M.V., Huertas, F., 2009. A full-scale in situ heating test for high-level nuclear waste disposal: observations, analysis and interpretation. Geotechnique 59, 377–399.

Guo, R., Martino, J., Dixon, D., 2005. Fully-coupled thermo-hydraulic modelling of the sealing bulkheads and adjacent rock in a full-scale underground test. Can. Geotech. J. 42 (2005), 1318–1329.

ITASCA, 2009. FLAC3D V4.0, Fast Lagrangian Analysis of Continua in 3 Dimensions, User's Guide. Itasca Consulting Group, Minneapolis, Minnesota.

Komine, H., 2004. Simplified evaluation for swelling characteristics of bentonites. Modelling the mechanical behaviour of expansive clays. Eng. Geol. 71, 265–279.

Kristensson, O., Åkesson, M., 2008. Mechanical modeling of MX-80 – quick tools for BBM parameter analysis. Phys. Chem. Earth, Parts A/B/C 33 (Suppl. 1), S508–S515.

Lloret, A., Alonso, E.E., 1985. State surfaces for partially saturated soils. In: Proc. 11th Int. Conf. Soil Mech. Fdn Eng., San Francisco, vol. 2, pp. 557–562.

Noorishad, J., Tsang, C.-F., 1996. ROCMAS simulator; a thermohydromechanical computer code. In: Stephansson, O., Jing, L., Tsang, C.-F. (Eds.), Coupled Thermo-Hydro-Mechanical Processes of Fractured Media, Developments in Geotechnical Engineering, vol. 79. Elsevier, pp. 551–558 (Elsevier).

Nygård, R., Gutierrez, M., Bratli, R.K., Høeg, K., 2006. Brittle–Ductile transition, shear failure and leakage in shales and mudrocks. Mar. Pet. Geol. 23, 201–212.

Ohnishi, Y., Kobayashi, A., 1996. THAMES. In: Stephansson, O., Jing, L., Tsang, C.-F. (Eds.), Coupled Thermo-Hydro-Mechanical Processes of Fractured Media, Developments in Geotechnical Engineering, vol. 79. Elsevier, pp. 545–549.

Olivella, S., Carrera, J., Gens, A., Alonso, E.E., 1994. Nonisothermal multiphase flow of brine and gas through saline media. Transp. Porous Media 15, 271–293.

Olivella, S., Gens, A., 2000. Vapour transport in low permeability unsaturated soils with capillary effects. Transp. Porous Media 40, 219–241.

Philip, J.R., de Vries, D.A., 1957. Moisture movement in porous material under temperature gradients. EOS Trans. AGU 38, 222–232.

Pruess, K., Oldenburg, C., Moridis, G., 2012. TOUGH2 User's Guide, Version 2.1, LBNL-43134(revised). Lawrence Berkeley National Laboratory, Berkeley, California.

Roscoe, K.H., Burland, J.B., 1968. On the generalized stress-strain behaviour of the "wet" clay. In: Heyman, J., Leckic, F.A. (Eds.), Engineering Plasticity. Cambridge University Press, Cambridge, pp. 535–609.

Rutqvist, J., Börgesson, L., Chijimatsu, M., Nguyen, T.S., Jing, L., Noorishad, J., Tsang, C.-F., 2001a. Coupled thermo-hydro-mechanical analysis of a heater test in fractured rock and bentonite at kamaishi mine – comparison of field results to predictions of four finite element codes. Int. J. Rock Mech. Min. Sci. 38, 129–142.

Rutqvist, J., Börgesson, L., Chijimatsu, M., Kobayashi, A., Nguyen, T.S., Jing, L., Noorishad, J., Tsang, C.-F., 2001b. Thermohydromechanics of partially saturated geological media – governing equations and formulation of four finite element models. Int. J. Rock Mech. Min. Sci. 38, 105–127.

Rutqvist, J., Wu, Y.-S., Tsang, C.-F., Bodvarsson, G., 2002. A modeling approach for analysis of coupled multiphase fluid flow, heat transfer, and deformation in fractured porous rock. Int. J. Rock Mech. Min. Sci. 39, 429–442, 97.

Rutqvist, J., Tsang, C.-F., 2004. A fully coupled three-dimensional THM analysis of the FEBEX in situ test with the ROCMAS code: prediction of THM behaviour in a bentonite barrier. In: Stephansson, O., Hudson, J.A., Jing, L. (Eds.), Coupled T-H-M-C Processes in Geo-systems: Fundamentals, Modelling, Experiments and Applications. Elsevier Geo-Engineering Book Series, Oxford, pp. 143–148.

Rutqvist, J., Chijimatsu, M., Jing, L., De Jonge, J., Kohlmeier, M., Millard, A., Nguyen, T.S., Rejeb, A., Souley, M., Sugita, Y., Tsang, C.F., 2005. Numerical study of the THM effects on the near-field safety of a hypothetical nuclear waste repository – BMT1 of the DECOVALEX III project. Part 3: effects of THM coupling in fractured rock. Int. J. Rock Mech. Min. Sci. 42, 745–755.

Rutqvist, J., 2011. Status of the TOUGH-FLAC simulator and recent applications related to coupled fluid flow and crustal deformations. Comput. Geosci. 37, 739–750.

Rutqvist, J., Ijiri, Y., Yamamoto, H., 2011. Implementation of the Barcelona Basic Model into TOUGH–FLAC for simulations of the geomechanical behavior of unsaturated soils. Comput. Geosci. 37, 751–762.

Rutqvist, J., Chen, F., Birkholzer, J., Liu, H.H., Müller, H., Garitte, B., Vietor, T., 2013. Modeling of coupled thermo-hydro-mechanical processes at Mont Terri heater experiment in Opalinus clay using TOUGH-FLAC. In: Proceedings of the 2013 International High-level Radioactive Waste Management Conference (IHLRWM), April 28–May 2, 2013 (Albuquerque, New Mexico).

Rutqvist, J., Zheng, L., Chen, F., Liu, H.-H., Birkholzer, J., 2014a. Modeling of coupled thermo-hydro-mechanical processes with links to geochemistry associated with bentonite-backfilled repository tunnels in clay formations. Rock Mech. Rock Eng. 47, 167–186.

Rutqvist, J., Davis, J., Zheng, L., Vilarrasa, V., Housworth, J., Birkholzer, J., 2014b. Investigation of Coupled THMC Processes and Reactive Transport: FY14 Progress. Prepared for U.S. Department of Energy, Used Fuel Disposition Campaign, FCRD-UFD-2014–000497, Lawrence Berkeley National Laboratory, LBNL-6720E.

Sánchez, M., Gens, A., Guimarães, L., do, N., Olivella, S., 2005. A double structure generalized plasticity model for expansive materials. Int. J. Numer. Anal. Methods Geomech. 29, 751–787.

Sánchez, M., Gens, A., Guimarães, L., do, N., Olivella, S., 2008. Implementation algorithm of a generalised plasticity model for swelling clays. Comput. Geotech. 35, 860–871.

Sánchez, M., Gens, A., Olivella, S., 2012. THM analysis of a large-scale heating test incorporating material fabric changes. Int. J. Numer. Anal. Methods Geomech. 36, 391–421.

Thomas, H.R., He, Y., Onofrei, C., 1998. An examination of the validation of a model of the hydro/thermo/mechanical behaviour of engineered clay barriers. Int. J. Numer. Anal. Methods Geomech. 22, 49–71.

Thomas, H.R., Vardon, P.P.J., Cleall, P.J., 2014. Three-dimensional behaviour of a prototype radioactive waste repository in fractured granitic rock. Can. Geotech. J. 51, 246–259.

Tsang, C.F., Bernier, F., Davies, C., 2005. Geohydromechanical processes in the excavation damaged zone in crystalline rock, rock salt, and indurated and plastic clays—in the context of radioactive waste disposal. Int. Rock Mech. Min. Sci. 42, 109–125.

Wang, Q., Kolditz, O., 2007. Object-oriented finite element analysis of thermo-hydro-mechanical (THM) problems in porous media. Int. J. Numer. Meth. Engng. 69, 162–201.

Wang, Q., Cui, Y.-J., Tang, A.M., Barnichon, J.-D., Saba, S., Ye, W.-M., 2013. Hydraulic conductivity and microstructure changes of compacted bentonite/sand mixture during hydration. Eng. Geol. 164, 67–76.

Yu, L., Weetjens, E., Sillen, X., Vietor, T., Li, X., Delage, P., Labiouse, V., Charlier, R., 2014. Consequences of the thermal transient on the evolution of the damaged zone around a repository for heat-emitting high-level radioactive waste in a clay formation: a performance assessment perspective. Rock Mech. Rock Eng. 47, 3–19.

Zheng, L., Rutqvist, J., Liu, H.-H., Birkholzer, J.T., Sonnenthal, E., 2014. Model evaluation of geochemically induced swelling/shrinkage in argillaceous formations for nuclear waste disposal. Appl. Clay Sci. 97–98, 24–32.

Chapter 10

Transport Properties through Partially Saturated Charged Membranes and Geophysical Approaches

A. Revil
Department of Geophysics, Colorado School of Mines, Green Center, Golden, CO, USA; ISTerre, CNRS, Université de Savoie, Le Bourget du Lac, France

Chapter Outline

Developments in Clay Science, Vol. 6C. http://dx.doi.org/10.1016/B978-0-08-100027-4.00010-3

10.1 INTRODUCTION

Geophysical methods can be used to provide unique nonintrusive information regarding transport phenomena in clay barriers. This can be achieved in two distinct ways. The first concerns the development of methods that can be used to characterize the texture of the porous materials such as the tortuosity (Jougnot et al., 2009), the pore-size distribution (Revil et al., 2014a), and the porosity and water content (Revil, 2013a). The second type of approaches concerns the use of time-lapse geophysical methods to monitor, possibly in real time, transport phenomena taking place in the pore network of clayey materials or affecting their mineral surface (e.g., Vaudelet et al., 2011a,b for silicates). The most efficient methods to perform these two tasks are general passive and active electrical methods such as self-potential, resistivity, and induced polarization. A recent review of these methods can be found in Revil et al. (2012). A growing research objective is to be able to forward-model the geophysical response from the reactive transport multiphase simulations in clayey materials. Once this task will be achieved, stochastic methods can be used to gain some better tool to monitor complex transport processes occurring in clayey materials. A good example of such coupled joint inversion of geophysical data including the physics of the monitored process can be found, for instance, in Jardani et al. (2013).

The goal of this chapter is to provide a model describing the generalized constitutive equations of transport of water and ions in the case of partially saturated charged porous media. In the last part of the chapter, the implications of this model in terms of geophysical observables will be discussed. In the following, the water phase is considered to be the wetting phase (index w) for the solid grains, and the nonwetting phase (air) is considered to be very compressible and electrically insulating. Charged porous material means that the interface between the solid phase and the pore-water phase contains an electrical double layer coating the surface of the grains as it is the case for clay mineral particles. This electrical double layer includes a diffuse layer of counterions and coions (usually cations and anions at clay mineral surfaces) and a Stern layer of adsorbed counterions (e.g., Avena and De Pauli, 1998; Leroy and Revil, 2004).

The advantage of the transport and thermodynamic models developed in Revil and Linde (2006) and Revil (2007) is that they are easily testable since all the conductivity and coupling terms of the matrix **L** of material properties are written in terms of the same fundamental parameters (permeability and formation factor) that can be measured independently. The formation factor is a fundamental topological property of the pore space and thus enters into the electrical conductivity problem. It can be defined as the ratio between the bulk tortuosity of the pore space and the connected porosity. This is in contrast with the phenomenological models found in the literature and based on nonequilibrium thermodynamics without a micro−macro description of the coupling properties (e.g., Mitchell, 1993; Sherwood and Craster, 2000 for the saturated case, Chen and Hicks, 2013; for the unsaturated one). These "pure thermodynamic models" cannot be used to couple reactive transport modeling codes to geophysical observations since they do not offer any connections with the microphysics existing at the pore scale and the effect of the texture of the material upon the transport properties. Mechanistic models exist based on the Donnan model used to represent the concentrations in the pore space of the porous body (e.g., Katchalsky and Curran, 1965). They are, however, open to criticisms in the case of thin electrical double layers (e.g., at high salinities). Also a number of existing mechanistic models are not based on a correct representation of the electrical double layer, especially regarding the decomposition of the double layer in a Stern and diffuse layer and therefore do not account for the partition of the counterions between the Stern and diffuse layers (e.g., Pride, 1994). Other micro−macro models (e.g., Loret et al., 2002; Chen and Hicks, 2013) ignore, for instance, electrostatic effects in the pore space and cannot therefore pretend to represent transport phenomena in clay materials with ions present in a pore water influenced by a charged surface.

The goal of the present chapter is to test various aspects of the model developed in Revil and Linde (2006) and Revil (2007) against a broad range of experimental data. The Section 10.2 is an introduction to the specific notations used in the present chapter. The Section 10.3 presents the transport model in terms of a simplified constitutive law disregarding its filtration component. The predictions of the model will be discussed in terms of electrical conductivity, streaming potential coupling coefficient, and electroosmotic permeability at full saturation and at various saturations of the pore-water phase. Section 10.4 of this chapter is devoted to the modeling of reverse osmosis for saturated and partially saturated membranes such as bentonite used as permeability barriers in the shallow subsurface for environmental purposes (e.g., Kang and Shackelford, 2009, 2010, 2011; Bohnloff and Shackelford, 2013; Zhang et al., 2014) and the containment of radioactive wastes (e.g., Bonaparte et al., 2008; Gens, 2013).

10.2 NOTATIONS

The properties averaged over the pore-water phase will be overlaid with a bar, and the properties in a reservoir locally in equilibrium with this pore water or in the two pore-water reservoirs in contact with a cylindrical-jacketed core sample will be written without a bar. The pore water is not neutral in a membrane. There are three volumetric charge densities ($\overline{Q}_V$, $\widehat{Q}_V$, Q_V) in the pore-water phase that will need to be clearly identified in this chapter. The first quantity, $\overline{Q}_V$, represents the phase average (over the pore-water phase) of the charge density due to the diffuse layer only. The second quantity, $\widehat{Q}_V$, represents the effective charge density of the diffuse layer that is dragged by the flow of the pore water. Finally, the third charge density, Q_V, represents the total charge density of the diffuse and Stern layers per unit volume of the pore-water phase.

10.3 ELECTROKINETIC PHENOMENA WITHOUT FILTRATION

10.3.1 Summary of the Theory

This section summarizes the linear constitutive equations derived in Revil and Linde (2006) at the scale of a representative elementary volume of the porous medium and in absence of filtration effects, assuming that the solid phase is at rest (i.e., no displacement) and surface conductivity in the electrical diffuse layer is due to a single counterion of ionic mobility $\beta_{(+)}$ (in $m^2\ V^{-1}\ s^{-1}$). The counterions of the Stern layer are considered immobile for quasi-static (or slow) transport phenomena (Revil and Skold, 2011) by opposition to dynamic phenomena in a broad range of frequencies (e.g., Revil, 2013a,b). The electrokinetic coupling effects investigated in this section are (1) the streaming current associated with the drag of the excess of charge in the electrical diffuse layer by the flow of the pore water and (2) the electroosmotic flow of the pore water associated in presence of an electrical field. This second effect is due to the viscous drag on the water molecules associated with the movement of the counterions in the direction of the electrical field. Both effects are related to the existence of the electrical double layer coating the surface of the clay minerals and their cation exchange capacity (CEC, see Figure 10.1).

The macroscopic electrical field is defined as $\overline{\mathbf{E}} = -\nabla\overline{\varphi}$ and the macroscopic pressure field as $\overline{\mathbf{P}} = -\nabla(\overline{p}_w - \overline{\pi})$. In these equations, $\overline{p}_w$ denotes the mechanical pressure of the pore-water phase and $\overline{\pi}$ the swelling pressure of the water phase in a jacketed core sample between two reservoirs. With the assumptions described above, the generalized form of the constitutive equations for the total current density $\mathbf{J}$ and Darcy velocity $\dot{\mathbf{w}}$ can be written in a matrix form as follows (Revil and Linde, 2006; Revil and Lu, 2014):

$$\begin{bmatrix} \mathbf{J} \\ \dot{\mathbf{w}} \end{bmatrix} = -\mathbf{L}\begin{bmatrix} \nabla\overline{\varphi} \\ \nabla\left(\overline{p}_w - \overline{\pi}\right) \end{bmatrix}. \tag{10.1}$$

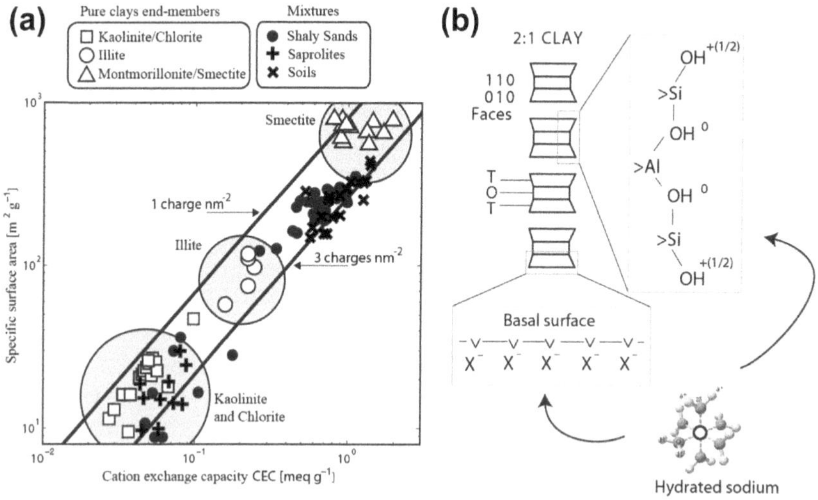

FIGURE 10.1 Cation exchange capacity (CEC) of the clay mineral surface. (a) The CEC denotes the maximum capacity of ion exchange on a mineral surface at a given pH. At first approximation, the specific surface area appears proportional to the CEC (expressed in meq g^{-1}). The two lines correspond to 1–3 elementary charges per unit surface area. (b) Hydrated cations like sodium are sorbed on the surface of the clay minerals (both on the basal surfaces and edges) in the Stern layer. The surface sites that are not compensated in the Stern layer are compensated in the diffuse layer of mobile counterions tied to the surface through Coulombic forces only. *Figure adapted from Revil et al. (2013) with the permission of the publisher.*

The matrix **L** is a 2 × 2 square matrix of material properties. These material properties are given by,

$$L_{11} = \sigma = \frac{1}{F} s_w^n \sigma_w + \frac{1}{F\phi} s_w^{n-1} \sigma_S, \tag{10.2}$$

$$L_{22} = \frac{k_w(s_w)}{\eta_w}, \tag{10.3}$$

$$L_{12} = L_{21} = L = \widehat{Q}_V(s_w) \frac{k_w(s_w)}{\eta_w}, \tag{10.4}$$

where η_w (in Pa s) denotes the dynamic viscosity of the pore water, $k_w(s_w)$ (in m^2) the saturation-dependent permeability, F denotes the electrical formation factor related to porosity by Archie's law $F = \phi^{-m}$ (Archie, 1942), $m > 1$ is the cementation exponent. As explained in the introduction, the formation factor can also be defined as the ratio between the bulk tortuosity of the pore space and the connected porosity. Note that the hydraulic conductivity (in m s^{-1}), if needed, is given by,

$$K_w(s_w) = \frac{k_w(s_w)}{\eta_w} \rho_w g. \tag{10.5}$$

The term,

$$\sigma_S = \beta_{(+)}\rho_s(1-f)\mathrm{CEC}, \tag{10.6}$$

entering Eqn (10.2) denotes the surface conductivity (in S m^{-1}) due to electrical conduction in the electrical diffuse layer. In Eqn (10.5), ρ_w denotes the mass density of the pore water and g the acceleration of the gravity. In Eqn (10.6), $\beta_{(+)}$ denotes the mobility of the counterions in the diffuse layer (e.g., $\beta_{(+)}(Na^+, 25°C) = 5.2 \times 10^{-8}\ m^2\ s^{-1}\ V^{-1}$), ρ_s is the density of the solid phase (typically 2700–2800 kg m^{-1} for the clay minerals layers, see Chapter 1), f is the fractions of counterions in the Stern layer (typically in the range 0.7–0.9, with 0.9 chosen by default for clay minerals with monovalent cations), and CEC is the cation exchange capacity expressed in meq g^{-1} (the conversion rate being 1 meq g^{-1} = 96 485 C kg^{-1} in SI units). The CEC can be eventually replaced by the specific surface area since the ratio of the CEC by the specific surface area corresponds to the surface charge density, which is pretty constant for clay minerals at a given pH (at least inside a factor 2, see Figure 10.1). The presence of a given cation can influence the value of the effective CEC that should be used in the model (see Appendix A).

The matrix $\mathbf{L}$ is symmetrical ($\mathbf{L} = \mathbf{L}^T$), a property known as Onsager's reciprocity in nonequilibrium thermodynamics. This relationship is valid only in the vicinity of thermodynamic equilibrium (Onsager, 1931a,b; Prigogine, 1947). The diagonal terms in the matrix $\mathbf{L}$ correspond to the electrical conductivity (classical Ohm's law) and hydraulic conductivity (classical Darcy's law). The off-diagonal terms correspond to the streaming current coupling term and electroosmotic coupling term, respectively. These terms are the electrokinetic terms related to the relative displacement between the electrical diffuse layer and the charged grains with their Stern layer.

The streaming current is associated with the drag of the excess of electrical charge of the diffuse layer by the flow of the pore water (Figure 10.2(a)), while the electroosmotic flow corresponds to the flow of the pore water under the effect of an electrical field (Figure 10.2(b)). In the first case, a source current density is generated (flux of a net charge amount per surface area per unit time). In the second case, the flow is due to the viscous coupling between the displacement of the excess of charge by the electrical field (electromigration) and the water molecules in the pore water.

In the following, the capillary pressure is at equilibrium (the deformation of the meniscus under the flow of the fluid phases is neglected). Hysteretic effects will be neglected in the variations of the material properties with saturation (see Vinogradov and Jackson, 2011; for a laboratory experiment showing clear hysteretic effects on the streaming potential coupling coefficient). Under these assumptions, the saturation can be related to the capillary pressure by the Brooks–Corey model (Brooks and Corey, 1964), or

$$s_w = \begin{cases} s_r + (1-s_r)\left(\dfrac{p_e}{p_c}\right)^{\lambda}, & p_c \geq p_e \\ 1, & p_c < p_e \end{cases}, \tag{10.7}$$

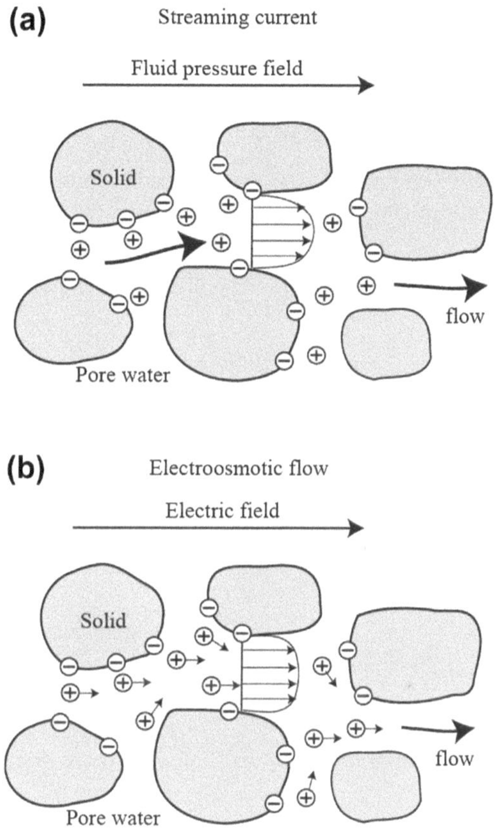

FIGURE 10.2 Electrokinetic effects in absence of filtration. Only the negative and positive excess of electrical charges are shown on the mineral surface and in the pore water, respectively. (a) Streaming current generated by the drag of the excess of charge of the pore water by the flow of the pore water under a pore-fluid pressure gradient. (b) Electroosmotic fluid flow generated by the application of an electrical field to the porous material and the viscous coupling between the movement of the counterions and the water molecules.

where $s_e = (s_w - s_r)/(1 - s_r)$ denotes the effective or reduced water saturation, s_r the irreducible water saturation, p_c the capillary pressure (in Pa), and p_e (in Pa) the capillary entry pressure (i.e., the critical pressure needed to displace the water phase by the gas phase when the porous material is fully water-saturated). The capillary pressure is the difference, at the scale of the representative elementary volume, between the pressure of the nonwetting phase (including the osmotic pressure component) and the pressure of the wetting phase. The wetting phase is water and the nonwetting phase is assumed to be connected to the atmosphere, so its pressure is constant. Therefore, the hydraulic driving force is $\overline{\mathbf{P}} = -\nabla(\overline{p}_w - \overline{\pi}) = \nabla p_c$.

The permeability at the partial saturation can be related to a relative permeability that depends only on the relative water saturation $k_r(s_w)$ and to the permeability at saturation k_S:

$$k_w(s_w) = k_S k_r(s_w). \tag{10.8}$$

In the Brooks–Corey model, the relative permeability is also given by a power law relationship (Purcell, 1949), or

$$k_r(s_w) = \left(\frac{p_c}{p_e}\right)^{-2-3\lambda} = s_e^r, \tag{10.9}$$

where

$$r = \frac{2+3\lambda}{\lambda}, \tag{10.10}$$

and λ is called the pore-size distribution index. Revil et al. (2014b) showed that the capillary entry pressure can be related to the permeability at saturation by the following expression,

$$p_e = \frac{2\gamma}{\sqrt{226F}} k_s^{-0.5}, \tag{10.11}$$

and that the Brooks–Corey index is related to n (the second Archie exponent entering Eqn (10.2) by $\lambda = 2/(n-1)$ such as $r = n+2$. These two relationships are useful for developing a correlation between the electrical and hydraulic parameters.

The streaming potential coupling coefficient C (here defined in V m^{-1}) is defined as follows,

$$C = \left(\frac{\partial \varphi}{\partial \psi}\right)_{\mathrm{J}=0} = -\frac{\widehat{Q}_V(s_w)K(s_w)}{\sigma(s_w)}. \tag{10.12}$$

The streaming potential coupling coefficient C can be written as a product of the streaming potential coupling coefficient at saturation C_S and a relative coupling coefficient C_r that is dimensionless. Neglecting the effect of surface conductivity (i.e., at the high salinity asymptotic limit), the relative coupling coefficient obtained with the Brooks–Corey model is given by,

$$C_r = s_w^{-(n+1)} s_e^{3+2/\lambda}. \tag{10.13}$$

The combination of Brooks–Corey model to the electroosmotic coupling term L (sometimes called the electroosmotic permeability) denotes

$$L(s_w) = \widehat{Q}_V(s_w)\frac{k_w(s_w)}{\eta_w} = L_r L_S, \tag{10.14}$$

where L_r (dimensionless) denotes the relative term and L_S the value of the electroosmotic permeability (m^3 V^{-1} s^{-1}) at saturation. The effective charge density dragged by the flow of the pore water depends on the saturation according to $\widehat{Q}_V(s_w) = \widehat{Q}_V^S/s_w$ (Revil et al., 2014b). This leads to,

$$L_r(s_w) = s_e^r s_w^{-1} = s_e^{3+2/\lambda} s_w^{-1}. \tag{10.15}$$

Note that, if the residual water saturation is very low, the relative permeability and the relative electroosmotic permeability are written as $k_r(s_w) = s_w^r$ and $L_r(s_w) = s_w^{r-1}$, respectively.

An alternative to the Brooks–Corey model is the popular Mualem–van Genuchten model (Mualem, 1976; van Genuchten (1980). This model can be written as follows,

$$s_e = \left(1 + \left|\frac{p_c}{p_e}\right|^{n_v}\right)^{-m_v}, \tag{10.16}$$

$$k_r(s_w) \approx \sqrt{s_e}\left[1 - \left(1 - s_e^{1/m_v}\right)^{m_v}\right]^2, \tag{10.17}$$

where $m_v \approx 1 - 1/n_v, p_c \geq p_e$, and n_v and m_v are the van Genuchten exponents. Therefore, the relative coupling coefficients are given by,

$$C_r = s_w^{-(n+1)} \sqrt{s_e}\left[1 - \left(1 - s_e^{1/m_v}\right)^{m_v}\right]^2, \tag{10.18}$$

$$L_r = s_w^{-1} \sqrt{s_e}\left[1 - \left(1 - s_e^{1/m_v}\right)^{m_v}\right]^2. \tag{10.19}$$

The present approach can easily be extended to other capillary pressure and permeability models in unsaturated conditions.

10.3.2 Application to Electrical Conductivity

In Figure 10.3, the electrical conductivity equation, developed in Revil (2013a), and including its saturation dependence, is fitted to the experimental data from Comparon (2005) obtained at 1.3 GHz for mixtures of MX80 bentonite and kaolinite. The measured conductivity of the pore water was 2.3 S m^{-1}. The theory describes well the water content dependence of this data set. The high surface conductivity of the mixture is also consistent with the fact that smectite has a high CEC (typically close to 1 meq g^{-1}, see Appendix A).

The second test concerns the influence of the CEC upon the surface conductivity of the material at full saturation. The surface conductivity ($\sigma_S/F\phi$) is here defined by the conductivity relationship,

$$\sigma = \frac{1}{F}\sigma_w + \frac{1}{F\phi}\sigma_S. \tag{10.20}$$

Surface conductivity is proportional to the CEC (see Figure 10.4) as predicted by Eqn (10.6).

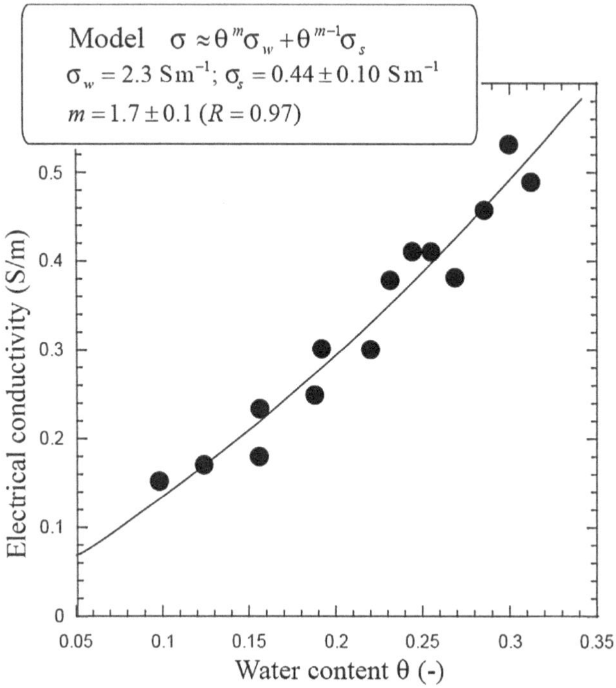

FIGURE 10.3 Fit of the simplified electrical conductivity models expressed as a function of the water content (see Appendix A) to the experimental data from Comparon (2005) at 1.3 GHz for mixtures of MX80 bentonite and kaolinite with a porosity ϕ of 0.40. In this figure, the surface conductivity at saturation is defined as $\phi^{m-1}\sigma_S$.

10.3.3 Application to the Streaming Potential Coupling Coefficient

In the following, the effect of salinity upon the streaming potential coupling coefficient C is discussed at first in saturated condition and then in unsaturated conditions. From Eqn (10.12) and the conductivity model corresponding to Eqn (10.20), the following expression of the streaming potential coupling coefficient can be obtained,

$$C = \frac{F\phi k_S \widehat{Q}_V^S}{\eta_w(\phi\sigma_w + \sigma_S)}. \tag{10.21}$$

Equation (10.21) is used to fit the data shown in Figure 10.5 for the Berea sandstone (with a minor amount of kaolinite) and performed at different salinities. Fitting the model to the data yields a surface conductivity $\sigma_S/F\phi = (1.2 \pm 0.3) \times 10^{-3}$ S m^{-1} and an excess charge density at saturation of $\widehat{Q}_V^S = 1.4 \pm 0.2$ C m^{-1}. These two values can be also predicted as follows.

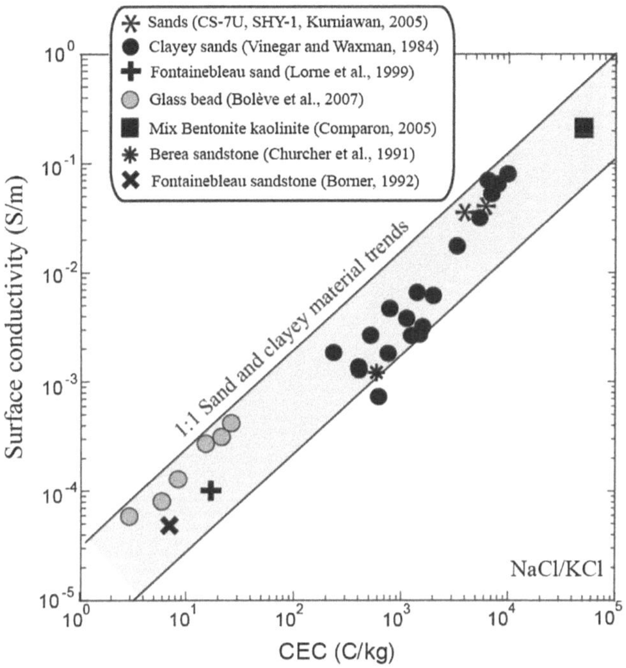

FIGURE 10.4 Surface conductivity ($\sigma_S/F\phi$) at saturation versus cation exchange capacity (CEC). Data from Vinegar and Waxman (1984, shaly sands, NaCl), Bolève et al. (2007, glass beads, NaCl), Lorne et al. (1999, Fontainebleau sand KCl), Churcher et al. (1991) (CEC for the Berea sandstone, surface conductivity from Figure 5), Kurniawan (2005, clean sand, sample CS-7U, porosity 0.1234, CEC = 4088 C kg^{-1}), Comparon (2005, mixtures of MX80 bentonite and kaolinite, porosity of 0.40, estimated CEC 0.5 meq g^{-1} from the CEC of smectite and the mass fraction of smectite), and Börner (1992, sample F3 Fontainebleau sandstone, porosity of 0.068, surface conductivity of 6.6×10^{-5} S m^{-1}, estimated CEC from the grain diameter, see Revil (2013b); CEC of 5.80 C kg^{-3}). The surface conductivity is mostly due to the contribution of the electrical diffuse layer.

First, the value of $\widehat{Q}_V^S$ can be obtained by $\log_{10}\widehat{Q}_V^S = -9.23 - 0.82 \log_{10}k_S$ (Revil et al., 2014b). This yields $\widehat{Q}_V^S = 2.0$ C m^{-1}, therefore in good agreement with the fitted value since $\widehat{Q}_V^S$ can vary over 11 orders of magnitude. Then, the surface conductivity can be compared to the value provided by Moore et al. (2004) using electrical conductivity measurements performed at different salinities. Moore et al. (2004) obtained $\sigma_S/F\phi = 2.7 \times 10^{-3}$ S m^{-1}. Here again, an acceptable agreement is obtained between the fitted surface conductivity value from the streaming potential coupling coefficient and the one determined from the electrical conductivity measurements.

The second test is provided by comparing the model to the data published recently by Mboh et al. (2012) who measured the relative streaming potential coupling coefficient of a clean (silica) sand characterized by the following

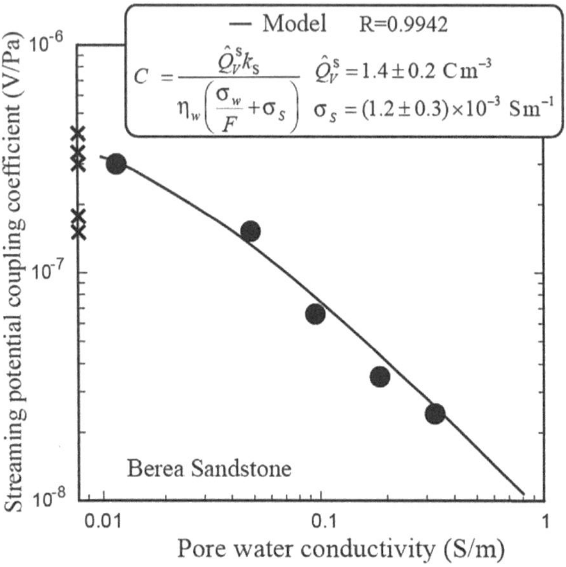

FIGURE 10.5 Streaming potential coupling coefficient at saturation versus salinity for the Berea sandstone. The black circles correspond to the experimental data obtained by Zhu and Toksöz (2012) (Berea sandstone, porosity $\phi = 0.23$, permeability $k_S = 450$ mD, NaCl). The crosses denote the experimental data obtained by Moore et al. (2004) (Berea sandstone, porosity 0.19, water). The plain line corresponds to fit the proposed model to the data of Zhu and Toksöz (2012).

properties: 99.3% silica, porosity $\phi = 0.41$, hydraulic conductivity at saturation $K_S = 8.25 \times 10^{-5}$ m s^{-1}, and mean grain diameter $d = 160$ μm (Table 10.1). Their column was equipped with ten nonpolarizable Ag/AgCl electrodes and six tensiometers, and the acquisition was done at 1 kHz. Mboh et al. (2012) measured a coupling coefficient at saturation of $C = -3.3$ mV m^{-1} for a pore-water conductivity (tap water) of $\sigma_W = 0.044$ S m^{-1} at 25°C. The second Archie exponent n (saturation exponent, Archie, 1942) was measured and found equal to 1.87. The van Genuchten exponent n_v was measured and found equal to 3.88 (by fitting the capillary pressure curve), which yields $m_v = 0.74$. A comparison between the data of Mboh et al. (2012) and Eqns (10.16) and (10.18) is shown in Figure 10.6 The best fit of the data yields $m_v = 0.69 \pm 0.05$, which is close to the value determined from the capillary pressure curve (Figure 10.6). As mentioned by Mboh et al. (2012), this implies that the values of the relative coupling coefficient bear information regarding the van Genuchten parameters as suggested previously by both Linde et al. (2007) and Revil et al. (2007).

In Figures 10.7 and 10.8, the data presented in Revil and Cerepi (2004) and Revil et al. (2007) are analyzed, especially the streaming potential coupling coefficient data of samples E3 and E39 (dolomitic limestones, see properties in Table 10.1). In both cases, the second Archie exponent n (saturation exponent)

TABLE 10.1 Petrophysical Properties of the Samples Discussed in the Main Text. M Corresponds to the Sand Sample Investigated by Mboh et al. (2012) While Samples E3 and E39 are two Dolomitic Samples Investigated by Revil et al. (2007)

Property	Symbol	M	E3	E39
Saturation exponent	n (−)	1.87	2.7	3.5
Cementation exponent	m (−)	—	1.93	2.49
Permeability	k_S (m^2)	8.4×10^{-12}	48.4×10^{-15}	23.8×10^{-15}
Porosity	ϕ (−)	0.41	0.203	0.159
Formation factor	F (−)	3.2[a]	21.7	97.4
Residual saturation	s_r (−)	0.09	0.31	0.34
Grain size	d (μm)	160	—	—
Pore size	r (μm)	—	1.18	0.17

[a]*Assuming a cementation exponent of 1.3, a classical value in clean sands.*

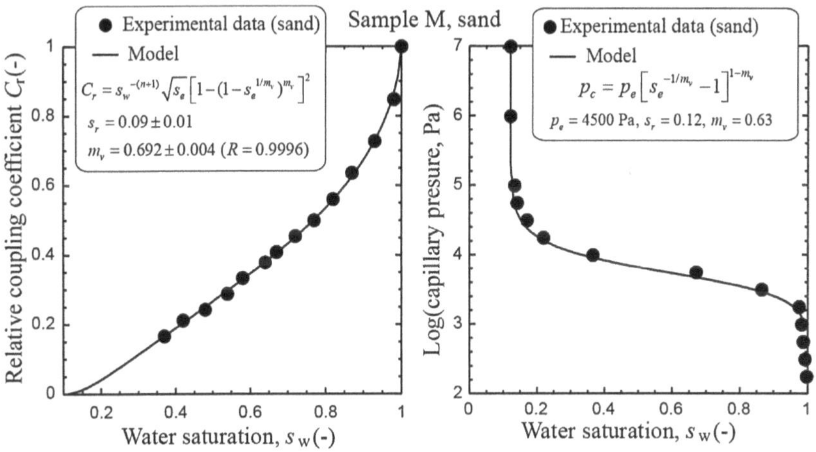

FIGURE 10.6 Streaming potential coupling coefficient versus water saturation. Comparison between experimental data and the prediction of our model associated with the van Genuchten model. The experimental data are from Mboh et al. (2012) (sample M, sand). Left panel: Relative streaming potential coupling coefficient versus saturation. The value of the saturation exponent (second Archie exponent) is the measured value $n = 1.87$. Right panel: Capillary pressure curve (the nonwetting fluid is air).

was independently determined using electrical resistivity measurements. The van Genuchten parameters were found to be approximately the same using the capillary pressure curves and the relative streaming potential coupling coefficient data, showing again the consistency of the model in terms of representation of the material properties with a fairly small number of fundamental parameters.

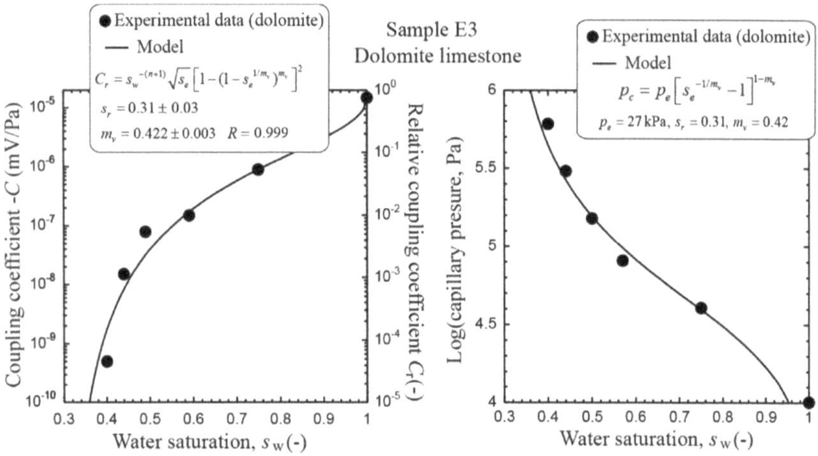

FIGURE 10.7 Streaming potential coupling coefficient versus water saturation. Comparison between experimental data and the prediction of our model based on the van Genuchten formulation (experimental data are from Revil and Cerepi, 2004; Revil et al., 2007; sample E3). Left panel: Absolute and relative streaming potential coupling coefficient versus saturation. The value of the saturation exponent (second Archie exponent) is the measured value $n = 2.7$. Right panel: Fit of the capillary pressure curve with the same van Genuchten parameters as obtained for the coupling coefficient (nonwetting fluid: mercury).

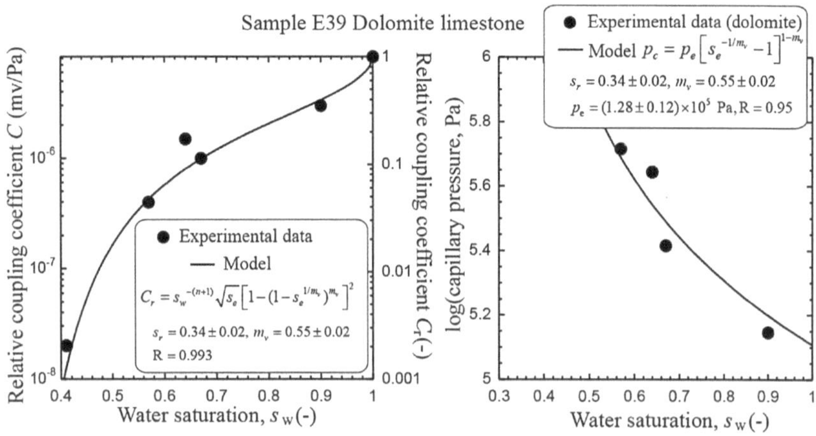

FIGURE 10.8 Streaming potential coupling coefficient versus water saturation. Comparison between experimental data and the prediction of our model based on the van Genuchten formulation (experimental data are from Revil and Cerepi, 2004; Revil et al., 2007; sample E39). Left panel: Absolute and relative streaming potential coupling coefficient versus saturation. The measured value of the saturation exponent (second Archie exponent) $n = 3.5$ is used. Right panel: Fit of the capillary pressure curve with the same van Genuchten parameters as obtained for the coupling coefficient (nonwetting fluid: mercury).

10.3.4 Application to the Electroosmotic Permeability

The experimental data of De Wet (1995) and Gabrieli et al. (2008) are used in order to test the model for the electroosmotic permeability (see Figure 10.9, Tamagnini et al., 2010). According to these authors, the material is a clayey silt from the site of Viadana, in the Po river valley in Italy. Tamagnini et al. (2010) reported other properties including the permeability at saturation ($k_S = 2 \times 10^{-17}$ m^2), the electroosmotic permeability at saturation ($L_S = -9 \times 10^{-10}$ m^3 V^{-1} s^{-1}), the electrical conductivity at saturation (0.109 S m^{-1}), and the inverse of the capillary entry pressure ($\alpha = 8.51 \times 10^{-3}$ Pa^{-1}). The pore water used for the experiment was distilled water. Indeed, the type of pore water does not matter since the relative electroosmotic permeability is independent on the pore-fluid composition. A good agreement was found between the experimental data point and both models, either based on the van Genuchten formulation (Figure 10.9) or based on the Brooks–Corey model (Figure 10.10).

We can try to compute the value of the electroosmotic permeability at saturation from the permeability at saturation. According to the relationship $\log_{10}\widehat{Q}_V^S = -9.23 - 0.82 \log_{10} k_S$, the charge density that can be dragged by the flow of the pore water at saturation, $\widehat{Q}_V^S$, is 30200 C m^{-3}. From Eqn (10.4), the electroosmotic permeability at saturation is given by $L_S = \widehat{Q}_V^S k_S / \eta_w$. Using

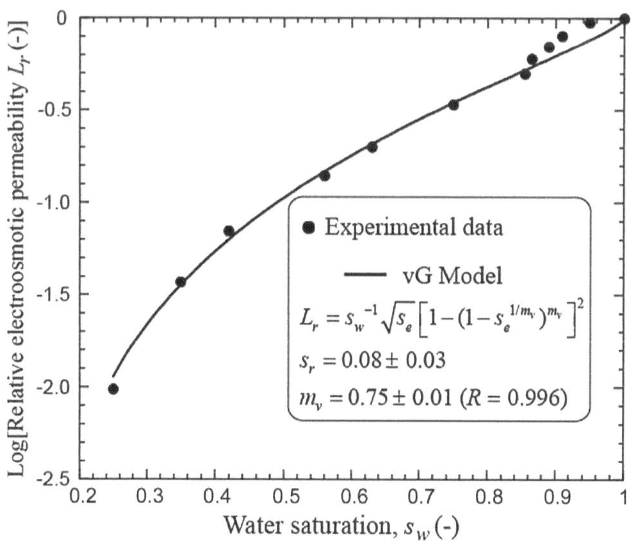

FIGURE 10.9 Variation of the relative electroosmotic permeability with saturation. Comparison between experimental data and the prediction of our model based on the van Genuchten formulation (experimental data are from De Wet, 1995; Gabrieli et al., 2008). Material: clayey silt from the site of Viadana, in the Po river valley, Italy. Pore water: distilled water.

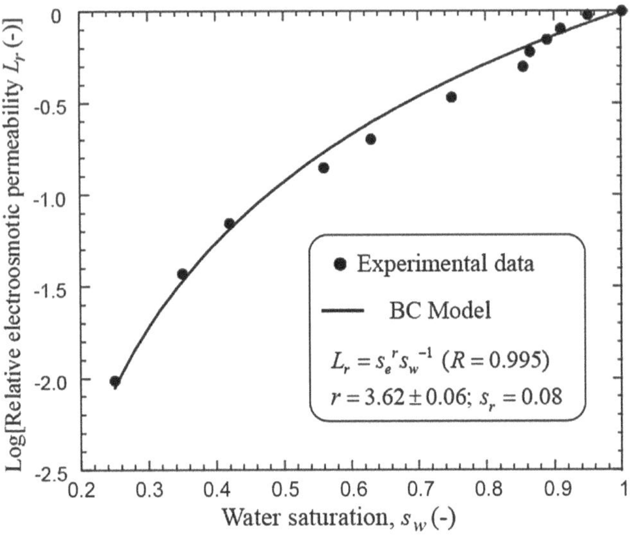

FIGURE 10.10 Same as Figure 7 but using the Brooks–Corey model to fit the data. The quality of the fit is very similar than for the van Genuchten model with the same amount of fitting parameters (two in both cases).

a dynamic viscosity of the pore water, η_W, of 10^{-3} Pa s and a permeability at saturation, k_S, of 2×10^{-17} m^2, L_S value amounts to 6×10^{-10} m^3 V^{-1} s^{-1}, which is close to the observed value $L_S = 9 \times 10^{-10}$ m^3 V^{-1} s^{-1}. Note that, because $\widehat{Q}_V^S$ is nearly inversely proportional to the permeability at saturation, the product $\widehat{Q}_V^S k_S$ is nearly independent on the porosity, which is also in agreement with the observation made by Tamagnini et al. (2010).

As explained in Section 3.5, the relative permeability and the relative electroosmotic permeability can be written as $k_r(s_W) = s_e^r$ and $L_r(s_W) \approx s_e^p$ with $p = r - 1$, respectively. In the case of Viadana silt, Tamagnini et al. (2010) reported $r = 5$ (using the data of Caruso and Jommi, 2005) and $p = 3.7$ (Figure 10.10) whereas the approximate relationship $p = r - 1$ would predict therefore $p = 4.0$, a value quite close to the previous one.

10.4 FILTRATION EFFICIENCY

The mechanism of ion filtration by reverse osmosis is explained in Figure 10.11 The charged porous material is behaving like a membrane for the salt and the salinity of the effluent is smaller than the salinity of the solution entering the porous material. For practical purposes, the theory is considered only to the case of a binary symmetric 1 : 1 electrolyte such as NaCl or KCl and the diffuse layer is assumed to be mostly populated by either Na^+ or K^+ counterions. The number of ionic species is therefore $N = 2$ (the anion and its associated cation). Then, this model is applied to the case of a saturated bentonite and then to an unsaturated bentonite.

10.4.1 Summary of the Theory

The generalized constitutive equation obtained in Revil and Lu (2014) is

$$\begin{bmatrix} \mathbf{J} \\ 2\mathbf{J}_d \\ \dot{\mathbf{w}} \end{bmatrix} = \begin{bmatrix} L_{11} & L_{12} & L_{13} \\ L_{21} & L_{22} & L_{23} \\ L_{31} & L_{32} & L_{33} \end{bmatrix} \begin{bmatrix} \overline{\mathbf{E}} \\ -\nabla\overline{\mu}_f \\ -\nabla\left(\overline{p}_w - \overline{\pi}\right) \end{bmatrix}, \tag{10.22}$$

with the following conductivity coefficients,

$$L_{11} = \sigma = \frac{1}{F} s_w^n \sigma_w + \left(\frac{1}{F\phi}\right) s_w^{n-1} \sigma_S, \tag{10.23}$$

$$L_{22} = \frac{1}{e^2}\left(\sigma_{(+)} + \sigma_{(-)}\right), \tag{10.24}$$

$$L_{33} = \frac{k_w(s_w)}{\eta_w}. \tag{10.25}$$

In these equations, the ionic contributions to the electrical conductivity and the free pore water Hittorf numbers are defined by

$$\sigma_{(+)} = \frac{1}{F} s_w^n t_{(+)}^w \sigma_w + \left(\frac{1}{F\phi}\right) s_w^{n-1} \sigma_S, \tag{10.26}$$

$$\sigma_{(-)} = \frac{1}{F} s_w^n \left(1 - t_{(+)}^w\right) \sigma_w, \tag{10.27}$$

$$t_{(+)}^w = \frac{\beta_{(+)}}{\beta_{(+)} + \beta_{(-)}}, \tag{10.28}$$

where the surface conductivity σ_S is given by Eqn (10.6) and is proportional to the CEC. The question of the CEC is therefore crucial to the accurate determination of the transport properties. The three independent coupling terms obey Onsager reciprocity and are given by,

$$L_{12} = L_{21} = \frac{1}{e}\left(\sigma_{(+)} - \sigma_{(-)}\right), \tag{10.29}$$

$$L_{13} = L_{31} = \widehat{Q}_V(s_w) \frac{k_w(s_w)}{\eta_w}, \tag{10.30}$$

$$L_{23} = L_{32} = \frac{k_w(s_w)}{\eta_w}\left(\overline{c}_{(+)} + \overline{c}_{(-)}\right). \tag{10.31}$$

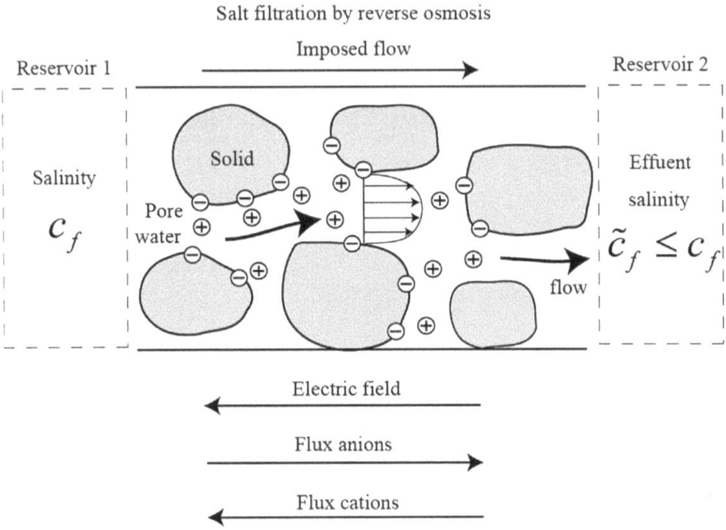

FIGURE 10.11 Concept of reverse osmosis or salt filtration. Though electrokinetic coupling, the salinity of the effluent is smaller than the salinity of the brine in the upstream reservoir.

In the next section, the present theory is used to determine some expressions for the osmotic coefficient in saturated and unsaturated conditions.

10.4.2 Determination of the Osmotic Efficiency

In the following, the total current density is zero (typically, the case of a cylindrical core sample between two reservoirs). From Eqn (10.22), the total current density is given by a generalized Ohm's law,

$$\mathbf{J} = \sigma\overline{\mathbf{E}} - \frac{1}{e}\left(\sigma_{(+)} - \sigma_{(-)}\right)\nabla\overline{\mu}_f - \widehat{Q}_V(s_w)\frac{k_w(s_w)}{\eta_w}\nabla(\overline{p}_w - \overline{\pi}). \tag{10.32}$$

The Hittorf numbers (comprised between 0 and 1) represent the fraction of electrical current carried by either the cations ($T_{(+)}$) or the anions ($T_{(-)}$):

$$T_{(+)} = \frac{\sigma_{(+)}}{\sigma} \tag{10.33}$$

$$T_{(-)} = \frac{\sigma_{(-)}}{\sigma} = 1 - T_{(+)}. \tag{10.34}$$

When the total current density is zero, the macroscopic electrical field is given by

$$\overline{\mathbf{E}}\big|_{\mathbf{J}=0} = \frac{1}{e}\left(T_{(+)} - T_{(-)}\right)\nabla\overline{\mu}_f + \frac{\widehat{Q}_V(s_w)k_w(s_w)}{\eta_w\sigma(s_w)}\nabla(\overline{p}_w - \overline{\pi}), \tag{10.35}$$

$$\mathbf{E}|_{\mathbf{J}=0} = \frac{1}{e}\left(2T_{(+)} - 1\right)\nabla\overline{\mu}_f + \frac{\widehat{Q}_V(s_w)k_w(s_w)}{\eta_w\sigma(s_w)}\nabla(\overline{p}_w - \overline{\pi}), \tag{10.36}$$

The streaming potential coupling coefficient (here in V/Pa) is defined as,

$$C \equiv \left.\frac{\Delta\overline{\varphi}}{\Delta(\overline{p}_w - \overline{\pi})}\right|_{\mathbf{J}=0,\nabla\mu_f=0} = -\frac{k_w\widehat{Q}_V}{\eta_w\sigma}. \tag{10.37}$$

Therefore, the macroscopic electrical field in Eqn (10.36) can be written as,

$$\overline{\mathbf{E}}\Big|_{\mathbf{J}=0} = \frac{1}{e}\left(2T_{(+)} - 1\right)\nabla\overline{\mu}_f - C\nabla(\overline{p}_w - \overline{\pi}), \tag{10.38}$$

Combining Eqn (10.38), for the electrical field $\overline{\mathbf{E}}$, with Eqn (10.22), for the Darcy velocity, the generalized constitutive Darcy equation is,

$$\dot{\mathbf{w}} = -\frac{k_w(s_w)}{\eta_w}\left[-\widehat{Q}_V(s_w)\overline{\mathbf{E}} + \left(\overline{c}_{(+)} + \overline{c}_{(-)}\right)\nabla\overline{\mu}_f + \nabla\left(\overline{p}_w - \overline{\pi}\right)\right], \tag{10.39}$$

$$\begin{aligned}\dot{\mathbf{w}} = -\frac{k_w}{\eta_w}\Big[&-\widehat{Q}_V\frac{1}{e}\left(2T_{(+)} - 1\right)\nabla\overline{\mu}_f + \widehat{Q}_V C\nabla\left(\overline{p}_w - \overline{\pi}\right)\\ &+ \left(\overline{c}_{(+)} + \overline{c}_{(-)}\right)\nabla\overline{\mu}_f + \nabla\left(\overline{p}_w - \overline{\pi}\right)\Big],\end{aligned} \tag{10.40}$$

$$\begin{aligned}\dot{\mathbf{w}} = -\frac{k_w}{\eta_w}\Bigg\{&\left(1 + C\widehat{Q}_V\right)\nabla\left(\overline{p}_w - \overline{\pi}\right)\\ &+ \left[-\frac{\widehat{Q}_V}{e}\left(2T_{(+)} - 1\right) + \left(\overline{c}_{(+)} + \overline{c}_{(-)}\right)\right]\nabla\overline{\mu}_f\Bigg\},\end{aligned} \tag{10.41}$$

According to Revil and Skold (2011), the product $C_S\widehat{Q}_V^S$ obeys to

$$C_S\widehat{Q}_V^S \ll 1, \tag{10.42}$$

in saturated conditions and can be therefore neglected in Eqn (10.41). In unsaturated conditions, C decreases faster with the saturation than $\widehat{Q}_V$ increases with the saturation, therefore $C\widehat{Q}_V$ can be safely neglected in Eqn (10.41) and the expression of the Darcy velocity can be expressed as,

$$\dot{\mathbf{w}} \equiv -\frac{k_w}{\eta_w}\left\{\nabla\left(\overline{p}_w - \overline{\pi}\right) + \left[-\frac{\widehat{Q}_V}{e}\left(2T_{(+)} - 1\right) + \left(\overline{c}_{(+)} + \overline{c}_{(-)}\right)\right]\nabla\overline{\mu}_f\right\}. \tag{10.43}$$

To account for the salinity of the pore water and the effect of the diffuse layer (the excess of positive charge in the pore water compensates for the surface charge and the charge in the Stern layer) Donnan model (Donnan, 1911) is used as described in Appendix B:

$$\overline{c}_{(+)} + \overline{c}_{(-)} = 2c_f\sqrt{1+\Theta^2}, \tag{10.44}$$

$$\Theta \equiv \frac{\overline{Q}_V^S}{2ec_f s_w}. \tag{10.45}$$

In addition, the gradient of the chemical potential of the salt between the two reservoirs can be related to the gradients in salinity c_f and osmotic pressure between the two reservoirs by Barbour and Fredlung (1989).

$$\nabla\overline{\mu}_f = \frac{\nabla\overline{\pi}}{2c_f}, \tag{10.46}$$

(see Appendix B). Using Eqns (10.44) and (10.46) in Eqn (10.43) yields the following generalized Darcy's law,

$$\dot{\mathbf{w}} \equiv -\frac{k_w}{\eta_w}\left\{\nabla\left(\overline{p}_w - \overline{\pi}\right) + \left[-\frac{\widehat{Q}_V}{2c_f e}\left(2T_{(+)} - 1\right) + \frac{\overline{c}_{(+)} + \overline{c}_{(-)}}{2c_f}\right]\nabla\overline{\pi}\right\}, \tag{10.47}$$

$$\dot{\mathbf{w}} \equiv -\frac{k_w}{\eta_w}\left\{\nabla\left(\overline{p}_w - \overline{\pi}\right) + \left[-\frac{\widehat{Q}_V}{2c_f e}\left(2T_{(+)} - 1\right) + \sqrt{1+\Theta^2}\right]\nabla\overline{\pi}\right\}, \tag{10.48}$$

$$\dot{\mathbf{w}} \equiv -\frac{k_w}{\eta_w}\left\{\nabla\overline{p}_w - \left[1 + \nu\left(2T_{(+)} - 1\right) - \sqrt{1+\Theta^2}\right]\nabla\overline{\pi}\right\}. \tag{10.49}$$

where the second dimensionless coefficient ν is defined as,

$$\nu \equiv \frac{\widehat{Q}_V}{2ec_f} = \frac{\widehat{Q}_V^S}{2ec_f s_w}. \tag{10.50}$$

The compact form of Eqn (10.49) is

$$\dot{\mathbf{w}} \equiv -\frac{k_w}{\eta_w}\left(\nabla\overline{p}_w - \varepsilon\nabla\overline{\pi}\right). \tag{10.51}$$

The coefficient ε is the (dimensionless) osmotic efficiency defined by (e.g., Groenevelt and Elrick, 1976),

$$\varepsilon \equiv \left(\frac{\partial\overline{p}_w}{\partial\overline{\pi}}\right)_{\mathbf{J},\dot{\mathbf{w}}=0}. \tag{10.52}$$

From Eqns (10.49) and (10.51), the osmotic efficiency is related to the three dimensionless numbers $T_{(+)}$, ν, and Θ by,

$$\varepsilon = 1 + \nu\left(2T_{(+)} - 1\right) - \sqrt{1+\Theta^2}. \tag{10.53}$$

Equation (10.53) is the same as the one obtained in Appendix C (Eqn 10.C21) for the filtration efficiency. As explained in Appendix B and Appendix C, the osmotic efficiency is given using the Donnan model by

$$\varepsilon = \frac{\sqrt{1+\Theta^2} - 1}{\sqrt{1+\Theta^2}}. \tag{10.54}$$

The same analysis in the context of Revil's model (see Appendix B) yields,

$$\varepsilon = \frac{\Theta}{1+\Theta} \tag{10.55}$$

Because Donnan's model, however, leads to a conductivity equation that is not in good agreement with experimental data, the use of Revil's formulation (Revil, 2013a) is preferred for the osmotic coefficient (and all the other transport properties). The dimensionless number Θ is related to its value at saturation by $\Theta = \Theta_s / s_w$ where the dimensionless parameter at saturation is given by,

$$\Theta_S \equiv \frac{\overline{Q}_V^S}{2ec_f} = \frac{(1-f)\rho_S}{2ec_f}\left(\frac{1-\phi}{\phi}\right)\text{CEC}. \tag{10.56}$$

In Eqn (10.56), the CEC is expressed in SI units (C kg^{-1}). It may be, however, more useful to use,

$$\Theta_S = \frac{10^{-3}(1-f)\rho_S}{2c_f}\left(\frac{1-\phi}{\phi}\right)\text{CEC}. \tag{10.57}$$

with c_f now expressed in mol L^{-1} and the CEC in meq g^{-1}.

10.4.3 Filtration Efficiency in Saturated Conditions

To test the model in saturated conditions, the data from Dominijanni et al. (2013) are used. The material is bentonite with 98% smectite ($\phi = 0.81$). The electrolyte is NaCl with concentrations from 5 to 100 mM. The measured CEC (methylene blue) is 1.05 meq g^{-1}. The permeability at saturation k_S is equal to 8.15×10^{-19} m^2. The model, Eqns (10.55) and (10.57), provides an excellent fit to the data (Figure 10.12). The porosity is assumed to be known, and the value of the partition coefficient to $f = 0.90$ is fixed (see discussion in Revil, 2012). The fitted CEC_{Na} value (0.38 ± 0.09 meq g^{-1}) is fairly consistent with the fact that $\text{CEC}_{\text{Na}} \approx 0.5$ CEC as discussed in Appendix A (and a total CEC = 1.05 meq g^{-1}).

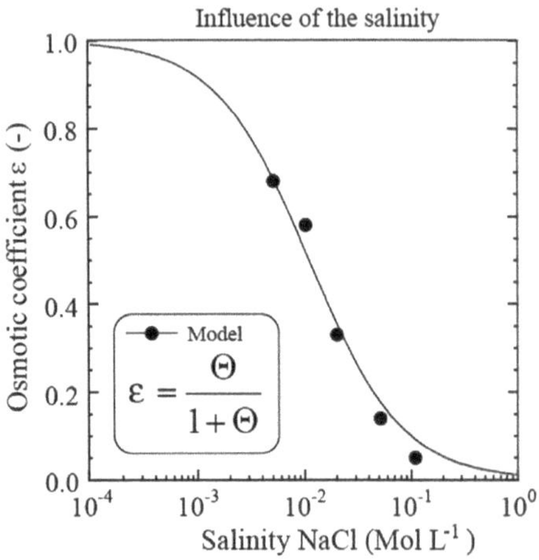

FIGURE 10.12 Comparison between the measurements of the osmotic coefficient as a function of the salinity (data from Dominijanni et al., 2013; bentonite with 98% smectite, NaCl solutions). The model use the porosity of the material ($\phi = 0.81$). The CEC_{Na} used for the model is 0.38 ± 0.09 meq g^{-1} (optimized, $R^2 = 0.975$) and a partition coefficient of $f = 0.90$.

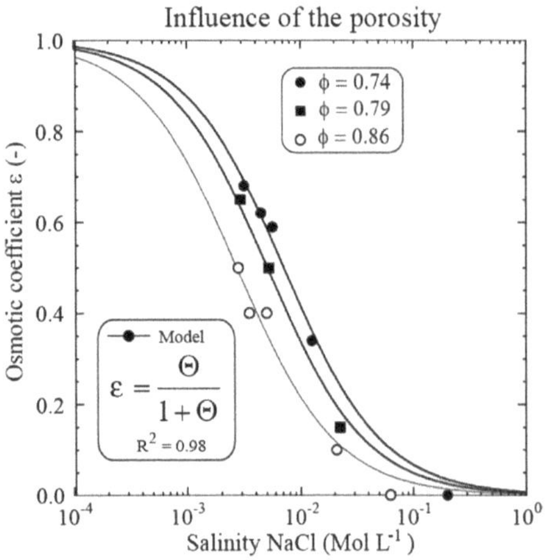

FIGURE 10.13 Comparison between the measurements of the osmotic coefficient as a function of the salinity for three different porosities (data from Malusis et al., 2003; bentonite, NaCl solutions). The model uses the porosity of the material. The CEC optimized by fitting the model to the data is 0.28 ± 0.03 meq g^{-1} (optimized, $R^2 = 0.98$) and a partition coefficient of $f = 0.95$ consistent with that found by Revil (2013a,b).

The sensitivity of the osmotic filtration to porosity is tested with the data from Malusis et al. (2003), who used bentonite, NaCl solutions, and made their measurements at three different porosities (see Figure 10.13). Here again the model fits the data very well with an optimized $CEC_{Na} = 0.28 \pm 0.03$ meq g^{-1} ($R^2 = 0.98$). The measured value is 0.48 meq g^{-1}. The model reproduces very well the increase of the membrane efficiency at a given salinity when the porosity decreases, keeping the CEC value constant.

The next test concerns the effect of the cation itself with bentonite and the influence of the cations Na^+ (data from Kemper and Rollins, 1966) and K^+

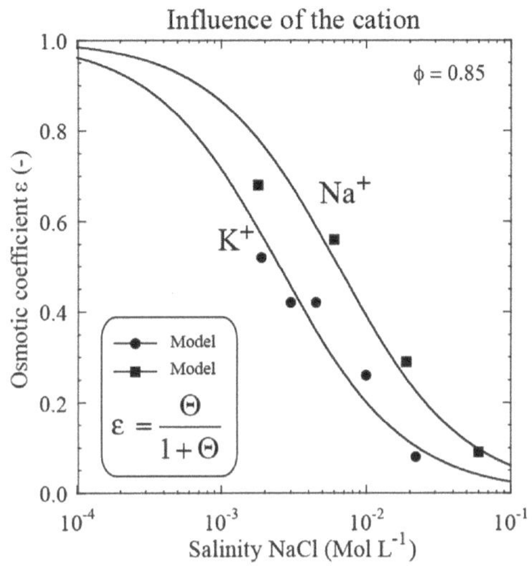

FIGURE 10.14 Comparison between the measurements of the osmotic coefficient as a function of the salinity for two different cations (Na^+: data from Kemper and Rollins, 1966; K^+: data from Malusis and Shackelford, 2002., in both case with bentonite). The model uses the porosity of the material (0.85 ± 0.01). The CEC optimized by fitting the model to the data ($R^2 = 0.94$ for Na^+, $R^2 = 0.90$ for K^+) and a partition coefficient of $f = 0.90$.

(data from Malusis and Shackelford, 2002). The mineralogy of the bentonite used by Malusis and Shackelford (2002) includes 71% smectite (montmorillonite), 15% quartz, 7% mixed layer illite/smectite minerals, and 7% other minerals. The CEC of this bentonite is 0.48 meq g^{-1}, with 53% of the exchange complex comprised of exchangeable sodium. In both cases, the porosity of the material is 0.85 ± 0.01. With $t^w_{(+)} = 0.38$ for NaCl and $t^w_{(+)} = 0.51$ for KCl, the optimized CEC values are $CEC_{Na} = 0.27 \pm 0.05$ meq g^{-1} and $CEC_K = 0.11 \pm 0.01$ meq g^{-1}, both with $f = 0.90$ (see Figure 10.14).

The last test is to see if the data from Meier et al. (2014) can be fitted at high salinities. Here again, a very good fit is obtained between the model and the data (Figure 10.15). It is likely that a better fit may be obtained if the salinity dependence of the partition coefficient with the salinity is taken into account (e.g., Revil and Skold, 2011).

10.4.4 Filtration Efficiency in Unsaturated Conditions

The data from Sample-Lord and Shackelford (2014) give insights concerning the impact of saturation on filtration efficiency. Two data points (for 100%- and 89%-saturated Na^+-bentonite specimens exposed to 20 mM) are used. For the first test (100% saturated Na^+-Bentonite), the porosity is 0.87 for a target source concentration of 20 mM KCl. The average final concentration in the specimen is 9.0 mM KCl. Its membrane efficiency reaches $\varepsilon = 0.612$.

The second test was done with 89%-saturated Na^+-Bentonite (note that in Sample-Lord and Shackelford, 2014; the saturation is 91% for this test, the value of 89% was actually confirmed upon disassembly). The porosity is

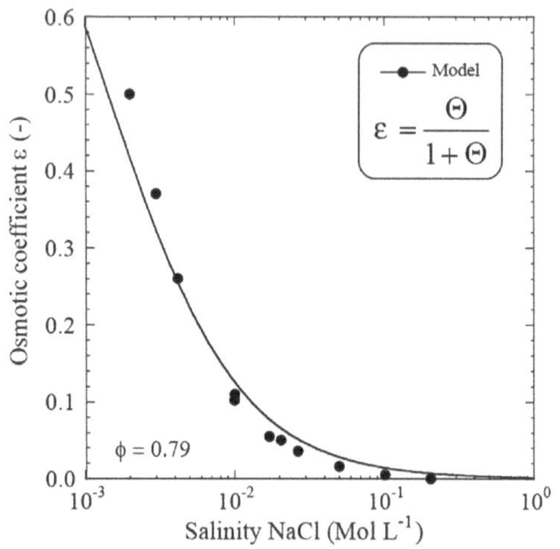

FIGURE 10.15 Comparison between the measurements of the osmotic coefficient as a function of the salinity at high salinities (data from Meier et al., 2014; KCl solutions). The model uses the porosity of the material ($\phi = 0.79 \pm 0.01$). The CEC optimized by fitting the model to the data ($R^2 = 0.97$) is 0.040 ± 0.003 meq g^{-1} and a partition coefficient of $f = 0.90$.

0.89 (also confirmed upon disassembly). The target source concentration is 20 mM KCl and the average final concentration in the specimen is 9.3 mM KCl. The measured membrane efficiency is $\varepsilon = 0.660$ (66.0%).

The CEC was expected to be about 80 cmol kg^{-1} (0.80 meq g^{-1}) based on previous test results. The measured membrane efficiencies for this material are higher than the values measured by Malusis and Shackelford (2002). This is expected, as for these tests the typical bentonite was treated via dialysis to increase the Na^+ content on the exchange complex (relative to Ca^{2+}), to enhance its membrane behavior.

The model to test is given by,

$$\varepsilon(s_w) = \frac{\Theta_S}{s_w + \Theta_S} \qquad (10.58)$$

The data are in excellent agreement with the model (Figure 10.16), using the measured porosity 0.87 ± 0.01 and a fitted CEC value of 0.72 meq g^{-1}, in close agreement with the value given above.

10.4.5 Diffusion Coefficient

In Figure 10.17, the prediction of the model for the diffusion coefficient of the salt (NaCl) through a bentonite (see Appendix C) yields,

$$D = \frac{D_f}{F}\left(\frac{1}{1+\Theta}\right), \qquad (10.59)$$

with the experimental data obtained by Dominijanni et al. (2013). The material is the same as discussed above for Figure 10.12 (bentonite with 98% smectite,

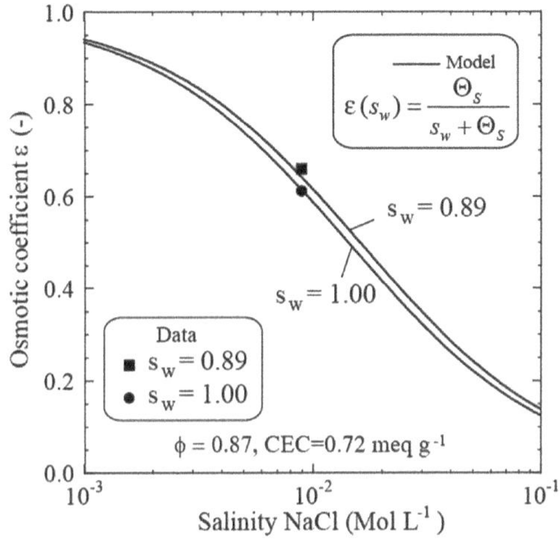

FIGURE 10.16 Comparison between the measurements of the osmotic coefficient as a function of the saturation (data from Sample-Lord and Shackelford, 2014; KCl solutions, two saturations investigated). The model uses the porosity of the material (0.87 ± 0.01). The CEC is 0.72 meq g^{-1} and a partition coefficient of $f = 0.90$.

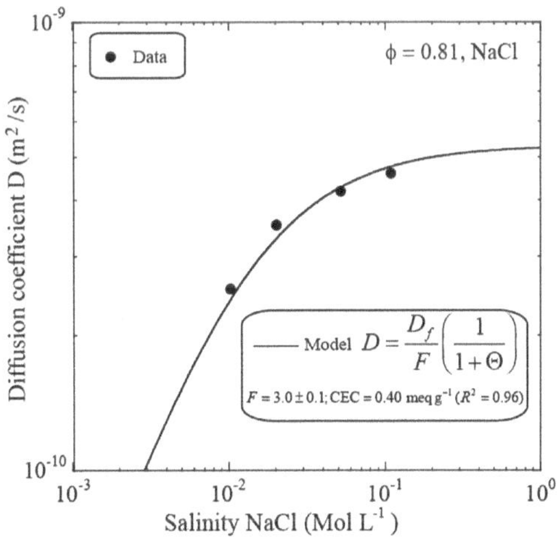

FIGURE 10.17 Comparison between the measurements of the diffusion coefficient as a function of the salinity (data from Dominijanni et al., 2013; bentonite with 98% smectite, NaCl solutions). The model uses the porosity of the material ($\phi = 0.81$). The formation factor is $F = 3.0 \pm 0.1$ CEC_{Na} used for the model is 0.38 ± 0.04 meq g^{-1} (optimized, $R^2 = 0.96$) and a partition coefficient of $f = 0.90$.

NaCl with concentrations from 5 mM to 100 mM, measured CEC with methylene blue 1.05 meq g^{-1}, porosity, $\phi = 0.81$). A test of the model is shown in Figure 10.12. Equation (10.59) provides an excellent fit to the data with the parameters already used above ($f = 0.90$, $\phi = 0.81$, CEC_{Na} of 0.38 ± 0.09 meq g^{-1}) and a formation factor of 3.0.

10.5 USE OF GEOPHYSICAL METHODS

The constitutive equations tested in this chapter can easily be combined with their associated continuity equations to provide the field equation of transport. These equations can be also combined with a rheology to account for the effect of the swelling pressure on the deformation of the material. The transport of salt and ions and the deformation of the materials can be remotely monitored through geophysical methods. Three methods are discussed in this chapter that can be cheaply used to monitor membranes used for the storage of wastes (Figure 10.18).

10.5.1 The Self-Potential Method

The self-potential method is a passive method in which an electrical field is passively measured using nonpolarizing electrodes. According to Eqn (10.32), the electrical field can be analyzed using the fact that the total current density is equal to Ohm's law plus source current terms including a streaming current contribution and a diffusion component:

$$\mathbf{J} = \sigma \overline{\mathbf{E}} + \mathbf{J}_S. \tag{10.60}$$

These source current densities act in turn as exciting terms inside the Maxwell equations and are responsible for the existence of electromagnetic

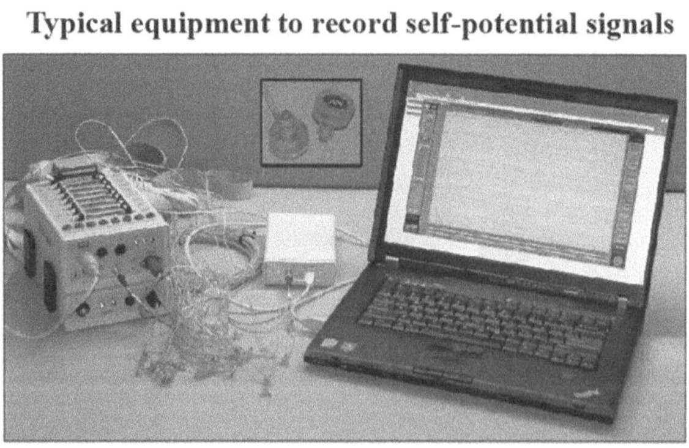

FIGURE 10.18 Description of the self-potential equipment used to perform self-potential monitoring including nonpolarization electrodes with preamplifiers connected to a multichannel high-precision voltmeter. This voltmeter is characterized by a high-impedance (100 MOhm) and is connected to a laptop computer used to record the data.

disturbances that can be remotely monitored. The conservation of charge in the low frequency of the Maxwell equations is given by,

$$\nabla \cdot \mathbf{J} = 0. \tag{10.61}$$

Combining Eqns (10.60) and (10.61), the electrical field is a solution of a Poisson equation:

$$\nabla \cdot \left(\sigma \overline{\mathbf{E}}\right) = -\nabla \cdot \mathbf{J}_S. \tag{10.62}$$

This equation is the same equation as solved in electroencephalography in which the electrical field and the magnetic field are measured on the scalp. In this case, the source of current is due to the existence of source current densities existing at the synapses between the neurons and corresponding to the opening of ionic channels. Recent examples of the use of self-potential methods including algorithms to perform the inverse problem can be found in Rittgers et al. (2013) (for the monitoring of corrosion) and Haas et al. (2013) (for the monitoring of hydromechanical disturbances). These two papers contain a description for the equipment and the electrodes that can be used for this monitoring purpose. Soueid Ahmed et al. (2014) developed a tomographic algorithm to image the distribution of the permeability of heterogeneous media (including aquifers) using pressure and self-potential measurements.

10.5.2 The Direct Current (DC) Resistivity Method

The self-potential method is categorized as a passive geoelectrical technique since the source electrical current density is produced inside the material in thermodynamic disequilibrium. The DC conductivity or resistivity method is based on the injection of electrical currents in the material using pair of electrodes and the measurements of the resulting electrical field. The goal of this method is to image the resistivity distribution in the material. Since the conductivity depends on the porosity, the saturation, the temperature and the ionic strength of the solution (Revil, 2013a), these parameters can be eventually monitored through time-lapse resistivity measurements. Recent techniques to jointly invert self-potential and DC resistivity data can be found in Jardani et al. (2013) using a stochastic technique and coupled inversion (i.e., the transport problem is included in the inversion of the geophysical data). With the massive development of multichannel resistive meters, this method can be used to take resistivity snapshots very quickly and image the progress of contaminants in a porous material.

10.5.3 The Induced Polarization Method

The induced polarization method is the sister of the DC resistivity method. In an alternating electrical field, the current and electrical field are not in phase. There is a small-phase lag between the two. This phase lag is related to polarization phenomena around the grains involving the electrical double layer. Using the resistivity method and measuring the phase lag, one can write the electrical conductivity as a complex number with an in-phase (real) component related to the electromigration of the charge carriers and an out-of-phase (quadrature or imaginary) component related to the polarization of the material, i.e., the reversible storage of electrical charges in the alternating electrical current. In theory, induced polarization can be performed in the frequency domain (i.e., the measurement of the phase lag is performed at different frequency of the exciting electrical current) and realized with the same equipment as used for DC resistivity. In practice, the impedance meter requires a high resolution on the phase (0.1 mrad) and few equipments have this phase resolution.

Generally, induced polarization measurements are very complementary to resistivity measurements. Resistivity measurements are mostly sensitive to what is occurring in the water phase while induced polarization measurements are very sensitive to adsorption/desorption phenomena in the electrical double layer (Vaudelet et al., 2011a,b).

10.6 CONCLUSIONS

The theory developed by Revil and Linde (2006) and Revil (2013a) is successful at describing various experimental data from the literature including electrical conductivity, streaming and electroosmotic coupling coefficient measurements, and osmotic and salt diffusion coefficient measurements. The constitutive equations seem to explain an unprecedented amount of experimental data in saturated and unsaturated conditions of clay material. The theory remains simple since it is based essentially on the knowledge of the connected porosity, the CEC, and the saturation of the material. This theory can be used in transport numerical codes coupled with hydromechanical codes to predict coupled transport phenomena in deformable partially saturated membranes. These models can be also used to forward-model the geophysical response of three geoelectrical methods (self-potential, DC resistivity, and induced polarization). In turn, time-lapse tomography using these methods can be efficiently used to monitor the performance of membranes over time.

GLOSSARY

Variable	Unit	Meaning
C	V/Pa	Streaming potential coupling coefficient
C_r	–	Relative streaming potential coefficient ($0 \leq C_r \leq 1$)
CEC	C/kg	Cation exchange capacity
$c_{(\pm)}$	m^{-3}	Concentration of cations and anions in the reservoir
$\overline{c}_{(\pm)}$	m^{-3}	Concentration of cations and anions in the pore space
c_w	m^{-3}	Concentration of water molecules in the reservoir

Variable	Unit	Meaning
$\overline{c}_w$	m^{-3}	Concentration of water molecules in the pore space
D	$m^2\ s^{-1}$	Diffusion coefficient of the salt in the porous material
D_f	$m^2\ s^{-1}$	Diffusion coefficient of the salt in the reservoir
$\overline{\mathbf{E}}$	$V\ m^{-1}$	Macroscopic electrical field
e	C	Elementary charge, 1.6×10^{-19}
f	–	Partition coefficient between the Stern and diffuse layers
F	–	Electrical formation factor
$\mathbf{J}$	$A\ m^{-2}$	Macroscopic total current density
$\mathbf{J}_d$	$m^{-2}\ s^{-1}$	Macroscopic diffusion flux of the salt
$\mathbf{J}_{(\pm)}$	$m^{-2}\ s^{-1}$	Macroscopic flux of the cations and anions
k_b	$m^2\ kg\ s^{-2}\ K^{-1}$	Boltzmann constant, $1.3806503 \times 10^{-23}$
k_w	m^2	Permeability at partial saturation
k_r	–	Relative permeability ($0 \leq k_r \leq 1$)
k_S	m^2	Hydraulic conductivity
K_w	$m\ s^{-1}$	Hydraulic conductivity at saturation
K_S	$m\ s^{-1}$	Matrix of material properties
$\mathbf{L}$	variable	Components of the matrix **L**
L_{ij}	variable	First (porosity) Archie exponent
L	$A\ m^{-1}\ Pa^{-1}$	Streaming current coupling coefficient
L_r	–	Relative streaming current coupling coefficient ($0 \leq L_r \leq 1$)
L_S	$A\ m^{-1}\ Pa^{-1}$	Streaming current coupling coefficient at full saturation
m	–	Second (saturation) Archie exponent
m_v	–	van Genuchten exponent
n	–	Second Archie's exponent
n_v	–	van Genuchten exponent
N	–	Number of ionic species
$\overline{\mathbf{P}}$	$Pa\ m^{-1}$	Capillary pressure field
p_w	Pa	Mechanical pore water pressure in the reservoir
$\overline{p}_w$	Pa	Mechanical pore water pressure in the pore-water phase
p_c	Pa	Capillary pressure at the representative elementary volume
p_e	Pa	Capillary entry pressure
$\widehat{Q}_V$	$C\ m^{-3}$	Effective charge density dragged by the pore water flow
$\widehat{Q}_V^S$	$C\ m^{-3}$	Value of $\widehat{Q}_V$ at saturation
Q_V	$C\ m^{-2}$	Total charge density per unit volume of the pore-water phase
$\overline{Q}_V$	$C\ m^{-2}$	Total charge density of the diffuse layer per unit pore volume
$\overline{Q}_V^S$	$C\ m^{-2}$	Value of $\overline{Q}_V$ at saturation
r	–	Relative permeability exponent
s_w	–	Water saturation
s_r	–	Irreducible water saturation
s_e	–	Effective water saturation
S_{Sp}	$m^2\ kg^{-1}$	Specific surface area
T	K	Absolute temperature

Variable	Unit	Meaning
$T_{(\pm)}$	–	Macroscopic Hittorf number of the cations and anions
$t^{W}_{(\pm)}$	–	Hittorf number of cations and anions in the reservoirs
$\dot{\mathbf{w}}$	m s^{-1}	Darcy velocity
$\dot{\mathbf{w}}$	m s^{-1}	Mechanical contribution to the Darcy velocity
ϕ	–	Connected porosity
$\beta_{(\pm)}$	m^2 s^{-1} V^{-1}	Mobility of the cations and anions in the bulk pore water
ε	–	Osmotic coefficient (filtration efficiency)
λ	–	Brooks–Corey saturation exponent
$\mu_{(\pm)}$	J	Chemical potential of cations and anions
$\tilde{\mu}_{(\pm)}$	J	Electrochemical potential of cations and anions
μ_W	J	Chemical potential of water
μ_f	J	Chemical potential of the salt
σ_S	S m^{-1}	Surface conductivity
σ_W	S m^{-1}	Conductivity of the pore water
$\sigma_{(\pm)}$	S m^{-1}	Ionic contribution to the total conductivity of the material
σ	S m^{-1}	Total electrical conductivity of the porous material
ψ	V	Electrical potential due to the electrical double layer
φ	V	Electrical potential in the reservoir
$\overline{\varphi}$	V	Electrical potential in the pore water
ρ_S	kg m^{-3}	Mass density of the solid
ρ_W	kg m^{-3}	Mass density of the pore water
π	Pa	Osmotic pressure in the reservoirs
$\overline{\pi}$	Pa	Osmotic pressure in the pore space
$\delta\pi$	Pa	Difference in osmotic pressure between the reservoir and the pore water
η_W	Pa s	Dynamic viscosity of the pore-water phase
Ω_W	m^3	Molecular volume of water
$\Omega_{(\pm)}$	m^3	Molecular volume of the anions and cations
Θ	–	Dimensionless variable
Θ_S	–	Value of Θ at saturation
ν	–	Dimensionless variable

APPENDIX A: CATION-DEPENDENT CEC

According to Caroll (1959), the cation exchange capacity (CEC) of kaolinite is in the range 0.03–0.10 meq g^{-1}, of chlorite and illite in the range 0.10–0.40 meq g^{-1}, of smectite in the range 0.70–1.5 meq g^{-1}, and of vermiculite in the range 1.0–2.0 meq g^{-1}. However, these values of the CEC correspond to the complete titration of the mineral surface using a cation with a strong affinity for the active sites on the mineral surface. For other cations, the apparent CEC depends on the type of cations. Woodruff and Revil (2011) proposed that the measured CEC for a cationic species i is scaled to the absolute CEC (measured using a cation that has a very strong affinity with clay mineral surface like ammonium or cobalt), according to

$$\mathrm{CEC}_i = \delta_i \mathrm{CEC}, \tag{10.A1}$$

where $\delta_i \leq 1$ (therefore $\mathrm{CEC}_i \leq \delta_i\mathrm{CEC}$) denotes a scaling parameter that is dependent on the type of cation populating the mineral surface. Revil and Lu (2013) obtained the following values: $\delta(\mathrm{Li}) = 0.4$, $\delta(\mathrm{Mg}) = 0.8$, $\delta(\mathrm{Ca}) = 0.8$, and $\delta(\mathrm{K}) = 0.4$.

APPENDIX B: OSMOTIC PRESSURE WITH THE DONNAN AND REVIL MODELS

The aim of this appendix is to provide, at first, the basics to understand the electrostatic component of the osmotic pressure in charged porous media, and then, both classical and new relationships in terms of the Donnan and Revil models (Donnan, 1911; Revil, 2013a) expressing the ionic concentrations in the pore space as a function of the CEC. The ionic concentration in the (neutral) fictitious reservoir of ions in local equilibrium with the pore water is $c_{(\pm)} = c_f$. The expressions of the electrochemical potentials of the cations and anions in the fictitious reservoir and in the pore space are (e.g., Helfferich, 1995),

$$\tilde{\mu}_{(\pm)} = \mu^0_{(\pm)} + k_b T \ln c_{(\pm)}, \tag{10.B1}$$

$$\bar{\tilde{\mu}}_{(\pm)} = \mu^0_{(\pm)} + k_b T \ln \bar{c}_{(\pm)} \pm e\psi, \tag{10.B2}$$

respectively. The bar is used to express that the quantity is considered in the pore space. In these expressions, ψ is the mean (microscopic) electrical potential distribution associated with the electrical diffuse layer in the capillary ($\psi = 0$ in the reservoir). Thermodynamic equilibrium between the fictitious reservoir and the pore space implies,

$$\bar{\tilde{\mu}}_{(\pm)} = \tilde{\mu}_{(\pm)}, \tag{10.B3}$$

which leads directly to Boltzmann distributions for the concentrations of the cations and anions in the pore space of the charged material. Solving these equations together with charge balance equations in the pore space of the material and in the reservoirs yields well-known expressions for the ionic concentrations in the pore space of the material and an expression for the mean electrostatic potential (Teorell, 1935; Meyer and Sievers, 1936),

$$\bar{c}_{(\pm)} = c_f\left(\sqrt{1+\Theta^2} \pm \Theta\right), \tag{10.B4}$$

$$\psi = \frac{k_b T}{2e} \ln\left[\frac{\sqrt{1+\Theta^2}+\Theta}{\sqrt{1+\Theta^2}-\Theta}\right], \tag{10.B5}$$

which corresponds to the Donnan model and where $\Theta \equiv \overline{Q}_V^S/(2ec_f s_w)$. The potential ψ is the mean electrostatic potential of the porous material as a whole. This generalized the original Teorell-Sievers-Meyer model to unsaturated materials.

The osmotic pressure in a reservoir and the chemical potential of the water in this reservoir are given by (Kedem and Katchalsky, 1958),

$$\pi = -\frac{k_b T}{\Omega_w} \ln c_w = -\frac{k_b T}{\Omega_w} \ln(1 - 2\Omega_f c_f) \approx 2k_b T c_f, \tag{10.B6}$$

$$\mu_w = \mu_w^0 + \Omega_w p_w + k_b T \ln c_w, \tag{10.B7}$$

respectively, where p_w is the mechanical pressure in the pore-water phase. The chemical potential of the water in the reservoir can be written as $\mu_w = \mu_w^0 + \Omega_w(p_w - \pi)$, and the chemical potential gradient in water is $\nabla \mu_w = \Omega_w \nabla(p_w - \pi)$. The chemical potential of the water in the pore space of the material is,

$$\overline{\pi} = -\frac{k_b T}{\Omega_w} \ln \overline{c}_w, \tag{10.B8}$$

$$\overline{\mu}_w = \mu_w^0 + \Omega_w \overline{p}_w + k_b T \ln \overline{c}_w, \tag{10.B9}$$

where $\overline{c}_w$ and $\overline{p}_w$ are the water concentration and internal fluid pressure in the pore space of the porous material, respectively. The chemical potential of water in the pore space can be written as,

$$\overline{\mu}_w = \mu_w^0 + \Omega_w(\overline{p}_w - \overline{\pi}). \tag{10.B10}$$

The conservation of mass in the reservoir and in the aqueous pore water is described as,

$$\Omega_w c_w + (\Omega_{(+)} + \Omega_{(-)})c_f = 1, \tag{10.B11}$$

$$\Omega_w \overline{c}_w + \Omega_{(+)}\overline{c}_{(+)} + \Omega_{(-)}\overline{c}_{(-)} = 1, \tag{10.B12}$$

respectively. Local thermodynamic equilibrium between the fictitious reservoir and the pore water implies $\overline{\mu}_w = \mu_w$. Equations (10.B6) to (10.B12) yield the following expression for the internal pore pressure in the charged porous material $\overline{p}_w = p_w - (\pi - \overline{\pi})$. The difference of osmotic pressure between the reservoir and the pore water is

$$\delta\pi = \pi - \overline{\pi} = -\frac{k_b T}{\Omega_w} \ln\left(\frac{c_w}{\overline{c}_w}\right), \tag{10.B13}$$

$$\delta\pi = -\frac{k_b T}{\Omega_w} \ln\left(\frac{1 - 2\Omega_w c_f}{1 - \Omega_w \overline{c}_{(+)} - \Omega_w \overline{c}_{(-)}}\right), \tag{10.B14}$$

$$\delta\pi \approx -k_b T\left(\overline{c}_{(+)} + \overline{c}_{(-)} - 2c_f\right), \tag{10.B15a}$$

using a first-order Taylor expansion of the argument of the logarithm and assumed that the molar volume of the different molecules are roughly the same. The combination of Eqns (10.B6) and (10.B8) with the mass conservation equations yields the osmotic pressure in the reservoir, $\pi = 2k_bTc_f$ (classical van't Hoff equation in a neutral electrolyte), and the osmotic pressure in the pore space of the porous material and $\overline{\pi} = k_bT(\overline{c}_{(+)} + \overline{c}_{(-)})$ (extension of the van't Hoff equation in the pore-water phase). The difference in osmotic pressure between the reservoir and the local pore space is further obtained as a function of the mean potential or as a function of the key-dimensionless variable Θ:

$$\delta\pi = -2c_fk_bT\left(\cosh\left(\frac{e\psi}{k_bT}\right) - 1\right), \tag{10.B15b}$$

$$\delta\pi = -2c_fk_bT\left(\sqrt{1+\Theta^2} - 1\right). \tag{10.B16}$$

Similarly, the osmotic pressure in the aqueous phase in the pore space can be written as,

$$\overline{\pi} = 2c_fk_bT\cosh\left(\frac{e\psi}{k_bT}\right). \tag{10.B17}$$

This is consistent with the integration of the excess of charge with respect to the mean potential. Indeed,

$$\delta\pi = \int_0^{\psi} \overline{Q}_V d\psi' = e\int_0^{\psi} \left(\overline{c}_{(+)} - \overline{c}_{(-)}\right) d\psi', \tag{10.B18}$$

$$\delta\pi = -2c_fk_bT\left[\cosh\left(\frac{e\psi}{k_bT}\right) - 1\right] = \pi - \overline{\pi}. \tag{10.B19}$$

The pressure difference $\delta\overline{\pi}$ is the difference between the osmotic pressure of the reservoir and of the pore water, and this difference arises because of the electrostatic force inside the pore space of the charged material.

The model proposed by Revil is quite different. It is based on the fact that the excess of cations contained in the pore space is just balancing the deficiency of charge on the surface of the clay minerals (Stern layer comprised). The cationic and anionic concentrations are given by,

$$\overline{c}_{(+)} = c_{(+)} + \frac{\overline{Q}_V^S}{es_w} = c_f(1 + 2\Theta), \tag{10.B20}$$

$$\overline{c}_{(+)} = c_{(-)} = c_f, \tag{10.B21}$$

which can be summarized by,

$$\overline{c}_{(\pm)} = c_f(1 + \Theta \pm \Theta). \tag{10.B22}$$

This equation replaces Eqn (10.B4) of the Donnan model. The osmotic pressure in the porous material is given by $\overline{\pi} = k_b T(\overline{c}_{(+)} + \overline{c}_{(-)})$, and therefore Revil's model yields $\overline{\pi} = 2k_b T c_f(1 + \Theta)$. The difference of osmotic pressure between the reservoir and the pore-water phase is therefore given by $\delta\pi = \pi - \overline{\pi} = -2k_b T c_f \Theta$. An expression of the osmotic efficiency in the Revil model, starting from its definition, is obtained as:

$$\varepsilon \equiv \left(\frac{\partial \overline{p}_w}{\partial \overline{\pi}}\right)_{\mathbf{J},\dot{\mathbf{w}}=0} = -\left(\frac{\partial(\pi - \overline{\pi})}{\partial \overline{\pi}}\right)_{\mathbf{J},\dot{\mathbf{w}}=0}, \tag{10.B23}$$

$$\varepsilon = 1 - \left(\frac{\partial \pi}{\partial \overline{\pi}}\right)_{\mathbf{J},\dot{\mathbf{w}}=0}, \tag{10.B24}$$

$$\varepsilon = 1 - \frac{1}{(1 + \Theta)} = \frac{\Theta}{1 + \Theta}. \tag{10.B25}$$

Similarly, the Hittorf number is given in the context of the Revil's model by,

$$T_{(+)}(s_w) = \frac{\beta_{(+)}\overline{c}_{(+)}}{\beta_{(+)}\overline{c}_{(+)} + \beta_{(-)}\overline{c}_{(-)}}, \tag{10.B26}$$

$$T_{(+)}(s_w) = \frac{\beta_{(+)}(1 + 2\Theta)}{\beta_{(+)}(1 + 2\Theta) + \beta_{(-)}}. \tag{10.B27}$$

Similarly in the Donnan model, the osmotic efficiency and the macroscopic Hittorf number are given by (Revil and Skold, 2011),

$$\varepsilon = \frac{\sqrt{1 + \Theta^2} - 1}{\sqrt{1 + \Theta^2}}, \tag{10.B28}$$

$$T_{(+)} = \frac{\beta_{(+)}\left(\sqrt{1 + \Theta^2} + \Theta\right)}{\beta_{(+)}\left(\sqrt{1 + \Theta^2} + \Theta\right) + \beta_{(-)}\left(\sqrt{1 + \Theta^2} - \Theta\right)}. \tag{10.B29}$$

Note that all the results of the Revil model can be deduced from the Donnan model by using the following change of variables $\sqrt{1 + \Theta^2}(\text{Donnan}) \Leftrightarrow 1 + \Theta(\text{Revil})$.

APPENDIX C: OSMOTIC COEFFICIENT, REVERSE OSMOSIS, AND SALT DIFFUSIVITY

The salinity flux is obtained by combining Eqns (10.22), (10.24), (10.29), and (10.31),

$$2\mathbf{J}_d = \frac{1}{e}\left(\sigma_{(+)} - \sigma_{(-)}\right)\overline{\mathbf{E}} - \frac{1}{e^2}\left(\sigma_{(+)} + \sigma_{(-)}\right) \times \nabla\overline{\mu}_f - \frac{k_w(s_w)}{\eta_w}\left(\overline{c}_{(+)} + \overline{c}_{(-)}\right)\nabla\left(\overline{p}_w - \overline{\pi}\right). \quad (10.C1)$$

For an experiment with a core sample comprised between two ionic reservoirs, the total current density is null and therefore the electrical field is given by,

$$\overline{\mathbf{E}}\Big|_{\mathbf{J}=0} = \frac{1}{e}\left(2T_{(+)} - 1\right)\nabla\overline{\mu}_f - C\nabla\left(\overline{p}_w - \overline{\pi}\right). \quad (10.C2)$$

Combining Eqns (10.C1) and (10.C2) yields, after some algebraic manipulations,

$$2\mathbf{J}_d = -\frac{2}{e^2}\left(\frac{\sigma_{(+)}\sigma_{(-)}}{\sigma_{(+)} - \sigma_{(-)}}\right)\nabla\overline{\mu}_f + \frac{k_w\widehat{Q}_V}{\eta_w e}\left(T_{(+)} - T_{(-)}\right)\nabla\left(\overline{p}_w - \overline{\pi}\right) - \frac{k_w(s_w)}{\eta_w}\left(\overline{c}_{(+)} + \overline{c}_{(-)}\right)\nabla\left(\overline{p}_w - \overline{\pi}\right), \quad (10.C3)$$

where Eqn (10.37) has been used for the expression of the streaming potential coupling coefficient. The gradient of the chemical potential of the pore water is given by,

$$\nabla\overline{\mu}_f = k_b T\left(\nabla\ln\overline{c}_{(+)} + \nabla\ln\overline{c}_{(-)}\right), \quad (10.C4)$$

$$\nabla\overline{\mu}_f = k_b T\nabla\ln\left(\overline{c}_{(+)}\overline{c}_{(-)}\right), \quad (10.C5)$$

$$\nabla\overline{\mu}_f = 2k_b T\nabla\ln\left(\overline{c}_f\right), \quad (10.C6)$$

where the mean salinity of the pore water is defined by

$$\overline{c}_f = \sqrt{\overline{c}_{(+)}\overline{c}_{(-)}}. \quad (10.C7)$$

Using also the property of the macroscopic Hittorf number, it follows:

$$\mathbf{J}_d = -\frac{2k_b T}{e^2\overline{c}_f}\left(\frac{\sigma_{(+)}\sigma_{(-)}}{\sigma_{(+)} - \sigma_{(-)}}\right)\nabla\overline{c}_f + \frac{k_w\widehat{Q}_V}{2\eta_w e}\left(2T_{(+)} - 1\right)\nabla\left(\overline{p}_w - \overline{\pi}\right) - \frac{k_w(s_w)}{2\eta_w}\left(\overline{c}_{(+)} + \overline{c}_{(-)}\right)\nabla\left(\overline{p}_w - \overline{\pi}\right). \quad (10.C8)$$

If the advective component of the diffusional flux is zero, the first term of Eqn (10.C8) is nothing else than the first Fick's law term:

$$\mathbf{J}_d = -D\nabla\overline{c}_f, \tag{10.C9}$$

$$D = \frac{2k_bT}{e^2\overline{c}_f}\left(\frac{\sigma_{(+)}\sigma_{(-)}}{\sigma_{(+)} - \sigma_{(-)}}\right), \tag{10.C10}$$

and D is the effective diffusion coefficient of the salt through the porous material. The opposite case corresponds to the case where the diffusional flux is dominated by advective effect. This case is the one for which the filtration coefficient of reverse osmosis can be defined. In this case, the diffusional flux of the salt is given by,

$$\mathbf{J}_d = -\frac{k_w}{2\eta_w}\left[\overline{c}_{(+)} + \overline{c}_{(-)} + \frac{\widehat{Q}_V}{e}\left(1 - 2T_{(+)}\right)\right]\nabla(\overline{p}_w - \overline{\pi}). \tag{10.C11}$$

The salinity of the effluent $\tilde{c}_f$ is defined by,

$$\mathbf{J}_d = \tilde{c}_f\dot{\mathbf{w}}_m, \tag{10.C12}$$

where $\dot{\mathbf{w}}_m$ denotes the mechanical component of the Darcy velocity given by

$$\dot{\mathbf{w}}_m = -\frac{k_w}{\eta_w}\nabla(\overline{p}_w - \overline{\pi}). \tag{10.C13}$$

It follows that the salinity of the effluent is given by,

$$\tilde{c}_f = \frac{1}{2}\left[\overline{c}_{(+)} + \overline{c}_{(-)} + \frac{\widehat{Q}_V}{e}\left(1 - 2T_{(+)}\right)\right], \tag{10.C14}$$

$$\tilde{c}_f = c_f\left[\frac{\overline{c}_{(+)} + \overline{c}_{(-)}}{2c_f} + \frac{\widehat{Q}_V}{2ec_f}\left(1 - 2T_{(+)}\right)\right], \tag{10.C15}$$

$$\tilde{c}_f = c_f\left[\sqrt{1 + \Theta^2} + \nu\left(1 - 2T_{(+)}\right)\right], \tag{10.C16}$$

where:

$$\overline{c}_{(+)} + \overline{c}_{(-)} = 2c_f\sqrt{1 + \Theta^2}, \tag{10.C17}$$

$$\Theta \equiv \frac{\overline{Q}_V}{2ec_f}, \tag{10.C18}$$

$$\nu \equiv \frac{\widehat{Q}_V}{2ec_f}. \tag{10.C19}$$

The filtration coefficient is properly defined by,

$$\varepsilon = \frac{c_f - \tilde{c}_f}{c_f} = 1 - \frac{\tilde{c}_f}{c_f}, \tag{10.C20}$$

$$\varepsilon = 1 - \sqrt{1 + \Theta^2} + \nu\left(2T_{(+)} - 1\right). \tag{10.C21}$$

Assuming $\Theta \approx \nu$ for bentonite and using the expression of the macroscopic Hittorf number (assuming the mobilities between the cations and anions are very close),

$$T_{(+)} \approx \frac{\sqrt{1 + \Theta^2} + \Theta}{2\sqrt{1 + \Theta^2}}, \tag{10.C22}$$

and

$$\varepsilon = 1 - \sqrt{1 + \Theta^2} + \frac{\Theta^2}{\sqrt{1 + \Theta^2}}, \tag{10.C23}$$

$$\varepsilon = \frac{\sqrt{1 + \Theta^2} - 1}{\sqrt{1 + \Theta^2}}. \tag{10.C24}$$

The same analysis in the context of Revil's model yields,

$$\varepsilon = \frac{\Theta}{1 + \Theta}. \tag{10.C25}$$

Both formula, Eqns (10.24) and (10.25), are consistent with what was found in Appendix B.

Following the same approach as Revil and Skold (2011), the diffusion coefficient can be written as,

$$D = \frac{D_f}{F}\gamma, \tag{10.C26}$$

where D_f is the mutual diffusion coefficient of the salt in the pore water and γ is a correction coefficient. Donnan's model with the same mobility for the cations and anions yields (Revil and Skold, 2011):

$$\gamma = \frac{1}{\sqrt{1 + \Theta^2}}, \tag{10.C27}$$

while Revil's model yields:

$$\gamma = \frac{1}{1 + \Theta}. \tag{10.C28}$$

In the limit $\Theta = 0$ (no electrical diffuse layer), $\gamma = 1$ and Eqn (10.C26) reduces to the classical expression of the diffusion coefficient $D = D_f/F$ while in the case $\Theta \gg 1$, $\gamma = 0$ and the clay material is behaving as a perfect membrane. In this case, there is no concentration gradient in the pore space since the concentration of cation is controlled only by the CEC while the concentration of anions is vanishingly small. The following general property can be checked $\gamma = 1 - \varepsilon$ (e.g., Dominijanni et al., 2013, Eqn (10.65)).

REFERENCES

Archie, G.E., 1942. The electrical resistivity log as an aid in determining some reservoir characteristics. Trans. Am. Inst. Min., Metall. Pet. Eng. 146, 54–62.

Avena, M.J., De Pauli, C.P., 1998. Proton adsorption and electrokinetics of an Argentinean montmorillonite. J. Colloid Interface Sci. 202, 195–204. http://dx.doi.org/10.1006/jcis.1998.5402.

Barbour, S., Fredlund, D., 1989. Mechanisms of osmotic flow and volume change in clay soils. Can. Geotech. J. 26 (4), 551–562.

Bolève, A., Crespy, A., Revil, A., Janod, F., Mattiuzzo, J.L., 2007. Streaming potentials of granular media: influence of the Dukhin and Reynolds numbers. J. Geophys. Res. 112, B08204. http://dx.doi.org/10.1029/2006JB004673.

Bohnhoff, G.L., Shackelford, C.D., 2013. Improving membrane performance via bentonite polymer nanocomposite. Appl. Clay Sci. 86, 83–98.

Bonaparte, R., Beech, J., Griffin, L., Phillips, D., Kumthekar, U., Reising, J., 2008. Design, construction, and performance of low-level radioactive waste disposal facility. In: Proceedings of 6th International Conference on Case Histories in Geotechnical Engineering, August 11–16, 2008, Arlington, VA.

Börner, F.D., 1992. Complex conductivity measurements of reservoir properties. In: Proc Third European Core Analysis Symposium, Paris, pp. 359–386.

Brooks, R.H., Corey, A.T., 1964. Hydraulic properties of porous media. In: Hydrology Papers 3, Colorado State University, Ft. Collins, Colorado.

Carroll, D., 1959. Ion exchange in clays and other minerals. Geol. Soc. Am. Bull. 70 (6), 749–780. http://dx.doi.org/10.1130/0016-7606.

Caruso, M., Jommi, C., 2005. An evaluation of indirect methods for the estimation of hydraulic properties of unsaturated soils. In: Bilsen, H., Nalbantoglu, Z. (Eds.), Problematic Soils, 1. Eastern Mediterranean University Press, pp. 183–191.

Chen, X., Hicks, M.A., 2013. Unsaturated hydro-mechanical-chemo coupled constitutive model with consideration of osmotic flow. Comput. Geotech. 54, 94–103.

Churcher, R.L., French, P.R., Shaw, J.C., Schramm, L.L., 1991. In: Paper SPE 21041 Presented at the 1991 Int. Symp. On Oilfield Chemistry, Anaheim, February 20-22, 1991.

Comparon, L., 2005. Etude expérimentale des propriétés électriques et diélectriques des matérieux argileux consolidés, PhD Thesis, Institut de Physique du Globe de Paris, 400 pp.

Dominijanni, A., Manassero, M., Puma, S., 2013. Coupled chemical-hydraulic-mechanical behaviour of bentonites. Géotechnique 63 (3), 191–205.

Donnan, F.G., 1911. Theorie der Membrangleichgewichte und Membranpotentiale bei Vorhandensein von nicht dialysierenden Elektrolyten. Ein Beitrag zur physikalisch-chemischen Physiologie. Theory of membrane equilibria and membrane potentials in the presence of non-dialysing electrolytes. A contribution tophysical-chemical physiology Z. Elektrochem.

Angew. Phys. Chem. 17 (14), 572–581. English translation republished 1995 in J. Membrane Sci. 100 (1), 45–55.

Gabrieli, L., Jommi, C., Musso, G., Romero, E., 2008. Electrokinetic treatment of a natural silt in saturated and unsaturated conditions. In: Shao, J.F., Burlion, N. (Eds.), Thermo-hydromechanical and Chemical Coupling in Geomaterials and Applications: Proceedings of the 3rd International Symposium GeoProc2008. Wiley, ISTE, pp. 203–210.

Gens, A., July 1–3, 2013. Coupled modelling of barriers for radioactive waste disposal. In: Manassero, M., Dominijanni, A., Foti, S., Musso, G. (Eds.), Proceedings of Coupled Phenomena in Environmental Geotechnics. CRC Press/Balkema, Torino, Italy, pp. 21–34. Taylor & Francis Group, London.

van Genuchten, M.T., 1980. A closed-form equation for predicting the hydraulic conductivity of unsaturated soils. Soil Sci. Soc. Am. J. 44, 892–898.

Groenevelt, P., Elrick, D., 1976. Coupling phenomena in saturated homo-ionic montmorillonite: II. Theoretical. Soil Sci. Soc. Am. J. 40, 820–823.

Haas, A.K., Revil, A., Karaoulis, M., Frash, L., Hampton, J., Gutierrez, M., Mooney, M., 2013. Electrical potential source localization reveals a borehole leak during hydraulic fracturing. Geophysics 78 (2), D93–D113. http://dx.doi.org/10.1190/GEO2012-0388.1.

Helfferich, F., 1995. Ion Exchange. Dover Publications, Inc., New York.

Jardani, A., Revil, A., Dupont, J.P., 2013. Stochastic joint inversion of hydrogeophysical data for salt tracer test monitoring and hydraulic conductivity imaging. Adv. Water Resour. 52, 62–77. http://dx.doi.org/10.1016/j.advwatres.2012.08.005.

Jougnot, D., Revil, A., Leroy, P., 2009. Diffusion of ionic tracers in the Callovo-Oxfordian clay-rock using the Donnan equilibrium model and the electrical formation factor. Geochimica et Cosmochimica Acta 73, 2712–2726.

Kang, J., Shackelford, C., 2009. Clay membrane testing using a flexible-wall cell under closed-system boundary conditions. Appl. Clay Sci. 44 (1–2), 43–58.

Kang, J., Shackelford, C., 2010. Membrane behavior of compacted clay liners. J. Geotech. Geoenviron. Eng. 136 (10), 1368–1382.

Kang, J., Shackelford, C., 2011. Consolidation enhanced behavior of a geosynthetic clay liner. Geotext. Geomembranes 29 (6), 544–556.

Katchalsky, A., Curran, P.F., 1965. Nonequilibrium Thermodynamics in Biophysics. Harvard University Press, Cambridge, MA.

Kedem, O., Katchalsky, A., 1958. Thermodynamics analysis of the permeability of biological membranes to non-electrolytes. Biochimica et Biophysica Acta 27, 229–246.

Kemper, W., Rollins, J., 1966. Osmotic efficiency coefficients across compacted clays. Soil Sci. Soc. Am. J. 30 (5), 529–534.

Kurniawan, B., 2005. Shaly Sand Interpretation Using CEC-dependent Petrophysical Parameters, PhD Thesis, Louisiana State University, 201 pp.

Leroy, P., Revil, A., 2004. A triple layer model of the surface electrochemical properties of clay minerals. J. Colloid Interface Sci. 270 (2), 371–380.

Linde, N., Jougnot, D., Revil, A., Matthaï, S.K., Arora, T., Renard, D., Doussan, C., 2007. Streaming current generation in two-phase flow conditions. Geophys. Res. Lett. 34 (3), L03306. http://dx.doi.org/10.1029/2006GL028878.

Loret, B., Hueckel, T., Gajo, A., 2002. Chemo-mechanical coupling in saturated porous media: elastic–plastic behaviour of homoionic expansive clays. Int. J. Solids Struct. 39, 2773–2806.

Lorne, B., Perrier, F., Avouac, J.-P., 1999. Streaming potential measurements. 1. Properties of the electrical double layer from crushed rock samples. J. Geophys. Res. 104, 17857–17877.

Malusis, M., Shackelford, C., 2002. Chemico-osmotic efficiency of a geosynthetic clay liner. J. Geotech. Geoenviron. Eng. 128 (2), 97–106.

Malusis, M.A., Shackelford, C.D., Olsen, H.W., 2003. Flow and transport through clay membrane barriers. Eng. Geol. 70, 235–248.

Mboh, C.M., Huisman, J.A., Zimmermann, E., Vereecken, H., 2012. Coupled hydrogeophysical inversion of streaming potential signals for unsaturated soil hydraulic properties. Vadose Zone J. 11 (2). http://dx.doi.org/10.2136/vzj2011.0115.

Meier, A., Sample-Lord, K., Castelbaum, D., Kallase, S., Moran, B., Ray, T., Shackelford, C., 2014. Persistence of Semipermeable Membrane Behavior for a Geosynthetic Clay Liner, 7th International Congress on Environmental Geotechnics. Melbourne, Australia.

Meyer, K.H., Sievers, J.F., 1936. La perméabilité des membranes. I. Théorie de la perméabilité ionique. Helv. Chem. Acta 19, 649.

Mitchell, J.K., 1993. Fundamentals of Soil Behavior, second ed. Wiley, New York.

Moore, J.R., Glaser, S.D., Morrison, H.F., Hoversten, G.M., 2004. The streaming potential of liquid carbon dioxide in Berea sandstone. Geophys. Res. Lett. 31, L17610. http://dx.doi.org/10.1029/2004GL020774.

Mualem, Y., 1976. A new model for predicting hydraulic conductivity of unsaturated porous media. Water Resour. Res. 12, 513–522.

Onsager, L., 1931a. Reciprocal relations in irreversible processes I. Phys. Rev. 37 (4), 405–426.

Onsager, L., 1931b. Reciprocal relations in irreversible processes II. Phys. Rev. 38 (12), 2265–2279.

Pride, S., 1994. Governing equations for the coupled electromagnetics and acoustics of porous media. Phys. Rev. B. 50, 15,678–15,696.

Prigogine, I., 1947. Etude Thermodynamique des Phénomènes Irréversibles. Desoer, Liège, Belgium.

Purcell, W.R., 1949. Capillary pressures—their measurement using mercury and the calculation of permeability. Trans. AIME 186, 39.

Revil, A., Cerepi, A., 2004. Streaming potential in two-phase flow condition. Geophys. Res. Lett. 31 (11), L11605. http://dx.doi.org/10.1029/2004GL020140.

Revil, A., Linde, N., 2006. Chemico-electromechanical coupling in microporous media. J. Colloid Interface Sci. 302, 682–694.

Revil, A., 2007. Thermodynamics of transport of ions and water in charged and deformable porous media. J. Colloid Interface Sci. 307 (1), 254–264.

Revil, A., Linde, N., Cerepi, A., Jougnot, D., Matthäi, S., Finsterle, S., 2007. Electrokinetic coupling in unsaturated porous media. J. Colloid Interface Sci. 313 (1), 315–327. http://dx.doi.org/10.1016/j.jcis.2007.03.037.

Revil, A., Skold, M., 2011. Salinity dependence of spectral induced polarization in sands and sandstones. Geophys. J. Int. 187, 813–824. http://dx.doi.org/10.1111/j.1365-246X.2011.05181.x.

Revil, A., 2012. Spectral induced polarization of shaly sands: influence of the electrical double layer. Water Resour. Res. 48, W02517. http://dx.doi.org/10.1029/2011WR011260.

Revil, A., Karaoulis, M., Johnson, T., Kemna, A., 2012. Review: some low-frequency electrical methods for subsurface characterization and monitoring in hydrogeology. Hydrogeol. J. 20 (4), 617–658. http://dx.doi.org/10.1007/s10040-011-0819-x.

Revil, A., Lu, N., 2013. Unified water sorption and desorption isotherms for clayey porous materials. Water Resour. Res. 49 (9), 5685–5699. http://dx.doi.org/10.1002/wrcr.20426.

Revil, A., Eppehimer, J.D., Skold, M., Karaoulis, M., Godinez, L., Prasad, M., 2013. Low-frequency complex conductivity of sandy and clayey materials. J. Colloid Interface Sci. 398, 193–209.

Revil, A., 2013a. Effective conductivity and permittivity of unsaturated porous materials in the frequency range 1 mHz–1GHz. Water Resour. Res. 49, 306–327. http://dx.doi.org/10.1029/2012WR012700.

Revil, A., 2013b. On charge accumulations in heterogeneous porous materials under the influence of an electrical field. Geophysics 78 (4), D271–D291. http://dx.doi.org/10.1190/GEO2012-0503.1.

Revil, A., Lu, N., 2014. Couplings in partially water-saturated porous media. 1. Electrokinetic theory with filtration. submitted to Adv. Water Resour.

Revil, A., Florsch, N., Camerlynck, C., 2014a. Spectral induced polarization porosimetry. Geophys. J. Int. 198, 1016–1033. http://dx.doi.org/10.1093/gji/ggu180.

Revil, A., Barnier, G., Karaoulis, M., Sava, P., 2014b. Seismoelectric coupling in unsaturated porous media: theory, petrophysics, and saturation front localization using an electroacoustic approach. Geophys. J. Int. 196 (2), 867–884. http://dx.doi.org/10.1093/gji/ggt440.

Rittgers, J.B., Revil, A., Karaoulis, M., Mooney, M.A., Slater, L.D., Atekwana, E.A., 2013. Self-potential signals generated by the corrosion of buried metallic objects with application to contaminant plumes. Geophysics 78 (5), EN65–EN82. http://dx.doi.org/10.1190/GEO2013-0033.1.

Sample-Lord, K.M., Shackelford, C.D., 2014. Membrane behavior of unsaturated bentonite barriers. In: Abu-Farsakh, M., Yu, X., Hoyos, L.R. (Eds.), Geo-Congress 2014: Geo-characterization and Modeling for Sustainability, Geotechnical Special Publication, 234. ASCE, Reston, VA, pp. 1900–1909.

Sherwood, J., Craster, B., 2000. Transport of water and ions through a clay membrane. J. Colloid Interface Sci. 230 (2), 349–358.

Soueid Ahmed, A., Jardani, A., Revil, A., Dupont, J.P., 2014. Hydraulic conductivity field characterization from the joint inversion of hydraulic heads and self-potential data. Water Resour. Res. 50. http://dx.doi.org/10.1002/2013WR014645.

Tamagnini, C., Jommi, C., Cattaneo, F., 2010. A model for coupled electro-hydro-mechanical processes in fine grained soils accounting for gas generation and transport. Annals of the Brazilian Academy of Sciences Anais da Academia Brasileira de Ciências 82 (1), 169–193.

Teorell, T., 1935. An attempt to formulate a quantitative theory of membrane permeability. Proc. Soc. Exp. Biol. Med. 33, 282.

Vaudelet, P., Revil, A., Schmutz, M., Franceschi, M., Bégassat, P., 2011a. Induced polarization signature of the presence of copper in saturated sands. Water Resources Research 47, W02526. http://dx.doi.org/10.1029/2010WR009310.

Vaudelet, P., Revil, A., Schmutz, M., Franceschi, M., Bégassat, P., 2011b. Changes in induced polarization associated with the sorption of sodium, lead, and zinc on silica sands. J. Colloid Interface Sci. 360, 739–752.

Vinegar, H.J., Waxman, M.H., 1984. Induced polarization of shaly sands. Geophysics 49, 1267–1287.

Vinogradov, J., Jackson, M.D., 2011. Multiphase streaming potential in sandstones saturated with gas/brine and oil/brine during drainage and imbibition. Geophys. Res. Lett. 38, L01301. http://dx.doi.org/10.1029/2010GL045726.

de Wet, M., 1995. Electro-kinetics, infiltration and unsaturated flow. In: Alonso, E., Delage, P. (Eds.), Unsaturated Soils, 1. A.A. Balkema, Rotterdam, pp. 283–291.

Woodruff, W.F., Revil, A., 2011. CEC-normalized clay-water sorption isotherm. Water Resources Res. 47, W11502. http://dx.doi.org/10.1029/2011WR010919.

Zhang, J., Wang, P., Huang, F., Tang, J.X., 2014. Preparation and characterization of polyacrylonitrile/bentonite blend membrane for wastewater treatment. Fresenius Environ. Bull. 23 (2), 455–462.

Zhu, Z., Toksöz, N., 2012. Experimental measurements of streaming potential and seismoelectric conversion in Berea sandstone. Geophys. Prospect. 78 (3), D143–D150.

Chapter 11

Upscaling Strategies for Modeling Clay-Rock Properties

Virginie Marry[a,b] and Benjamin Rotenberg[a,b]
[a] *Sorbonne Universités, UPMC Univ. Paris 06, UMR 8234 PHENIX, Paris, France;*
[b] *CNRS, UMR 8234 PHENIX, Paris, France*

Chapter Outline

11.1 INTRODUCTION

Clay-rocks are intrinsically multiscale materials, with heterogeneities over length scales covering several orders of magnitude. Macroscopic laws used to predict transport or mechanical properties of clays for practical applications, e.g., in the contexts of the geological disposal or nuclear waste and of carbon dioxide sequestration, reflect an average behavior of the material. However, some macroscopic phenomena such as swelling and transport properties, which are of prime importance in the role played by clay barriers, originate from processes that occur on the molecular scale and can only be captured by starting from a molecular point of view. It is then necessary to link the

Developments in Clay Science, Vol. 6C. http://dx.doi.org/10.1016/B978-0-08-100027-4.00011-5

molecular properties to the macroscopic average behavior observed on the sample scale, taking into account the complex heterogeneous structure of the material on intermediate scales.

Experimentally, it is not possible to probe all the structural, mechanical, and dynamical properties of the solid clay matrix and of the confined fluids and solutes on microscopic and mesoscopic scales. In addition, although experiments can be performed at various scales, the use of real samples (millimeter to centimeter sized), combined with the limitations inherent in the experimental techniques, often results in difficulties in the choice of the models used for data interpretation. When such information is missing, one may then resort to simulations to complement experimental investigation. Numerical simulations are usually performed on ideal and simplified systems, thereby allowing to dissect and decouple phenomena and to discriminate between the various hypotheses made in the analysis of experimental data.

Section 11.2 gives an overview of the simulations tools available for the description of clays on various scales, from the molecular scale to the pore scale. Section 11.3 then discusses how the complex structure of the material can be taken into account in upscaling particle- and pore-scale properties to the macroscopic behavior of the sample.

11.2 FROM THE ATOMIC SCALE TO THE MESOSCALE

Modeling clays at the atomic scale can seem remote from the prediction of macroscopic phenomena. Nevertheless, a molecular description is useful both to understand the underlying microscopic processes and to validate and improve the hypotheses made in models used at the sample scale in order to interpret or predict macroscopic phenomena. This section explains how insights gained at the atomic scale can be transferred into models at the mesoscale, i.e., a scale where the atomic structure of the molecules and crystalline structure can be omitted, without loosing the important molecular-level features: starting from the interlayer pores, the small dimensions of which renders essential description at the atomic scale, toward a multiporous description where the clay mineral particles are in contact with larger pores. As an illustration, the question of how the exchanges and the interfacial processes deduced from molecular modeling can be taken into account at the mesoscale will be addressed.

11.2.1 Interlayer Porosity

11.2.1.1 Molecular Details Matter

The modeling of interlayer porosity is of prime importance to understand the behavior of fluids such as water in clays, and their impact on mechanical properties. Clays can swell or not, depending on their crystalline structure, the type of compensating cations, the localization of the clay charge, and the characteristics of the surrounding solution. Moreover, swelling is modified at high compacting pressures, which can also impact the mechanical stability of the material. Given the narrow aperture of clay interlayer nanopores at high compactions, the behavior of compacted clays needs to be described at the atomic level to get a clear picture of the swelling properties (Delville, 2013). For the same reason, the behavior of the interlayer fluid is mainly controlled by interfacial phenomena at the nanometer scale, especially the solvent structure at the surface.

Classical molecular simulations such as Monte Carlo (MC) and molecular dynamics (MD) are appropriate modeling methods to explore the interlayer scale (Skipper et al., 1995; Shroll and Smith, 1999; Park and Sposito, 2000; Hensen et al., 2001; Marry et al., 2002; Tambach et al., 2004; Smith et al., 2006). They rely on the calculation of interatomic interactions via classical force-fields stemming from ab initio simulations or experimental data (Skipper et al., 1989; Delville, 1992; Teppen et al., 1997; Sainz-Diaz et al., 2001; van Duin et al., 2003; Cygan et al., 2004; Heinz et al., 2005). As an example, MD simulations yield direct predictions of the trajectories of the atoms by solving Newton's equations of motion (Frenkel and Smit, 2002). While atomic positions and velocities are not directly comparable to experiments, statistical mechanics provides the link with macroscopic observables as averages of microscopic quantities.

Molecular simulations have been extensively used to dissect clay swelling phenomena, especially at the first stages of hydration (crystalline swelling). For references, see the review (Anderson et al., 2010) or the updated general reference (Greathouse and Cygan, 2013). In particular, molecular simulations have been used to foresee the stable states of various clays and fluids, at pressures and temperatures corresponding to burial conditions relevant for landfill sites (Odriozola et al., 2004; de Lourdes Chavez et al., 2005; de Pablo et al., 2005; Odriozola and Aguilar, 2005; Botan et al., 2010; Rao et al., 2013). Molecular simulations do not only provide a global picture of the interlayer behavior in extreme conditions, but also give access to averaged quantities which can be used in meso- and macroscopic modeling. Apart from structural properties such as ionic distributions, that can be used to improve continuum descriptions (see the next section), molecular simulations allow calculating mechanical quantities such as the elastic tensor, which give quantitative information on the response of clays to mechanical constraints. This tensor is highly anisotropic due to the lamellar structure and the presence of water between the clay layers. It can be calculated using MD simulations, from the strain fluctuations of the system at equilibrium (Ebrahimi et al., 2012; Carrier et al., 2014) or from the response of the system to an external stress (Manevitch and Rutledge, 2004; Schmidt et al., 2005; Mazo et al., 2008; Suter and Coveney, 2009). The layer-scale elastic tensor can then be introduced in larger scale models (see Section 11.3.2).

MD simulations also allow computing dynamical properties of the confined fluid. A particularly important example is the diffusion coefficients of mobile species (solvent, ions, or dissolved gases), which can be derived from mean-square displacements or velocity correlation functions (Frenkel and Smit, 2002) and are not always accessible from experiments probing the same scale. Indeed, while quasielastic neutron scattering experiments give access to the water dynamics in interlayer pores (Martins et al., 2014), local diffusion coefficients for ions are not easily obtained experimentally. In swelling clays, confinement reduces the diffusion coefficients of mobile species in interlayer pores, by a factor that depends on the type of clay and the cation and, to some extent, on the force-field used in the simulations (Malikova et al., 2006; Marry et al., 2011; Michot et al., 2012). Typically, a decrease by a factor of 5–10 and 2–3 is observed in the monolayer and bilayer states, respectively. These diffusion coefficients, which are the key transport properties in pores where diffusion processes dominate, can then be used as input parameters in simulations (Churakov and Gimmi, 2011) or models (Ichikawa et al., 2004; Bourg and Sposito, 2010) on larger scales accounting for the global geometry of the material, as described in Section 11.3.

11.2.1.2 From Molecular to Continuous Descriptions

The computational cost associated with molecular modeling prevents their use for the simulation of large systems (much larger than the molecular size) and long times (much larger than the ns) associated with the scale of clay mineral particles. In order to upscale the transport or mechanical properties determined at the atomic scale (clay layer or interlayer pore), one can use insights from molecular simulations to parameterize coarser models. In fact, taking one step back one can consider the use of ab initio calculations to parameterize classical force-field for molecular simulations (Cygan et al., 2004) as the first step of a multiscale "bottom-up" strategy. The next step is to build simple models of transport (diffusion, advection) or mechanical properties based on molecular simulations. Several attempts in this direction have been carried out.

Computation of the swelling pressure as a function of distance between clay layers to determine the stable states is a time-consuming process and should rigorously be done in the Grand Canonical ensemble, i.e., with a fixed water chemical potential instead of fixed water content. Nevertheless, it is already possible to use simpler models to go beyond the limits of Derjaguin–Landau–Verwey–Overbeek theory, which only captures the electrostatic interactions between the charged layers at the mean-field level, ignores the finite size of ions and, in general, the roughness of the surface and the discrete charge distribution. The primitive model of electrolytes treats interlayer ions as charged hard spheres in a continuous solvent defined by its permittivity (Kjellander and Marcelja, 1986). It captures some ionic correlations leading, e.g., to an attraction between like-charged clay layers in the presence of multivalent compensating cations (Kjellander et al., 1988; Delville and Pellenq, 2000; Paillusson and Trizac, 2011). Such correlations within the primitive model can be described either using simulation (e.g., MC) or integral equations (Greberg et al., 1996, 1997). The oscillatory swelling pressure profiles observed in molecular simulations are due to the discreteness of the solvent molecules, which is not captured by the primitive model. Carrier (2014) recently proposed an extension, the Explicit Solvent Primitive Model, calibrated on molecular simulations, where solvent molecules are treated as Lennard-Jones particles with an electric dipole. This model reproduces quantitatively the swelling pressure curves obtained by all-atom simulations, at a much smaller computational cost.

Due to its importance in the context of clays as barriers, the diffusion of interlayer solutes has also been investigated in the framework of continuous solvent models. The first ingredient is the free energy landscape experienced by the diffusing species, called the potential of mean force (PMF), which can be determined from molecular simulations. The dynamics of the solute can then be modeled using Langevin dynamics which involves the friction that the environment (solvent, surfaces and other ions) exerts on the solute. Using such models, it is possible to derive simple kinetic models similar to the one used on the macroscopic scale, coupling diffusion and adsorption/desorption at the surface (Rotenberg et al., 2007a,b; Dufrêche et al., 2010). Nevertheless, the simple Langevin equation has its own limitations, which could also be clarified using molecular simulations. In particular, the separation of timescales between the solute and its environment is not achieved in clay interlayer pores so that a generalized Langevin description would be more appropriate (Carof et al., 2014).

11.2.2 Clay Mineral Particle in Contact with a Reservoir

Clay mineral particles are in contact with pores larger than interlayer pores (the size of the latter is of the order of 1 nm), as illustrated in Figure 11.1. These larger pores surrounding the clay mineral particles can be viewed as external fluid reservoirs. The interparticle fluid is expected to behave as in bulk liquid water provided that the distance to the particle surfaces is sufficient. While the properties of a bulk fluid can be derived from fluid mechanics equations, the presence of the clay surfaces raises two questions: (1) How does the fluid reservoir exchange mobile species with the clay mineral particle and (2) What is the range of influence of the surfaces on the pore fluid? In order to answer these questions, a bottom-up approach starting from the atomic scale is again necessary.

11.2.2.1 Exchange between the Particle and the Mesopore

The origins of exchanges between the pore and the particle are twofold: on the one hand the chemical reactions of adsorption/desorption on the edge sites of the clay layers and on the other hand the entrance/exit into/from interlayer spaces. Both mechanisms are crucial since they determine the way pollutants

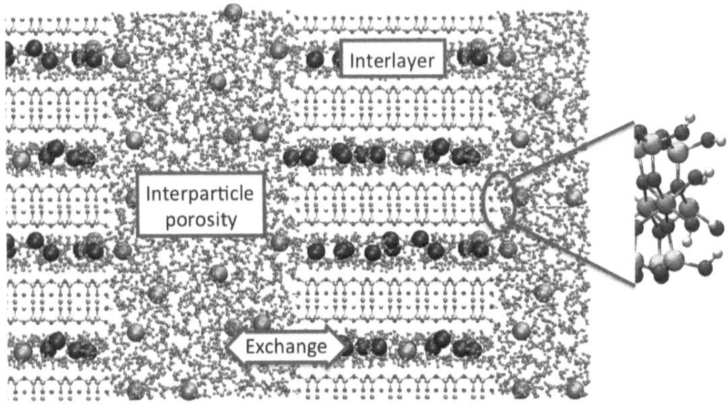

FIGURE 11.1 Interlayer and interparticle pores. Classical molecular simulations are necessary to correctly describe the behavior of water, ions, and solutes in clay interlayer pores as well as their exchange with interparticle pores. Adsorption onto clay edges requires the knowledge of the protonation state of the edge sites as a function of pH. Ab initio simulations provide a way to determine the corresponding relative acidities.

will transit through the material, even if the edge surfaces represent a small external surface compared to the basal ones.

Because chemical reactions, such as acid–base reactions, are involved in the fixation of mobile species on the edges of clay mineral particles, the upscaling should start from an ab initio (also referred to as "first principles") description, in which electrons involved in chemical bonds are taken into account. The acidity constants (pK_a values) assigned to different surface groups are usually obtained from titration experiments, which however cannot assign them precisely to specific sites. Such experiments can be interpreted in the light of semiempirical bond-valence models such as the multisite ion complexation model (Hiemstra et al., 1996; Tournassat et al., 2004; Machesky et al., 2008). However, molecular simulations at the quantum level are to date the only method to justify or invalidate the assumptions underlying such models and to bring new insights into the behavior of clay edges. Therefore the structure and reactivity of clay edges have been investigated using density functional theory (DFT)-based molecular modeling (Bickmore et al., 2003; Churakov, 2006, 2007; Kremleva et al., 2011). By coupling DFT with thermodynamic integration, acidic constants of the edge sites (SiOH, MOH, and MOH_2 with M = Al or Mg) could be calculated. For example, on the (010) surface of pyrophyllite the pK_a of SiOH and $AlOH_2$ groups are approximately 7 and 8, respectively, while that of $AlOH^-$ (~22) is outside of the water range (Tazi et al., 2012; Liu et al., 2013). Surrounding water needs to be taken into account in order to explain the protonation/deprotonation processes, and the acidity of a given site may be influenced by the proximity of other reactive sites. Predicted pK_a values are compatible with titration experiments (Liu et al., 2013). DFT calculations were also used to calculate the adsorption constant of ionic species on edge sites (Churakov and Dähn, 2012; Alexandrov and Rosso, 2013). This approach, although computationally intensive, is crucial since it gives a clear interpretation of the adsorption properties of clay edges and allows testing and improving the validity of the models used on larger scales where adsorption and acidity constants are needed.

When transport properties are considered, knowledge of adsorption thermodynamics on the edges, although important, is not sufficient. The way mobile species exchange between the clay mineral particles (interlayer pores) and the interparticle pores is crucial since it determines how species diffuse globally through the material. Few studies have considered explicitly the exchange properties at the atomic scale (Rotenberg et al., 2007c, 2009; Jardat et al., 2009; Hedström and Karnland, 2012). Although many features of radionuclide diffusion can be explained on the basis of Donnan theory, which predicts the relative concentrations of cations and anions between the particle and the reservoir (Birgersson and Karnland, 2009), refinements are needed to account for the excluded volume of the ions, their specific interactions with clay surfaces, and the correlations between them (Le et al., 2013). Molecular simulations can help to assess these effects but they are heavy and computer time-consuming. A more efficient approach consists in a multiscale strategy, whereby effective potentials between ions and surfaces are first calculated from molecular dynamics simulations. In a second step, Brownian dynamics simulations describing the solvent as a continuum and using the effective potentials are performed. This methodology allowed a systematic quantification of the effects of pore size and ionic concentrations on the Donnan equilibrium between particle and reservoir, taking into account ion size and ionic correlations (Jardat et al., 2009).

11.2.2.2 Influence of the Surfaces on the Fluid Properties

The presence of particle surfaces induces structural changes in the fluid at the interface. Moreover the clay charge and its location play important roles in the distribution of ions in the vicinity of the surfaces (de Carvalho and Skipper, 2001; Wang et al., 2005; Greathouse and Cygan, 2006; Wang et al., 2006; Marry et al., 2008). On the macroscopic scale, most of the models used to interpret tracer experiments rely on continuous transport equations such as Fick's and Darcy's laws in which several parameters are defined to account for the geometry of the material (e.g., porosity, tortuosity, see previous section) but also to explain the different behaviors of the mobile species depending on their charge and type and on experimental conditions (Shackelford and Moore, 2013). Thus, the description of the diffuse layer in the vicinity of the surfaces of the mesopores takes an important part in the interpretation of experimental data.

Molecular simulations can help to validate and improve the models of the electrical diffuse layer (Marry et al., 2003; Tournassat et al., 2009; Bourg and

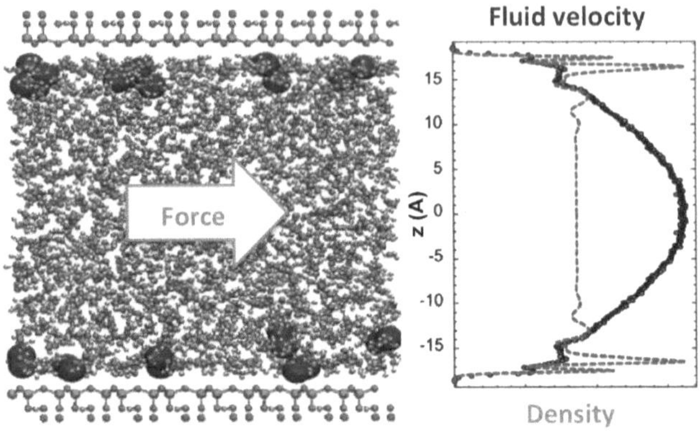

FIGURE 11.2 Hydrodynamics in interparticle pores. Molecular simulation provides a detailed description of the structure of the fluid in interparticle pores, including, e.g., its layering at the surface (oscillations in the density profiles). Nonequilibrium molecular dynamics techniques allow simulating hydrodynamic flows and investigating the validity of continuous hydrodynamics under extreme confinement.

Sposito, 2011) that are implemented in macroscopic modeling (Parkhurst and Appelo, 1999; Appelo and Wersin, 2007; Leroy et al., 2007). Due to the computational cost associated with all-atom simulations, mesoscopic models based on molecular simulations have been proposed to describe the interface more efficiently, while keeping track of the relevant microscopic information. In addition to the above-mentioned case of continuous solvent model simulations by Brownian dynamics (Jardat et al., 2009), a promising strategy can be found in classical molecular density functional theories, which manage to accurately render the molecular features of the interface (solvent structure) with a computational cost two to three orders of magnitude lower than all-atom simulations (Levesque et al., 2012; Jeanmairet et al., 2014).

Molecular details also play an important role in hydrodynamic and electrokinetic properties under confinement, as illustrated in Figure 11.2. All-atom simulations demonstrate that the hydrophobic/hydrophilic properties of clay surfaces depend in particular on the structure and charge density of the layers (Rotenberg et al., 2011). These properties are responsible for the degree of water adhesion on the surfaces. Importantly, the corresponding boundary conditions (slip vs stick) are only rarely taken into account when solving macroscopic transport equations (Marry et al., 2003; Botan et al., 2011, 2013).

11.3 FROM THE MESOSCOPIC TO THE MACROSCOPIC SCALE

Once a suitable model of the material and transport properties is available on the local scale, possibly derived from molecular simulations as discussed in the previous section, one should make the link between the local pore scale and the macroscopic scale that is relevant to the performance of engineered barriers. We discuss here the two main issues that need to be addressed for that purpose, namely how to represent the complex heterogeneous structure of the material and how to upscale the local constitutive equations to the macroscopic scale.

11.3.1 Representative Elementary Volume

The quantitative prediction of transport coefficients requires a realistic description of the poral space of the material. Figure 11.3 illustrates some model representation of clays discussed in the present section. Recent progress in imaging techniques has yielded detailed characterizations of the structure of clay materials on various scales from the angstrom (10^{-10} m) to the centimeter (10^{-2} m). The multiscale heterogeneous structure of clays has been investigated using a range of experimental tools, including transmission electron microscope, focused ion beam-scanning electron microscope, and X-ray computed tomography that provide complementary 2D or 3D images of the solid fraction. In turn, the analysis of these images provides either a direct picture of the solid structure and of the pore network. Some of these techniques further allow analyzing the local chemical composition and hence provide 2D or 3D mineralogical maps of the material (Robinet et al., 2012).

While it is possible to perform simulations of transport properties on numerical samples directly obtained from the experimental 3D images, this approach remains rather exceptional in the case of clayey media. Several difficulties exist in addition to the computational limitations associated with the simulation of complex systems. Importantly, images obtained at a coarse scale miss the fine pore network which plays an essential role in the transport properties of clays. Conversely, images with a very high resolution can be obtained only for a small fraction of the material, raising the question of the representativity of the selected zone, because fluctuations of local properties may occur on larger scales. In order to obtain realistic numerical samples for further simulations, it is therefore essential to quantify the statistical properties of the porous solid and the associated pore network structure. Even the definition of the pore size distribution, an apparently simple quantity, is not without its ambiguity. The geometrical analysis of the pore network may rather require chord distribution functions or other topological and morphological quantities such as two-point correlation functions or Fourier analysis (Levitz, 2007; Brisard et al., 2012).

Once the geometrical features of the real medium are quantified, it becomes possible to generate model numerical samples to be used for the simulation of various properties. From the most simple geometries such as parallel slit pores or regular sphere packings, various strategies have been proposed to account for the complex, heterogeneous, and anisotropic structure of clays for the simulation of mechanical or transport properties. For example,

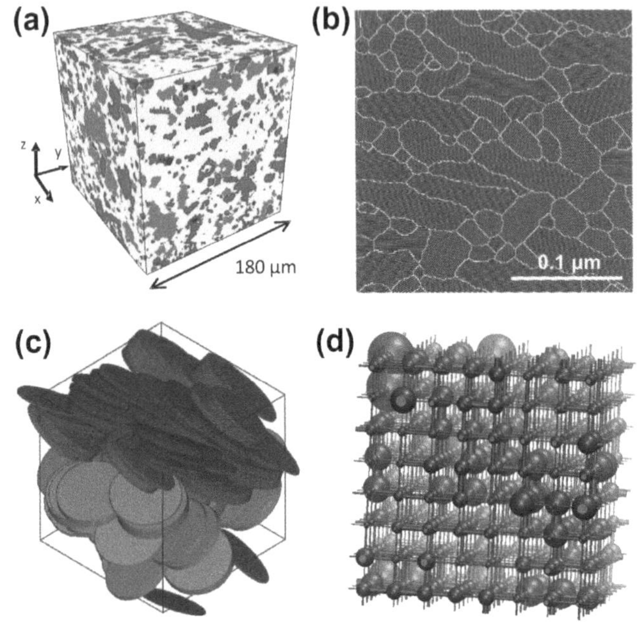

FIGURE 11.3 Models of representative elementary volume. (a) 3D mineralogical map of Callovo-Oxfordian clay obtained from X-ray tomography experiments. *(Reprinted from Robinet et al. (2012), Copyright 2012, with permission of John Wiley & Sons, Inc.)* (b) 2D grain structure used to model diffusion in montmorillonite. *(Reprinted from Churakov et al. (2014), with permission from Elsevier.)* (c) 3D packing of ellipsoids obtained using a potential of mean force from Molecular simulations. *(Reprinted from Ebrahimi et al. (2014), Copyright 2014, AIP Publishing LLC.)* (d) Pore network model *(Copyright Amaël Obliger.)*.

Coelho et al. (1997) used a deposition process mimicking sedimentation to generate packings of grains of arbitrary shapes. Thovert et al. (2001) proposed a grain reconstruction method derived from the two-point correlation function. Tyagi et al. (2013) introduced a lattice Monte Carlo method allowing the generation of 2D and 3D ensemble of grains with prescribed interlayer and interparticle porosities, anisotropic shape, and distribution of orientations. A promising approach for the generation of realistic structures has been introduced recently by Ebrahimi et al. (2014): In a first step, the PMF between clay layers is computed from molecular simulations; then, this PMF can be used to predict the aggregation state for various external pressures.

11.3.2 Upscaling from the Particle and Pore Scales to the Sample Scale

For practical applications, the mechanical and transport properties are characterized by a limited set of parameters, such as the effective diffusion coefficient or the permeability defined by macroscopic equations related to an effective homogeneous medium such as Fick's law, $J_i = -D_i^{eff} \nabla C_i$, for the diffusion of solute i under a concentration gradient or Darcy's law, $J_s = -K\nabla P$, for the flux of solvent s under a pressure gradient. From a fundamental point of view, one would like to understand how these macroscopic behaviors emerge from the underlying microscopic mechanisms.

A first approach consists in treating the complex material as an effective homogeneous medium and introducing empirical parameters relating the microscopic properties, such as the molecular diffusion coefficient D^m, to macroscopic ones, such as the effective diffusion coefficient D^{eff}. For example, one often introduces a tortuosity coefficient to account for the longer path that tracers follow through the pore network compared to the path they would follow in an identical unconfined fluid (Bourg et al., 2003; Gimmi and Kosakowski, 2011). This quantity can be defined from geometrical considerations, but is in fact more often determined empirically from D^m, D^{eff}, and the porosity ϕ. Other parameters, such as the constrictivity or the accessible porosity have been introduced to account for the fact that various species may explore the pores differently, in particular if they bear different charges. Nevertheless, the definition of such quantities from microscopic considerations is not always straightforward Leroy et al. (2006), Van Loon et al. (2007), Glaus et al. (2007).

It is possible to rigorously upscale the local constitutive equations to the macroscopic ones using the mathematical theory of homogenization. This theory not only provides a route to the macroscopic coefficient, but also to the exact form of the transport equation on the macroscopic scale, which may depend on the flow regime. For example, Darcy's law can be *derived* from Stokes law on the pore scale and an expression for the permeability is given as an average over the pore volume of fields that satisfy partial differential equations, which can then be solved numerically for a given geometry. In the case of (model) clays, homogenization theory has been used to predict mechanical properties (Brisard and Dormieux, 2010; Cariou et al., 2012, 2013), as well as tracer diffusion including surface diffusion or hydrodynamics in various flow regimes (Auriault and Adler, 1995; Mikelic and Van Duijn, 2011; Allaire and Hutridurga, 2012). Moyne and Murad (2006a,b) proposed the complete analysis of the electrokinetic and mechanical couplings. Recently, Allaire et al. (2014) introduced nonideality in the local transport equations, to extend the domain of validity in terms of electrolyte concentration, and obtained the corresponding homogenized macroscopic equations for a rigid porous medium. In practice, apart from a few exceptions (Allaire et al., 2013) the numerical solution of the homogenization problem has been only applied to simple geometries such as slit pores describing the interlayer porosity.

In order to deal with more realistic geometries, a straightforward albeit challenging and computationally intensive strategy consists in the direct numerical simulation of the process of interest in a complex representative elementary volume. Of course, the simulation technique depends on the considered phenomena. The effective diffusion coefficient can be computed from random walks of explicit noninteracting particles, propagated using a Brownian dynamics algorithm, by measuring their mean-square displacement as a function of time. Churakov and Gimmi (2011) used this approach to investigate the effect of the heterogeneity (interlayer/interparticle) and anisotropy of the local diffusion tensor on the effective diffusion tensor through the clay matrix. On a larger scale, Robinet et al. (2012) used the time-domain diffusion method to compute the effective coefficient of tracers in a chemically heterogeneous Callovo-Oxfordian clay sample. Archie's law for diffusion on the sample scale ($D^{eff} \propto \phi^m$, where ϕ is the porosity and m is called the cementation exponent) could be observed in such simulations. Random walk (Valfouskaya et al., 2005) or moment propagation simulations (Rotenberg et al., 2008) also yield information on the time-dependent diffusion coefficients, which encode information on how the tracers explore the pore space and can be determined experimentally using nuclear magnetic resonance.

For hydrodynamic flows, the permeability can be obtained by solving the Stokes equation, usually with no-slip boundary conditions at the solid surface. Similarly, the full electrokinetic coupling matrix can be computed by solving the Stokes equation (for the solvent) coupled to the Poisson–Nernst–Planck equations (for charged solutes). The equations are solved in the presence of a pressure, electric potential or concentration gradient and the induced solvent, charge and mass fluxes are computed, thereby providing the coefficients of the coupling matrix. This can be achieved numerically either directly using finite element or finite volume methods (Coelho et al., 1996) as well as mesoscopic lattice simulations such as the Lattice Boltzmann method (Rotenberg et al., 2010; Pazdniakou and Adler, 2013). An important outcome of these simulations was the observation that the electrokinetic transport properties can be rationalized, including for large surface potentials, by using an appropriate renormalization length (Gupta et al., 2006) initially introduced by Johnson et al. (1986).

Due to the computational cost of direct numerical simulation, an alternative approach to the numerical upscaling of local transport equations has been introduced to capture the effect of geometrical heterogeneity on transport properties. The pore network model (PNM) allows dealing with complex porous networks at the price of a simplified treatment of local transport properties. Initially developed to predict multiphase flow in porous media (Fatt, 1956), it describes the porosity as a network of pores connected by channels. In a nutshell, the PNM approach amounts to solving a set of conservation equations on the nodes of the network (in analogy with Kirchhoff's law for a network of resistors), on the basis of local fluxes through the channels connecting the nodes, under the effect of external macroscopic gradients. It has recently been extended to capture the effect of the surface charge of the porous solid, including all electrokinetic couplings (Obliger et al., 2014), thereby opening its use for clay materials. It was found in particular that the combination of electrokinetic effects in the smaller pores with the heterogenity of the sample may result in behaviors that cannot be inferred from the properties of a channel with the average diameter.

11.4 CONCLUSION

The combination of molecular, coarse-grained, and continuous modeling and simulations provides an efficient way to investigate coupled transport and mechanical phenomena in clays on various scales. In addition, several strategies can be adopted to generate representative numerical samples accounting for the complex structure of clay materials. The continuous equations describing the couplings on the particle and pore scales can then be upscaled, for a given representation of the complex pore structure, by complementary theoretical and numerical tools, thereby providing the link between the macroscopic behavior and its microscopic origin. This very efficient multiscale modeling approach is of course not limited to clays and has also been applied to other materials involved in the use of clays as barriers, such as cement, glasses, or metal alloys.

Due to the simplifying assumptions introduced to describe the system from the point of view of both structure and underlying interactions or processes, the results obtained from numerical simulations may not always be quantitatively comparable with experimental ones. Nevertheless, the constant progress in computing resources and efficiency allows the development of ever more accurate descriptions over larger length and timescales, thereby providing more realistic models of complex systems such as clay-rocks. As a result, multiscale simulation has now become a standard tool for the understanding of condensed matter in general.

REFERENCES

Alexandrov, V., Rosso, K., 2013. Insights into the mechanism of Fe(II) adsorption and oxidation at Fe-clay mineral surfaces from first-principles calculations. J. Phys. Chem. C 117, 22880–22886.

Allaire, G., Hutridurga, H., 2012. Homogenization of reactive flows in porous media and competition between bulk and surface diffusion. IMA J. Appl. Math. 77, 788–815.

Allaire, G., Brizzi, R., Dufrèche, J.F., Mikelic, A., Piatnitski, A., 2013. Ion transport in porous media: derivation of the macroscopic equations using upscaling and properties of the effective coefficients. Comput. Geosci. 17, 479–495.

Allaire, G., Brizzi, R., Dufrêche, J.F., Mikelic, A., Piatnitski, A., 2014. Role of non-ideality for the ion transport in porous media: derivation of the macroscopic equations using upscaling. Physica D 282, 39–60.

Anderson, R., Ratcliffe, I., Greenwell, H., Williams, P., Cliffe, S., Coveney, P., 2010. Clay swelling—a challenge in the oilfield. Earth-Sci. Rev. 98, 201–216.

Appelo, C.A.J., Wersin, P., 2007. Multicomponent diffusion modeling in clay systems with application to the diffusion of tritium, iodide, and sodium in Opalinus clay. Environ. Sci. Technol. 41, 5002–5007.

Auriault, J.L., Adler, P.M., 1995. Taylor dispersion in porous media: analysis by multiple scale expansions. Adv. Water Resour. 18, 217–226.

Bickmore, B.R., Rosso, K.M., Nagy, K.L., Cygan, R.T., Tadanier, C.J., 2003. Ab initio determination of edge surface structures for dioctahedral 2:1 phyllosilicates: implications for acid-base reactivity. Clays Clay Min. 51, 359–371.

Birgersson, M., Karnland, O., 2009. Ion equilibrium between montmorillonite interlayer space and an external solution—consequences for diffusional transport. Geochim. Cosmochim. Acta 73, 1908–1923.

Botan, A., Rotenberg, B., Marry, V., Turq, P., Noetinger, B., 2010. Carbon dioxide in montmorillonite clay hydrates: thermodynamics, structure, and transport from molecular simulation. J. Phys. Chem. C 114, 14962–14969.

Botan, A., Rotenberg, B., Marry, V., Turq, P., Noetinger, B., 2011. Hydrodynamics in clay nanopores. J. Phys. Chem. C 115, 16109–16115.

Botan, A., Marry, V., Rotenberg, B., Turq, P., Noetinger, B., 2013. How electrostatics influences hydrodynamic boundary conditions: Poiseuille and electro-osmostic flows in clay nanopores. J. Phys. Chem. C 117, 978–985.

Bourg, I.C., Bourg, A.C.M., Sposito, G., 2003. Modeling diffusion and adsorption in compacted bentonite: a critical review. J. Contam. Hydrol. 61, 293–302.

Bourg, I., Sposito, G., 2010. Connecting the molecular scale to the continuum scale for diffusion process in smectite-rich porous media. Environ. Sci. Technol. 44, 2085–2091.

Bourg, I., Sposito, G., 2011. Molecular dynamics simulations of the electrical double layer on smectite surfaces contacting concentrated mixed electrolyte ($NaCl$-$CaCl_2$) solutions. J. Colloid Interface Sci. 360, 701–715.

Brisard, S., Dormieux, L., 2010. FFT-based methods for the mechanics of composites: a general variational framework. Comput. Mater. Sci. 49, 663–671.

Brisard, S., Chae, R.S., Bihannic, I., Michot, L., Guttmann, P., Thieme, J., Schneider, G., Monteiro, P.J.M., Levitz, P., 2012. Morphological quantification of hierarchical geomaterials by X-ray nano-CT bridges the gap from nano to micro length scales. Am. Mineral. 97, 480–483.

Cariou, S., Duan, Z., Davy, C., Skoczylas, F., Dormieux, L., 2012. Poromechanics of partially saturated COx argillite. Appl. Clay Sci. 56, 36–47.

Cariou, S., Dormieux, L., Skoczylas, F., 2013. An original constitutive law for Callovo-Oxfordian argillite, a two-scale double-porosity material. Appl. Clay Sci. 80–81, 18–30.

Carof, A., Marry, V., Salanne, M., Hansen, J.P., Turq, P., Rotenberg, B., 2014. Coarse graining the dynamics of nano-confined solutes: the case of ions in clays. Mol. Simulat. 40, 237–244.

Carrier, B., 2014. Influence of Water on the Short-Term and Long-Term Mechanical Properties of Swelling Clays: Experiments on Self-supporting Films and Molecular Simulations (PhD. thesis), Université Paris-Est.

Carrier, B., Vandamme, M., Pellenq, R.J.M., Van Damme, H., 2014. Elastic properties of swelling clay particles at finite temperature upon hydration. J. Phys. Chem. C 118 (17), 8933–8943. http://dx.doi.org/10.1021/jp412160e.

de Carvalho, R., Skipper, N., 2001. Atomistic computer simulation of the clay-fluid interface in colloidal laponite. J. Chem. Phys. 114, 3727–3733.

de Lourdes Chavez, M., de Pablo, L., de Pablo, J., 2005. Monte Carlo molecular simulation of the hydration of K-montmorillonite at 353 K and 625 bar. Langmuir 20, 10764–10770.

Churakov, S.V., 2006. Ab initio study of sorption on pyrophyllite: structure and acidity of the edge sites. J. Phys. Chem. B 110, 4135–4146.

Churakov, S.V., 2007. Structure and dynamics of the water films confined between edges of pyrophyllite: a first principle study. Geochim. Cosmochim. Acta 71, 1130–1144.

Churakov, S.V., Gimmi, T., 2011. Up-scaling of molecular diffusion coefficients in clays: a two-step approach. J. Phys. Chem. C 115, 6703–6714.

Churakov, S., Dähn, R., 2012. Zinc adsorption on clays inferred from atomistic simulations and EXAFS spectroscopy. Environ. Sci. Technol. 46, 5713–5719.

Churakov, S.V., Gimmi, T., Unruh, T., Van Loon, L.R., Juranyi, F., 2014. Resolving diffusion in clay minerals at different time scales: combination of experimental and modeling approaches. Appl. Clay Sci. 96, 36–44.

Coelho, D., Shapiro, M., Thovert, J.F., Adler, P.M., 1996. Electroosmotic phenomena in porous media. J. Colloid Interface Sci. 181, 169–190.

Coelho, D., Thovert, J.F., Adler, P.M., 1997. Geometrical and transport properties of random packings of spheres and aspherical particles. Phys. Rev. E 55, 1959–1978.

Cygan, R.T., Liang, J.J., Kalinichev, A.G., 2004. Molecular models of hydroxide, oxyhydroxide, and clay phases and the development of a general force field. J. Phys. Chem. B 108, 1255–1266.

Delville, A., 1992. Structure of liquids at a solid interface: an application to the swelling of clay by water. Langmuir 8, 1796–1805.

Delville, A., Pellenq, R.J.M., 2000. Electrostatic attraction and/or repulsion between charged colloids: a (NVT) Monte-Carlo study. Mol. Simulat. 24, 1–24.

Delville, A., 2013. Beyond the diffuse layer theory: a molecular analysis of the structural, dynamical, and mechanical properties of charged solid/liquid interfaces. J. Phys. Chem. C 117, 14558–14569.

Dufrêche, J.F., Rotenberg, B., Marry, V., Turq, P., 2010. Bridging molecular and continuous descriptions: the case of dynamics in clays. An. Acad. Bras. Cienc. 82, 61–68.

van Duin, A., Strachan, A., Stewman, S., Zhang, Q., Xu, X., Goddard, W., 2003. Reaxff(sio) reactive force field for silicon and silicon oxide systems. J. Phys. Chem. A 107, 3803–3811.

Ebrahimi, D., Pellenq, R.M., Whittle, A., 2012. Nanoscale elastic properties of montmorillonite upon water adsorption. Langmuir 28, 16855–16863.

Ebrahimi, D., Whittle, A., Pellenq, R.M., 2014. Mesoscale properties of clay aggregates from potential of mean force representation of interactions between nanoplatelets. J. Chem. Phys. 140, 154309.

Fatt, I., 1956. The network model of porous media, part I: capillary characteristics. Petrol Trans. AIME 144–159.

Frenkel, D., Smit, B., 2002. Understanding Molecular Simulations, from Algorithms to Applications. Academic Press.

Gimmi, T., Kosakowski, G., 2011. How mobile are sorbed cations in clays and clay rocks? Environ. Sci. Technol. 45, 1443–1449.

Glaus, M.A., Baeyens, B., Bradbury, M.H., Jakob, A., Van Loon, L.R., Yaroshchuk, A., 2007. Diffusion of ^{22}Na and ^{85}Sr in montmorillonite: evidence of interlayer diffusion being the dominant pathway at high compaction. Environ. Sci. Technol. 41, 478–485.

Greathouse, J.A., Cygan, R.T., 2006. Water structure and aqueous uranyl(VI) adsorption equilibria onto external surfaces of beidellite, montmorillonite, and pyrophyllite: results from molecular simulations. Environ. Sci. Technol. 40, 3865–3871.

Greathouse, J.A., Cygan, R.T., 2013. Chapter 3—Molecular simulation of clay minerals. In: Bergaya, F., Lagaly, G. (Eds.), Developments in Clay Science, Volume 5 of Handbook of Clay Science. Elsevier, pp. 405–423.

Greberg, H., Kjellander, R., Akesson, T., 1996. Ion-ion correlations in electric double layers from Monte Carlo simulations and integral equation calculations. Mol. Phys. 87, 407–422.

Greberg, H., Akesson, T., Kjellander, R., 1997. Ion-ion correlations in electric double layers from Monte Carlo simulations and integral equation calculations part 2. Case of added salt. Mol. Phys. 92, 35–48.

Gupta, A.K., Coelho, D., Adler, P.M., 2006. Electroosmosis in porous solids for high zeta potentials. J. Colloid Interface Sci. 303, 593–603.

Hedström, M., Karnland, O., 2012. Donnan equilibrium in Na-montmorillonite from a molecular dynamics perspective. Geochim. Cosmochim. Acta 77, 266–274.

Heinz, H., Koerner, H., Anderson, K.L., Vaia, R.A., Farmer, B., 2005. Force field for mica-type silicates and dynamics of octadecylammonium chains grafted to montmorillonite. Chem. Mater. 17, 5658–5669.

Hensen, E., Tambach, T., Bliek, A., Smit, B., 2001. Adsorption isotherms of water in Li-, Na- and K-montmorillonite by molecular simulation. J. Phys. Chem. 115, 3322–3329.

Hiemstra, T., Venema, P., Riemsdijk, W.H.V., 1996. Intrinsic proton affinity of reactive surface groups of metal (hydr)oxides: the bond valence principle. J. Colloid Interface Sci. 184, 680–692.

Ichikawa, Y., Kawamura, K., Fujii, N., Kitayama, K., 2004. Microstructure and micro/macro-diffusion behavior of tritium in bentonite. Appl. Clay Sci. 26, 75–90.

Jardat, M., Dufrêche, J.F., Marry, V., Rotenberg, B., Turq, P., 2009. Salt exclusion in charged porous media: a coarse-graining strategy in the case of montmorillonite clays. Phys. Chem. Chem. Phys. 11, 2023–2033.

Jeanmairet, G., Marry, V., Levesque, M., Rotenberg, B., Borgis, D., 2014. Hydration of clays at the molecular scale: the promising perspective of classical density functional theory. Mol. Phys. 112, 1320–1329.

Johnson, D.L., Koplik, J., Schwartz, L.M., 1986. New pore-size parameter characterizing transport in porous media. Phys. Rev. Lett. 57, 2564–2567.

Kjellander, R., Marcelja, S., 1986. Double-layer interaction in the primitive model and the corresponding Poisson-Boltzmann description. J. Phys. Chem. 90, 1230–1232.

Kjellander, R., Marcelja, S., Quirk, J., 1988. Attractive double-layer interactions between calcium clay particles. J. Colloid Interface Sci. 126, 194–211.

Kremleva, A., Kruger, S., Rosch, N., 2011. Uranyl adsorption at (010) edge surfaces of kaolinite: a density functional study. Geochim. Cosmochim. Acta 75, 706–718.

Le, T., Moyne, C., Murad, M., Lima, S., 2013. A two-scale non-local model of swelling porous media incorporating ion size correlation effects. J. Mech. Phys. Solids 61, 2493–2521.

Leroy, P., Revil, A., Coelho, D., 2006. Diffusion of ionic species in bentonite. J. Colloid Interface Sci. 296, 248–255.

Leroy, P., Revil, A., Altmann, S., Tournassat, C., 2007. Modeling the composition of the pore water in a clay-rock geological formation (Callovo-Oxfordian, France). Geochim. Cosmochim. Acta 71, 1087–1097.

Levesque, M., Marry, V., Rotenberg, B., Jeanmairet, G., Vuilleumier, R., Borgis, D., 2012. Solvation of complex surfaces via molecular density functional theory. J. Chem. Phys. 137, 224107.

Levitz, P., 2007. Toolbox for 3D imaging and modeling of porous media: relationship with transport properties. Cement Concrete Res. 37, 351–359.

Liu, X., Lu, X., Sprik, M., Cheng, J., Meijer, E.J., Wang, R., 2013. Acidity of edge surface sites of montmorillonite and kaolinite. Geochim. Cosmochim. Acta 117, 180–190.

Machesky, M.L., Predota, M., Wesolowski, D.J., Vlcek, L., Cummings, P.T., Rosenqvist, J., Ridley, M.K., Kubicki, J.D., Bandura, A.V., Kumar, N., Sofo, J.O., 2008. Surface protonation at the rutile (110) interface: explicit incorporation of solvation structure within the refined MUSIC model framework. Langmuir 24, 12331–12339.

Malikova, N., Cadène, A., Marry, V., Dubois, E., Turq, P., 2006. Diffusion of water in clays on the microscopic scale: modeling and experiment. J. Phys. Chem. B 110, 3206–3214.

Manevitch, L., Rutledge, G., 2004. Molecular dynamics simulation of thermomechanical properties of montmorillonite crystal. 1. Isolated clay nanoplate. J. Phys. Chem. B 108, 1428–1435.

Marry, V., Turq, P., Cartailler, T., Levesque, D., 2002. Microscopic simulation of structure and dynamics of water and counterions in a monohydrated montmorillonite. J. Chem. Phys. 117, 3454.

Marry, V., Dufrêche, J.F., Jardat, M., Turq, P., 2003. Equilibrium and electrokinetic phenomena in charged porous media from microscopic and mesoscopic models: electro-osmosis in montmorillonite. Mol. Phys. 101, 3111–3119.

Marry, V., Rotenberg, B., Turq, P., 2008. Structure and dynamics of water at a clay surface from molecular dynamics simulation. Phys. Chem. Chem. Phys. 10, 4802–4813.

Marry, V., Dubois, E., Malikova, N., Durand-Vidal, S., Longeville, S., Breu, J., 2011. Water dynamics in hectorite clays: influence of temperature studied by coupling neutron spin echo and molecular dynamics. Environ. Sci. Technol. 45, 2850–2855.

Martins, M., Gates, W., Michot, L., Ferrage, E., Marry, V., Bordallo, H., 2014. Neutron scattering, a powerful tool to study clay minerals. Appl. Clay Sci. 96, 22–35.

Mazo, M., Manevitch, L., Gusarova, E., Shamaev, M., Berlin, A., Balabaev, N., Rutledge, G., 2008. Molecular dynamics simulation of thermomechanical properties of montmorillonite crystal. 1. Isolated clay nanoplate. J. Phys. Chem. B 112, 2964–2969.

Michot, L., Ferrage, E., Jimenez-Ruiz, M., Boehm, M., Delville, A., 2012. Anisotropic features of water and ion dynamics in synthetic Na- and Ca-smectites with tetrahedral layer charge. A combined quasi-elastic neutron-scattering and molecular dynamics simulations study. J. Phys. Chem. C 116, 16619–16633.

Mikelic, A., Van Duijn, C.J., 2011. Rigorous derivation of a hyperbolic model for Taylor dispersion. Math. Models Meth. Appl. Sci. 21, 1095–1120.

Moyne, C., Murad, M.A., 2006a. A two-scale model for coupled electro-chemo-mechanical phenomena and Onsager's reciprocity relations in expansive clays: I homogenization analysis. Transport Porous Med. 62, 333–380.

Moyne, C., Murad, M.A., 2006b. A two-scale model for coupled electro-chemo-mechanical phenomena and Onsager's reciprocity relations in expansive clays: II computational validation. Transport Porous Med. 63, 13–56.

Obliger, A., Jardat, M., Coelho, D., Békri, S., Rotenberg, B., 2014. Pore network model of electrokinetic transport through charged porous media. Phys. Rev. E 89, 043013.

Odriozola, G., Aguilar, J., Lopez-Lemus, J., 2004. Na-montmorillonite hydrates under ethane rich reservoirs: Npzzt and μpzzt simulations. J. Chem. Phys. 121, 4266–4275.

Odriozola, G., Aguilar, J., 2005. Stability of Ca-montmorillonite hydrates: a computer simulation study. J. Chem. Phys. 123, 174708.

de Pablo, L., Chavez, M., de Pablo, J., 2005. Stability of Na-, K-, and Ca-montmorillonite at high temperatures and pressures: a Monte Carlo simulation. Langmuir 21, 10874–10884.

Paillusson, F., Trizac, E., 2011. Interaction regimes for oppositely charged plates with multivalent counterions. Phys. Rev. E 84, 011407.

Park, S.H., Sposito, G., 2000. Monte Carlo simulation of total radial distribution functions for interlayer water in Li-, Na- and K-montmorillonite. J. Phys. Chem. B 104, 4642–4648.

Parkhurst, D.L., Appelo, C.A.J., 1999. User's Guide to PHREEQC (Version 2)—A Computer Program for Speciation, Batch-reaction, One-dimensional Transport, and Inverse Geochemical Calculations. U.S. Geological Survey. Water-Resources Investigations Report 99–4259.

Pazdniakou, A., Adler, P.M., 2013. Dynamic permeability of porous media by the lattice Boltzmann method. Adv. Water Resour. 62 (Part B), 292–302.

Rao, Q., Xiang, Y., Leng, Y., 2013. Molecular simulations on the structure and dynamics of water-methane fluids between Na-montmorillonite clay surfaces at elevated temperature and pressure. J. Phys. Chem. C 117, 14061–14069.

Robinet, J.C., Sardini, P., Coelho, D., Parneix, J.C., Prêt, D., Sammartino, S., Boller, E., Altmann, S., 2012. Effects of mineral distribution at mesoscopic scale on solute diffusion in a clay-rich rock: example of the Callovo-Oxfordian mudstone (Bure, France). Water Resour. Res. 48, W05554.

Rotenberg, B., Marry, V., Dufrêche, J.F., Giffaut, E., Turq, P., 2007a. A multiscale approach to ion diffusion in clays: building a two-state diffusion-reaction scheme from microscopic dynamics. J. Colloid Interface Sci. 309, 289–295.

Rotenberg, B., Marry, V., Dufrêche, J.F., Malikova, N., Giffaut, E., Turq, P., 2007b. Modelling water and ion diffusion in clays: a multiscale approach. C. R. Chim. 10, 1108–1116.

Rotenberg, B., Marry, V., Vuilleumier, R., Malikova, N., Simon, C., Turq, P., 2007c. Water and ions in clays: unraveling the interlayer/micropore exchange using molecular dynamics. Geochim. Cosmochim. Acta 71, 5089–5101.

Rotenberg, B., Pagonabarraga, I., Frenkel, D., 2008. Dispersion of charged tracers in charged porous media. EPL (Europhys. Lett.) 83, 34004.

Rotenberg, B., Morel, J.P., Marry, V., Turq, P., Morel-Desrosiers, N., 2009. On the driving force of cation exchange in clays: insights from combined microcalorimetry experiments and molecular simulation. Geochim. Cosmochim. Acta 73, 4034–4044.

Rotenberg, B., Pagonabarraga, I., Frenkel, D., 2010. Coarse-grained simulations of charge, current and flow in heterogeneous media. Farad. Discuss. 144, 223–243.

Rotenberg, B., Patel, A.J., Chandler, D., 2011. Molecular explanation for why talc surfaces can be both hydrophilic and hydrophobic. J. Am. Chem. Soc. 133, 20521–20527.

Sainz-Diaz, C., Hernandez-Laguna, A., Dove, M., 2001. Modeling of dioctahedral 2:1 phyllosilicates by means of transferable empirical potentials. Phys. Chem. Miner. 28, 130–141.

Schmidt, S., Katti, D., Ghosh, P., Katti, K., 2005. Evolution of mechanical response of sodium montmorillonite interlayer with increasing hydration by molecular dynamics. Langmuir 21, 8069–8076.

Shackelford, C., Moore, S., 2013. Fickian diffusion of radionuclides for engineered containment barriers: diffusion coefficients, porosities, and complicating issues. Eng. Geol. 152, 133–147.

Shroll, R., Smith, D., 1999. Molecular dynamics simulations in the grand canonial ensemble: application to clay mineral swelling. J. Chem. Phys. 111, 9025–9033.

Skipper, N., Refson, K., McConnell, J., 1989. Computer calculation of water-clay interactions using atomic pair potentials. Clay Miner. 24, 411–425.

Skipper, N., Chang, F.R., Sposito, G., 1995. Monte Carlo simulation of interlayer molecular structure in swelling clay minerals. 1. Methodology. Clays Clay Miner. 43, 285–293.

Smith, D., Wang, Y., Chaturvedi, A., Whitley, H., 2006. Molecular simulations of the pressure, temperature, and chemical potential dependencies of clay swelling. J. Phys. Chem. B 110, 20046–20054.

Suter, J., Coveney, P., 2009. Materials properties of clay nanocomposites: onset of negative Poisson ratio in large-scale molecular dynamics simulation. Soft Matter 5, 3896–3904.

Tambach, T., Hensen, E., Smit, B., 2004. Molecular simulations of swelling clay minerals. J. Phys. Chem. B 108, 7586–7596.

Tazi, S., Rotenberg, B., Salanne, M., Sprik, M., Sulpizi, M., 2012. Absolute acidity of clay edge sites from ab-initio simulations. Geochim. Cosmochim. Acta 94, 1–11.

Teppen, B., Rasmussen, K., Bertsch, P., Miller, D., Schafer, L., 1997. Molecular dynamics modeling of clay minerals. 1. Gibbsite, kaolinite, pyrophyllite, and beidellite. J. Phys. Chem. B 101, 1579–1587.

Thovert, J.F., Yousefian, F., Spanne, P., Jacquin, C.G., Adler, P.M., 2001. Grain reconstruction of porous media: application to a low-porosity Fontainebleau sandstone. Phys. Rev. E 63, 061307.

Tournassat, C., Ferrage, E., Poinsignon, C., Charlet, L., 2004. The titration of clay minerals: II. Structure-based model and implications for clay reactivity. J. Colloid Interface Sci. 273, 234–246.

Tournassat, C., Chapron, Y., Leroy, P., Bizi, M., Boulahya, F., 2009. Comparison of molecular dynamics simulations with triple layer and modified Gouy-Chapman models in a 0.1 m NaCl-montmorillonite system. J. Colloid Interface Sci. 339, 533–541.

Tyagi, M., Gimmi, T., Churakov, S.V., 2013. Multi-scale micro-structure generation strategy for up-scaling transport in clays. Adv. Water Resour. 59, 181–195.

Valfouskaya, A., Adler, P.M., Thovert, J.F., Fleury, M., 2005. Nuclear-magnetic-resonance diffusion simulations in porous media. J. Appl. Phys. 97, 083510.

Van Loon, L.R., Glaus, M.A., Müller, W., 2007. Anion exclusion effects in compacted bentonites: towards a better understanding of anion diffusion. Appl. Geochem. 22, 2536–2552.

Wang, J.W., Kalinichev, A.G., Kirkpatrick, R.J., Cygan, R.T., 2005. Structure, energetics, and dynamics of water adsorbed on the muscovite (001) surface: a molecular dynamics simulation. J. Phys. Chem. B 109, 15893–15905.

Wang, J.W., Kalinichev, A.G., Kirkpatrick, R.J., 2006. Effects of substrate structure and composition on the structure, dynamics, and energetics of water at mineral surfaces: a molecular dynamics modeling study. Geochim. Cosmochim. Acta 70, 562–582.

Summary and Perspective

Christophe Tournassat[a,b], Carl I. Steefel[b], Ian C. Bourg[b,c] and Faïza Bergaya[d]

[a] *Water, Environment and Ecotechnology Division, French Geological Survey (BRGM), Orléans, France;* [b] *Earth Sciences Division, Lawrence Berkeley National Laboratory, Berkeley, CA, USA;* [c] *Department of Civil and Environmental Engineering, Princeton University, Princeton, NJ, USA;* [d] *Centre de Recherche sur la Matière Divisée, Centre National de la Recherche Scientifique (CNRS), Orléans, France*

Clay barriers are at the heart of several of today's most important energy-related economic and societal challenges, including solid waste management, carbon capture and storage, geologic storage of high-level radioactive waste, exploration for unconventional gas and oil resources, and, by extension, climate change. The science of clay barriers is a fascinating research area for the range of techniques, theory, and applications that it encompasses. The authors of the chapters in this volume have striven to emphasize the most important knowledge related to clay barrier properties. The resulting overview is certainly not comprehensive, because of the number of processes and scales that must be taken into account. However, the editors believe that the present volume provides a thorough introduction to the key questions and challenges that must be overcome in order to ensure future breakthroughs in the science of clay barriers.

The primary function of clay barriers—confinement—is ensured by three remarkable properties: their low permeability to gas and water, the two carrier phases by which pollutants migrate; the low diffusivity of dissolved species in the pore network, which limits mass transfer and; the high retention capacity of clay minerals, which further limits the mobility of many inorganic and organic pollutants. The last property has been extensively studied in order to obtain quantitative information regarding the amount of pollutant that can be sequestered in clay barriers at in situ physical and chemical conditions. In parallel, spectrometric measurements have made it possible to unravel various uptake mechanisms that range from adsorption on the clay mineral surfaces, to cation exchange and surface complexation, to the incorporation of metals in the structure of existing or neoformed clay mineral layers. In the case of organic and inorganic anionic compounds, there is a need for predictive models of adsorption that can be generalized to different clay minerals and chemical conditions. Effective modeling tools to describe the adsorption of organic compounds, whether empirical or theoretical, are still lacking, especially with regard to situations where adsorption is characterized by significant hysteresis or strong competition of different species for the same adsorption sites. For inorganic cationic compounds, effective and predictive adsorption models have been developed that cover a large range of chemical conditions (pH and ionic strength). However, these models remain empirical in nature and their link to the fundamental properties of clay mineral surfaces is not entirely clear. Additional improvement in the characterization of reactive surface sites and their immediate environment, including the structure of clay mineral edge surfaces, the properties of interfacial water, and the details of surface charge screening by counter-ion adsorption and co-ion exclusion in the diffuse layer, is likely necessary to reconcile existing adsorption models with the properties of clay mineral surfaces. Knowledge of in situ chemical conditions in clay barriers (i.e., the chemical conditions in pore water) can be combined with retention models in order to determine whether conditions are favorable or unfavorable for the mitigation of a given contaminant. Fortunately, reliable experimental and modeling methods exist to characterize and understand pore water chemistry. However, among the parameters of interest, in situ redox conditions remain challenging to characterize, and the models available now are not yet fully capable of predicting the thermodynamic or kinetic controls on redox reactions in clay media. The uncertainties in redox conditions limit the precision with which predictions can be made on the retention of redox sensitive species or on the fate of redox mediated degradable compounds.

Diffusive mass fluxes of solute species in clay barriers are usually interpreted within the framework of Fickian diffusion laws, which state that the diffusion rate is linearly proportional to the concentration gradients of individual species. In porous clay materials, these diffusion rates are augmented with empirical parameters determined experimentally (diffusion coefficient, rock capacity factor). In principle, these macroscopic parameters can be predicted from knowledge of the microstructure of clay materials and of the fundamental phenomena that occur in individual pores (such as the structure of the electrical double layer and the mobility of water and ions near clay mineral surfaces). Important advances have been made on both of these topics with the aid of high-resolution microscopy and high-performance computational techniques (molecular, coarse-grained, and continuum scale modeling). However, much remains to be done. The characterization of the pore network at the nanometer scale is a significant challenge and the numerical simulation techniques available at the smallest scale remain limited to very simple systems compared to the natural systems of interest. These limitations partly explain the coexistence of many models of diffusion in clay barriers based on very different representations of the processes at the pore scale. Similarly, complex coupled processes must be taken into account to explain why Darcy's law does not consistently describe fluid flow in clay-rocks. Numerical methods allowing for the coupling between chemical potential, electrical potential, and thermal gradient are clearly necessary for the purposes of accurate fluid and solute transport calculations in this context. In addition, multiscale experimental and modeling approaches are necessary to build conceptually robust and well-parameterized models. With regard to gas migration in clay materials, our understanding of the fundamental physics controlling gas transport has increased in the last decade, but the complex interplay between gas diffusion, viscous flow, and mechanically induced changes in pore structure makes it a challenge to develop mechanistic numerical models. Recent development in pore network nanoscale imaging will certainly facilitate future breakthroughs in modeling gas transport at pore to macroscopic scales. At the larger scale of a geologic formation or a waste disposal site, upscaling is necessary that also takes into account the large-scale heterogeneity and anisotropy of the media of interest.

In addition to their short-term confinement properties, clay barriers must have a demonstrated stability (both chemical and mechanical) over long time scales. Clay minerals are resistant to alteration over a broad range of conditions because of their low dissolution kinetics but also because of the slow rate of diffusive mass transport in clay materials and because of the high chemical buffering capacity of clay minerals and accessory minerals in clay-rocks that limits the perturbations of the local chemical system. Dissolution rates are enhanced at low and high pH values and are strongly dependent on the deviation from equilibrium. Although empirical relationships are readily available to predict dissolution rates as a function of conditions, a fully consistent theory of crystal dissolution/precipitation for clay minerals remains to be developed. In this regard, the definition of a reliable proxy for the reactive surface area of dissolving and precipitating clay mineral remains a challenging issue. In the final analysis, it will be necessary to tackle the issue of the applicability to the field scale of dissolution rates obtained in laboratory conditions that are not fully representative of in situ conditions (dispersed particles in laboratory experiment vs compact material at in situ conditions). The development of microscopic and spectrometric techniques with atomic resolution should help identify fundamental mechanisms (at the scale of the reactive sites) in order to overcome these challenges and build more mechanistic models. At the scale of the barrier itself, observations and the results of larger scale modeling have begun to converge on the identification of mineralogical alteration fronts that are highly localized regardless of the alteration process involved (e.g., clay/cement or clay/iron), with the exception of redox perturbations that remain poorly understood. Significant pH variations also should be restricted to a narrow zone given the high pH buffer capacity of the clay materials. These observations suggest that large scale disaggregation of clay barriers due to chemical alteration is unlikely. However, the hydrological and mechanical properties of the altered zone could change dramatically with dissolution of primary and precipitation of secondary minerals. These changes can lead to porosity opening or clogging, a phenomenon that has been repeatedly observed in laboratory samples and natural analogues at all of the spatial and temporal scales investigated. The prediction of these processes at the larger spatial and temporal scales is made difficult by the scale dependence of the results predicted by the numerical models developed to date. The prediction in space and time of the coupling of chemical, hydrologic, and thermomechanical clay barrier properties is a formidable challenge that can be overcome only by integrating scientists from various disciplinary fields in the pursuit of these common objectives.

For all of the challenges discussed above, multiscale modeling is necessary to overcome the current limitations in process understanding and prediction. The development of new microscopic and spectrometric techniques together with innovative sample preparation procedures is a necessary step to characterize clay sample structures and heterogeneities at the relevant scale where fundamental processes must be understood and modeled. For large-scale applications, the development of highly coupled continuum scale models taking into account large scale heterogeneities is essential for progress to be made. These models must be based on rigorous upscaling procedures that take advantage of recent advances in computational methods and hardware, building on a range of macroscopic experimental observables and measurements that will serve as the basis for predicting the scale and time dependence of clay barrier properties.

Index

Note: Page numbers followed by "f" and "t" indicate figures and tables respectively.

D

Q

R

S

FSC
www.fsc.org
MIX
Papier | Fördert
gute Waldnutzung
FSC® C083411

Kolibri 360 GmbH
Ferdinand-Jühlke-Straße 7
99095 Erfurt, Deutschland
produktsicherheit@kolibri360.de